AF352889

Monitoring and Promoting Physical Activity, Physical Fitness and Motor Competence in Children

Monitoring and Promoting Physical Activity, Physical Fitness and Motor Competence in Children

Editors

Georgian Badicu
Ana Filipa Silva
Hugo Miguel Borges Sarmento

MDPI • Basel • Beijing • Wuhan • Barcelona • Belgrade • Manchester • Tokyo • Cluj • Tianjin

Editors

Georgian Badicu
Physical Education and
Special Motricity
Transilvania University
of Brasov
Brasov
Romania

Ana Filipa Silva
Instituto Politécnico de Viana
do Castelo
Escola Superior Desporto
e Lazer
Viana do Castelo
Portugal

Hugo Miguel Borges
Sarmento
Faculty of Sport Sciences and
Physical Education
University of Coimbra
Coimbra
Portugal

Editorial Office
MDPI
St. Alban-Anlage 66
4052 Basel, Switzerland

This is a reprint of articles from the Special Issue published online in the open access journal *Children* (ISSN 2227-9067) (available at: www.mdpi.com/journal/children/special_issues/Physical_Motor_ Competence_Children).

For citation purposes, cite each article independently as indicated on the article page online and as indicated below:

LastName, A.A.; LastName, B.B.; LastName, C.C. Article Title. *Journal Name* **Year**, *Volume Number*, Page Range.

ISBN 978-3-0365-6651-1 (Hbk)
ISBN 978-3-0365-6650-4 (PDF)

Cover image courtesy of Georgian Badicu

© 2023 by the authors. Articles in this book are Open Access and distributed under the Creative Commons Attribution (CC BY) license, which allows users to download, copy and build upon published articles, as long as the author and publisher are properly credited, which ensures maximum dissemination and a wider impact of our publications.

The book as a whole is distributed by MDPI under the terms and conditions of the Creative Commons license CC BY-NC-ND.

Contents

About the Editors

Georgian Badicu

Badicu Georgian is an associate professor at Transilvania University of Brasov, Faculty of Physical Education and Mountain Sports, Department of Physical Education and Special Motricity. He has published about 69 papers in JCR journals and is the author or co-author of 51 papers published by MDPI. His main interests are physical activity, fitness, education, obesity, well-being, recreation, football, public health, quality of life, sports activities, physical education, didactics of physical education, and sports.

Ana Filipa Silva

Ana Filipa Silva currently works as Assistant Professor in the Polytechnic Institute of Viana do Castelo (IPVC) in Melgaco, Portugal and as a researcher in Research Centre in Sports Sciences, Health Sciences and Human Development (CIDESD, Portugal). She has a European Ph.D. in Sport Sciences, Sports Training in Faculty of Sport, University of Porto, Porto, Portugal (2017, Portugal). Among others, the main publications are within the following topics: (i) motor control, (ii) youth sports performance, (iii) decision making in sports, and (iv) cognitive performance in sports. She is currently guest associate editor for *Frontiers* in Physiology, Frontiers in Sport, and Active Living and Human Movement. She has been guest editor of many Special Issues: (i) Training and Performance in Youth Sports at International Journal Environmental Research and Public Health; (ii) In Search of Individually Optimal Movement Solutions in Sport: Learning between Stability and Flexibility at Frontiers in Physiology and Frontiers in Sport and Active Living and Human Movement; (iii) Decision Making in Youth Sport Flexibility at Frontiers in Physiology and Frontiers in Sport and Active Living and Human Movement; and (iv) Children's Exercise Physiology, Volume II at Frontiers in Physiology and Frontiers in Sport and Active Living and Human Movement.

Hugo Miguel Borges Sarmento

Hugo Sarmento is professor at Faculty of Sport Sciences and Physical Education, University of Coimbra, Portugal. His main interests are: performance analysis and talent indentification and development; sport psychology; and physical activity and health.

Article

Body Image Perception in High School Students: The Relationship with Gender, Weight Status, and Physical Activity

Stefania Toselli [1], Luciana Zaccagni [2], Natascia Rinaldo [2], Mario Mauro [1,*], Alessia Grigoletto [3,*], Pasqualino Maietta Latessa [1] and Sofia Marini [1]

[1] Department for Life Quality Studies, University of Bologna, 47921 Rimini, Italy
[2] Department of Neuroscience and Rehabilitation, Faculty of Medicine, Pharmacy and Prevention, University of Ferrara, 44121 Ferrara, Italy
[3] Department of Biomedical and Neuromotor Sciences, University of Bologna, 40126 Bologna, Italy
* Correspondence: mario.mauro4@unibo.it (M.M.); alessia.grigoletto2@unibo.it (A.G.)

Citation: Toselli, S.; Zaccagni, L.; Rinaldo, N.; Mauro, M.; Grigoletto, A.; Maietta Latessa, P.; Marini, S. Body Image Perception in High School Students: The Relationship with Gender, Weight Status, and Physical Activity. *Children* **2023**, *10*, 137. https://doi.org/10.3390/children10010137

Academic Editors: Ferdinand Salonna and Tonia Vassilakou

Received: 24 October 2022
Revised: 26 December 2022
Accepted: 6 January 2023
Published: 10 January 2023

Copyright: © 2023 by the authors. Licensee MDPI, Basel, Switzerland. This article is an open access article distributed under the terms and conditions of the Creative Commons Attribution (CC BY) license (https://creativecommons.org/licenses/by/4.0/).

Abstract: Body image perception includes body size assessment, body desirability estimation, and perceptions concerning one's own body shape and size. Adolescence is a period of intense and prompt physical transformation, which changes the perception of one's body. This represents a critical period for the development of body image. Therefore, the present cross-sectional study aimed to evaluate body image perception and investigate the relationships between it, weight status, sex, and physical activity in a sample of high school students living in Italy. General demographic information and details about physical activity were collected. Body image perception was measured with a body silhouette and two indexes were calculated: the FID (Feel minus Ideal Discrepancy) to evaluate the discrepancy between the perceived current figure and the ideal figure; and the FAI (Feel weight status minus Actual weight status Inconsistency) to observe improper perception of weight status. In addition, body shape concerns were evaluated with the Body Shape Questionnaire (BSQ), in which participants reported the frequency of experiencing negative thoughts about their body shape in the last four weeks. Two hundred and four students were included in the study (155 = female, mean age = 17.13 ± 1.70; 49 = male, mean age = 17.25 ± 1.69). Females felt more concerned about body shape than males ($\chi^2 = 11.347$, $p = 0.001$). Distinctions emerged in terms of body mass index, the scores of Feel minus Ideal Discrepancy (FID), Feel weight status minus Actual weight status Inconsistency (FAI), the Body Shape Questionnaire (BSQ), and of the silhouette mean comparisons due to sex, weight status, and PA interaction effects ($p < 0.001$). Additionally, 94% of the BSQ variability could be explained by sex, weight status, and PA. Although no direct effects were observed on body image perception, healthy habit promotion, such as physical activity, could positively affect adolescent lives.

Keywords: body image; self-perception; physical activity; body shape questionnaire

1. Introduction

Body image is a multidimensional construct, which therefore encompasses perceptual, cognitive, and affective elements concerning one's own body and the bodies of others [1–3]. Body image perception includes body size assessment (how a person perceives his or her body), body attractiveness estimation (what is the type of body that a person considers most attractive), and perceptions related to one's own body shape and size [4]. Thus, interactions among physiological, cognitive, and sociocultural factors contribute to body image development [5]. During life, people are subjected to physical and psychological changes that affect the perception of their image. Thus, body image is not a static concept but rather a dynamic characteristic influenced by the perceptions of the individual about him- or herself [6]. Adolescence represents a period of intense and prompt physical changes and it therefore also changes the perception of one's body characteristics, and it can be a critical time for the development of body image [2,6–8]. Acceptance of and adaptation to these changes are essential for adolescents' feelings about their body image [9]. However, these transformations can represent a risk factor in adolescents' body image perception [6], thereby causing a distorted image of one's body. Moreover, distorted perception of one's body image can result from inaccurate evaluations of one's body, undesirable effects regarding one's body, miscognition of body-related stimuli, and specific body-related behaviours such as checking of body weight or avoidance of the consideration of body [10]. Worse still, it can result in unhealthy behaviours and adverse psychological outcomes such as an unhealthy diet, low self-esteem, and psychological disturbances [1,11,12].

There are several factors that influence the risk of body image disturbance during adolescence. The body image perception of adolescents has been reported to be associated with bio-social factors such as sex, age, socioeconomic status, and opportunities for health education [13,14]; environmental factors such as media exposure, peers, and school [1,14–17]; and behavioural factors such as weight control behaviour, physical activity, sexual behaviour, and eating patterns [11,18,19].

Regarding biological factors, the influence of sex on body image is well-known; body image dissatisfaction is more frequent among girls, but it is also present in thin boys [20,21].

Behavioural factors play a fundamental role, as they can be modified. Adolescents are responsive to the acquisition of habits and routines that are often influenced by their environment, which can be significant for their future life and of great importance in promoting healthy lifestyles during this time [22]. Among these lifestyle factors, proper nutrition and regular physical activity are of particular importance. Unfortunately, the data in the literature suggest a decline of optimal eating patterns and a decrease in physical activity in adolescents, who frequently assume sedentary behaviour [23].

Furthermore, adolescents are influenced by markets and advertising, which contribute to making their diet unbalanced and excessively caloric; rich in refined fats and sugars; poor in fruit, vegetables, legumes, and fish; and in many cases omitting breakfast [23]. Thus, in 2018, an international study conducted by the WHO found that one in five adolescent (21%) respondents was overweight or obese [24]. In Italy, in 11–15-year-old adolescents, the prevalence of overweight was 16.6% and obesity was 3.2%; excess weight decreased slightly with age and was greater in males (HBSC 2018: https://www.epicentro.iss.it/hbsc/indagine-2018, accessed on 22 February 2022). Distorted perception of weight status is quite prevalent in adolescents [25,26]. High-school-aged students tend to overrate height and underreport weight, thereby decreasing the prevalence of overweight and obesity estimates with lower average self-reported BMI [27]. Proper perception of weight is the key to determining the nutritional habits and weight management of adolescents. Indeed, it has been demonstrated that many overweight students are unlikely to participate in weight control practices [26,28–31].

As reported above, sports participation and physical activity (PA) can affect body image in adolescents [32–35]. In general, female and male adolescents who play sports present a lower body dissatisfaction; this can be connected to the role of sports participation in contributing to lower adiposity levels and, consequently, to lower body dissatisfaction in the adolescent population [36–38]. However, the associations between body image and sports participation in adolescents are not consistent, especially according to sex. Tebar et al. [39] reported that sports participation was associated with low body dissatisfaction in boys but not in girls. In sports contexts, females and males are not encouraged in the same way due to sex-stereotyped beliefs [40]. This appears in sex disparities in sports participation: in Italy, scant physical activity was higher among females and older adolescents [41]. Morano et al. [40] reported that 41% of girls and 59% of boys 11–19 years old practice sports. Despite the large importance of sports participation, the sport experience of adolescents has not been widely investigated, and the studies on body image concerns in adolescents involved in sports are rare and inconclusive.

Given this information, the main goal of the present study was to evaluate body image perception and to explore the relationships among it, weight status, sex, and physical activity in a sample of high school students living in Italy. Our hypothesis was that there would be a significant interaction effect between weight status, sex, and physical activity on body image perception in the sample. Specifically, we assumed that overweight/obese adolescents with a lower level of reported physical activity would have the lowest body perception and that these effects would be especially evident in females.

2. Materials and Methods

A cross-sectional study design was chosen, and data were collected and analysed in a sample of 204 adolescents attending secondary school in Piemonte (northern Italy). All adolescents attending the school could answer the questionnaires. The inclusion criteria were attending the school, completing all the parts of the questionnaires, having parental written consent, and agreeing to participate. The exclusion criteria were not having parental written consent and not completing all the parts of the questionnaires. So, we analysed only the fully completed questionnaires. The study was approved by the Bioethics Committee of the University of Bologna (approval no. 25027).

2.1. Outcome Assessment

General information about demographic variables (e.g., sex and age) and physical activity participation, both structured sport (e.g., volleyball with a team and coach) and independent physical activity (e.g., walking, yoga, etc.), was collected by an ad hoc questionnaire that was created for this purpose. The sports participation frequency of each subject was determined by the number of days per week of physical activity during a typical week as declared by the subject. Concerning anthropometric evaluation, participants' self-reported height and weight were converted to meters and kilograms to calculate body mass index (BMI).

2.2. Body Image Perception

The perception of body image was evaluated using the validated Body Silhouette Chart [42,43]. Nine male or female silhouettes, ordered in morphology from emaciation to obesity, were shown to the students, who were asked to select the one which they most desired (ideal figure) as well as the silhouette which they believed was most similar to their own (perceived current figure) [42]. The silhouette series was divided into underweight, normal weight, overweight, and obese [42]. The discrepancy between the perceived current figure and the ideal figure represented the degree of body image dissatisfaction (FID or Feel minus Ideal Discrepancy) [44]. By subtracting the number of the figure selected by the students as the ideal figure from the one selected as their perceived current figure, it was possible to calculate the FID index. A negative FID score indicates that the perceived current figure was thinner than the ideal figure, and a positive score indicates that the perceived current figure was bigger than the ideal figure. An FID score of 0 indicates that the student has chosen the same figure as perceived current and perceived ideal, showing no discrepancy.

The FAI (Feel weight status minus Actual weight status Inconsistency) index was used to evaluated the improper perception of weight status [45]. FAI inconsistency was calculated by subtracting the conventional code assigned to the real weight status of the participant from the code of her or his perceived weight status, which assigns a specific weight status category to each silhouette. The conventional code of real weight status of the participant and the perceived weight were classified according to the BMI assessed by Cole cut-off values by sex and age [46,47] (1 = underweight; 2 = normal weight; 3 = overweight; and 4 = obese). Instead, the classification recommended by Sànchez-Villegas was used for the silhouettes obtained [42]. In this classification method, silhouettes one, two, and three correspond to underweight (=1), silhouettes four and five correspond to normal weight (=2), silhouettes six and seven correspond to overweight (=3), and silhouettes eight and nine correspond to obesity (=4). A positive FAI score indicates that weight status is overestimated, a negative score indicates that weight status is underestimated, and a score of zero indicates no inconsistency in weight status perception.

2.3. Body Shape Concerns

Body shape concerns were assessed with a 14-item version [48] of the Body Shape Questionnaire (BSQ) [49], which was validated in Italian by [50]. The BSQ utilizes a six-point Likert scale ranging from one to six (one = never, six = always) to assess how frequently in the past four weeks a participant reported experiencing negative thoughts or feelings about their body shape (e.g., "Have you felt ashamed of your body?"). In addition, the questionnaire investigated if they had tried to control or lose weight with the same six-point Likert scale. Scores from all items are added to get a total score. Higher scores reflected greater body shape concerns. The total score was calculated on the sum of all the values was multiplied by 34/14 and subsequently related to the specific thresholds that refer to the complete form of the questionnaire: a score below 80 indicated "no concern", a score between 80 and 110 indicated a "slight concern", a score between 111 and 140 indicated a "moderate concern", and a score above 140 indicated a "marked concern".

2.4. Statistical Analysis

The statistical analysis was performed with STATA® software, version 17 (Publisher: StataCorp. 2021. Stata Statistical Software: Release 17. College Station, TX, USA, StataCorp LP). Descriptive statistics were calculated and reported such as mean $\pm$ standard deviation (SD) for quantitative variables (age, height, weight, BMI, silhouette think to look, silhouette want to be, FID, FAI, and BSQ) and frequency (number of observations) and percentage (%) for qualitative variables (weight status, weight control, body shape concern, and trying to lose weight). The chi-squared test (χ^2) was used to assess discrepancies in frequencies among categorial and binomial variables. The Shapiro–Wilk test was performed to check the variables' normal distribution. Due to the presence of predictors with different data levels, an analysis of covariances and the interaction effect between sex, weight status, and physical activity frequency were tested. A stepwise procedure with backward selection was carried out to perform the multiple regression analysis and to assess possible predictors of FID, FAI, and BSQ. Stepwise estimation included only predictors with a significance level for removal from the model of at most $p < 0.05$.

3. Results

A total of 204 students, consisting of 155 females (mean age = 17.13 $\pm$ 1.70) and 49 males (mean age = 17.25 $\pm$ 1.69), were included in the final analysis (Figure 1). Considering weight status, normal weight was the most represented category in both females and males.

Table 1 shows no statistically significant differences in frequencies between sexes except for body shape concern, which displayed more trouble in females. Additionally, male adolescents were less worried about weight loss.

Table 2 shows the mean comparisons and effects of the interaction between sex, weight status, and frequencies of PA. Sex differences were significant only for the desired figure and body shape questionnaire, while all variables related to body perception were influenced by weight status. In the underweight category, FID and FAI scores showed negative discrepancies, indicating a misperception tending to underestimation. Regarding effects of PA frequency, it did not show significant outcomes. However, interaction effects of sex, weight status, and PA frequency were relevant for every variable.

To assess possible predictors of body image dissatisfaction (assessed by FID score), inconsistency of weight status (assessed by FAI score), and body shape concern (assessed by BSQ score), three backward multiple regressions were performed. Table 3 shows the predictive model results. Regarding FID, weight and all the categories of PA frequency were positive predictors, whereas the conditions of being younger, male, and having tried to lose weight sometimes, were found to be associated with better body image satisfaction.

Although the whole model explained 60% of the FID variance, adolescent weight seemed to be the most influential variable ($\eta^2 = 0.417$).

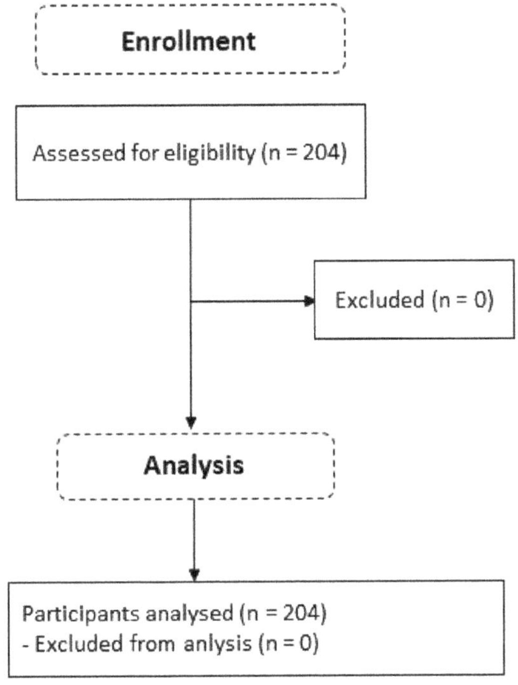

Figure 1. Sample flowchart.

Table 1. Frequencies of weight status, body shape concern, weight control and trying to lose weight by sex.

	Sex [n (%)]		Statistics	
	F = 155 (75.98)	M = 49 (24.02)	χ^2	p
Weight Status			2.945	0.400
Underweight	14 (9.03)	4 (8.16)		
Normal weight	97 (62.58)	26 (53.06)		
Overweight	33 (21.29)	12 (24.49)		
Obese	11 (7.10)	7 (14.29)		
Body shape concern (BSQ)			11.347	0.010
No	56 (36.13)	26 (53.06)		
Some	41 (26.45)	17 (34.69)		
Moderate	29 (18.71)	4 (8.16)		
Marked	29 (18.71)	2 (4.08)		
Weight control			1.083	0.298
No	36 (23.23)	15 (30.61)		
Yes	119 (76.77)	34 (69.39)		
Tried to lose weight			4.293	0.231
Never	42 (27.1)	16 (33.33)		
Sometimes	54 (34.84)	21 (43.75)		
Often	45 (29.03)	7 (8.33)		
Always	14 (9.03)	4 (8.33)		

Note: F, female; M, male; χ2, chi-squared test; *p*, *p*-value; %, percentage; N, number of observations.

Table 2. Mean comparisons and interaction effects between sex, weight status, and physical activity frequency.

Variable	Sex				Weight Status						Physical Activity Frequency						Interaction Effect			
	Male	Female	Stats		Under	Normal	Over	Obese	Stats		≤ 1 d·w^{-1}	2–3 d·w^{-1}	4–5 d·w^{-1}	>5 d·w^{-1}	Stats		Sex # WS # PA Frequencies			
	Mean (±SD)	Mean (±SD)	$F_{(1, 202)}$	p	Mean (±SD)	Mean (±SD)	Mean (±SD)	Mean (±SD)	$F_{(3, 200)}$	p	Mean (±SD)	Mean (±SD)	Mean (±SD)	Mean (±SD)	$F_{(3, 200)}$	p	$F_{(25, 178)}$	p	adj-R^2	η^2
Age	17.25 (1.69)	17.13 (1.70)	0.18	0.67	17.85 (1.52)	16.99 (1.74)	17.48 (1.56)	16.83 (1.62)	2.20	0.09	17.14 (1.67)	17.04 (1.72)	17.76 (1.82)	17.28 (1.72)	0.51	0.67	1.17	.27	0.02	0.14
Height	174.96 (7.59)	163.57 (6.72)	100.16	†	48.44 (3.78)	57.78 (8.00)	74.99 (10.8)	87.56 (13.79)	0.80	0.49	165.59 (8.17)	166.03 (8.88)	167.67 (9.86)	169.58 (7.94)	1.67	0.18	4.71	†	0.31	0.40
Weight	73.00 (16.86)	60.34 (11.75)	34.56	†	166.5 (6.69)	165.62 (8.26)	167.6 (9.43)	167.82 (9.11)	99.34	†	63.20 (14.52)	62.10 (14.30)	67.62 (13.36)	65.53 (12.96)	0.63	0.60	18.41	†	0.68	0.72
BMI	23.75 (4.82)	22.52 (3.95)	3.23	0.07	17.46 (0.70)	20.99 (1.79)	26.58 (1.85)	31.71 (2.58)	293.07	†	22.96 (4.29)	22.36 (4.03)	24.12 (4.79)	22.75 (4.09)	0.56	0.64	37.92	†	0.82	0.84
STL	4.27 (1.66)	4.06 (1.55)	0.64	0.42	2.50 (0.86)	3.50 (1.08)	5.78 (1.01)	6.17 (1.50)	73.98	†	4.17 (1.63)	4.10 (1.52)	4.22 (1.56)	3.81 (1.50)	0.39	0.76	9.34	†	0.51	0.57
SWB	3.51 (1.02)	2.97 (1.01)	10.38	§	2.72 (0.75)	2.89 (1.01)	3.64 (0.96)	3.61 (0.98)	9.08	†	3.17 (1.03)	3.14 (1.10)	3.11 (0.78)	2.73 (0.96)	1.30	0.28	2.30	†	0.14	0.24
FID	0.76 (1.56)	1.08 (1.41)	1.92	0.17	−0.22 (0.73)	0.62 (1.28)	1.93 (0.91)	2.56 (1.76)	28.41	†	1.00 (1.41)	0.97 (1.61)	1.11 (1.54)	1.08 (1.29)	0.05	0.99	4.02	†	0.27	0.36
FAI	0.14 (0.50)	0.14 (0.51)	0.01	0.93	−0.50 (0.51)	0.12 (0.42)	0.38 (0.49)	0.28 (0.57)	16.40	†	0.14 (0.55)	0.12 (0.42)	0.11 (0.33)	0.15 (0.54)	0.04	0.99	2.49	†	0.16	0.26
BSQ	82.17 (27.78)	101.03 (34.55)	12.10	†	71.37 (11.24)	89.13 (29.52)	112.2 (35.20)	132.76 (32.86)	19.42	†	99.05 (35.84)	95.17 (31.17)	84.73 (29.53)	92.66 (33.49)	0.71	0.55	3.54	†	0.24	0.33

Note: BMI, body mass index; STL, silhouette think to look; SWB, silhouette want to be; FID, Feel minus Ideal Discrepancy; FAI, Feel weight status minus Actual weight status Inconsistency; BSQ, Body Shape Questionnaire; SD, standard deviation; WS, weight status; PA, physical activity; F, statistic test of Snedecor–Fisher; d, days; w, week; p, p-value; adj-R^2, adjusted R-squared; η^2, eta-squared effect size; #, interaction; §, $p < 0.01$; †, $p < 0.001$.

Concerning FAI, being underweight, normal weight, and never trying to lose weight appeared as negative predictors of weight status estimation. Contrarily, body weight overestimation was associated with all the categories of PA frequency. The model explained 24% of the variance and had a highly significant R^2. Being male and not obese was significantly associated with a lower score of body shape concern. The BSQ score increased as body image dissatisfaction and PA frequency increased. The model explained 94% of the variance.

Finally, Figures 2–4 shows the linear prediction of the BSQ influenced by PA frequency categories and the weight status (A), PA frequency categories, FID values (B), and PA frequency categories and sex (C). Lower BSQ values corresponded to 4–5 days per week of PA in each plot except for overweight subjects (Figures 2–4, green line).

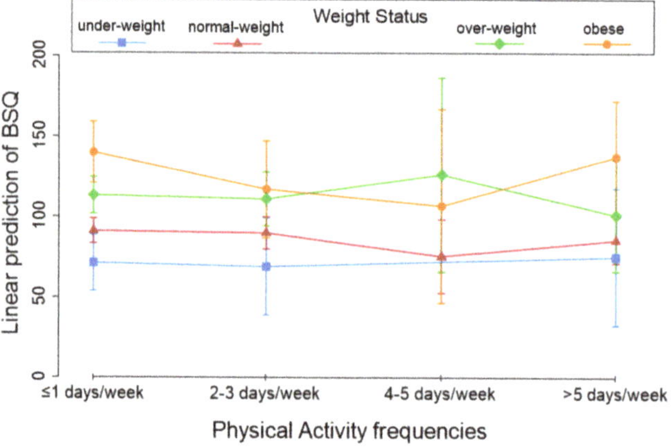

Figure 2. Linear prediction of BSQ by Weight Status for four Physical Activities frequencies.

Table 3. Predictors of FID, FAI, and BSQ: multiple regression analysis results.

FID (N = 202)						
Source	SS	df	MS	$F_{(9, 193)}$	p	adj-R^2
Model	388.62	9	43.18	35.26	<0.001	0.604
Residual	236.39	193	1.22			
Total	625	202	3.09			
Variable	β	SE	t	p		η^2
Height	−0.633	0.013	−4.89	<0.001		0.11
Weight	0.795	0.007	11.75	<0.001		0.417
Age	−0.12	0.047	−2.58	0.01		0.033
PA frequency						0.092
$\leq$1 d·w^{-1}	0.878	0.208	4.22	<0.001		
2–3 d·w^{-1}	0.887	0.208	4.26	<0.001		
4–5 d·w^{-1}	0.911	0.208	4.38	<0.001		
>5 d·w^{-1}	0.893	0.213	4.2	<0.001		
Sex (male)	−0.619	0.231	−2.68	<0.01		0.036
TLW (sometimes)	−0.315	0.166	−1.91	0.05		0.019
FAI (N = 203)						
Source	SS	df	MS	$F_{(7, 196)}$	p	adj-R^2
Model	15.17	7	2.17	10.4	<0.001	0.24
Residual	40.83	196	0.21			
Total	53	203	0.28			
Variable	β	SE	t	p		η^2
Weight Status						0.202
Underweight	−0.858	0.122	−7.01	<0.001		
Normal weight	−0.239	0.072	−3.34	0.001		
PA frequency						0.178
$\leq$1 d·w^{-1}	0.415	0.068	6.13	<0.001		
2–3 d·w^{-1}	0.376	0.081	4.63	<0.001		
4–5 d·w^{-1}	0.314	0.162	1.93	0.05		
>5 d·w^{-1}	0.438	0.108	4.06	<0.001		
TLW (never)	−0.152	0.071	−2.13	<0.05		0.022
BSQ (N = 202)						
Source	SS	df	MS	$F_{(9, 193)}$	p	adj-R^2
Model	1997339	9	221926.56	344.42	<0.001	0.939
Residual	124357.5	193	644.34			
Total	2121696.4	202	10503.45			
Variable	β	SE	t	p		η^2
Sex (male)	−16.159	4.45	−3.63	<0.001		0.064
FID	10.688	1.565	6.83	<0.001		0.194
Weight Status						0.081
Underweight	−38.557	10.059	−3.83	<0.001		
Normal weight	−29.164	7.732	−3.77	<0.001		
Overweight	−20.033	7.593	−2.64	0.01		
PA frequency						0.516
$\leq$1 d·w^{-1}	116.868	8.218	14.22	<0.001		
2–3 d·w^{-1}	115.445	8.752	13.19	<0.001		
4–5 d·w^{-1}	108.536	12.186	8.91	<0.001		
>5 d·w^{-1}	110.967	9.377	11.83	<0.001		

Note: SS, sum of squares; MS, mean of squares; df, degree of freedom; F, Snedecor–Fisher's statistical test; β, regression coefficient; SE, standard error; t, student's statistic test p, p-value; adj-R^2, adjusted R^2; η^2, eta-squared.

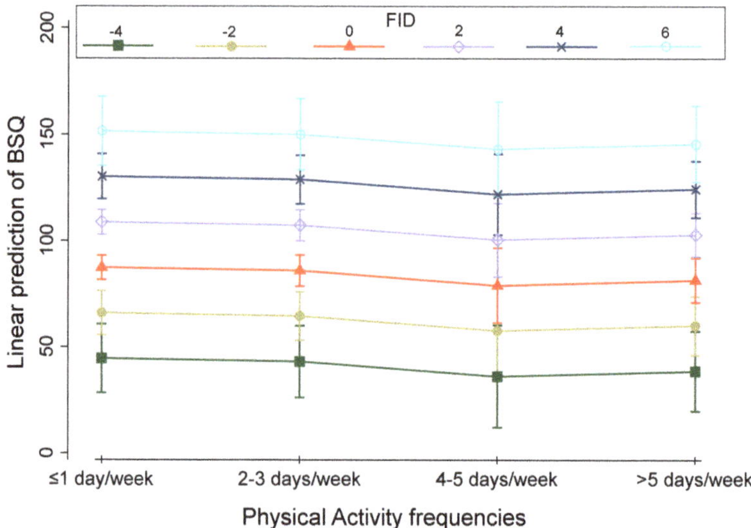

Figure 3. Linear prediction of BSQ by FID for four Physical Activities frequencies.

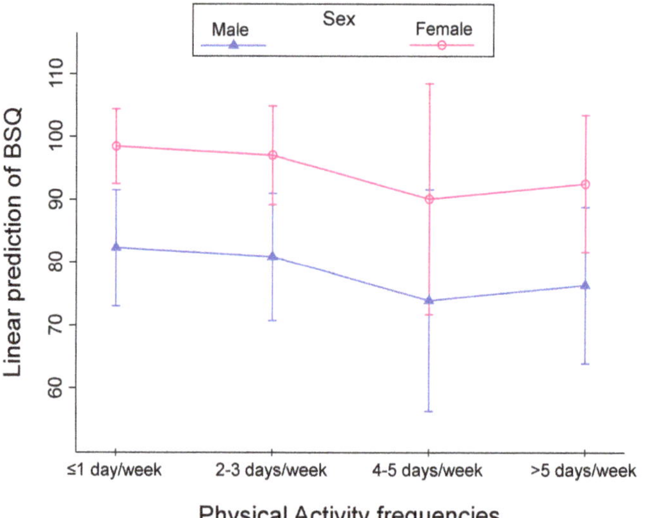

Figure 4. Linear prediction of BSQ by sex for four Physical Activities frequencies.

4. Discussion

The present study aimed to evaluate body image perception in a sample of high school students living in Italy and to explore the relationships between it, sex, weight status, and physical activity.

In the present study, no gender differences were observed in terms of weight status and weight control. Nonetheless, girls showed higher values of weight concerns. In addition, this study confirmed the presence of sex differences in body image perception. Admittedly, the absence of differences in weight status was unexpected since in Italy, a significantly higher proportion of males were overweight/obese compared with females [51,52]. These data confirmed that girls in this age range present significant weight concerns, as seen in similar studies [53–55]. It is noteworthy that even though the present study showed a greater prevalence of weight concern among girls, males also reported it. This result is of particular interest, and it cannot be neglected given the established sex-specific patterns, such as muscularity-oriented disordered eating [54,56]. Relatedly, a longitudinal study by Bucchianeri et al. [57] showed that in adolescence, body dissatisfaction increased with time in both sexes, but the levels of body dissatisfaction were remarkably higher in girls. Generally, body image dissatisfaction and weight control behaviour are greater issues for girls compared with males [55]. In our study, FID was higher in females, although it was not a significant difference between the sexes. Girls were found to be more inclined than boys to emphasize the aesthetic values of their bodies rather than functional ones. Moreover, females reported more dissatisfaction with both values than boys [58]. Consistent with this, even in childhood, girls seemed to be more conscious about how their body weight affects their appearance compared with boys [59]. The difference between real and ideal silhouettes with a higher incidence in girls may result in a differential diagnosis of body dysmorphic disorder, since it is more prevalent in women, beginning at the age of between 15 and 30 years [55]. The beauty standard currently imposed by media that values thinness reinforces this condition, which may lead to the adoption of restrictive diets regardless of the real need and may contribute to the occurrence and maintenance of low weight as well as the compromise of eating habits [60] and the presence of mental disorders [61]. For this purpose, it is noteworthy that even though the weight control outcome did not show significant differences between sexes, the percentage of individuals who reported putting attention on weight control is high in both sexes (76.77% in females and 69.39% in males), although a higher percentage of females declared that they tried to lose weight often or always compared to males (38.06% vs. 16.66%). Undoubtedly, the role of gender in body image has been changing; a study by Dzielska et al. [62] targeted adolescents from 26 countries from 2001/2002 to 2017/2018 and reported that the overall age-adjusted prevalence rates of weight-reduction behaviours (WRB) were 10.2% among boys and 18.0% among girls. The prevalence was higher for girls, but in more recent surveys, sex differences in WRB decreased, and a significant increase in the percentage of WRB among boys was observed in most countries. Therefore, from the present results of our study, being a male seems to be still a protective factor. Additionally, being male and having sometimes tried to lose weight was found to be associated with better body image satisfaction. Likewise, being male and not obese was significantly associated with a lower score of body shape concern.

Regarding the second investigated factor, weight status, overweight and obese students showed higher concerns, while no concern and slight concern were observed in underweight and normal weight subjects. FID was found to be related to a growing dissatisfaction with body image or to an increase in weight status. Thus, being underweight is the only category whose members would desire to be fatter, whereas in the other categories, the desire to be thinner prevailed. A relationship with weight also emerged from the regression analysis since weight was significantly associated with body image dissatisfaction. Relatedly, de Pinho et al. [55] reported a significant association between nutritional status and body image in adolescents; in particular, underweight and overweight adolescents were dissatisfied with their body image. Admittedly, weight status is associated with body dissatisfaction, weight concerns, and dietary restraints [63], potentially resulting in both restrictive and binge-purge disordered eating [64,65]. For this reason, studies regarding body image perception in this age group may be valuable for the early detection of disorders. Considering FAI, overweight subjects overestimate their weight status more than normal weight and obese subjects, while a misperception tending to underestimation is observed in underweight students. In addition, the regression analysis revealed that being underweight, normal weight, and having never trying to lose weight are negative predictors, and are thus protective factors, of weight status estimation. The number of teenagers who perceive themselves as overweight and adopt unhealthy strategies to lose weight is constantly increasing, as was previously reported by epidemiological surveys carried out in Italy [66].

The frequency of PA did not show significant associations with body image perception. The influence of sport participation and physical activity (PA) on body image in adolescents is not consistent. In general, adolescents of both sexes who play sports present a lower body dissatisfaction; however, associations between body image and sports practice in adolescents are not consistent, especially according to sex [32–39]. Our study showed that only intense physical activity effected perception [21]. From the results of the regression analysis, PA was associated with FID, FAI, and BSQ. In particular, all the categories of PA frequency were significantly associated with body image dissatisfaction and with body weight overestimation. Similarly, being physically active seemed to be associated with body shape concerns. Arguably, this is related to wider body awareness, which might contribute to generating more concern about body shape.

From the results of the present study, sex and weight status were found to be the prevailing factors influencing body image perception, while physical activity seemed to have a less decisive influence. Therefore, interventions should be directed at factors that influence the acceptance of their own body and the promotion of a healthy lifestyle. The sex-specific patterns in body image investigated in the present study underlined the need to focus on the risk factors for body image misperception and concerns by sex and weight status categories. Moreover, specific health promotion initiatives taking these aspects into account should be implemented.

The primary limitation of the study was the cross-sectional nature of the data, which limited causal inferences. A longitudinal study design during the whole adolescent period would be valuable to determine the direction of the relationship among the variables considered. In addition, height and weight were self-reported due to the COVID-19 public health emergency period, and this could have influenced the accuracy of the results. Information about sociodemographic variables was not collected, so it was not possible to evaluate the influence of these factors in explaining some of the possible links between physical activity participation and body perception. The sample size was limited, and further research with a larger sample is needed to verify the validity of the present results.

5. Conclusions

Although the prevalence of high school students who are of normal weight is high, many adolescents exhibit an altered body image and body shape perception, especially in the female sex. The daily practice of physical activity should reduce concerns related to body shape and facilitate better self-awareness. Despite more evidence being needed, the promotion of healthy habits such as physical activity could be an optimal strategy to improve mental wellness in adolescents.

Author Contributions: Conceptualization, S.T., N.R. and L.Z.; methodology, M.M. and S.M.; software, A.G. and P.M.L.; validation, S.T., S.M. and L.Z.; formal analysis, M.M., S.M. and N.R.; investigation, A.G. and S.M.; resources, P.M.L.; data curation, M.M. and S.T.; writing—original draft preparation, S.T., S.M., A.G. and M.M.; writing—review and editing, L.Z. and N.R.; visualization, A.G.; supervision, P.M.L.; project administration, S.T., L.Z. and N.R. All authors have read and agreed to the published version of the manuscript.

Funding: This research received no external funding.

Institutional Review Board Statement: The study was conducted in accordance with the Declaration of Helsinki, and approved by the Bioethics Committee of the University of Bologna (approval no. 25027 and date of approval 10 July 2020).

Informed Consent Statement: Informed consent was obtained from all subjects involved in the study.

Data Availability Statement: Not applicable.

Conflicts of Interest: The authors declare no conflict of interest.

References

1. Goonapienuwala, B.L.; Agampodi, S.B.; Kalupahana, N.S.; Siribaddana, S. Body Image Construct of Sri Lankan Adolescents. *Ceylon Med. J.* **2017**, *62*, 40–46. [CrossRef] [PubMed]
2. Voelker, D.K.; Reel, J.J.; Greenleaf, C. Weight Status and Body Image Perceptions in Adolescents: Current Perspectives. *AHMT* **2015**, *6*, 149–158. [CrossRef]
3. Chae, H. Factors Associated with Body Image Perception of Adolescents. *Acta. Psychol.* **2022**, *227*, 103620. [CrossRef] [PubMed]
4. Grogan, S. Promoting Positive Body Image in Males and Females: Contemporary Issues and Future Directions. *Sex Roles* **2010**, *63*, 757–765. [CrossRef]
5. Irvine, K.R.; McCarty, K.; McKenzie, K.J.; Pollet, T.V.; Cornelissen, K.K.; Tovée, M.J.; Cornelissen, P.L. Distorted Body Image Influences Body Schema in Individuals with Negative Bodily Attitudes. *Neuropsychologia* **2019**, *122*, 38–50. [CrossRef]
6. Markey, C.N. Invited Commentary: Why Body Image is Important to Adolescent Development. *J. Youth Adolesc.* **2010**, *39*, 1387–1391. [CrossRef]
7. De Araujo, T.S.; Barbosa Filho, V.C.; Gubert, F.d.A.; de Almeida, P.C.; Martins, M.C.; Carvalho, Q.G.d.S.; Costa, A.C.P.d.J.; Vieira, N.F.C. Factors Associated with Body Image Perception Among Brazilian Students from Low Human Development Index Areas. *J. Sch. Nurs.* **2018**, *34*, 449–457. [CrossRef]
8. Toselli, S.; Grigoletto, A.; Zaccagni, L.; Rinaldo, N.; Badicu, G.; Grosz, W.R.; Campa, F. Body Image Perception and Body Composition in Early Adolescents: A Longitudinal Study of an Italian Cohort. *BMC Public Health* **2021**, *21*, 1381. [CrossRef]
9. Yara, L.F.; Diego, A.S.S.; Andreia, P.; Adelson, F.d.S.; Edio, L.P. Insatisfação Com a Imagem Corporal Em Adolescentes de Uma Cidade de Pequeno Porte: Associação Com Sexo, Idade e Zona de Domicílio. *Rev. Bras. Cineantropometria Desempenho Hum.* **2011**, *13*, 202–207. [CrossRef]
10. Schuck, K.; Munsch, S.; Schneider, S. Body Image Perceptions and Symptoms of Disturbed Eating Behavior among Children and Adolescents in Germany. *Child Adolesc. Psychiatry Ment. Health* **2018**, *12*, 10. [CrossRef]
11. Bibiloni, M.d.M.; Pich, J.; Pons, A.; Tur, J.A. Body Image and Eating Patterns among Adolescents. *BMC Public Health* **2013**, *13*, 1104. [CrossRef]
12. Lee, J.; Lee, Y. The Association of Body Image Distortion with Weight Control Behaviors, Diet Behaviors, Physical Activity, Sadness, and Suicidal Ideation among Korean High School Students: A Cross-Sectional Study. *BMC Public Health* **2016**, *16*, 39. [CrossRef]
13. Chung, A.E.; Perrin, E.M.; Skinner, A.C. Accuracy of Child and Adolescent Weight Perceptions and Their Relationships to Dieting and Exercise Behaviors: A NHANES Study. *Acad. Pediatr.* **2013**, *13*, 371–378. [CrossRef]
14. Shin, A.; Nam, C.M. Weight Perception and Its Association with Socio-Demographic and Health-Related Factors among Korean Adolescents. *BMC Public Health* **2015**, *15*, 1292. [CrossRef]
15. Carey, R.N.; Donaghue, N.; Broderick, P. Body Image Concern among Australian Adolescent Girls: The Role of Body Comparisons with Models and Peers. *Body Image* **2014**, *11*, 81–84. [CrossRef]
16. Joeng, H.E.; Lee, I.S. Multilevel analysis of factors associated with subjective weight perception among normal body weight adolescents based on the 2017 Korean Youth's risk behavior survey (KYRBS). *J. Korean Acad. Community Health Nurs.* **2018**, *29*, 476–487. [CrossRef]
17. Latino, F.; Greco, G.; Fischetti, F.; Cataldi, S. Multilateral Training Improves Body Image Perception in Female Adolescents. *J. Hum. Sport Exerc.* **2019**, *14*, S927–S936. [CrossRef]
18. Alwan, H.; Viswanathan, B.; Paccaud, F.; Bovet, P. Is Accurate Perception of Body Image Associated with Appropriate Weight-Control Behavior among Adolescents of the Seychelles. *J. Obes.* **2011**, *2011*, 817242. [CrossRef]
19. Gualdi-Russo, E.; Rinaldo, N.; Masotti, S.; Bramanti, B.; Zaccagni, L. Sex Differences in Body Image Perception and Ideals: Analysis of Possible Determinants. *Int. J. Environ. Res. Public Health* **2022**, *19*, 2745. [CrossRef]
20. Jiménez Flores, P.; Jiménez Cruz, A.; Bacardi Gascón, M. Body-image dissatisfaction in children and adolescents: A systematic review. *Nutr. Hosp.* **2017**, *34*, 479–489. [CrossRef]
21. Toselli, S.; Spiga, F. Sport Practice, Physical Structure, and Body Image among University Students. *J. Eat. Disord.* **2017**, *5*, 31. [CrossRef] [PubMed]
22. Duno, M.; Acosta, E.; Duno, M.; Acosta, E. Body Image Perception among University Adolescents. *Rev. Chil. De Nutr.* **2019**, *46*, 545–553. [CrossRef]
23. Gallego, R.; Mera-Gallego, I.; Pérez, J.A.; Rodríguez, P.; Fernández Cordeiro, M.; Reneda, Á.; Verez, N.; Floro, N.; Andrés-Rodríguez, N.F.; Echevarría, R. Análisis de Hábitos Nutricionales y Actividad Física de Adolescentes Escolarizados. RIVA-CANGAS Analysis of Nutritional Habits and Physical Activity among Adolescent Students. RIVACANGAS. Bachelor's Thesis, Uppsala University, Uppsala, Sweden, 2017.
24. Inchley, J.; Budisavljevic, S.; Currie, D.; Torsheim, T.; Jåstad, A.; Cosma, A.; Kelly, C.; Arnarsson, Á.M.; Barnekow, V.; Weber, M.M. *Spotlight on Adolescent Health and Well-Being*; HBSC Study; WHO: Geneva, Switzerland, 2018; pp. 150–160.
25. Zarychta, K.; Mullan, B.; Luszczynska, A. Am I Overweight? A Longitudinal Study on Parental and Peers Weight-Related Perceptions on Dietary Behaviors and Weight Status among Adolescents. *Front. Psychol.* **2016**, *7*, 83. [CrossRef] [PubMed]
26. Brener, N.D.; Eaton, D.K.; Lowry, R.; McManus, T. The Association between Weight Perception and BMI among High School Students. *Obes. Res.* **2004**, *12*, 1866–1874. [CrossRef] [PubMed]

27. Brener, N.D.; Mcmanus, T.; Galuska, D.A.; Lowry, R.; Wechsler, H. Reliability and Validity of Self-Reported Height and Weight among High School Students. *J. Adolesc. Health* **2003**, *32*, 281–287. [CrossRef]
28. Khanna, D.; Mutter, C.M.; Kahar, P. Perception of Overall Health, Weight Status, and Gaining Weight in Relationship with Self-Reported BMI Among High School Students. *Cureus* **2021**, *13*, e19637. [CrossRef]
29. Coletta, A.; Baetge, C.; Murano, P.; Galvin, E.; Rasmussen, C.; Greenwood, M.; Levers, K.; Lockard, B.; Simbo, S.Y.; Jung, Y.P.; et al. Efficacy of Commercial Weight Loss Programs on Metabolic Syndrome. *FASEB J.* **2016**, *30*, lb216. [CrossRef]
30. Baetge, C.; Earnest, C.P.; Lockard, B.; Coletta, A.M.; Galvan, E.; Rasmussen, C.; Levers, K.; Simbo, S.Y.; Jung, Y.P.; Koozehchian, M.; et al. Efficacy of a Randomized Trial Examining Commercial Weight Loss Programs and Exercise on Metabolic Syndrome in Overweight and Obese Women. *Appl. Physiol. Nutr. Metab.* **2017**, *42*, 216–227. [CrossRef]
31. Levers, K.; Simbo, S.; Lockard, B.; Baetge, C.; Galvan, E.; Byrd, M.; Jung, Y.; Jagim, A.; Oliver, J.; Koozehchian, M.; et al. Effects of Exercise and Diet-Induced Weight Loss on Markers of Inflammation I: Impact on Body Composition and Markers of Health and Fitness. *J. Int. Soc. Sports Nutr.* **2013**, *10*, P15. [CrossRef]
32. Campbell, A.; Hausenblas, H.A. Effects of Exercise Interventions on Body Image: A Meta-Analysis. *J. Health Psychol.* **2009**, *14*, 780–793. [CrossRef]
33. Gomez-Baya, D.; Mendoza, R.; Matos, M.G.d.; Tomico, A. Sport Participation, Body Satisfaction and Depressive Symptoms in Adolescence: A Moderated-Mediation Analysis of Gender Differences. *Eur. J. Dev. Psychol.* **2019**, *16*, 183–197. [CrossRef]
34. Smolak, L.; Murnen, S.K.; Ruble, A.E. Female Athletes and Eating Problems: A Meta-Analysis. *Int. J. Eat. Disord.* **2000**, *27*, 371–380. [CrossRef]
35. Soulliard, Z.A.; Kauffman, A.A.; Fitterman-Harris, H.F.; Perry, J.E.; Ross, M.J. Examining Positive Body Image, Sport Confidence, Flow State, and Subjective Performance among Student Athletes and Non-Athletes. *Body Image* **2019**, *28*, 93–100. [CrossRef]
36. Añez, E.; Fornieles-Deu, A.; Fauquet-Ars, J.; López-Guimerà, G.; Puntí-Vidal, J.; Sánchez-Carracedo, D. Body Image Dissatisfaction, Physical Activity and Screen-Time in Spanish Adolescents. *J. Health Psychol.* **2018**, *23*, 36–47. [CrossRef]
37. Neumark-Sztainer, D.; Goeden, C.; Story, M.; Wall, M. Associations between Body Satisfaction and Physical Activity in Adolescents: Implications for Programs Aimed at Preventing a Broad Spectrum of Weight-Related Disorders. *Eat. Disord.* **2004**, *12*, 125–137. [CrossRef]
38. Tambalis, K.D.; Panagiotakos, D.B.; Psarra, G.; Sidossis, L.S. Prevalence, Trends and Risk Factors of Thinness among Greek Children and Adolescents. *J. Prev. Med. Hyg.* **2019**, *60*, E386–E393. [CrossRef]
39. Tebar, W.R.; Gil, F.C.S.; Werneck, A.O.; Delfino, L.D.; Santos Silva, D.A.; Christofaro, D.G.D. Sports Participation from Childhood to Adolescence is Associated with Lower Body Dissatisfaction in Boys—A Sex-Specific Analysis. *Matern. Child. Health J.* **2021**, *25*, 1465–1473. [CrossRef]
40. Morano, M.; Robazza, C.; Ruiz, M.C.; Cataldi, S.; Fischetti, F.; Bortoli, L. Gender-Typed Sport Practice, Physical Self-Perceptions, and Performance-Related Emotions in Adolescent Girls. *Sustainability* **2020**, *12*, 8518. [CrossRef]
41. Stival, C.; Lugo, A.; Barone, L.; Fattore, G.; Odone, A.; Salvatore, S.; Santoro, E.; Scaglioni, S.; van den Brandt, P.A.; Gallus, S.; et al. Prevalence and Correlates of Overweight, Obesity and Physical Activity in Italian Children and Adolescents from Lombardy, Italy. *Nutrients* **2022**, *14*, 2258. [CrossRef]
42. Sánchez-Villegas, A.; Madrigal, H.; Martínez-González, M.A.; Kearney, J.; Gibney, M.J.; de Irala, J.; Martínez, J.A. Perception of Body Image as Indicator of Weight Status in the European Union. *J. Hum. Nutr. Diet.* **2001**, *14*, 93–102. [CrossRef]
43. Stunkard, A.J.; Sørensen, T.; Schulsinger, F. Use of the Danish Adoption Register for the Study of Obesity and Thinness. *Res. Publ. Assoc. Res. Nerv. Ment. Dis.* **1983**, *60*, 115–120. [PubMed]
44. Mciza, Z.; Goedecke, J.H.; Steyn, N.P.; Charlton, K.; Puoane, T.; Meltzer, S.; Levitt, N.S.; Lambert, E.V. Development and Validation of Instruments Measuring Body Image and Body Weight Dissatisfaction in South African Mothers and Their Daughters. *Public Health Nutr.* **2005**, *8*, 509–519. [CrossRef] [PubMed]
45. Zaccagni, L.; Masotti, S.; Donati, R.; Mazzoni, G.; Gualdi-Russo, E. Body Image and Weight Perceptions in Relation to Actual Measurements by Means of a New Index and Level of Physical Activity in Italian University Students. *J. Transl. Med.* **2014**, *12*, 42. [CrossRef] [PubMed]
46. Cole, T.J.; Lobstein, T. Extended International (IOTF) Body Mass Index Cut-Offs for Thinness, Overweight and Obesity. *Pediatr. Obes.* **2012**, *7*, 284–294. [CrossRef] [PubMed]
47. Cole, T.J.; Flegal, K.M.; Nicholls, D.; Jackson, A.A. Body Mass Index Cut Offs to Define Thinness in Children and Adolescents: International Survey. *BMJ* **2007**, *335*, 194. [CrossRef]
48. Dowson, J.; Henderson, L. The Validity of a Short Version of the Body Shape Questionnaire. *Psychiatry Res.* **2001**, *102*, 263–271. [CrossRef]
49. Cooper, P.J.; Taylor, M.J.; Cooper, Z.; Fairburn, C.G. The Development and Validation of the Body Shape Questionnaire. *Int. J. Eat. Disord.* **1987**, *6*, 485–494. [CrossRef]
50. Matera, C.; Nerini, A.; Stefanile, C. Assessing Body Dissatisfaction: Validation of the Italian Version of the Body Shape Questionnaire-14 (BSQ-14). *Counseling* **2013**, *6*, 235–244.
51. Galfo, M.; D'Addezio, L.; Censi, L.; Roccaldo, R.; Martone, D. Overweight and Obesity in Italian Adolescents: Examined Prevalence and Socio-Demographic Factors. *Cent. Eur. J. Public Health* **2016**, *24*, 262–267. [CrossRef]
52. Bianco, A.; Filippi, A.R.; Breda, J.; Leonardi, V.; Paoli, A.; Petrigna, L.; Palma, A.; Tabacchi, G. Combined Effect of Different Factors on Weight Status and Cardiometabolic Risk in Italian Adolescents. *Ital. J. Pediatr.* **2019**, *45*, 32. [CrossRef]

53. Lichtenstein, M.B.; Hemmingsen, S.D.; Støving, R.K. Identification of Eating Disorder Symptoms in Danish Adolescents with the SCOFF Questionnaire. *Nord. J. Psychiatry* **2017**, *71*, 340–347. [CrossRef]
54. D'Anna, G.; Lazzeretti, M.; Castellini, G.; Ricca, V.; Cassioli, E.; Rossi, E.; Silvestri, C.; Voller, F. Risk of Eating Disorders in a Representative Sample of Italian Adolescents: Prevalence and Association with Self-Reported Interpersonal Factors. *Eat. Weight. Disord.* **2022**, *27*, 701–708. [CrossRef]
55. Pinho, L.d.; Brito, M.F.S.F.; Silva, R.R.V.; Messias, R.B.; Silva, C.S.d.O.e.; Barbosa, D.A.; Caldeira, A.P. Perception of Body Image and Nutritional Status in Adolescents of Public Schools. *Rev. Bras. Enferm.* **2019**, *72*, 229–235. [CrossRef]
56. Nagata, J.M.; Ganson, K.T.; Murray, S.B. Eating Disorders in Adolescent Boys and Young Men: An Update. *Curr. Opin. Pediatr.* **2020**, *32*, 476–481. [CrossRef]
57. Bucchianeri, M.M.; Arikian, A.J.; Hannan, P.J.; Eisenberg, M.E.; Neumark-Sztainer, D. Body Dissatisfaction from Adolescence to Young Adulthood: Findings from a 10-Year Longitudinal Study. *Body Image* **2013**, *10*, 1–7. [CrossRef]
58. Abbott, B.D.; Barber, B.L. Embodied Image: Gender Differences in Functional and Aesthetic Body Image among Australian Adolescents. *Body Image* **2010**, *7*, 22–31. [CrossRef]
59. Shriver, L.H.; Harrist, A.W.; Page, M.; Hubbs-Tait, L.; Moulton, M.; Topham, G. Differences in Body Esteem by Weight Status, Gender, and Physical Activity among Young Elementary School-Aged Children. *Body Image* **2013**, *10*, 78–84. [CrossRef]
60. Fortes, L.d.S.; Cipriani, F.M.; Coelho, F.D.; Paes, S.T.; Ferreira, M.E.C. Does self-esteem affect body dissatisfaction levels in female adolescents? *Rev. Paul Pediatr.* **2014**, *32*, 236–240. [CrossRef]
61. Lewer, M.; Bauer, A.; Hartmann, A.S.; Vocks, S. Different Facets of Body Image Disturbance in Binge Eating Disorder: A Review. *Nutrients* **2017**, *9*, 1294. [CrossRef]
62. Dzielska, A.; Kelly, C.; Ojala, K.; Finne, E.; Spinelli, A.; Furstova, J.; Fismen, A.-S.; Ercan, O.; Tesler, R.; Melkumova, M.; et al. Weight Reduction Behaviors Among European Adolescents—Changes From 2001/2002 to 2017/2018. *J. Adolesc. Health* **2020**, *66*, S70–S80. [CrossRef]
63. Goldfield, G.S.; Moore, C.; Henderson, K.; Buchholz, A.; Obeid, N.; Flament, M.F. Body Dissatisfaction, Dietary Restraint, Depression, and Weight Status in Adolescents. *J. Sch. Health* **2010**, *80*, 186–192. [CrossRef] [PubMed]
64. Gonçalves, S.; Machado, B.C.; Martins, C.; Hoek, H.W.; Machado, P.P.P. Retrospective Correlates for Bulimia Nervosa: A Matched Case—Control Study. *Eur. Eat. Disord. Rev.* **2016**, *24*, 197–205. [CrossRef] [PubMed]
65. Lebow, J.; Sim, L.A.; Kransdorf, L.N. Prevalence of a History of Overweight and Obesity in Adolescents with Restrictive Eating Disorders. *J. Adolesc. Health* **2015**, *56*, 19–24. [CrossRef] [PubMed]
66. Guarino, R.; Pellai, A.; Bassoli, L.; Cozzi, M.; Di Sanzo, M.A.; Campra, D.; Guala, A. Overweight, Thinness, Body Self-Image and Eating Strategies of 2,121 Italian Teenagers. Available online: https://www.hindawi.com/journals/tswj/2005/873485/ (accessed on 12 September 2022).

Disclaimer/Publisher's Note: The statements, opinions and data contained in all publications are solely those of the individual author(s) and contributor(s) and not of MDPI and/or the editor(s). MDPI and/or the editor(s) disclaim responsibility for any injury to people or property resulting from any ideas, methods, instructions or products referred to in the content.

Systematic Review

Recreational Soccer Training Effects on Pediatric Populations Physical Fitness and Health: A Systematic Review

Filipe Manuel Clemente [1,2,]*, Jason Moran [3], Rodrigo Ramirez-Campillo [4], Rafael Oliveira [5,6,7], João Brito [5,6,7], Ana Filipa Silva [1,6,8], Georgian Badicu [9], Gibson Praça [10] and Hugo Sarmento [11]

[1] Escola Superior Desporto e Lazer, Instituto Politécnico de Viana do Castelo, Rua Escola Industrial e Comercial de Nun'Álvares, 4900-347 Viana do Castelo, Portugal
[2] Instituto de Telecomunicações, Delegação da Covilhã, 1049-001 Lisboa, Portugal
[3] School of Sport, Rehabilitation and Exercise Sciences, University of Essex, Colchester CO4 3WA, UK
[4] Exercise and Rehabilitation Sciences Institute, School of Physical Therapy, Faculty of Rehabilitation Sciences, Universidad Andres Bello, Santiago 7591538, Chile
[5] The Research Centre in Sports Sciences, Health Sciences and Human Development (CIDESD), 5001-801 Vila Real, Portugal
[6] Sports Science School of Rio Maior, Polytechnic Institute of Santarém, 2040-413 Rio Maior, Portugal
[7] Life Quality Research Centre, 2040-413 Rio Maior, Portugal
[8] Research Center in Sports Performance, Recreation, Innovation and Technology (SPRINT), 4960-320 Melgaço, Portugal
[9] Department of Physical Education and Special Motricity, Faculty of Physical Education and Mountain Sports, Transilvania University of Brasov, 500068 Brasov, Romania
[10] Sports Department, Universidade Federal de Minas Gerais, Belo Horizonte 31270-901, Brazil
[11] University of Coimbra, Research Unit for Sport and Physical Activity (CIDAF), Faculty of Sport Sciences and Physical Education, 3000-248 Coimbra, Portugal
* Correspondence: filipe.clemente5@gmail.com

Citation: Clemente, F.M.; Moran, J.; Ramirez-Campillo, R.; Oliveira, R.; Brito, J.; Silva, A.F.; Badicu, G.; Praça, G.; Sarmento, H. Recreational Soccer Training Effects on Pediatric Populations Physical Fitness and Health: A Systematic Review. *Children* **2022**, *9*, 1776. https://doi.org/10.3390/children9111776

Academic Editor: Niels Wedderkopp

Received: 26 October 2022
Accepted: 16 November 2022
Published: 18 November 2022

Publisher's Note: MDPI stays neutral with regard to jurisdictional claims in published maps and institutional affiliations.

Copyright: © 2022 by the authors. Licensee MDPI, Basel, Switzerland. This article is an open access article distributed under the terms and conditions of the Creative Commons Attribution (CC BY) license (https://creativecommons.org/licenses/by/4.0/).

Abstract: This systematic review analyzed the effects of recreational soccer programs on physical fitness and health-related outcomes in youth populations. Studies were sought in the following databases: (i) PubMed, (ii) Scopus, (iii) SPORTDiscus, and (iv) Web of Science. The eligibility criteria included (1) population: youth (<18 years old) populations with no restrictions on sex or health condition; (2) intervention: exposure to a recreational soccer training program of at least four weeks duration; (3) comparator: a passive or active control group not exposed to a recreational soccer training program; (4) outcomes: physical fitness (e.g., aerobic, strength, speed, and change-of-direction) or health-related measures (e.g., body composition, blood pressure, heart rate variability, and biomarkers); (5) study design: a randomized parallel group design. The search was conducted on 6 September 2022 with no restrictions as to date or language. The risk of bias was assessed using the PEDro scale for randomized controlled studies. From a pool of 37,235 potentially relevant articles, 17 were eligible for inclusion in this review. Most of the experimental studies revealed the beneficial effects of recreational soccer for improving aerobic fitness and its benefits in terms of blood pressure and heart-rate markers. However, body composition was not significantly improved by recreational soccer. The main results revealed that recreational soccer training programs that are implemented twice a week could improve the generality of physical fitness parameters and beneficially impact cardiovascular health and biomarkers. Thus, recreational soccer meets the conditions for being included in the physical education curriculum as a good strategy for the benefit of the general health of children and young people.

Keywords: football; sports medicine; physical exercise; physical fitness; physical conditioning; child; adolescent

1. Introduction

Youth populations are experiencing rising levels of obesity and declining levels of physical activity in recent times [1,2]. Moreover, epidemiological studies indicate the rise of health issues related to inactivity and sedentarism [3]. Consequently, there is growing concern over the effects of sedentary lifestyles on the health-based parameters of young people [4] as this sedentary behavior might be related to reduced quality of life and life expectancy. This issue might be explained by the different cultural and social changes experienced in recent decades, such as the increase in time spent looking at screens (TVs and video games, for example) and the development of motorized transport systems, which reduce the demand for actively physical behaviors in the youth population [5]. Hence, the World Health Organization strongly recommended in their 2020 physical activity guidelines for children and adolescents that these populations should be exposed to a minimum of 60 min/day of moderate-to-vigorous intensity physical activity. In the particular case of vigorous activities, these should be performed at least three times per week, focusing on vigorous aerobic exercise and strength-related activities that strengthen muscles and bones [6]. Based on that recommendation, programs focusing on providing youths with physical activity have recently been proposed and tested. These programs can include physical education activities, which are usually offered to children during schooltime, and recreational practices, such as soccer, which are not necessarily related to scholarly activities. In the current study, we will review the effects of recreational soccer physical activity programs on the respective pediatric populations' physical fitness and health.

Studies have shown that the time spent in sedentary activities, defined as low-energy expenditure during waking hours, accounts for a large proportion of the day, between roughly 50 and 60% [2]. Physical activity has also been reported to reduce morbidity and mortality in adolescence [7]. Therefore, exercise—a non-sedentary behavior—might be seen as a non-pharmacological approach to decreasing the exposure to health risks for sedentary populations [5]. For example, aerobic training programs improved health parameters, such as insulin action and plasma lipids, in sedentary people [8]. Therefore, it seems that these other activities might also benefit other sedentary populations.

Different training programs could be offered to the general population when considering physical activities on a wider scale. For example, more prolonged-duration low-intensity aerobic exercises showed positive effects in sedentary subjects [3]. On the other hand, recreational sports participation—such as soccer practices—might offer higher motivation and enjoyment [9]. Indeed, a previous systematic review showed that activities with highlighted social, motivational, and competitive components were as effective and efficient as continuous running-based programs [10]. Another previous study showed that running and football-based training programs effectively improved the health-based parameters of sedentary women. However, training-induced cardiac adaptations appeared to be more consistent after football training than after running [11]. Furthermore, another study showed that recreational soccer training programs could also offer an opportunity to positively impact body composition, in comparison to people performing continuous running [12]. Hence, sports-based activities should increase adherence and improve upon the positive effects of physical activity on health parameters.

Previous studies have tested if recreational soccer practice [13,14] can positively impact physical fitness (e.g., aerobic capacity, strength, power, speed, and change of direction) health-based parameters, such as body composition (e.g., fat mass) and biomarkers (e.g., insulin) or other markers (e.g., blood pressure and heart rate variability). For example, a systematic review showed that recreational soccer increased cardiovascular and bone health and improved body composition [15]. Another study demonstrated the positive effects of recreational soccer practice on untrained women's risk factors for bone fractures [16]. With regard to pediatric populations, it has been shown that recreational soccer can reduce risk factors associated with obesity [17] and improve different health markers [18]. Indeed, in relation to pediatric populations, adherence to physical activity programs through recreational sports practice seems even more relevant. Soccer is a popular sport worldwide and many children support clubs and players regularly. Therefore, including recreational soccer instead of generic physical activity practices could facilitate long-term adherence to physical activity and enhance the positive effects of this practice on health-related parameters in youth. Indeed, a previous study showed that greater motivation was associated with increased recreational sports participation [19]. However, to our knowledge, no previous systematic review has examined the effects of recreational soccer on health-based parameters in pediatric populations, a gap in knowledge that is addressed by the current study.

The diversity of systematic reviews in recreational soccer is clear from the extant literature [20–27]. As can be seen from the available systematic reviews, some have focused on specific outcomes, such as bone health [26,27], fat mass [22], or maximal oxygen uptake [20], while others summarized the effects on different outcomes [21,23–25]. Additionally, half of these systematic reviews were exclusively dedicated to adult populations [21,22,26,27], while the remaining reviews focused on mixed populations, such as children, adolescents, adults, and older individuals (both healthy and unhealthy), thus making it difficult to determine a clear overview of the effects on children and youths [20,23–25]. Considering the relevance of physical exercise for promoting positive physical fitness adaptations in children and young people, and the opportunities offered by recreational soccer, there is scope for a systematic review of this population. A systematic review of recreational soccer's effects on health-related outcomes and the physical fitness of children and untrained youth populations can facilitate a precise understanding of the methodologies used in interventions and identify the adaptations induced by such interventions, in comparison to control groups who are not exposed to the same stimulus. Therefore, this systematic review aimed to analyze the effects of recreational soccer programs on physical fitness (e.g., aerobic capacity, strength, power, speed, and change of direction) and health-related markers (e.g., body composition, blood pressure, heart rate variability, and biomarkers) in untrained child and youth populations. This systematic review mainly focuses on comparisons with the control groups, aiming to understand the real effects of increased activity on health-related outcomes and physical fitness.

2. Materials and Methods

This systematic review followed the PRISMA 2020 guidelines [28]. The systematic review protocol was first submitted and then published on the Open Science Framework, with the registration number 10.17605/OSF.IO/FY4PX on 6 September 2022. The protocol can be accessed via the web address https://osf.io/nuebg/?view_only=11a89f39e2b34516 a482953db17d2acf (accessed on 5 September 2022).

2.1. Eligibility Criteria

Original articles published in peer-reviewed journals or "ahead of print" were eligible for consideration. We imposed no restrictions on the language in which the gathered articles were written [29]. Table 1 presents the eligibility criteria, based on the PICOS criteria.

2.2. Information Sources

The following databases were searched: (i) PubMed; (ii) Scopus, (iii) SPORTDiscus, and (iv) Web of Science (core collection). The searches were performed on 6 September 2022. Additionally, manual searches were performed on the included studies' reference lists to identify potentially relevant titles. The abstracts of these articles were checked for the relevant inclusion criteria, and, if necessary, the full text was investigated. Moreover, a consultation of two external experts (as recognized by Expertscape at the Worldwide level: https://expertscape.com/ex/soccer) (accessed on 5 September 2022) was performed, aiming to strengthen the search. Errata and article retractions were searched for each of the included articles, to identify possible sources of bias [31].

Table 1. Eligibility criteria for the current study.

	Inclusion Criteria	**Exclusion Criteria**
Population	Youth populations (under 18) with no restriction on sex or clinical conditions. Populations were included in Tier 0, indicating sedentary behavior, or Tier 1, indicating recreationally active, of the participant classification framework [30]; this means that: (Tier 0) they do not meet minimum activity guidelines, and, thus, can be considered inactive; or (Tier 1) they meet the World Health Organization minimum activity target and/or may participate in multiple sports/forms of activity.	Adults (>18 years old) or youths enrolled in Tiers 2 to 5 of the participant classification framework [30].
Intervention	Players were exposed to a structured recreational soccer training program for a minimum of four weeks, with no restrictions on the maximum length. Similarly, there were no restrictions on training volume, intensity, or weekly training frequency.	Exposed to less than four weeks of training intervention. Exposed to training programs for other sports than soccer.
Comparator	Passive control groups (not exposed to other training interventions, while retaining their regular physical activity levels and lifestyle) or active control groups (exposed to other exercise programs, not including recreational soccer training)	Exposed to training programs, which included recreational soccer.
Outcomes	Physical fitness outcomes (e.g., cardiorespiratory measures, speed or change-of-direction measures, muscular strength and power measures, and balance measures) and/or body characteristics or composition (e.g., body mass index, fat mass, and lean mass) and/or health-related markers (e.g., biochemical markers and inflammatory markers)	Acute physiological and/or physical responses (i.e., responses to a single training session or those experienced during exercise). Socio/psychological factors. Technical/tactical factors.
Study design	Randomized parallel group design.	No randomized designs. No controlled designs.

2.3. Search Strategy

The search was conducted using the Boolean operators AND/OR. No filters or limitations were applied to the publication date or language, to increase the chances of identifying appropriate studies [32]. The search strategy presented the following codes:

[Title/Abstract] "Soccer*" OR "Football*"
AND
[All fields/Full text] "recreation*" OR "untrain*" OR "health"
The entire search strategy can be found in Table 2.

2.4. Selection Process

The retrieved records (title, abstracts, and full texts) were independently screened by two of the authors (F.M.C. and H.S.). Disagreements between the two authors were discussed in a joint reanalysis. In cases where no consensus could be reached, a third author (A.F.S.) participated in a collaborative meeting to come to a final decision. The EndNote™

20.3 software (Clarivate^TM) was used to manage the records, including the removal of duplicates, either automatically or manually.

Table 2. The complete search strategy for each database.

Database	Specificities of the Databases	Search Strategy	Number of Articles in Automatic Search
PubMed	None to report	(recreation * OR untrain * OR health) AND (Soccer [Title/Abstract] OR Football [Title/Abstract])	7700
Scopus	Search for title and abstract also includes keywords	(TITLE-ABS-KEY (soccer OR football) AND ALL (recreation * OR untrain * OR health))	18,909
SPORTDiscus	None to report	TI (soccer or Football) AND TX (recreation * OR untrain * OR health)	4752
Web of Science	Search for title and abstract also includes keywords and its designated "topic"	Soccer OR Football (Title) and recreation * OR untrain * OR health (Topic)	5874

*: is the code for extension of the word.

2.5. Data Extraction Process

Two authors (F.M.C. and A.F.S.) independently extracted the data from the included studies. Information about study methods, results, and principal conclusions was extracted. A third author (H.S.) verified the collected information and helped in the case of any disagreements. A Microsoft® Excel worksheet was designed to collect the data. If relevant information was omitted from an article, the corresponding author was contacted to help to obtain the required information. If no reply was received after establishing the first contact, we sent a second message after three days. Following that, the author was contacted twice in two days, using the same message to achieve a response or attain one two weeks after the first contact.

2.6. Data Items

Participant-related and context information was obtained for the following items: the date of publication, the main goal of the research; sample size; country of origin; age; sex and clinical information.

The intervention-related information included: the timing of the academic season; program duration; training frequency; level of adherence to training; dose (e.g., duration, repetitions, rest, intensity, frequency, and density); rules of play; format of play and pitch size.

The physical fitness outcomes included (but were not restricted to) cardiorespiratory-related measures (e.g., maximal oxygen uptake, maximal aerobic speed, maximal heart rate), neuromuscular-related measures (e.g., muscular power and strength), speed and change-of-direction-related measures (e.g., sprint performance, change-of-direction performance), and balance and mobility-related measures (e.g., dynamic and static balance). Health-related outcomes focused on body characteristics and body composition (e.g., body mass index, lean mass, and fat mass), blood pressure (e.g., systolic and diastolic pressure), echocardiographic measures (e.g., cardiac output), bone health (e.g., bone mineral content), biochemical parameters (e.g., total cholesterol and glucose tolerance), and inflammatory parameters (e.g., leptin).

The comparators included information about passive control groups or active control groups (namely, type of exercise, intensity, and volume).

2.7. Study Risk of Bias Assessment

The physiotherapy evidence database (PEDro) scale was utilized to assess the risk of bias in the included studies. This scale has previously been tested for validity and reliability [33]. The scale facilitates the rating of eleven specific study criteria, ten of which are used to classify the overall score of the article, which ranges from 0 (lowest quality) to 10 (highest quality). Usually, score thresholds provide a qualitative classification of "poor" (<4 points), "fair" (4–5 points), "good" (6–8 points), and "excellent" (9–10 points). The scale assesses the following items: C1 means that the eligibility criteria were specified; C2 means that subjects were randomly allocated to groups; C3 means that allocation was concealed; C4 means that the groups were similar at baseline regarding the most important prognostic indicators; C5 means that blinding was applied to all subjects; C6 means that there was blinding of all therapists who administered the therapy; C7 means that there was blinding of all assessors who measured at least one key outcome; C8 means that the measures of at least one key outcome were obtained from more than 85% of the subjects who were initially allocated to groups; C9 means that all subjects for whom outcome measures were available received the treatment or control condition as allocated or, where this was not the case, the data for at least one key outcome were analyzed by "intention to treat"; C10 means that the results of between-group statistical comparisons are reported for at least one key outcome; C11 means that the study provides both point measures and measures of variability for at least one key outcome. Two authors (F.M.C. and R.O.) independently reviewed and rated the included articles, based on the PEDro scale. After that, two authors (F.M.C. and R.O.) shared the scores and discussed them on a point-by-point basis. In cases where a consensus could not be reached, a third author (A.F.S.) was invited to provide their own score and make a final decision.

3. Results

3.1. Study Identification and Selection

The initial search resulted in the identification of 37,235 titles (Figure 1). Duplicates (10,444 titles) were subsequently removed, either automatically or manually. The remaining 26,791 titles were screened for relevance, based on their titles and abstracts. Of those papers, 26,727 titles were removed. The full texts of the remaining 64 titles were then inspected, and from there, 48 more were removed, based on the eligibility criteria. Four of the potentially included articles [34–37] reported the same clinical trial registration number (NCT02000492). Based on that finding and considering that multiple reports of the same study should be collated so that each study, rather than each report, is the unit of interest in the review, as suggested by Cochrane, we chose only article [38] to remain. From the manual searches, five articles were retrieved. Of those, one was considered eligible for inclusion in the systematic review. Following the full search, 17 articles were included in the final analysis.

3.2. Study Characteristics

The current study's characteristics can be found in Table 3. Among the included studies, nine were conducted in Denmark, with two performed in Brazil. Germany, China, the Faroe Islands, and Saudi Arabia yielded one study each. The remaining articles did not present any details about each study's country of origin.

The characteristics of the recreational soccer training programs can be found in Table 4. The range of training program durations provided was between a minimum of eight weeks and a maximum of ten months. Weekly training frequency varied from two to five days. Training duration varied between 12 and 90 min. Small-sided games were used as training drills in most of the studies.

The characteristics of the control groups included in the studies can be found in Table 5. Most control groups comprised youths who were only enrolled in regular physical education classes.

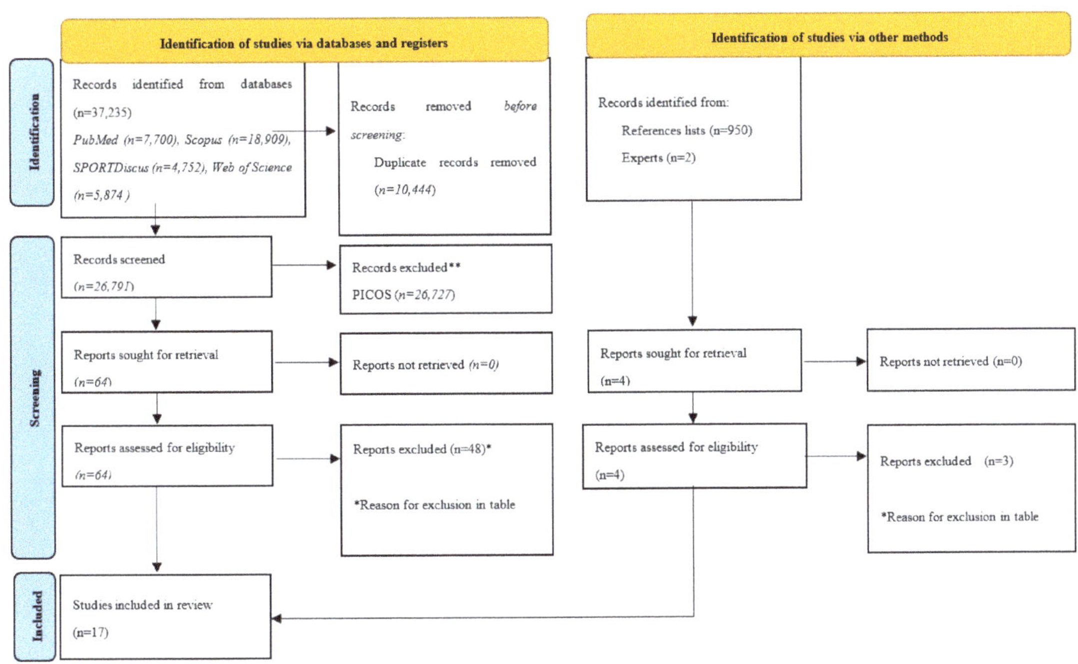

Figure 1. PRISMA 2020 flow diagram. * Reason for exclusion in table ** records excluded.

Table 3. Characteristics of the included studies.

Study	Clinical Registration	Country	N	Age (Years)	Sex	Assessments (Number)	Tests Applied	Outcomes Presented in the Study
[39]	Not reported	Germany	22	10.8 ± 1.2	Both	3 (pre, mid, post)	Anthropometry; cycling ergometry; CMJ; sit-and-reach; OLS; agility run; 20-m shuttle run test	BMI; POmax; VO$_2$max; lactate max; HRmax; CMJ height; change-of-direction time; 20-m shutle run time; HRmax at 20-m shutle run test; OLS time; sit-and-reach distance;
[34]	NCT01711892	Denmark	97	9.4 ± 0.4	Both	2 (pre, post)	Echocardiography; Anthropometry	BMI; systolic and diastolic BP; resting HR; LVDD; LVSD; LV; LA; LVEF; IVT; LVPWD; CO; DT; IVRT; RVDD; TAPSE
[40]	Not reported	Denmark	526	11.1 ± 0.4 and 11.0 ± 0.5	Both	2 (pre, post)	Anthropometry; X-ray absorptiometry; arterial blood pressure; 20-m sprint test; horizontal jump length; YYIRT; flamingo balance test	BMI; body fat; lean mass; systolic and diastolic BP; mean arterial pressure; resting HR; YYIRT distance; 20-m sprint time; flaming balance; horizontal jump length.
[41]	TCTR20150512001	Brazil	30	14.1 ± 1.3 and 14.8 ± 1.4	Both	2 (pre, post)	Anthropometry; X-ray absorptiometry; arterial blood pressure; cardiopulmonary exercise testing; heart rate variability; biochemical markers; endothelial function assessment; inflammatory biomarkers	BMI; body fat; fat-free mass; systolic and diastolic BP; mean blood pressure; maximal oxygen uptake; HRmax; total cholesterol; HDL; LDL; triglycerides; C-reactive protein; fasting glucose; glucose tolerance; insulin; HOMA-IR; leptin; IL-6; resistin; TNF-α; adiponectin; ET-1; NEFA.
[42]	Not reported	NA	22	15.9 ± 0.6	Boys	2 (pre, post)	Anthropometry; arterial blood pressure; heart rate variability; resting heart rate; YYIRT; 20-m sprint test; bilateral standing long jump; stork balance test.	Sum of skinfolds; systolic and diastolic BP; HR rest; *Ln SDNN*; *Ln rMSSD*; *Ln HF*; *HF*; *Ln LF*; *LF*; *LF/HF*; *SD1*; YYIRT distance; balance time; sprint time; jump distance.

Table 3. *Cont.*

Study	Clinical Registration	Country	N	Age (Years)	Sex	Assessments (Number)	Tests Applied	Outcomes Presented in the Study
[35]	Not reported	NA	35	11 to 13	Boys	2 (pre, post)	CMJ test; 10- and 30-m sprints; leg spreading, lying on the back; flexibility of the body when bending; flexibility of the body when stretching; biochemical markers.	CMJ; Sprint time at 10 and 30 m; leg spreading lying on the back; flexibility of the body when bending; flexibility of the body when stretching; leukocytes; erythrocytes; hemoglobin; glucose; cholesterol; triglycerides.
[36]	Not reported	NA	35	11 to 13	Boys	2 (pre, post)	Bioimpedance; CMJ test; agility *t*-test; sit-and-reach test; YYIRT; blood pressure.	BMI; body fat; lean body mass; muscle mass; CMJ; agility test; sit-and-reach test; YYIRT; resting and maximal HR; systolic and diastolic BP.
[43]	Not reported	NA	20	15.9 ± 0.6	Boys	2 (pre, post)	Anthropometry; YYIRT; 10- and 20-m sprint; sit-and-reach test; CMJ; standing long jump; stork balance test.	BMI; body fat; lean mass; sprint time; CMJ height; standing long-jump; sit-and-reach distance; balance; YYIRT distance.
[38]	NCT02000492	Denmark	295	9.3 ± 0.4	Both	2 (pre, post)	Anthropometry; X-ray absorptiometry; flamingo balance test; horizontal jump test; 20-m sprint test; coordination wall with three stages of increased difficulty.	Bone mineral content; lean mass; areal bone mineral density; the number of falls in the balance test; sprint time; coordination ability.
[44]	H-16026885	Denmark	931	11.9 ± 0.4 11.8 ± 0.2	Both	2 (pre, post)	Cognitive test battery, including detection, identification, and one-back and one-card learning tasks.	Psychomotor function; attention; working memory; visual memory.
[45]	Not reported	Faroe Islands	491	11.1 ± 0.3	Both	2 (pre, post)	Anthropometry; X-ray absorptiometry; blood pressure; stork balance test; horizontal jump test; YYIRT.	Systolic and diastolic BP; mean arterial pressure; resting HR; BMI; body fat; lean body mass; horizontal jump; postural balance; YYIRT distance.
[37]	Not reported	China	38	9 to 10	Boys	2 (pre, post)	Anthropometry; X-ray absorptiometry; 20-m shuttle run test; 50-m sprint; standing long jump; handgrip; 1-min sit up; sit-and-reach; single-leg standing,	BMI; body fat; fat mass and fat-free mass; maximal oxygen uptake; sprint time; standing long jump distance; 1-min sit-up; core muscle function; body balance; heart rate index.
[46]	No reported	NA	105	15.7 ± 0.6	Both	2 (pre, post)	Anthropometry; backward overhead medicine ball (3 kg) throw test; vertical jump test; YYIRT.	BMI; backward overhead medicine ball throw distance; vertical jump; YYIRT distance
[47]	TCTR20150512001	Brazil	13	13.9 ± 1.6 14.7 ± 2.3	Both	2 (pre, post)	Anthropometry; blood pressure; biochemical markers.	BMI; systolic and diastolic BP; body fat; HDL: triglycerides; fasting blood glucose level.
[48]	Not reported	Saudi Arabia	30	14.4 ± 2.0 15.6 ± 1.8 17.8 ± 0.4	Boys	2 (pre, post)	Blood pressure; blood glucose monitoring; biochemical markers.	LDL; HDL; triglyceride; systolic and diastolic blood pressure; total day insulin; fasting blood glucose; HbA1c.
[49]	REC-010712	NA	53	17.0 ± 0.6 16.7 ± 0.4 16.7 ± 0.4	Boys	2 (pre, post)	Anthropometry; multistage fitness test; push-up test; abdominal curl conditioning test; blood pressure.	BMI; body fat; systolic and diastolic BP; resting HR; VO2max; sit-ups; push-ups.
[50]	H-16026885	Denmark	1122	11.6 ± 0.5 11.4 ± 0.5	Both	2 (pre, post)	Anthropometry; bioimpedance; blood pressure; YYIRT; stork balance stand test; standing long jump.	YYIRT distance; VO2max; BMI; standing forward jump; balance; body fat; muscle mass; systolic and diastolic BP; resting HR.

Abbreviations: CMJ: counter-movement jump; BMI: body mass index; POmax: maximal power output; VO2max: maximal oxygen uptake; HR: heart rate; OLS: one-leg-standing; BP: blood pressure; CO, cardiac output; DT, transmitral deceleration time; IVRTglobal, global isovolumetric relaxation time; IVT, interventricular septum thickness; LA, left atrial; LVDD, left ventricular diastolic diameter; LVEF, left ventricular ejection fraction; LVPWD, left ventricular posterior wall diameter; LVSD, left ventricular systolic diameter; RVDD, right ventricular diastolic diameter; TAPSE, tricuspid annular plane systolic excursion; YYIRT: yo-yo intermittent recovery test level 1; IL-6, interleukin-6; TNF-α, tumoral necrosis factor-α; ET-1, endothelin-1; NEFA, non-esterified fatty acids; HDL, high-density lipoprotein; LDL, low-density lipoprotein; Ln, normal logarithm; SDNN, standard deviation of the normalized R–R intervals; RMSSD, root mean square of the standard deviation; HF, high frequency; LF, low frequency, SD1 geometric parameter of the Poincaré plot; LNSDNN: log-natural standard deviation of the NN (R-R) intervals; NA: not available.

Table 4. Characteristics of the recreational soccer training programs.

Study	Training Attendance	Duration	Days Per Week	Total Sessions	Training Duration (min)	Sets (n)	Recovery (min)	Work Duration (min)	Work Intensity	Training Drills
[39]	60 to 69%	6 months	3	54	60	NA	NA	NA	$80 \pm 8\%$ HRmax	Warm-up; SSGs (50%); technique (20%); fitness courses with a ball (20%)
[34]	$77 \pm 18\%$	10 weeks	3	21 ± 5	49	NA	NA	NA	$71 \pm 6\%$ HRmax	Warm-up; SSGs.
[40]	NA	11 weeks	2	22	45	NA	NA	NA	NA	Technique; SSGs
[41]	NA	12 weeks	3	36	52.1 ± 5.6	NA	NA	NA	$84.5 \pm 4.1\%$ HRmax	Warm-up (10 min); SSGs (40 min; cool-down (10 min)
[42]	NA	8 weeks	2	16	NA	NA	NA	NA	NA	Warm-up; SSGs (30–45 min)
[35]	>50%	12 weeks	NA	NA	60	4	2 min	8 min per set	$75.1 \pm 2.3\%$ HRmax	Warm-up (10 min); SSGs (32 min); cool-down (10 min)
[36]	>50%	12 weeks	NA	NA	60	4	2 min	8 min per set	$75.1 \pm 2.3\%$ HRmax	Warm-up (10 min); SSGs (32 min); cool-down (10 min)
[43]	NA	8 weeks	2	16	30–45	NA	NA	NA	$84.6 \pm 6.3\%$ HRpeak	Warm-up; SSGs (30–45 min)
[38]	NA	10 months	3	NA	NA	NA	NA	NA	0.48 ± 0.15 arbitrary units player load	Warm-up (3 to 5 min); SSGs
[44]	NA	11 weeks	2	22	45	NA	NA	NA	NA	NA
[45]	NA	11 weeks	2	22	45	NA	NA	NA	NA	SSGs.
[37]	NA	10 weeks	3	30	60	NA	NA	NA	NA	Warm-up (10 min); dribbling (10 min); dribbling and shooting (10 min); passing (10 min); running (10 min); cool-down (10 min).
[46]	>85%	32 weeks	2	64	45	4	3	5	NA	Warm-up (10 min); stretching (4 min); acceleration running (2 min); soccer (30 min); cool-down (5 min)
[47]	NA	12 weeks	3	36	60	NA	NA	NA	$84.5 \pm 4.1\%$ HRmax	Warm-up (10 min); SSGs (40 min); cool-down (10 min)
[48]	21–24 n	12 weeks	2	24	90	NA	NA	NA	~80% HRmax	Warm-up (5–10 min); game
[49]	NA	8 weeks	NA	28	60	NA	NA	NA	NA	NA
[50]	NA	11 weeks	2	22	45	NA	NA	NA	NA	NA

Abbreviations: SSGs: small-sided games; NA: not available; HRmax: maximal heart rate.

Table 5. Characteristics of control groups.

Study	Characteristic	Duration/Frequency	Attendance	Training Intensity	
[39]	One group performing standard classes	6 months/thrice a week	72%	$77 \pm 6\%$	Warm-up; aerobic endurance activities (40%); coordination and flexibility (20%); strength (15%); speed (15%).
[34]	One group performing standard classes	10 weeks/twice a week	NA	NA	40 min of physical education classes.
[40]	One group performing standard classes	11 weeks/twice a week	NA	NA	45 min of physical education classes.
[41]	NA	NA	NA	NA	NA
[42]	Inactive group	NA	NA	NA	Kept their regular physical activity level.

Table 5. *Cont.*

Study	Characteristic	Duration/Frequency	Attendance	Training Intensity	
[35]	One high-intensity interval training group and one control group	12 weeks/NA	NA	High-intensity interval training (80.0 ± 3.0% HRmax) Control (68.3 ± 2.2% HRmax)	Warm-up (10 min); 3 sets of high-intensity interval runs (100% maximal aerobic speed) interspaced by 3 min of passive rest; cool down (10 min) The control group performed the regular physical education classes.
[36]	One high-intensity interval training group and one control group	12 weeks/NA	NA	High-intensity interval training (80.0 ± 3.0% HRmax) Control (68.3 ± 2.2% HRmax)	Warm-up (10 min); 3 sets of high-intensity interval runs (100% maximal aerobic speed) interspaced by 3 min of passive rest; cool down (10 min) The control group performed the regular physical education classes.
[43]	Control group enrolled in regular physical education classes.	8 weeks/2 sessions a week	NA	NA	One hour of physical education classes per session.
[38]	One group performed circuit strength training and one acted as the control.	10 months/3 sessions	NA	0.34 ± 0.09 arbitrary units player load	Circuit strength training consisted of 30 s of all-out exercise with 45 s rest in between. Six to ten stations were used focusing on plyometric and dynamic or static strength (upper and lower body).
[44]	Control group enrolled in regular physical education classes.	11 weeks/2 session	NA	NA	Regular physical education classes of 45 min each.
[45]	Control group enrolled in regular physical education classes.	11 weeks/2 session	NA	NA	Regular physical education classes of 45 min each.
[37]	Inactive group	NA	NA	NA	Kept their regular physical activity level.
[46]	Control group enrolled in regular physical education classes.	32 weeks/ 2 sessions week	NA	NA	Regular physical education classes.
[47]	Inactive group	32 weeks	NA	NA	Kept their regular physical activity level.
[48]	Diet-only group and control group	12 weeks	NA	NA	One group had a nutritional program without exercise and the other acted as a control not receiving the program.
[49]	Control (inactive)	NA	NA	NA	Kept their regular physical activity level.
[50]	Control group enrolled in regular physical education classes.	11 weeks/ 2 sessions week	NA	NA	Regular physical education classes.

Abbreviations: NA: not available; HRmax: maximal heart rate.

3.3. Risk of Bias in Studies

Table 6 presents the assessment of the risk of bias. The criteria with the lowest scores were the ones associated with eligibility, the allocation being concealed, the blinding of participants and the person who administrated the protocol, and the blinding of the person who made the assessments.

Table 6. Assessment of the risk of bias.

Study	C1	C2	C3	C4	C5	C6	C7	C8	C9	C10	C11	Score
[39]	0	1	0	1	1	0	0	1	1	1	1	7
[34]	0	1	0	1	0	0	1	1	1	1	1	7
[40]	0	1	0	1	0	0	1	1	1	1	1	7
[41]	0	1	1	1	1	0	1	1	1	1	1	9
[42]	1	1	0	1	0	0	0	1	1	1	1	7
[35]	1	1	0	1	0	0	0	1	1	1	1	7
[36]	1	1	0	0	0	1	0	1	1	1	1	7
[43]	1	1	0	1	0	0	0	1	1	1	1	7
[38]	0	1	0	1	0	0	0	0	1	1	1	5
[44]	0	1	0	0	0	0	0	1	1	1	1	5
[45]	0	1	0	1	0	0	0	1	1	1	1	6
[37]	1	1	1	1	0	0	0	1	1	1	1	8
[46]	0	1	0	1	0	0	0	1	1	1	1	6
[47]	1	1	0	1	0	0	1	1	1	1	1	7
[48]	1	1	1	1	0	1	1	1	1	1	1	10
[49]	0	1	0	1	0	0	0	1	1	1	1	6
[50]	0	1	0	0	0	0	0	1	1	1	1	5

C1: eligibility criteria were specified; C2: subjects were randomly allocated to groups; C3: allocation was concealed; C4: the groups were similar at baseline regarding the most important prognostic indicators; C5: there was blinding of all subjects; C6: there was blinding of all therapists who administered the therapy; C7: there was blinding of all assessors who measured at least one key outcome; C8: measures of at least one key outcome were obtained from more than 85% of the subjects initially allocated to groups; C9: all subjects for whom outcome measures were available received the treatment or control condition as allocated, or, where this was not the case, data for at least one key outcome were analyzed according to "intention to treat"; C10: the results of between-group statistical comparisons are reported for at least one key outcome; C11: the study provides both point measures and measures of variability for at least one key outcome.

3.4. Results of Individual Studies

Table 7 shows the main results for the physical fitness variables. The main findings showed improvements in cardiorespiratory fitness.

Table 7. Main findings of the physical fitness outcomes.

Study	Physical Fitness—Evidence of the Main Findings (Differences or Not after Training Programs in Terms of the Main Physical Fitness Outcomes)	General Effects of Soccer Training
[39]	6-month soccer training improved maximal power output, balance, flexibility, jump ability, agility, and cardiorespiratory fitness in overweight children.	Favorable
[40]	11-week soccer training improved the ability of a 20-m sprint and cardiorespiratory fitness performance in children.	Favorable
[41]	12-week soccer training increased the VO2peak in obese adolescents.	Favorable
[42]	8-week soccer training increased cardiorespiratory fitness and 20-m sprint performance in untrained adolescents.	Favorable
[35]	12-week soccer training increased explosive power and flexibility of lower extremities in overweight children.	Favorable
[36]	12-week soccer training increased agility and cardiorespiratory fitness in overweight and obese children.	Favorable
[43]	8-week soccer training decreased time in sprint performance (10 and 20 m), and increased jump ability, balance, and cardiorespiratory fitness in adolescents.	Favorable

Table 7. *Cont.*

Study	Physical Fitness—Evidence of the Main Findings (Differences or Not after Training Programs in Terms of the Main Physical Fitness Outcomes)	General Effects of Soccer Training
[38]	10-month soccer training improved cardiorespiratory fitness, although without significant differences in children.	Favorable
[45]	11 weeks of "FIFA 11 for Health" improved balance and cardiorespiratory fitness in both genders. Specifically, when only girls were analyzed, cardiorespiratory fitness and jump ability were improved, while boys only improved cardiorespiratory fitness in children.	Favorable
[37]	10-week soccer training improved cardiorespiratory fitness, VO2peak, 50-m sprinting ability, jump ability, core muscle strength, and balance in children.	Favorable
[46]	8-month soccer training improved resistance in the upper body and cardiorespiratory fitness in adolescents.	Favorable
[49]	8-week soccer training improved abdominal strength and cardiorespiratory fitness in adolescents.	Favorable
[50]	An 11-week study of "11 for Health in Denmark" showed positive cardiorespiratory fitness and VO2max levels in children.	Favorable

Abbreviations:VO2max: maximal oxygen uptake; VO2peak: the peak of oxygen uptake.

Table 8 shows the main results for the health variables. The main findings showed improvements in blood pressure or heart rate variables.

Table 8. The main findings regarding health-related outcomes.

Study	Health-Related Outcomes—Evidence of the Main Findings (Differences or Not after Training Programs in Terms of Health-Related Outcomes)	General Effects of Soccer Training
[39]	6-month soccer training did not improve any biochemical or inflammatory marker in overweight children.	No significant effect
[34]	10-week soccer training improved the posterior wall diameter, interventricular septum thickness, and global isovolumetric relaxation in pre-adolescent children.	Favorable
[40]	11-week soccer training decreased systolic blood pressure and mean arterial blood pressure in children.	Favorable
[41]	12-week soccer training decreased systolic blood pressure, total cholesterol, triglycerides, C-reactive protein, insulin resistance, sympathetic activity, and vascular resistance. At the same time, parasympathetic activity, high-density lipoprotein cholesterol, and vascular conductance increased in obese adolescents.	Favorable
[42]	8-week soccer training increased high-frequency power, the root mean squared value of the standard deviation (rMSSD), and decreased sympathetic activity in untrained adolescents.	Favorable
[35]	12-week soccer training led to a positive effect on biochemical parameters, such as the increased number of erythrocytes in overweight children.	Favorable
[36]	12-week soccer training decreased the resting and maximal heart rate in overweight and obese children.	Favorable
[38]	10-month soccer training decreased the diastolic blood pressure and elicited discrete cardiac adaptations, such as interventricular septum thickness, cuspid annular plane systolic excursion, and left-atrial volume index in children.	Favorable
[44]	11 weeks of "FIFA 11 for Health" improved cognitive performance by reducing reaction time in terms of psychomotor function, attention, and working memory in children.	Favorable
[45]	11 weeks of "FIFA 11 for Health" decreased the systolic blood pressure in children.	Favorable
[37]	10-week soccer training improved heart function in children.	Favorable
[51]	10-month soccer training improved interventricular septum thickness and peak transmitral flow velocity in early diastole, while no other changes were observed in children.	
[47]	12-week soccer training was effective in reducing metabolic syndrome in obese adolescents.	Favorable
[46]	8-month soccer training improved the physical aggression subscale (physical aggression, verbal aggression, hostility, and anger) in adolescents.	Favorable

Table 8. *Cont.*

Study	Health-Related Outcomes—Evidence of the Main Findings (Differences or Not after Training Programs in Terms of Health-Related Outcomes)	General Effects of Soccer Training
[48]	12-week soccer training with diet restriction decreased the glycated hemoglobin, while no other changes were shown for this group or soccer training without diet restriction, which means that diet was important to improving glycemia in adolescents with type 1 diabetes.	Favorable
[49]	8-week soccer training did not cause any significant changes in blood pressure or heart rate health in adolescents.	No significant effect

Table 9 shows the main results for body composition changes. Most of the studies did not find a significant impact of recreational soccer on body composition.

Table 9. Main findings of body composition.

Study	Body Composition—Evidence of the Main Findings (Differences or Not after Training Programs in Terms of Body Composition)	General Effects of Soccer Training
[39]	During the 6-month soccer training, height and weight increased in overweight children.	No significant effect
[34]	During the 10-week soccer training, body composition variables did not change in children.	No significant effect
[40]	During the 11-week soccer training, body mass index and body fat percentage decreased in children.	Favorable
[41]	During the 12-week soccer training, body mass index, waist circumference, and percentage of body fat decreased in obese adolescents.	Favorable
[42]	During the 8-week soccer training, body composition variables did not change in untrained adolescents.	No significant effect
[36]	During the 12-week soccer training, body composition variables did not change in overweight and obese children.	No significant effect
[38]	During the 10-month soccer training, body composition variables did not change in any children.	No significant effect
[45]	During the 11-week period of "FIFA 11 for Health", height, weight, body mass index, and lean body mass increased, while body fat decreased in children of both genders.	Favorable
[37]	During the 10-week soccer training, body fat, fat mass, and abdominal fat decreased in children.	Favorable
[46]	During the 8-month soccer training, body composition variables did not change in adolescents.	No significant effect
[47]	The 12 weeks of soccer training were not enough to significantly decrease the fat percentage.	No significant effect
[49]	During 8-week soccer training, body composition variables did not change in adolescents.	No significant effect
[50]	The 11 weeks of "11 for Health in Denmark" showed a positive effect on body mass index in children.	Favorable

4. Discussion

The purpose of this systematic review was to analyze physical fitness and health-related markers in untrained children and youths exposed to recreational soccer (RS). The study mainly focused on comparisons with the control groups, aiming to understand the isolated effects on body composition, health-related outcomes, and physical fitness.

The main findings from this review are: (a) RS programs are appropriate for young people when an adequate and properly supervised program is followed; (b) youth RS programs spanning a period of eight to eleven weeks significantly improved cardiorespiratory fitness, blood pressure, and heart rate-related variables; (c) RS programs seem to be beneficial in improving body composition, although the results do not present a clearly discernible pattern.

4.1. Main Findings Regarding Physical Fitness

From the analyzed studies, the main findings showed improvements in cardiorespiratory fitness (CRF) in twelve studies. Although occurrences of cardiovascular disease (CVD) are rarely seen during childhood, the associated pathophysiological processes often begin as early as adolescence [52]. Since there is a correlation between childhood obesity and the eventual emergence of CVD risk factors in adulthood, recent studies [53–55] considered the importance of CVD risk factors, such as systolic blood pressure, body fat percentage, and aerobic fitness as the main outcomes related to the level of physical activity in children from 9 years of age and older. Moreover, CRF seems to prevent cardiovascular disease, regardless of body mass or composition [56,57], which highlights engagement in regular physical activity as an important behavior in controlling the aforementioned risk factors.

There is evidence from the original research in the present review that chronic exposure to RS has a high potential to improve CRF in children and adolescents with and without excess weight or obesity. RS seems to elicit high cardiovascular demands (endurance performance or VO_2max) in healthy children and clinical populations. Moreover, this type of exercise is considered safe, with positive long-term effects on physical fitness and health indices [58].

The improvements reported by studies in terms of cardiorespiratory fitness may be as a consequence of exposure to RS, with the most effective interventions occurring two to five times per week, with intensities above ~75% HRmax, for eight weeks to ten months. Indeed, improvements in cardiorespiratory fitness for young schoolchildren have been shown to be closely related to time spent in the highest aerobic intensity zone [59].

The greater levels of daily physical activity observed in children who regularly play RS [57] may also partly explain these increases in physical fitness. In overweight and obese children, together with large improvements in terms of motor skill performance, increased exercise capacity may also facilitate greater participation in everyday activities.

Along with improvements in cardiorespiratory fitness, the studies in this review report similar improvements in 10-m [43], 20-m [40], 30-m, and 50-m [37] sprint times, increased explosive power [39], jump ability [60], agility [39], balance [36], and flexibility [36].

In contrast to the above results, Ørntoft et al. [40] report that during an 11-week period, balance and jump performance remained unchanged in both groups (IG, intervention group and CG, control group) in their study. Similarly, Larsen et al. [60] reported that ten months of soccer training did not cause significant changes in physical fitness in children. Differences in the magnitude of adaptive responses across the various studies are possibly related to the different methodologies used and the baseline values of the groups of participants in those investigations.

4.2. Main Findings in Relation to Body Composition

The early prevention of overweight and related diseases in children and adolescents is crucial, meaning that regular physical activity has been accepted as a means to reduce the incidence of obesity and related comorbidities [61].

Although several school-based physical activity interventions, lasting between ten and fifty-two weeks, have been reported to have had positive effects on aerobic fitness and other fitness components, few have had positive effects on body composition and body fat percentage. Some studies [39] have reported that body composition (as expressed by BMI) remained largely similar over a six-month period. In two soccer-based interventions reported by Krustrup et al. [34], in normal weight and overweight children (aged 8–12 years) who were playing small-sided soccer matches for 3×40 and 3×60 min/week for 10 and 12 weeks, respectively, there were improvements in physical fitness, but there were no changes in BMI or fat percentage. Several other studies on RS also reported that body composition variables did not change in untrained, normal-weight, overweight and obese children, and adolescents. Based on these consistent results, it is perhaps the case that short-duration intervention programs may not promote changes in body composition and that longer studies are required to observe such changes. Conversely, another study [62] observed significant decreases in BMI z-scores over similar time periods to those described herein. However, these studies also included educational and nutritional advice programs, which may have contributed to the observed decreases in BMI. It should, additionally, be noted that the use of BMI z-scores, adjusted for age and sex and using a WHO reference population, increases the sensitivity of the index [63].

Vasconcellos et al. [47] reported between- and within-group differences for body weight, BMI, and waist circumference (WC) in obese adolescents (12–17 years) who played RS only during a 12-week physical activity intervention (3 times/week; 60-min/sessions). Additionally, the intervention group showed significant decreases in body-fat percentage, while no changes were observed in the control group. Interestingly, shorter soccer-based programs (for an 11-week intervention period) such as the "FIFA 11 for Health" program, applied in several studies [40,44], and the "11 for Health in Denmark" used by Ryom et al. [50] also had positive effects on body composition, BMI ($\Delta-0.15\text{kg}/\text{m}^2$) and fat percentage ($\Delta-0.8\%$) when compared with a CG. There were within-group decreases of 23.1 ± 8.4 to $22.5 \pm 8.3\%$ in terms of body fat percentage in the intervention group, as well as an increase in lean body mass (1.0 ± 1.7 vs 0.7 ± 1.6 kg) and a lower BMI when compared to the control group.

Since changes in body fat percentage and blood pressure can be achieved through dietary manipulation, physical exercise, and other daily behaviors, an additive effect can be obtained when such methods are combined [64]. It could be speculated that there are additional effects of the "FIFA 11 for Health" program, other than the football activity itself. For example, Ørntoft et al. [40] report that education nutritional habits, when combined with high-intensity physical training, may positively influence children's behavior. However, further studies are required to elucidate whether the increases in awareness of a healthier lifestyle that were observed following the "FIFA 11 for Health" program can result in the desired behavioral changes [65].

Another approach when analyzing the study results shows that the effects of PA on body composition parameters, such as body weight or body mass index, are inconsistent in young age groups and can distort any changes in body composition. The effects of PA on body weight are controversial, since such assessments may not consider the body composition (e.g., fat mass) [66]. The calculation of BMI uses two measures (height and weight) with a high level of variability in terms of pediatric age, which can decrease the accuracy in assessing obesity in the early stages of life.

The peri-pubertal period, in which the majority of participants in the studies herein are found, may also be a confounding factor.

Another plausible consideration is that the increase in average energy expenditure during intervention training programs can also contribute to an increase in appetite and, thus, body composition may remain constant as a result [39]. However, in most studies, nutrition was not controlled for during the intervention period and, therefore, a definitive conclusion cannot be made.

The meta-analysis conducted by Atlantis et al. [67] recommended 155–180 min of aerobic exercise per week to positively impact the fat-mass levels of overweight children. Thus, the nearly unchanged body composition reported in some studies [39,60] might be at least partly explained by an insufficient training volume (~120 min/wk). Possibly, two hours of moderate to vigorous exercise, or an energy expenditure of about 700 kcal per week, can be recommended based on the American College of Sports Medicine and the American Heart Association, aiming to maintain cardiovascular health in adults [68].

As already mentioned above regarding the study by Ørntoft et al. [40], BMI and the body-fat percentage were reduced during the 11-week period of the "FIFA 11 for Health" program. It is noteworthy that this reduction was achieved with 90 min of activity per week over an 11-week intervention period, which is much less than in previous investigations observing similar reductions in BMI and body fat percentage, as reported in interventions with higher training volume and longer duration. For example, Faude et al. [60] prescribed

180 min over 6 months, while Cvetković et al. [36] prescribed 180 min over 12 weeks, and Hadjicharalambous et al. [49] recommended 180 min/week over 8 weeks.

To allow more definite conclusions with regard to the long/short term and higher/low intensity and volume effectiveness of exercise programs in improving the body composition of normal and overweight children, further studies appear to be necessary.

However, it is important to consider that several studies [69–72] suggest that exercise improves health, even if no weight is lost, and that improved health and fitness may increase daily physical activity levels and compliance with exercise programs.

4.3. Main Findings Regarding Health-Related Outcomes

Of the studies in this review that analyzed the health-related outcomes of RS, two did not find any significant improvement [49,60]. The main findings showed improvements in blood pressure [34,51,73], cardiac function and HR variables [34,40], and mental and cognitive health [43,60].

Physical inactivity and lifestyle-related diseases during childhood and adolescence are associated with an increase in the risk of cardiovascular disease and largely contribute to disease and disability during adulthood. Early intervention with excessively overweight children seems mandatory for a healthier adult life [39]. This is of enormous importance to the current and future health of children and adolescents since it has been observed that obesity tends to track from childhood into adulthood [74]. Likewise, mental health is one of the current major concerns in children and adolescents [75], which may also have consequences in the future; however, in the present review, only three studies addressed this crucial issue [39].

The relationship between higher fitness levels and physical activity and the achievement of better cognitive health and performance is well established [76–78]. Faude et al. [60] report that the self-esteem of overweight children was considerably improved through training, with a larger effect found in the soccer-playing group. This may be explained not only by the social interactions experienced but also because of the competitive nature of the game, leading to feelings of success and coherence among the team [75]. In particular, emotional support may be enhanced when sports are conducted together with peers, an effect that is likely to be more pronounced in team sports such as soccer [79].

The study by Lind et al. [44] included an RS program ("FIFA 11 for Health for Europe") on cognitive performance in pre-adolescent children. The authors reported a reduction in reaction time in terms of psychomotor function, attention, and working memory. Several studies have highlighted the relationship between cognitive functioning and soccer, suggesting the importance of cognitive functions for performance in this sport.

4.4. Behavioral Area

In the behavioral area, only the authors of [46] report that the implementation of RS in regular physical education classes seems to be a potentially appropriate stimulus for reducing aggression (physical aggression, verbal aggression, hostility, and anger) in adolescents. A recent systematic review states that the prosocial behavior of RS plays a key role in interpersonal relationships concerning the growth of children and adolescents [80].

4.5. Cardiovascular Adaptations

Other health-related outcomes of RS programs, such as cardiovascular adaptations, have been reported in studies showing significant structural and functional effects on the cardiovascular system. Thus, studies of normal and overweight children have reported considerable cardiovascular adaptations to medium- and short-term soccer interventions [34,38,40]. Only Hadjicharalambous et al. [49] reported that an 8-week soccer training intervention did not cause any significant changes in cardiometabolic health in adolescents.

The analyzed studies report structural and functional cardiovascular changes, such as: (a) increased left ventricular posterior wall diameter; (b) improvements in right ventricular systolic function and increased global isovolumetric relaxation time; (c) improved interventricular septum thickness, cuspid annular plane systolic excursion, and left atrial volume index and peak transmitral flow velocity in early diastole; (d) beneficial changes in endothelial function and vascular conductance; (e) decreased submaximal HR and resting and maximal HR. Such cardiac adaptations to physical exercise, known as the "athlete's heart", can be elicited in obese and non-obese children, who are heterogeneous in terms of fitness levels and sports participation [34]. These observations demonstrate that in childhood, the heart adapts quickly to the physiological changes induced by physical training [81]. Despite the relatively long-term duration (10 months) in the study by Larsen, Nielsen, et al. [51], short-term interventions report similar adaptations (over 10 weeks) from the application of an RS program.

It has been shown that regular and extensive training in adults is associated with changes in cardiac morphology, namely, increased left and right ventricular cavity dimension, wall thickness, and mass [82–84]. Cardiac adaptations to exercise also occur in children and adolescents, independent of the influence of growth and maturation [85]; a recent systematic review of echocardiographic studies concludes that these adaptations are more pronounced in structural left ventricular parameters, with the functional parameters being preserved or slightly improved by exercise [86].

However, there are still very few studies on the effects of RS training on cardiac morphology and function that use more accurate measurement methods, such as cardiac magnetic resonance or 3-D echocardiography [81].

As a consequence of the above-described cardiac adaptations to RS, in the current review, although Ørntoft et al. [40] state that there seem to be no changes in terms of resting HR, studies revealed that RS was effective in decreasing resting, submaximal, and maximal HR in normal weight, overweight, and obese children [35]. Wang et al. [37] also reported improvements in the HR index, which is determined via resting HR, exercise HR, and recovery HR (1 min post-exercise), after 30 squats in 30 s.

Autonomic heart rate regulation was evaluated via heart rate variability in two included studies [39,51]. Studies show that autonomic modulation, both at rest and during exercise, increases when a positive adaptation to training occurs, leading to an improvement in physical performance [87].

Changes that were reported in cardiac autonomic activity, namely, higher parasympathetic outflow and lower sympathetic outflow after RS programs, are concomitant with a lowered systolic blood pressure. Altogether, an improvement in vagal modulation and a reduction in sympathetic activity reflect the enhancement of hemodynamic and cardiac autonomic function in short-term programs, which can be considered a cardio-protective effect of an RS program [39,51].

4.6. Blood Pressure

Blood pressure (BP) is reported in some studies as showing significant reductions after the application of intervention programs, namely, the "FIFA 11 for Health" program and other RS programs [51,73]. The positive effects of physical activity programs on BP in children aged 6–12 years have been reported in a meta-analysis of physical activity intervention studies. Despite the positive effects of the interventions, there was no consensus on a reduction in all blood pressure parameters in the studies analyzed. Some studies reported a decrease only in systolic blood pressure (SBP), while others reported it only in diastolic pressure or mean blood pressure. Interestingly, some studies reported that programs of shorter duration (≤8 weeks) appear to have no effect on BP [49].

Despite the relative ambiguity of the pooled results, the beneficial effects of high levels of physical activity on blood pressure in children are clear [88,89]; even 2 mmHg reductions in systolic and diastolic blood pressure are associated with a reduction in coronary heart disease in adults [90–92]. It is well established that a reduction in BP, if sustained, is associated with lowered arterial stiffness and a decreased atherosclerosis progression rate in adulthood [93,94].

Thus, the present findings suggest that RS programs could be an efficient strategy for BP regulation, even in overweight and obese children, with positive health implications as one progresses through the various developmental stages to adulthood.

4.7. Biochemical Parameters

Besides the various tests for assessment of the cardiorespiratory system and body composition, blood biochemical parameters are a significant indicator of health status. In the present review, the studies by Vasconcellos et al. [41] and Cvetković et al. [36] report that RS promoted a statistically significant increase in the number of erythrocytes at the end of the intervention period. Both studies also verified a decrease in blood glucose. Glucose was assessed using insulin concentration, following the homeostasis model assessment of insulin resistance (HOMA-IR). The index of HOMA-IR takes under consideration both the fasting insulin values and fasting glucose values. Overweight and obese adolescents often present higher negative health outcomes than those of normal weight. This can be explained by the higher values of insulin concentration and HOMA-IR [95,96].

Concerning triglyceride blood levels, RS was shown to have favorable effects (small effect size) according to Vasconcellos et al. [41]. These results are not consistent across the different studies, perhaps due to the baseline fitness levels, the duration of interventions, and the cardiometabolic profiles of the children and adolescents [49].

It is also important to understand individual responses, due to the complexity of the participants' levels of overweight and obesity in some studies. Studies reporting the effects of individual and group interventions on cardiometabolic risk factors in overweight or obese pubescent girls and boys are still lacking.

The completion of considerably larger and longer-term studies, including groups undergoing high and low volume and intensity soccer, will shed more light on the possible temporal developments and dose-response relationships, as well as the associated blood biochemical parameter adaptations.

4.8. Limitations

One of this study's limitations is associated with the sample size, related to the included studies. Although there is no explicit information in the articles, some similar aspects were reported in papers from the same research groups that could be associated with using partial data from the same dataset (i.e., data slicing). This can artificially inflate the sample size of the current systematic review. However, since there is no reported information about that fact (in the original articles), all the included studies were considered for selection, based on the assumption of trust, and based on the fact that the articles researched different outcomes. Another limitation of this review is the limited number of subjects investigated in the studies, as well as the relatively short intervention periods, and, in this regard, the likely suboptimal statistical power. Furthermore, the studies were analyzed on a "per-protocol" basis; therefore, only the efficacy under ideal conditions can be assessed. Another limitation is that activity and caloric intake in daily living were not controlled, while seasonal influences on the investigated parameters may have occurred.

In addition, hormonal changes resulting from the onset of puberty cannot be ruled out as an influential confounding factor. Specifically, some children in the various studies were at the beginning of puberty and rapid hormonal changes may have occurred.

4.9. Practical Implications

Based on the current findings, it is verified that recreational soccer can effectively improve the physical fitness and health of children and young people. A two-week course of training sessions of about 60 min each, using small-sided games, can be a recommended strategy for achieving a beneficial impact on participants. However, regarding fat mass, it is recommended that this should be complemented by dietary and nutritional advice, which allows for increasing the effect since recreational soccer seems insufficient for achieving a good impact.

5. Conclusions

The 11-week intervention of a football-based health education program, as used in some studies, presents itself as an effective program to be used within a school's curriculum to induce improvements in psychosocial and physiological health profiles, along with an increase in health knowledge. We recommend the tentative use of such programs alongside a call for greater efforts to engage in more studies on this particular issue. However, we would like to highlight the point that a major limitation of this review is the limited number of subjects investigated in the studies and the relatively short intervention periods, and, in this regard, the possibly suboptimal statistical power. Thus, generalizations from the findings should be made with caution.

Author Contributions: F.M.C. conceived the idea, drafted the project, wrote and revised the article, and approved the article; J.M. wrote and revised the article and approved the article; R.R.-C. wrote and revised the article and approved the article; R.O. wrote and revised the article and approved the article; J.B. wrote and revised the article and approved the article; A.F.S., wrote and revised the article and approved the article; G.B., wrote and revised the article and approved the article; G.P., wrote and revised the article and approved the article; H.S. wrote and revised the article and approved the article. All authors have read and agreed to the published version of the manuscript.

Funding: This work is funded by the Fundação para a Ciência e Tecnologia/Ministério da Ciência, Tecnologia e Ensino Superior, through national funds and, when applicable, co-funded EU funds, under the project UIDB/50008/2020.

Institutional Review Board Statement: Not applicable.

Informed Consent Statement: Not applicable.

Data Availability Statement: Not applicable.

Conflicts of Interest: The authors declare no conflict of interest.

References

1. Biddle, S.J.H.; García Bengoechea, E.; Wiesner, G. Sedentary Behaviour and Adiposity in Youth: A Systematic Review of Reviews and Analysis of Causality. *Int. J. Behav. Nutr. Phys. Act.* **2017**, *14*, 43. [CrossRef] [PubMed]
2. Saunders, T.J.; Chaput, J.P.; Tremblay, M.S. Sedentary Behaviour as an Emerging Risk Factor for Cardiometabolic Diseases in Children and Youth. *Can. J. Diabetes* **2014**, *38*, 53–61. [CrossRef] [PubMed]
3. le Roux, E.; de Jong, N.P.; Blanc, S.; Simon, C.; Bessesen, D.H.; Bergouignan, A. Physiology of Physical Inactivity, Sedentary Behaviours and Non-exercise Activity: Insights from the Space Bedrest Model. *J. Physiol.* **2022**, *600*, 1037–1051. [CrossRef]
4. Duvivier, B.M.F.M.; Schaper, N.C.; Bremers, M.A.; van Crombrugge, G.; Menheere, P.P.C.A.; Kars, M.; Savelberg, H.H.C.M. Minimal Intensity Physical Activity (Standing and Walking) of Longer Duration Improves Insulin Action and Plasma Lipids More than Shorter Periods of Moderate to Vigorous Exercise (Cycling) in Sedentary Subjects When Energy Expenditure Is Comparable. *PLoS ONE* **2013**, *8*, e55542. [CrossRef] [PubMed]
5. Biddle, S.J.H.; Gorely, T.; Marshall, S.J.; Murdey, I.; Cameron, N. Physical Activity and Sedentary Behaviours in Youth: Issues and Controversies. *J. R. Soc. Promot. Health* **2004**, *124*, 29–33. [CrossRef]
6. Carson, V.; Hunter, S.; Kuzik, N.; Gray, C.E.; Poitras, V.J.; Chaput, J.P.; Saunders, T.J.; Katzmarzyk, P.T.; Okely, A.D.; Connor Gorber, S.; et al. Systematic Review of Sedentary Behaviour and Health Indicators in School-Aged Children and Youth: An Update. *Appl. Physiol. Nutr. Metab.* **2016**, *41*, S240–S265. [CrossRef] [PubMed]
7. Chaput, J.-P.; Willumsen, J.; Bull, F.; Chou, R.; Ekelund, U.; Firth, J.; Jago, R.; Ortega, F.B.; Katzmarzyk, P.T. 2020 WHO Guidelines on Physical Activity and Sedentary Behaviour for Children and Adolescents Aged 5–17 Years: Summary of the Evidence. *Int. J. Behav. Nutr. Phys. Act.* **2020**, *17*, 141. [CrossRef]
8. Hallal, P.C.; Victora, C.G.; Azevedo, M.R.; Wells, J.C.K. Adolescent Physical Activity and Health: A Systematic Review. *Sports Med.* **2006**, *36*, 1019–1030. [CrossRef]
9. Castagna, C.; de Sousa, M.; Krustrup, P.; Kirkendall, D.T. Recreational Team Sports: The Motivational Medicine. *J. Sport Health Sci.* **2018**, *7*, 129. [CrossRef]
10. Milanović, Z.; Pantelić, S.; Čović, N.; Sporiš, G.; Mohr, M.; Krustrup, P. Broad-Spectrum Physical Fitness Benefits of Recreational Football: A Systematic Review and Meta-Analysis. *Br. J. Sports Med.* **2019**, *53*, 926–939. [CrossRef]
11. Andersen, L.J.; Hansen, P.R.; Søgaard, P.; Madsen, J.K.; Bech, J.; Krustrup, P. Improvement of Systolic and Diastolic Heart Function after Physical Training in Sedentary Women. *Scand. J. Med. Sci. Sports* **2010**, *20* (Suppl. S1), 50–57. [CrossRef] [PubMed]

12. Milanović, Z.; Pantelić, S.; Kostić, R.; Trajković, N.; Sporiš, G. Soccer vs. Running Training Effects in Young Adult Men: Which Programme Is More Effective in Improvement of Body Composition? Randomized Controlled Trial. *Biol. Sport* **2015**, *32*, 301–305. [CrossRef] [PubMed]
13. Montealegre Suárez, D.; Lerma Castaño, P.; Rojas Calderón, M.; Perdomo Trujillo, J.; Torres Méndez, M. Condición Física de Niños Futbolistas En Función de La Posición de Juego (Physical Condition of Children Footballers Depending on the Playing Position). *Rev. Iberoam. De Cienc. De La Act. Física Y El Deporte* **2020**, *9*, 23. [CrossRef]
14. Ayala Hernández, H.; Rivera Girón, A.; Castineyra Mendoza, S.; Gómez Figueroa, J. Influencia de La Educación Física en Jugadores de Fútbol Asociación Sub-13 y Sub-15 (INFLUENCE OF PHYSICAL EDUCATION IN U-13 AND U-15 ASSOCIATION SOCCER PLAYERS). *Rev. Iberoam. De Cienc. De La Act. Física Y El Deporte* **2021**, *10*, 37–46. [CrossRef]
15. Sarmento, H.; Manuel Clemente, F.; Marques, A.; Milanovic, Z.; David Harper, L.; Figueiredo, A. Recreational Football Is Medicine against Non-Communicable Diseases: A Systematic Review. *Scand. J. Med. Sci. Sports* **2020**, *30*, 618–637. [CrossRef] [PubMed]
16. Helge, E.W.; Aagaard, P.; Jakobsen, M.D.; Sundstrup, E.; Randers, M.B.; Karlsson, M.K.; Krustrup, P. Recreational Football Training Decreases Risk Factors for Bone Fractures in Untrained Premenopausal Women. *Scand. J. Med. Sci. Sports* **2010**, *20* (Suppl. S1), 31–39. [CrossRef]
17. Mielke, G.I.; Bailey, T.G.; Burton, N.W.; Brown, W.J. Participation in sports/recreational activities and incidence of hypertension, diabetes, and obesity in adults. *Scand. J. Med. Sci. Sports* **2020**, *30*, 2390–2398. [CrossRef] [PubMed]
18. Pinho, C.D.F.; Farinha, J.B.; Lisboa, S.D.C.; Bagatini, N.C.; Leites, G.T.; Voser, R.D.C.; Gaya, A.R.; Reischak-Oliveira, A.; Cunha, G.D.S. Effects of a small-sided soccer program on health parameters in obese children. *Rev. Brasileira Med. Esporte* **2022**, *29*. [CrossRef]
19. Tsorbatzoudis, H.; Alexandris, K.; Zahariadis, P.; Grouios, G. Examining the Relationship between Recreational Sport Participation and Intrinsic and Extrinsic Motivation and Amotivation. *Percept. Mot. Ski.* **2006**, *103*, 363–374. [CrossRef]
20. Milanović, Z.; Pantelić, S.; Čović, N.; Sporiš, G.; Krustrup, P.; Milanovic, Z.; Pantelic, S.; Covic, N.; Sporis, G.; Krustrup, P. Is Recreational Soccer Effective for Improving VO2max ? A Systematic Review and Meta-Analysis. *Sports Med.* **2015**, *45*, 1339–1353. [CrossRef]
21. Luo, H.; Newton, R.U.; Ma'ayah, F.; Galvão, D.A.; Taaffe, D.R. Recreational Soccer as Sport Medicine for Middle-Aged and Older Adults: A Systematic Review. *BMJ Open Sport Exerc. Med.* **2018**, *4*, e000336. [CrossRef] [PubMed]
22. Clemente, F.; González-Fernández, F.; Ceylan, H.; Silva, R.; Ramirez-Campillo, R. Effects of Recreational Soccer on Fat Mass in Untrained Sedentary Adults: A Systematic Review with Meta-Analysis. *Hum. Mov.* **2022**, *23*, 15–32. [CrossRef]
23. Krustrup, P.; Aagaard, P.; Nybo, L.; Petersen, J.; Mohr, M.; Bangsbo, J. Recreational football as a health promoting activity: A topical review. *Scand. J. Med. Sci. Sports* **2010**, *20*, 1–13. [CrossRef]
24. Castillo-Bellot, I.; Mora-Gonzalez, J.; Fradua, L.; Ortega, F.B.; Gracia-Marco, L. Effects of Recreational Soccer on Health Outcomes: A Narrative Review. *J Sci Sport and Exercise* **2019**, *1*, 142–150. [CrossRef]
25. Hammami, A.; Chamari, K.; Slimani, M.; Shephard, R.; Yousfi, N.; Tabka, Z.; Bouhlel, E. Effects of Recreational Soccer on Physical Fitness and Health Indices in Sedentary Healthy and Unhealthy Subjects. *Biol. Sport* **2016**, *33*, 127–137. [CrossRef]
26. Clemente, F.M.; Ramirez-Campillo, R.; Sarmento, H.; Castillo, D.; Raya-González, J.; Rosemann, T.; Knechtle, B. Effects of Recreational Small-Sided Soccer Games on Bone Mineral Density in Untrained Adults: A Systematic Review and Meta-Analysis. *Healthcare* **2021**, *9*, 457. [CrossRef]
27. Milanović, Z.; Čović, N.; Helge, E.W.; Krustrup, P.; Mohr, M. Recreational Football and Bone Health: A Systematic Review and Meta-Analysis. *Sports Med.* **2022**. *Online ahead of print*. [CrossRef]
28. Page, M.J.; McKenzie, J.E.; Bossuyt, P.M.; Boutron, I.; Hoffmann, T.C.; Mulrow, C.D.; Shamseer, L.; Tetzlaff, J.M.; Akl, E.A.; Brennan, S.E.; et al. The PRISMA 2020 Statement: An Updated Guideline for Reporting Systematic Reviews. *BMJ* **2021**, *372*, n71. [CrossRef]
29. Rechenchosky, L.; Menegassi, V.M.; Jaime, M.D.O.; Borges, P.H.; Sarmento, H.; Mancha-Triguero, D.; Serra-Olivares, J.; Rinaldi, W. Scoping Review of Tests to Assess Tactical Knowledge and Tactical Performance of Young Soccer Players. *J. Sports Sci.* **2021**, *39*, 2051–2067. [CrossRef]
30. McKay, A.K.A.; Stellingwerff, T.; Smith, E.S.; Martin, D.T.; Mujika, I.; Goosey-Tolfrey, V.L.; Sheppard, J.; Burke, L.M. Defining Training and Performance Caliber: A Participant Classification Framework. *Int. J. Sports Physiol. Perform.* **2022**, *17*, 317–331. [CrossRef]
31. Higgins, J.; Thomas, J. *Cochrane Handbook for Systematic Reviews of Interventions*; Cochrane: London, UK, 2021.
32. Wong, S.S.-L.; Wilczynski, N.L.; Haynes, R.B. Developing Optimal Search Strategies for Detecting Clinically Sound Treatment Studies in EMBASE. *J. Med. Libr. Assoc.* **2006**, *94*, 41–47. [PubMed]
33. Maher, C.G.; Sherrington, C.; Herbert, R.D.; Moseley, A.M.; Elkins, M. Reliability of the PEDro Scale for Rating Quality of Randomized Controlled Trials. *Phys. Ther.* **2003**, *83*, 713–721. [CrossRef] [PubMed]
34. Larsen, M.N.; Nielsen, C.M.; Ørntoft, C.; Randers, M.B.; Helge, E.W.; Madsen, M.; Manniche, V.; Hansen, L.; Hansen, P.R.; Bangsbo, J.; et al. Fitness Effects of 10-Month Frequent Low-Volume Ball Game Training or Interval Running for 8-10-Year-Old School Children. *Biomed Res. Int.* **2017**, *2017*, 2719752. [CrossRef] [PubMed]
35. Larsen, M.N.; Nielsen, C.M.; Helge, E.W.; Madsen, M.; Manniche, V.; Hansen, L.; Hansen, P.R.; Bangsbo, J.; Krustrup, P. Positive Effects on Bone Mineralisation and Muscular Fitness after 10 Months of Intense School-Based Physical Training for Children Aged 8–10 Years: The FIT FIRST Randomised Controlled Trial. *Br. J. Sports Med.* **2018**, *52*, 254–260. [CrossRef] [PubMed]

36. Larsen, M.N.; Nielsen, C.M.; Madsen, M.; Manniche, V.; Hansen, L.; Bangsbo, J.; Krustrup, P.; Hansen, P.R. Cardiovascular Adaptations after 10 Months of Intense School-Based Physical Training for 8- to 10-Year-Old Children. *Scand. J. Med. Sci. Sports* **2018**, *28* (Suppl. S1), 33–41. [CrossRef]
37. Larsen, M.N.; Madsen, M.; Nielsen, C.M.; Manniche, V.; Hansen, L.; Bangsbo, J.; Krustrup, P.; Hansen, P.R. Cardiovascular Adaptations after 10 months of Daily 12-Min Bouts of Intense School-Based Physical Training for 8-10-Year-Old Children. *Prog. Cardiovasc. Dis.* **2020**, *63*, 813–817. [CrossRef]
38. Krustrup, P.; Hansen, P.R.; Nielsen, C.M.; Larsen, M.N.; Randers, M.B.; Manniche, V.; Hansen, L.; Dvorak, J.; Bangsbo, J. Structural and Functional Cardiac Adaptations to a 10-Week School-Based Football Intervention for 9-10-Year-Old Children. *Scand. J. Med. Sci. Sports* **2014**, *24* (Suppl. S1), 4–9. [CrossRef]
39. Hammami, A.; Kasmi, S.; Razgallah, M.; Tabka, Z.; Shephard, R.J.; Bouhlel, E. Recreational Soccer Training Improves Heart-Rate Variability Indices and Physical Performance in Untrained Healthy Adolescent. *Sport Sci. Health* **2017**, *13*, 507–514. [CrossRef]
40. Cvetković, N.; Stojanović, E.; Stojiljković, N.; Nikolić, D.; Milanović, Z. Effects of a 12 Week Recreational Football and High-intensity Interval Training on Physical Fitness in Overweight Children./Efekti Rekreativnog Fudbala od 12 Nedelja i Intervalnog Treninga Visokog Intenziteta (Hiit) Na Fizičku Spremnost Kod dece Sa Prek. *Facta Univ. Ser. Phys. Educ. Sport* **2018**, *16*, 435–450.
41. Cvetković, N.; Stojanović, E.; Stojiljković, N.; Nikolić, D.; Scanlan, A.T.; Milanović, Z. Exercise Training in Overweight and Obese Children: Recreational Football and High-Intensity Interval Training Provide Similar Benefits to Physical Fitness. *Scand. J. Med. Sci. Sports* **2018**, *28* (Suppl. S1), 18–32. [CrossRef]
42. Hammami, A.; Randers, M.B.; Kasmi, S.; Razgallah, M.; Tabka, Z.; Chamari, K.; Bouhlel, E. Effects of Soccer Training on Health-Related Physical Fitness Measures in Male Adolescents. *J. Sport Health Sci.* **2018**, *7*, 169–175. [CrossRef] [PubMed]
43. Lind, R.R.; Geertsen, S.S.; Ørntoft, C.; Madsen, M.; Larsen, M.N.; Dvorak, J.; Ritz, C.; Krustrup, P. Improved Cognitive Performance in Preadolescent Danish Children after the School-Based Physical Activity Programme "FIFA 11 for Health" for Europe-A Cluster-Randomised Controlled Trial. *Eur. J. Sport Sci.* **2018**, *18*, 130–139. [CrossRef] [PubMed]
44. Skoradal, M.B.; Purkhús, E.; Steinholm, H.; Olsen, M.H.; Ørntoft, C.; Larsen, M.N.; Dvorak, J.; Mohr, M.; Krustrup, P. "FIFA 11 for Health" for Europe in the Faroe Islands: Effects on Health Markers and Physical Fitness in 10- to 12-Year-Old Schoolchildren. *Scand. J. Med. Sci. Sports* **2018**, *28* (Suppl. S1), 8–17. [CrossRef] [PubMed]
45. Wang, J.; Cao, L.; Xie, P.; Wang, J. Recreational Football Training Improved Health-Related Physical Fitness in 9- to 10-Year-Old Boys. *J. Sports Med. Phys. Fitness* **2018**, *58*, 326–331. [CrossRef] [PubMed]
46. Trajkovi, N.; Madic, D.M.; Milanovic, Z.; MacAk, D.; Padulo, J.; Krustrup, P.; Chamari, K. Eight Months of School-Based Soccer Improves Physical Fitness and Reduces Aggression in High-School Children. *Biol. Sport* **2020**, *37*, 185–193. [CrossRef]
47. Vasconcellos, F.; Cunha, F.A.; Gonet, D.T.; Farinatti, P.T.V. Does Recreational Soccer Change Metabolic Syndrome Status in Obese Adolescents? A Pilot Study. *Res. Q. Exerc. Sport* **2021**, *92*, 91–99. [CrossRef]
48. Mohammed, M.H.H.; Al-Qahtani, M.H.H.; Takken, T. Effects of 12 Weeks of Recreational Football (Soccer) with Caloric Control on Glycemia and Cardiovascular Health of Adolescent Boys with Type 1 Diabetes. *Pediatr. Diabetes* **2021**, *22*, 625–637. [CrossRef]
49. Hadjicharalambous, M.; Zaras, N.; Apostolidis, A.; Tsofliou, F. Recreational Soccer, Body Composition and Cardiometabolic Health: A Training-Intervention Study in Healthy Adolescents. *Int. J. Hum. Mov. Sports Sci.* **2022**, *10*, 524–533. [CrossRef]
50. Ryom, K.; Christiansen, S.R.; Elbe, A.M.; Aggestrup, C.S.; Madsen, E.E.; Madsen, M.; Larsen, M.N.; Krustrup, P. The Danish "11 for Health" Program Raises Health Knowledge, Well-Being, and Fitness in Ethnic Minority 10- to 12-Year-Olds. *Scand. J. Med. Sci. Sports* **2022**, *32*, 138–151. [CrossRef]
51. Vasconcellos, F.; Seabra, A.; Cunha, F.; Montenegro, R.; Penha, J.; Bouskela, E.; Nogueira Neto, J.F.; Collett-Solberg, P.; Farinatti, P. Health Markers in Obese Adolescents Improved by a 12-Week Recreational Soccer Program: A Randomised Controlled Trial. *J. Sports Sci.* **2016**, *34*, 564–575. [CrossRef]
52. Abrignani, M.G.; Lucà, F.; Favilli, S.; Benvenuto, M.; Rao, C.M.; di Fusco, S.A.; Gabrielli, D.; Gulizia, M.M. Lifestyles and Cardiovascular Prevention in Childhood and Adolescence. *Pediatr. Cardiol.* **2019**, *40*, 1113–1125. [CrossRef] [PubMed]
53. Drozdz, D.; Alvarez-Pitti, J.; Wójcik, M.; Borghi, C.; Gabbianelli, R.; Mazur, A.; Herceg-čavrak, V.; Lopez-Valcarcel, B.G.; Brzeziński, M.; Lurbe, E.; et al. Obesity and Cardiometabolic Risk Factors: From Childhood to Adulthood. *Nutrients* **2021**, *13*, 4176. [CrossRef] [PubMed]
54. de Jesus, J.M. Expert Panel on Integrated Guidelines for Cardiovascular Health and Risk Reduction in Children and Adolescents: Summary Report. *Pediatrics* **2011**, *128*, S213–S256.
55. Lang, J.J.; Belanger, K.; Poitras, V.; Janssen, I.; Tomkinson, G.R.; Tremblay, M.S. Systematic Review of the Relationship between 20 m Shuttle Run Performance and Health Indicators among Children and Youth. *J. Sci. Med. Sport* **2018**, *21*, 383–397. [CrossRef] [PubMed]
56. Högström, G.; Nordström, A.; Nordström, P. High Aerobic Fitness in Late Adolescence Is Associated with a Reduced Risk of Myocardial Infarction Later in Life: A Nationwide Cohort Study in Men. *Eur. Heart J.* **2014**, *35*, 3133–3140. [CrossRef] [PubMed]
57. Ortega, F.B.; Ruiz, J.R.; Castillo, M.J.; Sjöström, M. Physical Fitness in Childhood and Adolescence: A Powerful Marker of Health. *Int. J. Obes.* **2008**, *32*, 1–11. [CrossRef]
58. Ortega, F.B.; Ruiz, J.R.; Labayen, I.; Martínez-Gómez, D.; Vicente-Rodriguez, G.; Cuenca-García, M.; Gracia-Marco, L.; Manios, Y.; Beghin, L.; Molnar, D.; et al. Health Inequalities in Urban Adolescents: Role of Physical Activity, Diet, and Genetics. *Pediatrics* **2014**, *133*, e884–e895. [CrossRef]

59. Soares, I.F.; Cunha, F.A.; Vasconcellos, F. Effects of a 12-Week Recreational Soccer Program on Resting Metabolic Rate Among Adolescents with Obesity. *J. Sci. Sport Exerc.* 2022. [CrossRef]
60. Faude, O.; Kerper, O.; Multhaupt, M.; Winter, C.; Beziel, K.; Junge, A.; Meyer, T. Football to Tackle Overweight in Children. *Scand. J. Med. Sci. Sports* **2010**, *20*, 103–110. [CrossRef]
61. Zouhal, H.; Hammami, A.; Tijani, J.M.; Jayavel, A.; de Sousa, M.; Krustrup, P.; Sghaeir, Z.; Granacher, U.; ben Abderrahman, A. Effects of Small-Sided Soccer Games on Physical Fitness, Physiological Responses, and Health Indices in Untrained Individuals and Clinical Populations: A Systematic Review. *Sports Med.* **2020**, *50*, 987–1007. [CrossRef]
62. Castagna, C.; Krustrup, P.; Póvoas, S. Cardiovascular fitness and health effects of various types of team sports for adult and elderly inactive individuals-a brief narrative review. *Progress in Cardiovascular Diseases* **2020**, *63*, 709–722. [CrossRef] [PubMed]
63. Koyuncuoğlu Güngör, N. Overweight and Obesity in Children and Adolescents. *J. Clin. Res. Pediatr. Endocrinol.* **2014**, *6*, 129–143. [CrossRef] [PubMed]
64. Weintraub, D.L.; Tirumalai, E.C.; Farish Haydel, K.; Fujimoto, M.; Fulton, J.E.; Robinson, T.N. Team Sports for Overweight Children the Stanford Sports to Prevent Obesity Randomized Trial (SPORT). *Arch. Pediatr. Adolesc. Med.* **2008**, *162*, 232–237. [CrossRef]
65. Seabra, A.; Brito, J.; Figueiredo, P.; Beirão, L.; Seabra, A.; Carvalho, M.J.; Abreu, S.; Vale, S.; Pedretti, A.; Nascimento, H.; et al. School-Based Soccer Practice Is an Effective Strategy to Improve Cardiovascular and Metabolic Risk Factors in Overweight Children. *Prog. Cardiovasc. Dis.* **2020**, *63*, 807–812. [CrossRef]
66. de Sousa, M.V.; Fukui, R.; Krustrup, P.; Pereira, R.M.R.; Silva, P.R.S.; Rodrigues, A.C.; de Andrade, J.L.; Hernandez, A.J.; da Silva, M.E.R. Positive Effects of Football on Fitness, Lipid Profile, and Insulin Resistance in Brazilian Patients with Type 2 Diabetes. *Scand. J. Med. Sci. Sports* **2014**, *24*, 57–65. [CrossRef] [PubMed]
67. Fuller, C.W.; Ørntoft, C.; Larsen, M.N.; Elbe, A.M.; Ottesen, L.; Junge, A.; Dvorak, J.; Krustrup, P. "FIFA 11 for Health" for Europe. 1: Effect on Health Knowledge and Well-Being of 10- to 12-Year-Old Danish School Children. *Br. J. Sports Med.* **2017**, *51*, 1483–1488. [CrossRef]
68. Jovanović, R.; Nikolovski, D.; Radulović, O.; Novak, S. Physical Activity Influence on Nutritional Status of Preschool Children. *Acta Med. Median.* **2010**, *49*, 17–21.
69. Atlantis, E.; Barnes, E.H.; Singh, M.A.F. Efficacy of Exercise for Treating Overweight in Children and Adolescents: A Systematic Review. *Int. J. Obes.* **2006**, *30*, 1027–1040. [CrossRef]
70. Haskell, W.L.; Lee, I.M.; Pate, R.R.; Powell, K.E.; Blair, S.N.; Franklin, B.A.; Macera, C.A.; Heath, G.W.; Thompson, P.D.; Bauman, A. Physical Activity and Public Health. *Med. Sci. Sports Exerc.* **2007**, *39*, 1423–1434. [CrossRef] [PubMed]
71. Haskell, W.L.; Lee, I.-M.; Pate, R.R.; Powell, K.E.; Blair, S.N.; Franklin, B.A.; Bauman, A. Physical Activity and Public Health: Updated Recommendation for Adults from the American College of Sports Medicine and the American Heart Association. *Circulation* **2007**, *116*, 1081–1093. [CrossRef]
72. Shaw, K.; Gennat, H.; O'Rourke, P.; del Mar, C. Exercise for Overweight or Obesity. *Cochrane Database Syst. Rev.* **2006**, *2006*, CD003817. [CrossRef] [PubMed]
73. Ørntoft, C.; Fuller, C.W.; Larsen, M.N.; Bangsbo, J.; Dvorak, J.; Krustrup, P. "FIFA 11 for Health" for Europe. II: Effect on Health Markers and Physical Fitness in Danish Schoolchildren Aged 10-12 Years. *Br. J. Sports Med.* **2016**, *50*, 1394–1399. [CrossRef] [PubMed]
74. Laskowski, E.R. The Role of Exercise in the Treatment of Obesity. *PM&R* **2012**, *4*, 840–844. [CrossRef]
75. Biddle, S.J.H.; Ciaccioni, S.; Thomas, G.; Vergeer, I. Physical Activity and Mental Health in Children and Adolescents: An Updated Review of Reviews and an Analysis of Causality. *Psychol. Sport Exerc.* **2019**, *42*, 146–155. [CrossRef]
76. Zolotarjova, J.; ten Velde, G.; Vreugdenhil, A.C.E. Effects of Multidisciplinary Interventions on Weight Loss and Health Outcomes in Children and Adolescents with Morbid Obesity. *Obes. Rev.* **2018**, *19*, 931–946. [CrossRef]
77. Okoniewski, W.; Lu, K.D.; Forno, E. Weight Loss for Children and Adults with Obesity and Asthma a Systematic Review of Randomized Controlled Trials. *Ann. Am. Thorac. Soc.* **2019**, *16*, 613–625. [CrossRef]
78. Herman, K.M.; Craig, C.L.; Gauvin, L.; Katzmarzyk, P.T. Tracking of Obesity and Physical Activity from Childhood to Adulthood: The Physical Activity Longitudinal Study. *Int. J. Pediatric Obes.* **2009**, *4*, 281–288. [CrossRef]
79. Scharfen, H.E.; Memmert, D. The Relationship between Cognitive Functions and Sport-Specific Motor Skills in Elite Youth Soccer Players. *Front. Psychol.* **2019**, *10*, 817. [CrossRef]
80. Sabarit, A.; Reigal, R.E.; Morillo-Baro, J.P.; de Mier, R.J.R.; Franquelo, A.; Hernández-Mendo, A.; Falcó, C.; Morales-Sánchez, V. Cognitive Functioning, Physical Fitness, and Game Performance in a Sample of Adolescent Soccer Players. *Sustainability* **2020**, *12*, 5245. [CrossRef]
81. Erikstad, M.K.; Martin, L.J.; Haugen, T.; Høigaard, R. Group Cohesion, Needs Satisfaction, and Self-Regulated Learning: A One-Year Prospective Study of Elite Youth Soccer Players' Perceptions of Their Club Team. *Psychol. Sport Exerc.* **2018**, *39*, 171–178. [CrossRef]
82. Price, M.S.; Weiss, M.R. Relationships among Coach Leadership, Peer Leadership, and Adolescent Athletes' Psychosocial and Team Outcomes: A Test of Transformational Leadership Theory. *J. Appl. Sport Psychol.* **2013**, *25*, 265–279. [CrossRef]
83. Li, J.; Shao, W. Influence of Sports Activities on Prosocial Behavior of Children and Adolescents: A Systematic Literature Review. *Int. J. Environ. Res. Public Health* **2022**, *19*, 6484. [CrossRef] [PubMed]

84. Barczuk-Falęcka, M.; Małek, Ł.A.; Krysztofiak, H.; Roik, D.; Brzewski, M. Cardiac Magnetic Resonance Assessment of the Structural and Functional Cardiac Adaptations to Soccer Training in School-Aged Male Children. *Pediatr. Cardiol.* **2018**, *39*, 948–954. [CrossRef] [PubMed]

85. Zdravkovic, M.; Perunicic, J.; Krotin, M.; Ristic, M.; Vukomanovic, V.; Soldatovic, I.; Zdravkovic, D. Echocardiographic Study of Early Left Ventricular Remodeling in Highly Trained Preadolescent Footballers. *J. Sci. Med. Sport* **2010**, *13*, 602–606. [CrossRef]

86. Scharhag, J.; Schneider, G.; Urhausen, A.; Rochette, V.; Kramann, B.; Kindermann, W. Athlete's Heart Right and Left Ventricular Mass and Function in Male Endurance Athletes and Untrained Individuals Determined by Magnetic Resonance Imaging. *J. Am. Coll. Cardiol.* **2002**, *40*, 1856–1863. [CrossRef]

87. Galderisi, M.; Cardim, N.; D'Andrea, A.; Bruder, O.; Cosyns, B.; Davin, L.; Donal, E.; Edvardsen, T.; Freitas, A.; Habib, G.; et al. Themulti-Modality Cardiac Imaging Approach to the Athletés Heart: An Expert Consensus of the European Association of Cardiovascular Imaging. *Eur. Heart J. Cardiovasc. Imaging* **2015**, *16*, 353–353r. [CrossRef]

88. Unnithan, V.B.; Rowland, T.; George, K.; Bakhshi, A.; Beaumont, A.; Sculthorpe, N.; Lord, R.N.; Oxborough, D.L. Effect of Long-Term Soccer Training on Changes in Cardiac Function during Exercise in Elite Youth Soccer Players. *Scand. J. Med. Sci. Sports* **2022**, *32*, 892–902. [CrossRef]

89. Weberruß, H.; Engl, T.; Baumgartner, L.; Mühlbauer, F.; Shehu, N.; Oberhoffer-Fritz, R. Cardiac Structure and Function in Junior Athletes: A Systematic Review of Echocardiographic Studies. *Rev. Cardiovasc. Med.* **2022**, *23*, 129. [CrossRef]

90. Bellenger, C.R.; Fuller, J.T.; Thomson, R.L.; Davison, K.; Robertson, E.Y.; Buckley, J.D. Monitoring Athletic Training Status Through Autonomic Heart Rate Regulation: A Systematic Review and Meta-Analysis. *Sports Med.* **2016**, *46*, 1461–1486. [CrossRef]

91. García-Hermoso, A.; Saavedra, J.M.; Escalante, Y. Effects of Exercise on Resting Blood Pressure in Obese Children: A Meta-Analysis of Randomized Controlled Trials. *Obes. Rev.* **2013**, *14*, 919–928. [CrossRef]

92. Knowles, G.; Pallan, M.; Thomas, G.N.; Ekelund, U.; Cheng, K.K.; Barrett, T.; Adab, P. Physical Activity and Blood Pressure in Primary School Children: A Longitudinal Study. *Hypertension* **2013**, *61*, 70–75. [CrossRef] [PubMed]

93. Lauer, R.M.; Clarke, W.R.; Mahoney, L.T.; Witt, J. Childhood Predictors for High Adult Blood Pressure: The Muscatine Study. *Pediatr. Clin. North Am.* **1993**, *40*, 23–40. [CrossRef]

94. Cook, N.R.; Cohen, J.; Hebert, P.R.; Taylor, J.O.; Hennekens, C.H. Implications of Small Reductions in Diastolic Pressure for Primary Prevention. *Arch. Intern. Med.* **1995**, *155*, 701–709. [CrossRef] [PubMed]

95. Rowland, T.W. The Role of Physical Activity and Fitness in Children in the Prevention of Adult Cardiovascular Disease. *Prog. Pediatric Cardiol.* **2001**, *12*, 199–203. [CrossRef]

96. Juonala, M.; Magnussen, C.G.; Venn, A.; Dwyer, T.; Burns, T.L.; Davis, P.H.; Chen, W.; Srinivasan, S.R.; Daniels, S.R.; Kähönen, M.; et al. Influence of Age on Associations between Childhood Risk Factors and Carotid Intima-Media Thickness in Adulthood: The Cardiovascular Risk in Young Finns Study, the Childhood Determinants of Adult Health Study, the Bogalusa Heart Study, and the Muscatine Study for the International Childhood Cardiovascular Cohort (I3C) Consortium. *Circulation* **2010**, *122*, 2514–2520. [CrossRef]

Article

Motor Competence Assessment (MCA) Scoring Method

Luis Paulo Rodrigues [1,2,3,*], Carlos Luz [3,4,5], Rita Cordovil [6,7], André Pombo [3,4] and Vitor P. Lopes [2,8]

1   Instituto Politécnico de Viana do Castelo, Escola Superior Desporto e Lazer de Melgaço, 4900-347 Viana do Castelo, Portugal
2   Research Center in Sports Sciences, Health and Human Development (CIDESD), 5000-801 Vila Real, Portugal
3   Research Center in Sports Performance, Recreation, Innovation and Technology (SPRINT), 4960-320 Melgaço, Portugal
4   Escola Superior de Educação de Lisboa, Instituto Politécnico de Lisboa, 1549-003 Lisboa, Portugal
5   Centro Interdisciplinar de Estudos Educacionais, 1549-003 Lisboa, Portugal
6   Faculdade de Motricidade Humana, Universidade de Lisboa, 1499-002 Cruz Quebrada-Dafundo, Portugal
7   CIPER, Faculdade de Motricidade Humana, Universidade de Lisboa, 1499-002 Cruz Quebrada-Dafundo, Portugal
8   Instituto Politécnico de Bragança, Campus de Santa Apolónia, 5300-223 Bragança, Portugal
*   Correspondence: lprodrigues@esdl.ipvc.pt

Abstract: The Motor Competence Assessment (MCA) is a quantitative test battery that assesses motor competence across the whole lifespan. It is composed of three sub-scales: locomotor, stability, and manipulative, each of them assessed by two different objectively measured tests. The MCA construct validity for children and adolescents, having normative values from 3 to 23 years of age, and the configural invariance between age groups, were recently established. The aim of this study is to expand the MCA's development and validation by defining the best and leanest method to score and classify MCA sub-scales and total score. One thousand participants from 3 to 22 years of age, randomly selected from the Portuguese database on MC, participated in the study. Three different procedures to calculate the sub-scales and total MCA values were tested according to alternative models. Results were compared to the reference method, and Intraclass Correlation Coefficient, Cronbach's Alpha, and Bland–Altman statistics were used to describe agreement between the three methods. The analysis showed no substantial differences between the three methods. Reliability values were perfect (0.999 to 1.000) for all models, implying that all the methods were able to classify everyone in the same way. We recommend implementing the most economic and efficient algorithm, i.e., the configural model algorithm, averaging the percentile scores of the two tests to assess each MCA sub-scale and total scores.

Keywords: human development; motor development; motor test; motor performance; lifespan

Citation: Rodrigues, L.P.; Luz, C.; Cordovil, R.; Pombo, A.; Lopes, V.P. Motor Competence Assessment (MCA) Scoring Method. *Children* **2022**, *9*, 1769. https://doi.org/10.3390/children9111769

Academic Editors: Hugo Miguel Borges Sarmento, Georgian Badicu and Ana Filipa Silva

Received: 10 September 2022
Accepted: 14 November 2022
Published: 17 November 2022

Publisher's Note: MDPI stays neutral with regard to jurisdictional claims in published maps and institutional affiliations.

Copyright: © 2022 by the authors. Licensee MDPI, Basel, Switzerland. This article is an open access article distributed under the terms and conditions of the Creative Commons Attribution (CC BY) license (https://creativecommons.org/licenses/by/4.0/).

1. Introduction

Motor competence (MC) relates to human movement, its development, and performance. It has been described as the ability to be skilful in a wide variety of gross motor skills (stability, locomotor, and manipulative) that are associated with multiple developmental outcomes including physical health [1–3], psychological, social-emotional, and cognition/achievement [4–7]. Importantly, it is expected that adequate levels of MC will facilitate the proficiency of novel motor tasks throughout the lifespan and the learning of new skills [8].

The Motor Competence Assessment (MCA) is an instrument developed to assess motor competence across a lifespan. For the first time, it is possible to assess MC from childhood to old age using the same instrument, without a developmental (age) ceiling effect and using a feasible and objective test battery [8–10]. Although there are other instruments specifically developed or adapted to evaluate motor competence throughout a lifespan, such as the Test of Motor Competence (TMC) [11] and the last form of the Körperkoordinationstest für Kinder (KTK3+) [12], the first includes a fine motor component that is outside our definition of MC, and the second shows a clear ceiling effect on one of the tests used (balancing backwards) and no locomotor component is included.

The MCA is a quantitative test battery composed of six tests divided into three sub-scales: stability, locomotor, and manipulative. Each sub-scale is calculated from two tests, objectively measured. All motor tests in the MCA are quantitative (product-oriented), with no developmental (age) ceiling effect, and are easy to execute even with little practice. After determining the construct validity of the MCA for children and adolescents [9], the normative values from 3 to 23 years of age were established [10]. Very recently, the MCA configural invariance between age groups was tested, proving the usefulness of the MCA model throughout growth and development periods from early childhood to young adulthood [12].

Typically, the scales or batteries used to assess motor development or motor competence of children use a classification system that involves transforming the tests' raw data to standard scores relative to age and sex, but with a limited range (from 1 to 10, or more usually 1 to 20). Examples are the KTK, the Peabody Developmental Motor Scales 2 (PDMS-2) [13], the Test of Gross Motor Development 3 (TGMD-3) [14], and the Movement Assessment Battery for Children (MABC-2) [15], where a singular value of a standard score can accommodate a range of results from the test. For example, at the age of six, all values between 15 and 23 s on the one-leg balance test of the MABC are transformed into a standard score of eight. The same reduction in information happens when assessing the sub-scales or components, and the total value, or motor quotient, of the child. In such a procedure, a chunk of information is lost each time a transformation to standard scores is performed, and consequently we lose possible discrimination between children's real values. Since the MCA always uses quantitative ratio scales to assess the six tests, we argue all this information relative to age and sex (normative percentile values) should be used to better discriminate between subjects.

Although the MCA has been used with different populations and cultures [16–18], the method for scoring each of the three sub-scales (locomotor, stability, and manipulative) and for total MCA score, still needs to be defined. The initial theoretical framework on the development of motor skills [19,20] proposed an equal participation of the locomotor, stability, and manipulative components in overall motor competence. However, this structural relationship needs to be tested relative to the circumstances of the tests used to mark each component, and to the developmental ages where the evaluation takes place. Furthermore, since each MCA test is not a perfect marker of the component but only a proxy of it, there is a need to interpret possible different weights of the tests in the classification method of the components (sub-scales) and total MCA scores. The previous work undertaken on the validation of the MCA established the metrics for these questions, i.e., the relative weight of each test to the sub-scale according to different age periods [8–10]. Hence, the aim of this study will be to expand the MCA's development and validation by using the previous findings on the developmental characteristics of the MCA tests to define the best and leanest method to score and classify MCA sub-scales and total score.

2. Materials and Methods

2.1. Participants

An initial pool of 2150 participants (1003 females; 1147 males) from the Portuguese database on MC, a convenience sample consisting of students from preschool to university, was utilized to select the sample for this study. A computer program was used to randomly select 1000 participants (250 per age group from 3 to 6, 7 to 10, 11 to 16, and 17 to 23 years of age, gender-balanced within age groups). No differences were found on any of the MCA tests (all p's > 0.50), between the selected and the non-selected participants. All participants showed no motor or cognitive impairment.

The study was approved by the Ethics Committee of the Faculty of Human Kinetics at the University of Lisbon and the Scientific Council of the School of Sports and Leisure at the Polytechnic Institute of Viana do Castelo. All school directors authorized the study, and the informed consent of adult participants or parents/tutors of underage children was obtained. All children gave their verbal assent prior to data collection. All procedures were in accordance with the 1964 Declaration of Helsinki and its later amendments [21].

2.2. Instruments and Procedures

The MCA is composed of three sub-scales, stability, locomotor, and manipulative. Each sub-scale has two tests, specifically: lateral jumps (LJs) and shifting platforms (SPs) for stability; standing long jump (SLJ) and 4 × 10 m shuttle run (SHR) for locomotor; and ball kicking velocity (BKV) and ball throwing velocity (BTV) for the manipulative sub-scale. All tests were assessed in a quantitative scale (i.e., distance, time, number of executions, or velocity), and do not have a ceiling effect related to age or sex (for full description see Rodrigues et al., 2019). The literature describes the tests' reliability as ranging from good to excellent [22–25], and the values of the Intraclass Correlation Coefficient (ICC)were 0.95, 0.99, 0.97, 0.99, 0.98, and 0.98 respectively for SP, JS, SHR, SLJ, BKV, and BTV.

Participants completed a 10 min general and standardized warm-up before the testing. The test setting was organized in small groups (5 participants per task). Trained examiners administered the tests. A proficient demonstration and a verbal explanation of each test were provided before each test. A test trial was always provided to the participant before the actual test administration. Instructions were given for participants to perform each task at their maximum. Only motivational feedback was given during the test. All data collection was supervised by one of the authors of this study. Testing always took place in a gymnasium.

For the calculation of the sub-scales and total MCA, participants' results in each of the MCA tests were transformed into age- and sex-normative values (percentiles) according to the process previously described [8]. Three different procedures to calculate the sub-scales values were tested according to three alternative models [25]: (1) the Weighted Age Group Model (WAGM) accounted for different weights (percentage) of each test's representation in the final sub-scale score, according to the age groups' previous results [8]; (2) the Overall Model (OM)used the same weight of each test, independent of the age, for calculating the sub-scale score, according to the loading coefficient of the model found for the overall sample [8]; and (3) the Configural Model (CM), in which all tests represented an equal value for the sub-scale's calculation (see Table 1).

Table 1. Loading coefficients for each sub-scale of the three tested models.

		Weighted Age Group Model				Overall Model	Configural Model
		3 to 6	**7 to 10**	**11 to 16**	**17 to 22**		
Stability	LJ	0.86	0.85	0.77	0.73	0.94	1
	SP	0.90	0.59	0.82	0.66	0.93	1
Locomotor	SLJ	0.89	0.63	0.94	0.88	0.94	1
	SHR	−0.82	−0.86	−0.86	−0.79	0.89	1
Manipulative	BKV	0.82	0.85	0.89	0.93	0.97	1
	BTV	0.78	0.76	0.91	0.91	0.95	1

LJs—lateral jumps; SPs—shifting platforms; SLJ—standing long jump; SHR—shuttle run; BKV. ball kicking velocity; BTV—ball throwing velocity.

According to the different methods used, the formula to calculate each sub-scale score was:

$$| ((LC\ test\ 1/(LC\ test\ 1 + LC\ test\ 2)) * P\ test\ 1) + ((LC\ test\ 2/(LC\ test\ 1 + LC\ teste\ 2)) * P\ test\ 2)| \tag{1}$$

where LC = Loading Coefficient and P = percentile value.

After defining the best method for calculating the sub-scales scores, identical methodological procedures were followed to calculate the total MCA score. The relative representation of each sub-scale was used for calculating the final MCA score according to the same three models previously used (WAGM, OM, and CM):

$$\textbf{TOTAL MCA} = (((LC1+LC2)/\text{sum LC}) * STAB) + (((LC3+LC4)/\text{sum LC}) * LOC) + (((LC5+LC6)/\text{sum LC}) * MAN) \quad (2)$$

where LC1 = Loading Coefficient test 1; STAB = stability sub-scale score; LOC = locomotor sub-scale score; MAN = manipulative sub-scale score.

2.3. Statistical Analysis

The results for the sub-scales and total MCA according to the three tested models were compared with the values produced by the Weighted Age Group Model (WAGM), the one selected as our reference method. This choice relates to the fact that the WAGM model provides more information on the weight of each test to the sub-scale score and the total score (configural, loading coefficients, and age-related information). The Intraclass Correlation Coefficient tested for the relationship between the results accounting for the absolute value; the Cronbach's Alpha was used as a measure of reliability. The Bland–Altman plot with limits of agreement was used to describe agreement between the three methods taking the WAGM as the reference method [26,27].

Subsequently, the chosen method for calculating the MCA sub-scales scores was then used with our data to find each sub-scale score, and the MCA total score was then assessed according to the three different models previously used (WAGM, OM, and CM).

Initial analyses were made using IBM SPSS Statistics for Windows, Version 26.0. (Armonk, NY, USA: IBM Corp). The Bland–Altman plot and subsequent analyses were calculated in MedCalc® Statistical Software version 20.106 (MedCalc Software Ltd., Ostend, Belgium; https://www.medcalc.org; accessed on 1 February 2022).

3. Results

As depicted in Table 2, an almost perfect reliability level was found between the tested methods when compared to the reference method, with values ranging from 0.999 to 1.000 in the ICC and the Cronbach's Alpha. Additionally, the mean difference (bias) resulting from the Bland–Altman technique always returned a very low value, very close to zero (not significantly different from zero in all cases except for WAGM * CM), with also a very narrow interval between 95% limits of agreement (Figure 1).

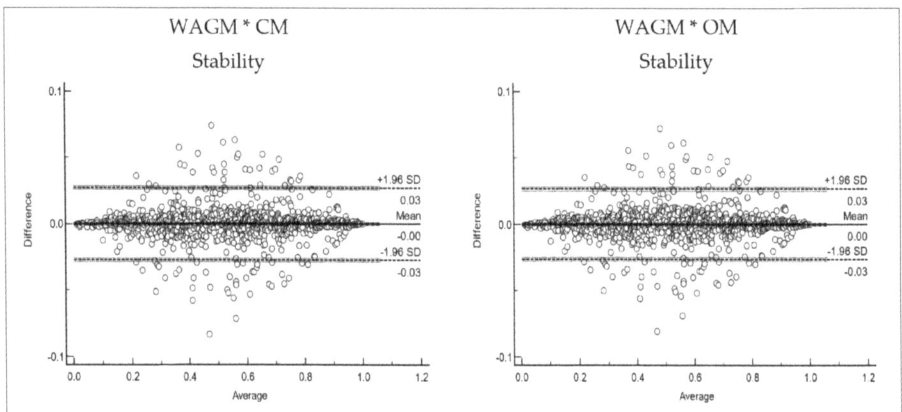

Figure 1. *Cont.*

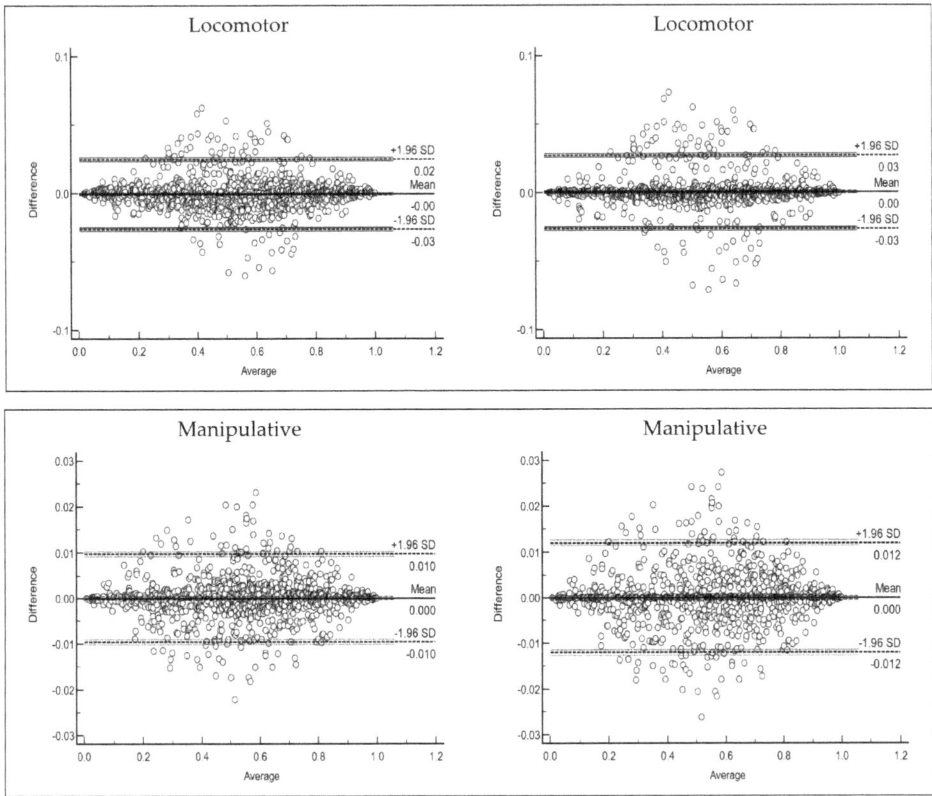

Figure 1. Bland–Altman graphs representing differences between the predicted models and the Weighted Age Group Model (WAGM) reference values for the MCA sub-scales, according to the percentile positions of the reference values of the WAGM vs. the mean of the two measurements, with limits of agreement and respective 95% confidence intervals.

Table 2. Reliability and Bland–Altman agreement analysis (mean bias and 95% limits of agreement) between the MCA sub-scale scores. Comparisons of the Configural Model and the Overall Model procedures with the Weighted Age Group Model.

	Cronbach's Alpha	Intraclass Correlation Coefficient	Bland–Altman		
			Mean Bias (95%CI)	Lower Limit	Upper Limit
Stability					
WAGM * CM	0.999	0.999	0.00001 (−0.0009 to 0.0009)	−0.0271	0.0271
WAGM * OM	0.999	0.999	−0.00006 (−0.0009 to 0.0008)	−0.0266	0.0264
Locomotor					
WAGM * CM	0.999	0.999	−0.0006 (−0.0015 to 0.0001)	−0.0261	0.0248
WAGM * OM	0.999	0.999	0.000500 (−0.0004 to 0.0013)	−0.0263	0.0272

Table 2. *Cont.*

	Cronbach's Alpha	Intraclass Correlation Coefficient	Bland–Altman		
			Mean Bias (95%CI)	Lower Limit	Upper Limit
Manipulative					
WAGM * CM	1.000	1.000	−0.00007 (−0.0004 to 0.0002)	−0.0097	0.0096
WAGM * OM	1.000	1.000	−0.000006 (−0.0004 to 0.0004)	−0.012	0.012

WAGM—Weighted Age Group Model; CM—Configural Model; OM—Overall Model.

The sub-scales scores according to the CM method were then used (Table 3) with our data to find the MCA total score according to the three different models previously used (WAGM, OM, and CM).

Table 3. Reliability and Bland–Altman agreement analysis (mean bias and 95% limits of agreement) between the MCA total scores). Comparisons of the Configural Model and the Overall Model procedures with the Weighted Age Group Model.

	Cronbach's Alpha	Intraclass Correlation Coefficient	Bland–Altman		
			Mean Bias (95%CI)	Lower Limit	Upper Limit
Total MCA					
WAGM * CM	0.999	0.999	−0.0006458 (−0.0010 to 0.000004)	−0.01779	0.01649
WAGM * OM	1.000	1.000	−0.0005122 (−0.0011 to −0.0001)	−0.01681	0.01578

WAGM—Weighted Age Group Model; CM—Configural Model; OM—Overall Model

When looking at Figure 1, which shows the Bland–Altman plot with the 95% limits of agreement, all differences between the tested (predicted) models and the reference values found when the WAGM model was used, were very small. Even the biggest differences were within a 0.1 percentile margin of error, meaning that any difference between the three tested models was negligible. Very few cases were beyond the 95% limits of agreement.

4. Discussion

The aim of this study was to determine the best algorithm for assessing the MCA sub-scales and total scores, and one representative of the developmental MC relationships. All test batteries for motor competence or motor development use sub-scales and total scores. In fact, the MCA has been used in recent years with each author using a composite of the tests to represent sub-scales and total MCA scores e.g., [16–18]. This practice has reflected the theoretical model assumed for MCA, long used in the field, that all tests and sub-scales have equal participation in the determination of the motor competence, but this had not been tested until now. To test this, we used three different models previously fitted to a sample of 1000 participants within 3 to 23 years of age, and their loading weights were used for calculation of the sub-scales and total MCA [26]. Since we hypothesized that the better-adjusted model should consider age-related constraints, the Weighted Age Group Model values were used as a reference model to compare to the other two: an Overall Model (using all the sample), and the Configural Model where all weights are assumed to be equal.

The analysis showed that no substantial differences could be found between the three methods used for assessing the sub-scales. The reliability values were perfect or in the vicinity of perfection (0.999 to 1.000), which implies that all the methods were able to classify everyone in the same way. The Bland–Altman plots represented in Figure 1 show the dimension of the differences between the reference model and the other two. The 95% limits of agreement found for both the CM and the OM resulted in a very narrow range, between −0.01 and 0.01 percentiles. That means that using either of the three methods would produce almost the same results for all individuals tested. Furthermore, the Bland–Altman bias analysis showed negligible values, indicating that all the methods can be used with similar results independent of the test scores' magnitude.

Given these results for the sub-scales, we suggest implementing the most economic and efficient algorithm, i.e., the configural model algorithm, averaging the percentile scores of the two tests to assess each sub-scale. This method was used with our data to find each sub-scale score, and the MCA total score was then assessed according to the three different models previously used (WAGM, OM, and CM). Again, the results proved very similar in outcome, independent of the method used, with reliability values ranging from 0.999 to 1.000, and the 95% confidence intervals for the differences found for both the CM (−0.0010 to 0.000) and the OM (−0.0011 to −0.0001) were very small.

Other scoring methods used in comparable test batteries usually include the transformation of individual test's raw values into z-scores related to age and sex (test z-score), and the transformation of the sum of all the z-scores into a new value that is then transformed into a new z-score (total or motor quotient z-score). That is the case for KTK3+ [12], or the TMC [11] (although the latter uses a general z-score value for each test, and not an age- and sex-related z-score). These methods are similar to the technique we used, where the mean and standard deviation used to find the z-scores results from a previous populational study. Other tests, such as the PDMS-2 [13], the TGMD-3 [14], or the MABC-2 [15], include a first transformation of the raw values into age- and sex-related standard scores (with a range of 1 to 10, or 1 to 20), and then its sum into a new standard score or percentile interval to represent the total motor quotient or percentile classification. In this case, each raw value is transformed two times, losing some information in each transformation, and resulting in less discriminating scores than the MCA.

A limitation of our results is the fact that no single gold standard is established for these specific measures of the MCA (sub-scales and total score). As explained previously, we choose the WAGM model as the most appropriate and complete model to determine the real value of the MCA scores. We also intend, in the future, to compare the MCA scores with other batteries used for assessing MC such as the KTK 3+, TMC, MABC-2, TGMD-3, or the Bruininks–Oseretsky.

5. Practical Implications

This study represents the final step for validating the MCA testing battery as a tool for assessing motor competence across all ages. MCA stability, locomotor, and manipulative sub-scales and MCA total score values can now be used when testing motor competence with any age and sex. The average of the normative age- and sex-related values for each test can be used to calculate the MCA sub-scales and total MCA scores. MCA is now ready to be used as a classification tool for MC throughout the human lifespan in different contexts (research, clinical, sports, education). Classification of the MCA cut-off values for the motor-impaired, underperformers, and high performers can now be established.

6. Conclusions

In conclusion, we advocate the use of the transformed percentile values to evaluate the performance on each test, and the average of these values to assess each sub-scale and the total MCA score, as the best and most effective method to score and classify MCA. With these insights, the MCA is ready to be used as a classification tool for MC throughout the lifespan in different contexts (research, clinical, sports, education). Further classification cut-off values for the motor-impaired, underperformers, and high performers can now be procured in future works.

Author Contributions: Conceptualization, L.P.R., C.L., R.C. and V.P.L.; Methodology, L.P.R., C.L., A.P. and V.P.L.; Formal analysis, L.P.R., C.L., R.C., A.P. and V.P.L.; Investigation, L.P.R., R.C. and A.P.; Data curation, C.L. and R.C.; Writing–original draft, L.P.R.; Writing–review & editing, C.L., R.C., A.P. and V.P.L.; Funding acquisition, L.P.R., R.C. and V.P.L. All authors have read and agreed to the published version of the manuscript.

Funding: This work was supported by the Portuguese Science Foundation (FCT) under Grant number UIDB/00447/2020 (unit 447); the European Regional Development Fund (ERDF) through the Regional Operational Program North 2020 under Project TECH—Technology, Environment, Creativity and Health, Grant number Norte-01-0145-FEDER-00004; and the Portuguese Science Foundation (FCT) under Grant number UID04045/2020.

Institutional Review Board Statement: The study was approved by the Scientific Council of the School of Sports and Leisure at the Polytechnic Institute of Viana do Castelo, and by the Ethics Committee of the Faculty of Human Kinetics at the University of Lisbon (Approval Number: CEFMH 26/2013; Date: 3 December 2013). School directors approved the study, adult participants and parents or tutors of underage children gave their informed consent, and children gave their verbal assent prior to data collection. All procedures were in accordance with the 1964 Declaration of Helsinki and its later amendments.

Informed Consent Statement: Adult participants and parents or tutors of underage children gave their informed consent, and children gave their verbal assent prior to data collection.

Data Availability Statement: Contact the correspondent author for data information.

Conflicts of Interest: The authors declare no conflict of interest.

References

1. Barnett, L.M.; Morgan, P.J.; van Beurden, E.; Beard, J.R. Perceived Sports Competence Mediates the Relationship between Childhood Motor Skill Proficiency and Adolescent Physical Activity and Fitness: A Longitudinal Assessment. *Int. J. Behav. Nutr. Phys. Act.* **2008**, *5*, 40. [CrossRef] [PubMed]
2. Chagas, D.V.; Marinho, B. Exploring the Importance of Motor Competence for Behavioral and Health Outcomes in Youth. *Percept. Mot. Ski.* **2021**, *128*, 2544–2560. [CrossRef] [PubMed]
3. Lima, R.A.; Bugge, A.; Ersbøll, A.K.; Stodden, D.F.; Andersen, L.B. The longitudinal relationship between motor competence and measures of fatness and fitness from childhood into adolescence. *J. Pediatr.* **2019**, *95*, 482–488. [CrossRef] [PubMed]
4. Barnett, L.M.; Webster, E.K.; Hulteen, R.M.; De Meester, A.; Valentini, N.C.; Lenoir, M.; Pesce, C.; Getchell, N.; Lopes, V.P.; Robinson, L.E.; et al. Through the Looking Glass: A Systematic Review of Longitudinal Evidence, Providing New Insight for Motor Competence and Health. *Sports Med.* **2022**, *52*, 875–920. [CrossRef] [PubMed]
5. Leonard, H.C.; Hill, E.L. Review: The Impact of Motor Development on Typical and Atypical Social Cognition and Language: A Systematic Review. *Child Adolesc. Ment. Health* **2014**, *19*, 163–170. [CrossRef]
6. Robinson, L.E.; Stodden, D.F.; Barnett, L.M.; Lopes, V.P.; Logan, S.W.; Rodrigues, L.P.; D'Hondt, E. Motor Competence and Its Effect on Positive Developmental Trajectories of Health. *Sports Med.* **2015**, *45*, 1273–1284. [CrossRef]
7. De Meester, A.; Barnett, L.M.; Brian, A.; Bowe, S.J.; Jiménez-Díaz, J.; Van Duyse, F.; Irwin, J.M.; Stodden, D.F.; D'Hondt, E.; Lenoir, M.; et al. The Relationship Between Actual and Perceived Motor Competence in Children, Adolescents and Young Adults: A Systematic Review and Meta-Analysis. *Sports Med.* **2020**, *50*, 2001–2049. [CrossRef]
8. Rodrigues, L.P.; Cordovil, R.; Luz, C.; Lopes, V.P. Model Invariance of the Motor Competence Assessment (MCA) from Early Childhood to Young Adulthood. *J. Sports Sci.* **2021**, *39*, 2353–2360. [CrossRef]
9. Luz, C.; Rodrigues, L.P.; Almeida, G.; Cordovil, R. Development and Validation of a Model of Motor Competence in Children and Adolescents. *J. Sci. Med. Sport* **2016**, *19*, 568–572. [CrossRef]
10. Rodrigues, L.P.; Luz, C.; Cordovil, R.; Bezerra, P.; Silva, B.; Camões, M.; Lima, R. Normative Values of the Motor Competence Assessment (MCA) from 3 to 23 Years of Age. *J. Sci. Med. Sport* **2019**, *22*, 1038–1043. [CrossRef]
11. Sigmundsson, H.; Lorås, H.; Haga, M. Assessment of Motor Competence Across the Life Span: Aspects of Reliability and Validity of a New Test Battery. *SAGE Open* **2016**, *6*, 2158244016633273. [CrossRef]
12. Coppens, E.; Laureys, F.; Mostaert, M.; D'Hondt, E.; Deconinck, F.J.A.; Lenoir, M. Validation of a Motor Competence Assessment Tool for Children and Adolescents (KTK3+) With Normative Values for 6- to 19-Year-Olds. *Front. Physiol.* **2021**, *12*, 652952. [CrossRef] [PubMed]
13. Folio, M.R.; Fewell, R.R. *Peabody Developmental Motor Scales*, 2nd ed.; Pearson's Clinical Assessment: San Antonio TX, USA, 2000.
14. Ulrich, D.A. *Test of Gross Motor Development*, 3rd ed.; Pro-Ed: Austin, TX, USA, 2016.
15. Henderson, S.E.; Sugden, D.A.; Barnett, A.L. *Movement Assessment Battery for Children-2*, 2nd ed.; The Psychological Corporation: London, UK, 2007.

16. Luz, C.; Cordovil, R.; Rodrigues, L.P.; Gao, Z.; Goodway, J.D.; Sacko, R.S.; Nesbitt, D.R.; Ferkel, R.C.; True, L.K.; Stodden, D.F. Motor competence and health-related fitness in children: A cross-cultural comparison between Portugal and the United States. *J. Sport Health Sci.* **2019**, *8*, 130–136. [CrossRef] [PubMed]
17. Silva, A.F.; Nobari, H.; Badicu., G.; Ceylan, H.I.; Lima, R.; Lagoa, M.J.; Luz, C.; Clemente, F.M. Reliability levels of motor competence in youth athletes. *BMC Pediatr.* **2022**, *20*, 430. [CrossRef]
18. Sá, C.D.S.C.D.; Luz, C.; Pombo, A.; Rodrigues, L.P.; Cordovil, R. Motor Competence in Children With and Without Ambliopia. *Percept. Mot. Skills* **2021**, *128*, 746–765. [CrossRef]
19. Gabbard, C. *Lifelong Motor Development*, 7th ed.; Wolters Kluwer: Philadelphia, PA, USA, 2018.
20. Goodway, J.D.; Ozmun, J.C.; Gallahue, D.L. *Understanding Motor Development: Infants, Children, Adolescents, Adults*; Jones & Bartlett Learning: Burlington, MA, USA, 2019.
21. World Medical Association. World Medical Association Declaration of Helsinki. Ethical Principles for Medical Research Involving Human Subjects. *Bull. World Health Organ.* **2001**, *79*, 373.
22. Iivonen, S.; Sääkslahti, A.; Laukkanen, A. A review of studies using the körperkoordinationstest für kinder (ktk). *Eur. J. Adapt. Phys. Act.* **2015**, *8*, 18–36. [CrossRef]
23. AAHPER. *Aapher Youth Fitness Test Manual*; American Alliance for Health, Physical Education, and Recreation: Washington DC, USA, 1975.
24. Fernandez-Santos, J.R.; Ruiz, J.R.; Cohen, D.D.; Gonzalez-Montesinos, J.L.; Castro-Piñero, J. Reliability and validity of tests to assess lower-body muscular power in children. *J. Strength Cond. Res.* **2015**, *29*, 2277–2285. [CrossRef]
25. Moreira, J.; Lopes, M.C.; Miranda-Júnior, M.V.; Valentini, N.C.; Lage, G.M.; Albuquerque, M.R. Körperkoordinationstest Für Kinder (KTK) for Brazilian Children and Adolescents: Factor Analysis, Invariance and Factor Score. *Front. Psychol.* **2019**, *10*, 2524. [CrossRef]
26. Nevill, A.M.; Atkinson, G. Assessing Agreement between Measurements Recorded on a Ratio Scale in Sports Medicine and Sports Science. *Br. J. Sports Med.* **1997**, *31*, 314–318. [CrossRef]
27. Bland, J.M.; Altman, D.G. Statistical Methods for Assessing Agreement between Two Methods of Clinical Measurement. *Antimicrob. Agents Chemother.* **1986**, *2*, 307–310. [CrossRef]

Systematic Review

Mixed-Methods Systematic Review to Identify Facilitators and Barriers for Parents/Carers to Engage Pre-School Children in Community-Based Opportunities to Be Physically Active

Rachel L. Knight [1], Catherine A. Sharp [1], Britt Hallingberg [2], Kelly A. Mackintosh [1] and Melitta A. McNarry [1,*]

[1] Applied Sports, Technology, Exercise and Medicine (A-STEM) Research Centre, Swansea University, Swansea SA1 8EN, UK
[2] Cardiff School of Sport and Health Sciences, Cardiff Metropolitan University, Cardiff CF5 2YB, UK
* Correspondence: m.mcnarry@swansea.ac.uk

Abstract: Background: Low physical activity levels in young children is a major concern. For children aged 0–5 years, engagement with opportunities to be physically active are often driven by the adults responsible for the child's care. This systematic review explores the barriers and facilitators to parents/caregivers engaging pre-school children in community-based opportunities for physical activity, within real-world settings, or as part of an intervention study. **Methods:** EBSCOhost Medline, CINHAL plus, EBSCOhost SPORTDiscus, Web of Science, ProQuest, and ASSIA were systematically searched for quantitative and qualitative studies published in English between 2015 and 16 May 2022. Data extracted from 16 articles (485 parents/carers; four countries) were quality-assessed using the Mixed Methods Assessment Tool and coded and themed via thematic analysis. **Results:** Nine themes (eight core, one minor) were identified and conceptualised into a socio-ecological model, illustrating factors over four levels: Individual—beliefs and knowledge (and parental parameters); Interpersonal—social benefits, social network, and family dynamic; Community—organisational factors and affordability; and Built and Physical Environment—infrastructure. **Discussion:** The findings provide valuable insights for practitioners and policy makers who commission, design, and deliver community-based physical activity opportunities for pre-school children. Developing strategies and opportunities that seek to address the barriers identified, as well as build on the facilitators highlighted by parents, particularly factors related to infrastructure and affordability, are imperative for physical activity promotion in pre-school children. The perspectives of fathers, socioeconomic and geographical differences, and the importance parents place on physical activity promotion all need to be explored further.

Keywords: physical activity; sedentary time; sedentary behaviour; youth; socio-ecological mode

Citation: Knight, R.L.; Sharp, C.A.; Hallingberg, B.; Mackintosh, K.A.; McNarry, M.A. Mixed-Methods Systematic Review to Identify Facilitators and Barriers for Parents/Carers to Engage Pre-School Children in Community-Based Opportunities to Be Physically Active. *Children* **2022**, *9*, 1727. https://doi.org/10.3390/children9111727

Academic Editor: Georgian Badicu

Received: 15 October 2022
Accepted: 6 November 2022
Published: 10 November 2022

Publisher's Note: MDPI stays neutral with regard to jurisdictional claims in published maps and institutional affiliations.

Copyright: © 2022 by the authors. Licensee MDPI, Basel, Switzerland. This article is an open access article distributed under the terms and conditions of the Creative Commons Attribution (CC BY) license (https://creativecommons.org/licenses/by/4.0/).

1. Introduction

The benefits of being physically active are irrefutable [1]. Unfortunately, physical activity (PA) levels of young children, including pre-school children (0–5-year-olds), are a major concern [2,3]. Specifically, whilst data on this age group are sparse, the emerging evidence is congruent with that of their older counterparts (5–18-year-olds), with high amounts of time in of sedentary behaviour and low levels of PA already established behaviours for many [4,5]. This is especially concerning given that childhood behaviours are well evidenced to track into adolescence and adulthood [6].

For young children, being insufficiently active can limit the protective effect of PA against cardiovascular disease [7] and obesity [8] and have negative connotations for cognitive development [9], psychosocial health [10], and the mastery of fundamental movement skills [11]. Multiple reviews have explored and identified correlates of PA and sedentary behaviour in children (i.e., [12,13]), highlighting that elements relating to parental practices and opinions are, at least to some degree, key determinants of a child's health behaviours, including PA.

The engagement of pre-school children in organised opportunities to be physically active has been shown to positively influence sedentary behaviour [14] and moderate-to-vigorous physical activity (MVPA) [15]; both important ways to facilitate the achievement of physical activity levels recommended for optimal health [1]. Indeed, Chen et al. [15] found that the participation of 3–5-year-olds in organised sport (outside of school time) was associated with 10% more MVPA throughout the day. Likewise, when evaluating the effects of a 10-week, family-driven active play programme in children under five years old, O'Dwyer et al. [14] discovered that attendance at structured activities away from the home was a significant predictor of lower weekend sedentary time. The authors suggest that this may, at least in part, be due to parental perceptions of PA, and their ability to facilitate the engagement of their child.

The United Nations Convention on the Rights of the Child [16] states that every child has the right to good health and the right to play. However, on an individual level, engagement of pre-school children with opportunities to be physically active is driven by the adults responsible for the child's care. Tackling physical inactivity and sedentary behaviour in early childhood therefore requires an in-depth understanding of the key factors that influence adults decision-making processes and behaviours. Such information is essential to assist policy makers and, where necessary, address the (mis)perceptions of those individuals who have the greatest influence during the early years period.

A review of global qualitative research synthesised and mapped the factors relating to parental, care provider, and child perceptions of PA behaviour influencers [17]. Seven core themes were outlined, with the most frequently reported barriers and facilitators being those on an interpersonal level, including the role that adults (parents, care providers, family) have in facilitating health behaviours. Given recent evidence surrounding the barriers and facilitators to physical activity engagement among young children [12,17], it is also important to understand how these present for parents/caregivers, specifically in accessing wider opportunities to support their children's physical activity within local communities. This can provide further recommendations for public health interventions. Consequently, this mixed-methods systematic review aims to build on prior findings by: (i) specifically exploring the barriers and facilitators to parents/caregivers living within developed economies engaging pre-school children (0–5-year-olds) in community-based opportunities for PA, either within real-world settings or as part of an intervention study, capturing both qualitative and quantitative evidence; (ii) categorising and discussing the findings at a socio-ecological level in line with the framework of Sallis et al. [18]; and (iii) establishing which barriers and facilitators may be central to future action planning. From here on, for ease, reference to *parents* will encompass parental, carer, and kinship relationships.

2. Methods

This mixed-methods systematic review of both quantitative and qualitative research, including peer-reviewed and non-peer-reviewed (grey) literature, was designed and conducted in line with the Joanna Briggs Institute (JBI) methodology [19]. This approach enabled the complexities of health-related research to be fully explored, ensuring the widest range of understanding was generated [20]. This review is reported in accordance with the Preferred Reporting Items for Systematic reviews and Meta-Analysis (PRISMA) [21] and registered on PROSPERO, registration number CRD42022331738.

2.1. Data Sources, Searches, and Study Selection Criteria

Six electronic databases (EBSCOhost Medline, CINHAL plus, EBSCOhost SPORTDiscus, Web of Science, ProQuest, ASSIA), limited to publications in English between 2015 (in line with publication of the United Nations Sustainable Development Goals [22]) and 16th May 2022, were searched by the first author (RLK). Search strategies of key terms were created based on previous reviews of similar topics (e.g., [17]) and refined with guidance from a subject Librarian and input from the review team. The Boolean terms and their variations used included, but were not limited to: (parent* OR carer* OR grandparent* OR guardian*) AND (preschool* OR "early years" OR "young children") AND (barrier* OR facilitator* OR perception* OR challeng*) AND ("physical activit* OR "active play" OR "fundamental movement" OR "motor skills").

Two authors (RLK and CS or MAM) independently reviewed all generated citations and abstracts to select eligible studies using Rayyan (QRCI, Qatar), with those coded 'include' subsequently full-text screened (independently by RK and MM) against the pre-defined inclusion/exclusion criteria (Table 1). Disagreements regarding eligibility of citations and abstracts or full-text articles were resolved through discussion between the authors (initially: k = 0.46, after discussion: k = 0.88), with a third reviewer (CS or BH) consulted where a consensus could not be reached (n = 3). Online File S1 provides full details of database-specific terms, restrictions and search strategies, and secondary and grey literature search processes.

Table 1. Study inclusion and exclusion criteria.

Variable	Inclusion Criteria	Exclusion Criteria
Population or participants	Parent/guardian/carer of a child(ren) aged 0–5 years Any gender Restricted to the 37 countries listed on the UN 2022 list of developed economies	Studies including the parent, guardian, or carer of children aged >5 years, where data cannot be separated Studies that include clinical populations (e.g., autism, cerebral palsy, cystic fibrosis), where data cannot be separated
Phenomena of interest	Perceived parental, guardian, or carer barriers/facilitators to engagement of children aged 0–5 years with PA driven opportunities	Studies including the perceptions of children or education providers Studies that involve data obtained relating specifically to the effects of COVID-19 pandemic curtailment restrictions
Context	Community-based PA opportunities—any public, private, or third-sector activity provided either freely or following payment within real-world settings or as part of an intervention study	Studies where a degree of activity offered (free or paid) is not based on or driven by PA Studies that report outcomes relating to physical education provision or interventions delivered during core nursery and school hours and therefore participation is not under the direct influence of a parent/guardian/carer
Study designs	Any quantitative, qualitative, or mixed-methods study design providing original results (including grey literature, such as conference proceedings) No restrictions on measurement type for assessing barriers/facilitators. Mixed-methods studies will be considered if the data from the qualitative and quantitative elements can be clearly extracted	Studies not providing original results, such as systematic reviews, meta-analysis, general reviews, or editorials Studies where barriers/facilitators are only reported anecdotally within the discussion, not as core results

PA = physical activity; UN = United Nations.

2.2. Data Extraction

A standardised form pre-developed by the JBI (mixed methods convergent integrated approach) [19] was used by RLK to extract data from the included studies for evidence synthesis. Extracted information included: authors; year of publication; study design, setting, and population, including sample size; participant demographics and characteristics; study methodology; inclusion/exclusion criteria; empirical and/or contextual results relating to parental/carer perceived barriers and facilitators to engagement with community-based PA opportunities (for quantitative data: outcomes of descriptive and interferential analyses; for qualitative: themes or sub-themes supported with illustrations, such as supporting data or direct quotations); type of community-based opportunity (if applicable); and quality-assessment information. A second reviewer (CS or BH) independently reviewed 50% of the extracted data. Two discrepancies were resolved through discussion. Supplementary data were consulted where necessary and available.

2.3. Quality Assessment

One author (RLK) independently assessed study quality using the Mixed Methods Assessment Tool (MMAT) [23]. A second reviewer (CS or BH) independently randomly checked 50% of the studies to ensure consistency. No disagreements occurred. Each study was subsequently attributed an overall quality score, ranging from 1*, where 20% of the quality criteria have been met, to 5*, where 100% of the quality criteria have been met [24]. No studies were excluded due to low quality; rather, all issues were considered when interpreting the results of each study. As per JBI recommendations for mixed-methods systematic reviews, an assessment of certainty was not conducted.

2.4. Data Synthesis

A convergent integrated approach to data synthesis was undertaken [19]. Quantitative data were transformed into textual descriptions or narrative interpretations to facilitate integration with extracted qualitative data. The transformed quantitative data were concurrently analysed and categorised with the qualitative data to form one set of themes and narratives. Thematic synthesis of the pooled data was conducted by RLK following the approach of Thomas and Harden [25]. The findings were coded line-by-line with 'free codes', organised into related areas, and 'descriptive' themes constructed. The descriptive themes were organised into overarching 'analytical' themes. This process was completed in collaboration with the review team; generated themes were discussed, reflected upon, and refined, with reference back to the original articles when required. The analytical themes were used to construct a conceptual framework based on the socioecological model categories outlined by Sallis et al. [18]. This process was undertaken by RLK with a 'critical friend' (CS) blindly crossmatching 10% of the studies against the framework to ensure the accuracy, rigour, transparency, and credibility of the undertaken process [26].

3. Results

Overall, 5856 articles were identified from electronic database searches; a further 282 were identified from secondary search strategies. Following duplicate removal, 2477 articles were screened, 2286 excluded, 91 retrieved for full-text eligibility screening, and 16 retained and included in the final analysis (Figure 1). The remaining articles included data from 485 parents of a child aged 0–5 years old [27–42], from four different countries (Australia, the United States of America (USA), the United Kingdom (UK), and Slovenia). Three articles did not provide details of the age of parents [28,32,40]; one article did not provide the sex of parents [38]. It was not possible to separate core demographic details from other study participants not relevant to this review within four articles [31,38,41,42]; however, given that it was possible to separate the facilitators and barriers, these studies were still included.

All articles collected qualitative data, predominantly via semi-structured interviews (n = 9/16) [27–34,36], with five articles adopting a mixed-methods design [28,29,34,40,42]. From these five articles, only the quantitative data from one article [28] met the criteria for inclusion in the data synthesis and was assessed for quality. Only 5 out of 16 articles (31%) had the primary aim to explore factors associated with the child's engagement in community-based opportunities for PA [27,28,31,34,42]. For the others, the aim was either focused on the general PA context (n = 8/16) [32,33,35–38,40,41] or on other health-related behaviours (n = 3/16) [29,30,39], with data pertinent to this review captured as a coincidental finding. Further individual study characteristics and overall MMAT quality scores are provided in Table 2, with a full breakdown of the quality assessments and supporting justification presented in Online Tables S1 and S2.

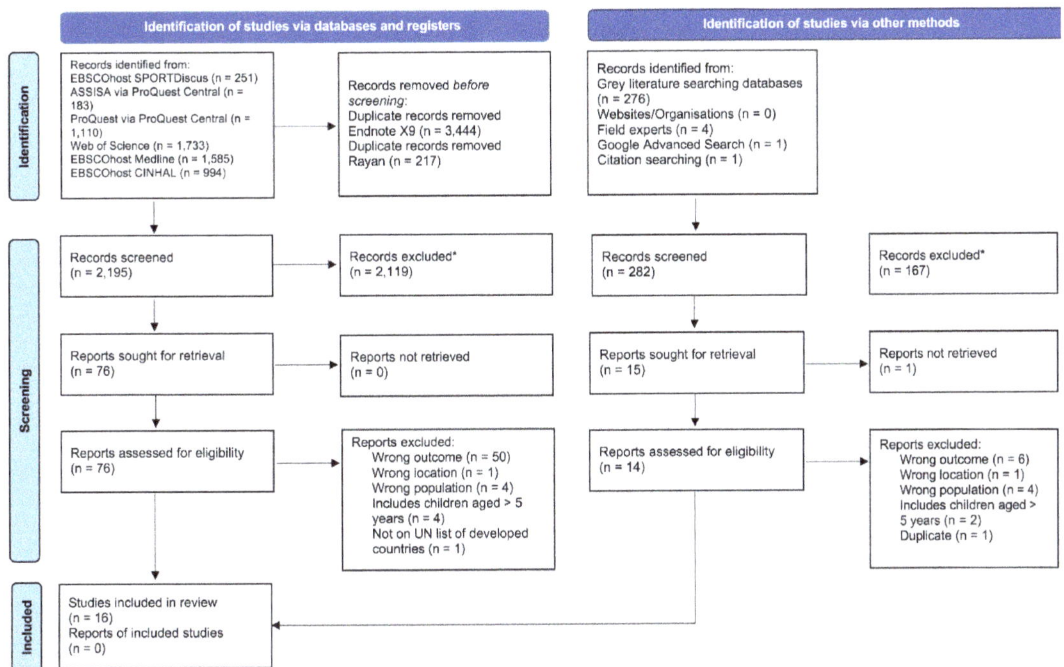

Figure 1. PRISMA 2020 flow diagram for new systematic reviews which included searches of databases, registers, and other sources. Note: * Automation tools were not used; *n* = number; UN = United Nation.

In concert with the dimensions of the socio-ecological framework of Sallis et al. [18], a narrative synthesis of the findings is outlined in the following section. To assist with framing potential interactions between individual-level parental influences and wider factors [18], this is accompanied by a model conceptualised from the nine themes (eight core, one minor) identified within the analysis process (Figure 2). An overview of the themes with indicative text quotes is presented in Online File S2.

Table 2. Study characteristics and MMAT scores.

Author	Study Design	Location and Context	Inclusion/ Exclusion Criteria	Phenomena of Interest	Participants (*n*)	Core Characteristics		Overall MMAT Grade
						Parent	Child	
Allen et al. (2021) [27]	Descriptive via semi-structured interviews (random sample)	UK: North-east of England Private /franchised swimming lessons	Aged ≥ 18 years; previously paid for private swimming lessons for child(ren) (0–4 years) for 12 months	The reasons why middle-class parents decide to pay for private/franchised swimming lessons for pre-school children in line with Bourdieu's triad of capital, habitus, and field	n = 8 (75% female) Aged 25–44 years	Education level 50% degree 25% college 25% secondary school Employment 50% employed 12.5% unemployed 12.5% homemaker 25% other Income 25% low-middle-class 75% mid-middle class	Aged 0–4 years	****
Andrews et al. (2019) [28]	Mixed methods Stage 1: Cross-sectional via individually constructed self-report questionnaire (stratified purposive sample) Stage 2: Descriptive via semi-structure interviews (self-selected from Stage 1)	Australia: Melbourne Selected inner and outer-suburban areas, one > 25 km and one < 10 km from the central business district	Lived in their suburb for ≥12 months; parent of at least one child aged 0–4 years attending pre-school	Who children played with and where they played in the two communities and the reasons for any differences in children's play experiences	Inner suburbs Questionnaire n = 72 (95.8% female) Interviews n = 10 Outer suburbs Questionnaire n = 26 (92.3% female) Interviews n = 10 No age range detailed	Inner suburbs 77.8% born in Australia Education level 34.4% postgraduate 37.5% degree 2.8% year 12 not completed Employment Hours/week paid work Mean = 16 (σ 14.6) Outer suburbs 46.2% born in Australia Education level 15.4% postgraduate 34.6% degree 11.5% year 12 not completed Employment Hours/week paid work Mean = 10.2 (σ 15.4)	Aged 0–4 years	**

Table 2. *Cont.*

Author	Study Design	Location and Context	Inclusion/ Exclusion Criteria	Phenomena of Interest	Participants (*n*)	Core Characteristics		Overall MMAT Grade
						Parent	Child	
Downing et al. (2016) [29]	Mixed methods. Descriptive via structured interviews (only qualitative data relevant; purposive sample)	Australia: Victoria Urban and regional local government areas covering the lowest decile for socioeconomic disadvantage	Mother of child(ren) aged 1–3 years; from low SES urban or regional area; able to speak, read, and write fluent English	The views of mothers from disadvantaged (low SES position) urban and regional areas (e.g., beyond major capital cities) as potential end users of child active play and screen time behaviour change interventions, with a focus on text messaging and web-based delivery platforms	Urban *n* = 22 (100% female) Mean age = 33.9 years (σ 6.4) Regional *n* = 10 (100% female) Mean age = 32.4 years (σ 5.7)	Urban Education level 46% degree or higher 46% year 12/trade/diploma 9% year 10 or equivalent Employment 60% employed 23% home duties 18% stu-dent/unemployed/other Regional Educational level 50% degree or higher 30% year 12/trade/diploma 20% Year 10 or equivalent Employment 30% employed 50% home duties 20% student/ unemployed/other	Mean age 2.5 years (σ 0.9) Lived with mother all/most of time	****
Fuller et al. (2019) [30]	Descriptive via semi-structured interviews (purposive sample)	Australia: Greater Brisbane Metropolitan area. Playgroup attendees	Parent of child attending a playgroup that agreed to hold a focus group	Barriers/facilitators to using parenting practices that support healthy obesity-related behaviour development in their child and what is acceptable in terms of intervention delivery mode and timing. Findings to inform the design of a community playgroup childhood obesity prevention intervention	*n* = 30 (93.3% female) 1 father 1 grandmother 13% aged > 30 years 37% aged 30–35 years 50% aged 36+ years	Education level 50% university 40% TAFE or trade 10% secondary school Employment 50% employed 50% unemployed	Median age = 24 months	*****

Table 2. *Cont.*

Author	Study Design	Location and Context	Inclusion/ Exclusion Criteria	Phenomena of Interest	Participants (*n*)	Core Characteristics		Overall MMAT Grade
						Parent	Child	
Gridley (2022) [31]	Descriptive via semi-structured interviews (purposive sample)	UK: England Indoor climbing centres	Parents with a child who had attended 2+ sessions of a developmentally adapted indoor bouldering programme	The perceptions of parents whose children took part in a developmentally adapted indoor bouldering programme designed for < 6-year-old children	*n* = 6 overall Mean age = 35.5 years (σ 6.4) Derived from text that *n* = 3 attended rock tots (33% female)	Unable to separate any details from those of parents who attended rock kids (children aged > 5 years)	Aged 1–4 years (rock tots)	*****
Grzywacz et al. (2016) [32]	Descriptive via semi-structured interviews (part of a larger study; purposive sample)	USA: North Carolina Latino farmworker families	Female with child aged 2–5 years; household member employed in agriculture in the last 12 months	The beliefs held by mothers in Latino farmworker families about the contribution of PA to pre-school-aged children's health and the perceived barriers or constraints that impose limits on pre-school-aged farmworkers' children's PA	*n* = 33 (100% female) No age range detailed	n = 16 migrant families n = 17 seasonal families Education No details Employment 40% worked on farms during past 12 months	Aged 2–5 years	*****
Hnatiuk et al. (2020) [33]	Descriptive via semi-structured interviews (purposive sample)	Australia: Western Sydney	Aged ≥ 18 years; parent of child aged 2–4 years	Parents' perceptions about: (i) PA and possible benefits of family-based co-participation in PA, (ii) the perceived facilitators and barriers to co-participation in Western Sydney, and (iii) recommendations for improving co-participation within their community	*n* = 15 (93% female) Mean age = 34.9 years (σ 5.4)	73% born in Australia Education level 20% postgraduate 40% undergraduate degree 27% trade certificate or diploma 13% year 12 or equivalent Employment 53% employed 6% student	Aged 2–4 years Mean age = 3.5 years (σ 0.56)	*****

Table 2. *Cont.*

Author	Study Design	Location and Context	Inclusion/ Exclusion Criteria	Phenomena of Interest	Participants (n)	Core Characteristics		Overall MMAT Grade
						Parent	Child	
Houghton et al. (2015) [34]	Mixed methods. Descriptive via structured interviews (only qualitative data relevant; purposive sample)	UK: Liverpool Sure Start Children's Centres	Fathers/male carers of child aged 3–5 years; living within catchment area of 26 Sure Start Children's Centres; attended at least 4 out of 6 sessions	The effectiveness and feasibility of a physically active-play-based programme on fathers' engagement with their pre-school-aged children across Liverpool	$n = 36$ 95.7% fathers; 3.2% grandads; and 1.1% other Mean age = 37.7 years (σ 8.7)	70.2% White British 26.6% had not used Sure Start service before Education level No details Employment 56.4% employed 33% unemployed 4.3% retired 3.2% full time students 3.2% unable to work	Mean age = 3.8 years (σ 1.2)	*****
Joseph et al. (2019) [35]	Descriptive via focus groups (part of a larger longitudinal mixed-methods study; purposive sample)	USA: Louisiana Early care education centres	Legal guardian/caretaker of child aged 3–5 years enrolled in ECC; understand and speak English; free from cognitive impairment; parent not an ECC provider; child not at ECC assessed for longitudinal study arm	The views and differences between parent and ECC provider perspectives on barriers/facilitators to children's PA and screen time, awareness of relevant regulations and recommendations, and caregiver-identified opportunities to leverage screen time to increase young children's PA	$n = 28$ (94% female) Age range 24–59 years	67% African American (Data for n = 18) Education level (Data for n = 18) 22% > college degree 22% some college/college degree 56% ≤ high school diploma Employment (Data for n = 16) 69% employed 31% unemployed/retired	Mean age = 3.7 years (σ 0.8)	*****
Lindsay et al. (2022) [36]	Descriptive via semi-structured interviews (part of a larger longitudinal mixed-methods study; purposive sample)	USA: Massachusetts Brazilian immigrant families	Father of at least one child aged 2–5 years; Brazilian ethnicity and born in Brazil; aged ≥ 21 years; lived in Massachusetts; lived in the USA ≥ 12 months	The perspectives and parenting practices of Brazilian immigrant fathers with regard to their pre-school-age children's PA and sedentary time	$n = 21$ (100% male) Mean age = 34.4 years (σ 2.8) Age range 27–43 years	Education level 4.7% college 71.5% high school/high school diploma 19.1% < high school Employment 100% employed	Aged 2–5 years	*****

Table 2. *Cont.*

Author	Study Design	Location and Context	Inclusion/ Exclusion Criteria	Phenomena of Interest	Participants (*n*)	Core Characteristics		Overall MMAT Grade
						Parent	Child	
Lindsay et al. (2019) [37]	Descriptive via focus groups (part of a larger longitudinal mixed-methods study) (convenience sample)	USA: Greater Boston area, Massachusetts Brazilian immigrant families	Mother of at least one child aged 2–5 years; Brazilian ethnicity and born in Brazil; lived in the USA for $\geq$12 months	PA parenting practices used by Brazilian immigrant mothers of pre-school-aged children	$n = 37$ (100% female) Mean age = 35.3 (σ 2.8) Age range 26–41 years	Education level 56.8% graduated high school Employment 92% owned their own housecleaning business	Aged 2–5 years	*****
Martin-Biggers et al. (2015) [38]	Descriptive via focus groups (part of a larger longitudinal mixed-methods study (purposive sample) Each focus group randomly asked about 2 different topics from a selection of 7)	USA: New Jersey and Arizona	Parent of child aged 2–5 years; primary language English or Spanish	Parents' cognitions associated with key obesity-prevention behaviours	$n = 28$ (active -play/barriers focus groups) No sex demographics supplied	Unable to extract focus-group-specific data	Aged 2–5 years	*****
Penilla et al. (2017) [39]	Descriptive via focus groups (purposive sample)	USA: San Francisco Community-based organization in the Mission District	Parent/guardian of a child between 2 and 5 years; English- or Spanish-speaking; of Mexican, Guatemalan, or Salvadoran descent	The broader social and environmental influences, namely, how urban neighbourhoods influence children's weight status through parents' child-feeding and PA routines	$n = 49$ (55% female) Mean age mothers 30 years (σ 5.8) Mean age fathers 35 years (σ 9.1)	Mothers Education level Mean years in education 12 (σ 3.7) Employment 33% currently employed Fathers Education level Mean years in education 11 (σ 2.3) Employment 68% currently employed	Aged 2–5 years	*****
Pišot (2020) [40]	Mixed methods Descriptive via structured interviews (only qualitative data relevant)	No details: Presume Slovenia	Mother of child aged 4 years Not stated	In mothers of pre-school children, the perceived barriers that enable or inhibit them to ensure their children are physically active in their leisure time	$n = 54$ (100% female)	No other demographic details supplied	Aged 4 years	**

Table 2. *Cont.*

Author	Study Design	Location and Context	Inclusion/ Exclusion Criteria	Phenomena of Interest	Participants (n)	Core Characteristics		Overall MMAT Grade
						Parent	Child	
Roscoe et al. (2017) [41]	Phenomenological via semi-structured interviews (purposive sample)	UK: North Warwickshire Area of high deprivation		Pre-school staff and parents' perceptions of pre-school children's PA and FMS, considering the environment, facilities, play, socio-economic status, and PA barriers	$n = 10$ (100% female)	Unable to separate any other details from those of staff also included in focus groups	Aged 2–4 years	*****
Stirrup et al. (2015) [42]	Mixed methods Ethnography via semi-structured interviews (part of a larger, ongoing study—only qualitative data relevant; purposive sample)	UK: Single unspecified local authority	Parent of child aged 0–5 years; attending pre-school facility; classified as middle class	Opportunities for pre-school children (aged 0–5 years) to participate in PA, the influence of social class and culture on parents' opportunities (and dispositions) to invest in and their choice of PA experiences. Plus, via Bourdieu's concepts of habitus and capital, the tendency for parents to provide copious developmental opportunities via 'intensive mothering'	$n = 3$ interviews (100% female; all selected from sample of $n = 79$ questionnaire respondents as examples of middle-class)	No demographic details supplied for interview participants	Aged 0–5 years	***

[a] σ = standard deviation; ECC = early childcare centre; FMS = fundamental movement skills; n = number; PA = physical activity; SES = socioeconomic status; TAFE = technical and further education; UK = United Kingdom; USA = United States of America. [b] Overall study quality was assessed using the Mixed Methods Assessment Tool (MMAT) and is reported using asterisks (*) as a descriptor, ranging from 1*, where 20% of the quality criteria have been met, to 5*, where 100% of the quality criteria have been met [14].

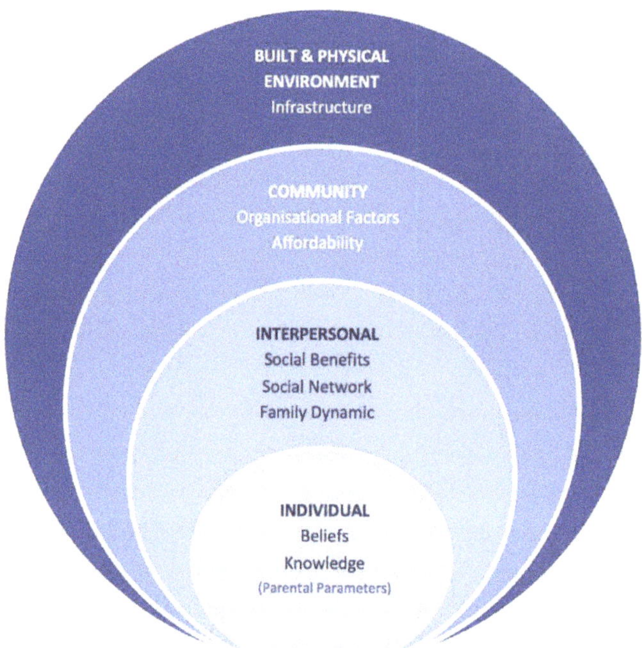

Figure 2. Socio-ecological model of factors that influence parental engagement of pre-school children in community-based opportunities for physical activity.

3.1. Individual

Beliefs and Knowledge (and Parental Parameters)

At an intrinsic level, parental *beliefs* regarding the value they place on PA, and the wider benefits they perceive it could bring, appear central to their decision-making processes. Structured PA was perceived to provide exposure to stimulating learning opportunities and environments [27,42] and to foster a broad range of positive and healthy child-development traits. In addition to being an approach to develop social and emotional skills [27,28,31,42], companionship [28], confidence [42], and teamwork [27] were also referred to. Specifically relating to enrolment in private swimming lessons, participation was viewed as a way for children to gain an essential life skill (personal safety), and an early childhood education that would be advantageous in the longer-term [27]. For parents classified by studies as middle-class, swimming lessons were also perceived as a way to demonstrate 'good parenting' by enabling them to comply with social norms and mitigate their own concerns regarding fear of drowning and water safety [27].

Perceived benefits were also deemed to be derived from professional instruction. Specifically, knowing a professional would be present in one study facilitated fathers' engagement with an active play intervention [34], whilst the anticipated development of strong, sport-specific skills due to instruction motivated attendance at bouldering classes [31]. Paying for skilled instruction and access to an appropriate role model that would enhance a child's behaviour and subsequent learning, facilitated enrolment in private swimming lessons [27]. Indeed, an individual, versus a group-based approach, was considered to be associated with even greater gains [27].

Closely linked to beliefs, parental *knowledge*, either prior knowledge or a lack of, influenced their decision making. At a basic level, not having the skills to teach their child to swim themselves facilitated attendance at private swimming lessons [27]. However, the recognition and understanding that early skill development can be beneficial for securing leisure opportunities in later life [27] and bring health and fitness benefits in the short term [31] may still be offset by other factors. For example, Houghton et al. [34] found a misconception among fathers that available sessions, including toddler groups, were purely for mothers, reflecting how lack of knowledge can be a barrier to attendance. Briefly, although not a major finding, it is pertinent to note that factors relating to *parental parameters*, namely, health and personality traits (shyness, laziness), were also highlighted as potential barriers to engagement by fathers attending community-based active play sessions as part of an intervention study [34].

3.2. Interpersonal
Social Benefits, Social Networks, and Family Dynamics

A key driver for attending playgroups [28] and parks [33] was the opportunity they provide for parents [28,42], particularly mothers [33,37], to *socialise*. Moreover, *social networks* provide a platform for peer support and influence. Playgroups provide a sense of togetherness, where comradeship can thrive, parenting practices can be shared in a safe space, and observations made of other parent–child interactions to inform one's own future parenting practices [30]. Social networks facilitated the use of parks [33] and other PA opportunities through the sharing of knowledge and resources, for example, attending in groups or sharing transport [37]. Where such peer influence had a predominantly positive impact, including participation on the basis that 'everyone else is doing it' [27], differing parenting approaches and/or attitudes, particularly towards choice of leisure time activity (café versus park), may act as a barrier to pre-schoolers' PA [33].

Challenges can also arise from within the family unit that influence the *family dynamic* and, subsequently, PA engagement. Parents in general, and fathers specifically, highlighted the difficulties that arose from having children of different ages. Specifically, competing child commitments [33] and having children outside the age-range of the targeted class [34] were raised as barriers to engagement. Moreover, while mothers often delegated the teaching of new sports to fathers or professionals [40], they were, however, often the 'gatekeeper' to engagement in PA opportunities, self-perceiving that they were the drivers of initiating new activities [40] and perceived by fathers to be the 'organiser' and reason for their attendance and co-participation at activities [34].

3.3. Community
Organisational Factors and Affordability

When targeting PA in pre-schoolers, the unique requirements of this population and their parents should be acknowledged at an *organisational* level. Age-group-targeted opportunities that are developmentally appropriate, based on the principles of learning through play, and provide a different experience (e.g., bouldering classes) inspired parents and facilitated attendance [31]. Moreover, a lack of professionally supervised activities for this age group acted as a barrier [37]. The timing of community-based opportunities was also important. Holding more events on weekends [33] could increase engagement; commitment and schedule clashes were barriers identified by both mothers [37] and fathers [34]. For opportunities specifically targeting the engagement of fathers, Saturday sessions led to the greatest attendance levels [34].

The engagement of parents could, in part, be facilitated or restricted by advertising strategies. Social media was identified as a tool for sharing PA ideas and local events [33]. However, it was noted that whilst improved advertising of community PA opportunities could facilitate greater co-participation of parents with their child(ren) [33], poor advertising was a barrier that led to a lack of understanding about available sessions [34]. Being used in an alternative format, marketing techniques delivered through social media that highlighted the perceived threat of not being able to swim was an effective way of encouraging parents to enrol their pre-schooler in private swimming lessons [27].

A major barrier for parents was the affordability of PA. The high cost of participation in organised sports or structured activities (e.g., ice skating) limited engagement [36–39,41].

For families in deprived areas, the cost of, for example, general swimming, let alone private swimming lessons, precluded participation [41]. Despite wanting to enrol their child(ren), for Brazilian immigrant mothers and fathers in the USA, it was simply not an activity they could readily afford [36,37]. For those with limited finances, a lack of disposable income was a significant barrier to pre-schoolers' PA [36,41], whereas for middle-class parents, being able to afford the additional cost of private swimming lessons was not something they were concerned about [27].

The provision of low-cost and free activities was viewed as an approach to incentivise participation, particularly for those living in deprived areas [41]. Improving access to resources already available in the community (school facilities) and greater support from community councils were highlighted as ways to improve PA opportunities for young families [33]. Additionally, where fathers identified the limited availability of affordable organised sport as a barrier to their pre-schoolers' PA [36], individuals embracing or local providers facilitating opportunities to engage in locality-specific activities (bushwalking) [33] could provide a viable solution.

3.4. Built and Physical Environment

Infrastructure

The *infrastructure* surrounding adequate access to parks, open spaces, general resources, their quality, and how easy it is to travel to them presented an important facilitator of play, and therefore PA, for pre-schoolers. A lack of available, well-resourced facilities and/or parks with age-appropriate equipment [28,33], all-weather provisions [29,33], and adequate green space [36] were mentioned as hinderances. Access to open space was somewhat more of an issue in outer, than inner, Australian suburbs [28], whilst, in some rural farmland areas of the USA, parks were reportedly non-existent [32]. In addition to a lack of access to outdoor facilities, the availability of indoor facilities, particularly in more rural areas [29], was also highlighted. Parents would like greater access to indoor play centres [29] and programmes and services, such as local events and playgroups [33].

Even when access is not an issue, the quality and safety of available facilities was often raised, particularly in parks [28,35,39]. Deteriorating conditions, where broken and removed equipment has not been replaced [28], safety hazards (broken glass and equipment) [39], the presence of dogs within park areas, and safety issues associated with traffic and park locations [28], all deterred parents from taking their preschool-aged children. Similarly, these issues were highlighted to a greater extent in outer rather than inner suburb areas [28].

Inherently linked are issues surrounding distance and transportation. Whether in rural or urban areas, the distance parents have to travel to attend parks, green spaces, and other facilities and opportunities that encourage play and PA had a significant impact on their frequency of attendance [29,32,33,36]. In more rural farmland areas [32], outer suburbs [28], and areas of high deprivation [41], attendance is often governed by access to transport, whether public or private. With this comes the associated cost of fuel [38] or fares, and sometimes excessive and unmanageable travel times [32].

3.5. Discussion

This mixed-methods systematic review examined the facilitators and barriers to parents engaging pre-school children in community-based PA opportunities and environments. To support the interpretation and translation of the findings, the results of the review were categorised using a socio-ecological approach (Figure 2). From the 16 included articles, eight core themes were identified over four levels: (i) *Individual*—belief, knowledge; (ii) *Interpersonal*—social benefits, family dynamic, and social network; (iii) *Community*—organisational factors and affordability; and (iv) *Built and Physical Environment*—infrastructure.

Parental beliefs regarding the value of PA and their knowledge around its benefits influence engagement with community opportunities. Beliefs, which focused on the positive development of traits and experiences that could be acquired by the child, act as a facilitator [27,28,31,42]. However, limited knowledge, whether through lack of skills to teach their child themselves or about specific PA opportunities, can simultaneously facilitate [27] and prohibit attendance [34]. At an individual level, although not a core theme, perhaps due to limited studies, the influence of parental parameters was also highlighted. Given the limited evidence to support this theme, more detailed discussions are precluded.

The importance of personal growth resonates within the existing literature on parental motivations for supporting pre-school children's leisure activities. For example, higher levels of behavioural engagement in leisure activities, which is frequently organised sport, presents more often among parents who value leisure activities for their ability to shape children's competencies and provide enjoyment [43]. There is also a strong relationship between greater participation in leisure opportunities for children and more favourable family economic factors [44]. Given, however, the strong relationship between the accessibility of opportunities in communities to be physically active and family resources, such as time, money, and location, as evidenced in this review, these are likely to present as more prominent barriers for less affluent families.

It is important to highlight that the values placed on leisure-time use may vary among parents of different social classes [45]. Previous research has identified less affluent parents as being less likely to believe that structured leisure activities may help children overcome social and behavioural difficulties [43] and, in contrast to more affluent families, to place more emphasis on safety and opportunities for free time (as opposed to structured activity) [46]. Considering parental motivations and values is central to shaping future opportunities for parents of young children, as well as reducing health inequalities. How these motivations and values are associated with family affluence within local communities warrants further research.

Social norms are evolving; whilst, historically, mothers have been the primary-care providers for children, changes in policy over recent years have advocated that it should no longer be the default. For example, in 2015 in the UK, Shared Parental Leave was introduced [47]. This change in the provision of primary childcare impacts the landscape of the household and family dynamic roles; with fathers playing a more central role at home, it is important to consider at whom opportunities that facilitate PA are targeted (mainly mothers) [34]. Indeed, the social networks and benefits that parents [28,42], and particularly mothers [33,37], obtain from taking their children to activities drive both their initial engagement with community-based PA opportunities for their child and their sustained attendance. Whether such factors equally motivate fathers is unclear from the current evidence base.

The Capability, Opportunity, Motivation, and Behaviour model (COM-B) [48] proposes that three factors (*Capability*, *Opportunity*, or *Motivation*), either combined or in isolation, can facilitate health-related behaviour. In this instance, considering *Motivation*, whilst parents might initially be driven to attend community-based opportunities for PA for their child's or for their own benefit (e.g., personal social benefits), observing their child's enjoyment through being active and having fun may subsequently lead to bi-directional effects, with positive changes in parental beliefs leading to more engaged participation from the child [43].

When considering *Opportunity*, the direct cost of activities or cost of getting to locations (e.g., parks) [36–39,41] and infrastructure-based limitations [28,29,32,33,35,36,39] both presented as major barriers to participation. In the UK, at a time when planning departments are now more heavily involved in creating infrastructures that encourage PA, these issues should be ones that can be alleviated. Additionally, current policies and developments in active education settings [49] support the use of pre-existing local community facilities, such as schools [33], as venues for PA opportunities, minimising both cost and travel requirements, but also building community cohesion.

3.6. Future Directions

Multiple factors across multiple different socio-ecological levels that parents identify as influencing the engagement of their pre-school child in community-based opportunities for PA were highlighted. Whilst these provide key areas for practitioner and policy-maker consideration, for some, more research is required to further understand their impact. First, the results indicate gender differences in parental perceptions of barriers and facilitators. Whilst the views and needs of fathers are discussed within some of the included articles, the percentage of female study participants is substantially greater (Table 2), and drivers of engagement potentially differ. For example, playgroups and parks provide opportunities for mothers to socialise [33,37]; however, this needs to be explored further from a male perspective. Dual-advertising campaigns may be needed to target the parenting roles simultaneously and widen participation and engagement to the whole family unit.

Second, it is apparent that the factors that may have the greatest influence on parents who are more or less affluent and/or live in rural, versus urban, locations might differ. Where affordability and access barriers present the primary challenge in areas with higher deprivation (i.e., [29,32,36,37,39]), for parents where cost is not perceived to be an issue, influencing factors appear to be situated at an individual or interpersonal level (i.e., [27,28,31,33,42]). Given the limited number of studies that have specifically set out to explore this in parents of pre-school children, studies that compare such differences in greater depth within local communities are required. Proportional universalism [50], where policy implementation strategies and provision are provided to all, but provision is graded dependent on need, is required.

Finally, concerning parental beliefs and knowledge, further work is needed to ascertain where on the spectrum of importance parents place engaging pre-schoolers in PA. Conventionally, it has been perceived that pre-schoolers are sufficiently active by default (as supported by the findings of Hesketh et al. [17]). However, with emerging evidence to show that this is not the case, with high levels of sedentary behaviour and low levels of PA potentially already established in some (i.e., [3,5]), this pre-conception needs to be urgently challenged.

3.7. Strengths and Limitations

This review was designed to take a systematic approach, using rigorous methods and validated tools, following a pre-defined protocol. Positively, the participant samples, in combination, represent a broad range of the target population encompassing males, females, ethnical diversity and different levels of socio-economic status. However, it is still possible that some studies containing pertinent data may not have been captured. It is important to note that only studies published in English from one of the 37 countries listed on the United Nations 2022 list of developed economies [51] were included, thus limiting the extrapolation of findings to other countries. Future research should explore and consider the specific environmental and cultural parameters and constraints of less developed economies. Additionally, the quality of included studies varied. Whilst 11 met 100% of the MMAT quality criteria [30–34,36–39,41], two articles only met 40% of the criteria [28,40], one met 60%, [42] and two met 80% [27,29]. It also became apparent during the latter stages of the screening processes that studies containing the views of expectant parents (terminology not agreed and included within the search terms by the team and its wider collaborators) may have provided additional insight. As per the pre-defined criteria, within this review, such studies (e.g., [52]) were duly excluded from the final analysis. It is important to additionally highlight that the results for the theme 'Beliefs and Knowledge' are largely based on the findings of a single study [27], potentially introducing an element of bias. Therefore, this needs to be considered when interpreting the results.

4. Conclusions

Physical inactivity is detrimental to health and well-being across the lifespan, making the embodiment of healthy behaviours from the early years phase essential. The message from PA guidelines is consistent and has not changed since 2011 [53]; therefore, identifying ways to help parents realise the potential of early PA is important. The findings within this evidence synthesis provide valuable insights for practitioners and policy makers involved in commissioning, designing, and delivering community-based PA opportunities for pre-school children. To tackle physical inactivity and promote PA, developing strategies and opportunities that acknowledge and seek to address the barriers identified and build on the facilitators highlighted by parents, particularly surrounding infrastructure and affordability, may be paramount.

Supplementary Materials: The following supporting information can be downloaded at: https://www.mdpi.com/article/10.3390/children9111727/s1, Table S1: Coding criteria for the Mixed Methods Assessment Tool; Table S2: Quality assessment results from Mixed Methods Assessment Tool; File S1: Database-specific terms, restrictions and search strategies, and secondary and grey literature search processes; File S2: Themes and indicative text quotes.

Author Contributions: The study was conceptualized by K.A.M., M.A.M. and C.A.S. Literature searching was undertaken by R.L.K. Data analysis was undertaken by R.L.K., with secondary assistance from C.A.S. and B.H. and critical interpretation from M.A.M. The first draft of the manuscript was written by R.L.K. All authors have read and agreed to the published version of the manuscript.

Funding: Funding was received from Sport Wales, which enabled the appointment of the research assistant (first author) who conducted this review.

Institutional Review Board Statement: No ethical approval needed; all data were available in the scientific literature.

Informed Consent Statement: Not applicable.

Data Availability Statement: The data that support the findings of this study are available from the corresponding author upon reasonable request.

Conflicts of Interest: Britt Hallingberg is a trustee for the Llanharan Community Development Project Ltd., which provides play opportunities for local communities in Wales. All other authors declare that they have no conflict of interest.

References

1. World Health Organization. Global Action Plan on Physical Activity 2018–2030: More Active People for a Healthier World. Geneva: World Health Organization. 2018. Available online: https://apps.who.int/iris/bitstream/handle/10665/272722/9789241514187-eng.pdf (accessed on 16 July 2022).
2. Bornstein, D.B.; Beets, M.W.; Byun, W.; McIver, K. Accelerometer-derived physical activity levels of preschoolers: A meta-analysis. *J. Sci. Med. Sport* **2011**, *14*, 504–511. [CrossRef] [PubMed]
3. O'Brien, K.T.; Vanderloo, L.M.; Bruijns, B.A.; Truelove, S.; Tucker, P. Physical activity and sedentary time among preschoolers in centre-based childcare: A systematic review. *Int. J. Behav. Nutr. Phys. Act.* **2018**, *15*, 117. [CrossRef]
4. Cardon, G.; Van Cauwenberghe, E.; De Bourdeaudhuij, I. What do we know about physical activity in infants and toddlers: A review of the literature and future research directions. *Sci. Sports* **2011**, *26*, 127–130. [CrossRef]
5. Downing, K.L.; Hnatiuk, J.; Hesketh, K.D. Prevalence of sedentary behavior in children under 2 years: A systematic review. *Prev. Med.* **2015**, *78*, 105–114. [CrossRef] [PubMed]
6. van der Zee, M.D.; van der Mee, D.; Bartels, M.; de Geus, E.J.C. Tracking of voluntary exercise behaviour over the lifespan. *Int. J. Behav. Nutr. Phys. Act.* **2019**, *16*, 17. [CrossRef]
7. Sääkslahti, A.; Numminen, P.; Varstala, V.; Helenius, H.; Tammi, A.; Viikari, J.; Välimäki, I. Physical activity as a preventive measure for coronary heart disease risk factors in early childhood. *Scand. J. Med. Sci. Sports* **2004**, *14*, 143–149. [CrossRef]
8. Jiménez-Pavón, D.; Kelly, J.; Reilly, J.J. Associations between objectively measured habitual physical activity and adiposity in children and adolescents: Systematic review. *Int. J. Pediatr. Obes.* **2010**, *5*, 3–18. [CrossRef] [PubMed]
9. LeBlanc, A.G.; Spence, J.C.; Carson, V.; Connor Gorber, S.; Dillman, C.; Janssen, I.; Kho, M.E.; Stearns, J.A.; Timmons, B.W.; Tremblay, M.S. Systematic review of sedentary behaviour and health indicators in the early years (aged 0–4 years). *Appl. Physiol. Nutr. Metab.* **2012**, *37*, 753–772. [CrossRef]
10. Burdette, H.L.; Whitaker, R.C. Resurrecting free play in young children: Looking beyond fitness and fatness to attention, affiliation, and affect. *Arch. Pediatr. Adolesc. Med.* **2005**, *159*, 46–50. [CrossRef]
11. Williams, C.L.; Carter, B.J.; Kibbe, D.L.; Dennison, D. Increasing physical activity in preschool: A pilot study to evaluate animal trackers. *J. Nutr. Educ. Behav.* **2009**, *41*, 47–52. [CrossRef] [PubMed]

12. Hesketh, K.R.; O'Malley, C.; Paes, V.M.; Moore, H.; Summerbell, C.; Ong, K.K.; Lakshman, R.; van Sluijs, E.M. Determinants of Change in Physical Activity in Children 0–6 years of Age: A Systematic Review of Quantitative Literature. *Sports Med.* **2017**, *47*, 1349–1374. [CrossRef] [PubMed]

13. Kim, H.; Akira, M.; Ma, J. Factors impacting levels of physical activity and sedentary behavior among young children: A literature review. *Int. J. Appl. Sports Sci.* **2017**, *29*, 1–12. [CrossRef]

14. O'Dwyer, M.V.; Fairclough, S.J.; Knowles, Z.; Stratton, G. Effect of a family focused active play intervention on sedentary time and physical activity in preschool children. *Int. J. Behav. Nutr. Phys. Act.* **2012**, *9*, 117. [CrossRef] [PubMed]

15. Chen, C.; Sellberg, F.; Ahlqvist, V.H.; Neovius, M.; Christiansen, F.; Berglind, D. Associations of participation in organized sports and physical activity in preschool children: A cross-sectional study. *BMC Pediatr.* **2020**, *20*, 328. [CrossRef]

16. United Nations. Convention on the Rights of the Child. Geneva: United Nations. 1989. Available online: https://www.ohchr.org/en/instruments-mechanisms/instruments/convention-rights-child (accessed on 4 September 2022).

17. Hesketh, K.R.; Lakshman, R.; Sluijs, E.M.F. Barriers and facilitators to young children's physical activity and sedentary behaviour: A systematic review and synthesis of qualitative literature. *Obes. Rev.* **2017**, *18*, 987–1017. [CrossRef]

18. Sallis, J.F.; Cervero, R.B.; Ascher, W.; Henderson, K.A.; Kraft, M.K.; Kerr, J. An ecological approach to creating active living communities. *Annu. Rev. Public Health* **2006**, *27*, 297–322. [CrossRef]

19. Lizarondo, L.; Stern, C.; Carrier, J.; Godfrey, C.; Rieger, K.; Salmond, S.; Apostolo, J.; Kirkpatrick, P.; Loveday, H. Chapter 8: Mixed methods systematic reviews. In *JBI Manual for Evidence Synthesis*; Aromataris, E., Munn, Z., Eds.; JBI: North Adelaide, Australia, 2020; Available online: https://jbi-global-wiki.refined.site/space/MANUAL/4687380/Chapter+8%3A+Mixed+methods+systematic+reviews (accessed on 10 May 2022).

20. Higgins, J.P.; Thomas, J.; Chandler, J.; Cumpston, M.; Li, T.; Page, M.J.; Welch, V.A. (Eds.) Cochrane Handbook for Systematic Reviews of Interventions Version 6.3 (Updated February 2022). Cochrane. 2022. Available online: https://training.cochrane.org/handbook (accessed on 10 May 2022).

21. Page, M.J.; McKenzie, J.E.; Bossuyt, P.M.; Boutron, I.; Hoffmann, T.C.; Mulrow, C.D.; Shamseer, L.; Tetzlaff, J.M.; Akl, E.A.; Brennan, S.E.; et al. The PRISMA 2020 statement: An updated guideline for reporting systematic reviews. *BMJ.* **2021**, *372*, n71. [CrossRef]

22. United Nations Department of Economic and Social Affairs. Transforming Our World: The 2030 Agenda for Sustainable Development. Geneva: United Nations. 2015. Available online: https://sdgs.un.org/2030agenda (accessed on 4 September 2022).

23. Hong, Q.N.; Pluye, P.; Fàbregues, S.; Bartlett, G.; Boardman, F.; Cargo, M.; Dagenais, P.; Gagnon, M.-P.; Griffiths, F.; Nicolau, B.; et al. The Mixed Methods Appraisal Tool (MMAT) version 2018 for information professionals and researchers. *Educ. Inf.* **2018**, *34*, 285–291. [CrossRef]

24. Hong, Q.N. Reporting the Results of the MMAT. 2020. Available online: http://mixedmethodsappraisaltoolpublic.pbworks.com/w/file/140056890/Reporting%20the%20results%20of%20the%20MMAT.pdf (accessed on 11 May 2022).

25. Thomas, J.; Harden, A. Methods for the thematic synthesis of qualitative research in systematic reviews. *BMC Med. Res. Methodol.* **2008**, *8*, 45. [CrossRef]

26. Smith, B.; McGannon, K.R. Developing rigor in qualitative research: Problems and opportunities within sport and exercise psychology. *Int. Rev. Sport Exerc. Psychol.* **2018**, *11*, 101–121. [CrossRef]

27. Allen, G.; Velija, P.; Dodds, J. 'We just thought everyone else is going so we might as well': Middle-class parenting and franchised baby/toddler swimming. *Leis. Stud.* **2021**, *40*, 169–182. [CrossRef]

28. Andrews, F.J.; Stagnitti, K.; Robertson, N. Social play amongst preschool-aged children from an inner and an outer metropolitan suburb. *J. Soc. Incl.* **2019**, *10*, 4–17. [CrossRef]

29. Downing, K.L.; Best, K.; Campbell, K.J.; Hesketh, K.D. Informing active play and screen time behaviour change interventions for low socioeconomic position mothers of young children: What do mothers want? *BioMed Res. Int.* **2016**, *2016*, 2139782. [CrossRef]

30. Fuller, A.B.; Byrne, R.A.; Golley, R.K.; Trost, S.G. Supporting healthy lifestyle behaviours in families attending community playgroups: Parents' perceptions of facilitators and barriers. *BMC Public Health* **2019**, *19*, 1740. [CrossRef] [PubMed]

31. Gridley, N. Parental perceptions of an indoor bouldering programme for toddlers and pre-schoolers in England: An initial exploratory study. *J. Adventure Educ. Outdoor Learn.* **2022**. [CrossRef]

32. Grzywacz, J.; Arcury, T.; Trejo, G.; Quandt, S. Latino mothers in farmworker families' beliefs about preschool children's physical activity and play. *J. Immigr. Minor. Health* **2016**, *18*, 234–242. [CrossRef]

33. Hnatiuk, J.A.; Dwyer, G.; George, E.S.; Bennie, A. Co-participation in physical activity: Perspectives from Australian parents of pre-schoolers. *Health Promot. Int.* **2020**, *35*, 1474–1483. [CrossRef]

34. Houghton, L.J.; O'Dwyer, M.; Foweather, L.; Watson, P.; Alford, S.; Knowles, Z.R. An impact and feasibility evaluation of a six-week (nine hour) active play intervention on fathers' engagement with their preschool children: A feasibility study. *Early Child Dev. Care* **2015**, *185*, 244–266. [CrossRef]

35. Joseph, E.D.; Kracht, C.L.; St Romain, J.; Allen, A.T.; Barbaree, C.; Martin, C.K.; Staiano, A.E. Young children's screen time and physical activity: Perspectives of parents and early care and education center providers. *Glob. Pediatr. Health* **2019**, *6*, 2333794X19865856. [CrossRef]

36. Lindsay, A.C.; de Sá Melo Alves, A.; Vianna Gabriela Vasconcellos de, B.; Arruda Carlos André, M.; Hasselmann, M.H.; Machado Márcia Maria, T.; Greaney, M.L. A qualitative study conducted in the United States exploring the perspectives of Brazilian

immigrant fathers about their preschool-age children's physical activity and screen time. *J. Public Health* **2022**, *30*, 1619–1632. [CrossRef]

37. Lindsay, A.C.; Arruda, C.A.M.; Machado, M.M.T.; Greaney, M.L. Parenting practices that encourage & discourage physical activity in preschool-age children of Brazilian immigrant families. *Ann. Behav. Med.* **2019**, *53*, e0214143. [CrossRef]
38. Martin-Biggers, J.; Spaccarotella, K.; Hongu, N.; Alleman, G.; Worobey, J.; Byrd-Bredbenner, C. Translating it into real life: A qualitative study of the cognitions, barriers and supports for key obesogenic behaviors of parents of preschoolers. *BMC Public Health* **2015**, *15*, 189. [CrossRef] [PubMed]
39. Penilla, C.; Tschann, J.M.; Sanchez-Vaznaugh, E.V.; Flores, E.; Ozer, E.J. Obstacles to preventing obesity in children aged 2 to 5 years: Latino mothers' and fathers' experiences and perceptions of their urban environments. *Int. J. Behav. Nutr. Phys. Act.* **2017**, *14*, 148. [CrossRef]
40. Pišot, S. Mother's perspective of maintaining an active outdoor leisure time for a preschool child. /Usklajevanje aktivnega preživljanja prostega časa predšolskih otrok iz perspektive mater. *Ann. Kinesiol.* **2020**, *11*, 83–97. [CrossRef]
41. Roscoe, C.M.P.; James, R.S.; Duncan, M.J. Preschool staff and parents' perceptions of preschool children's physical activity and fundamental movement skills from an area of high deprivation: A qualitative study. *Qual. Res. Sport. Exerc. Health* **2017**, *9*, 619–635. [CrossRef]
42. Stirrup, J.; Duncombe, R.; Sandford, R. 'Intensive mothering' in the early years: The cultivation and consolidation of (physical) capital. *Sport Educ. Soc.* **2015**, *20*, 89–106. [CrossRef]
43. Ooi, L.L.; Rose-Krasnor, L.; Shapira, M.; Coplan, R.J. Parental beliefs about young children's leisure activity involvement. *J. Leis. Res.* **2020**, *51*, 469–488. [CrossRef]
44. Snellman, K.; Silva, J.M.; Frederick, C.B.; Putnam, R.D. The engagement gap: Social mobility and extracurricular participation among American youth. *Ann. Am. Acad. Pol. Soc. Sci.* **2015**, *657*, 194–207. [CrossRef]
45. Holloway, S.L.; Pimlott-Wilson, H. Enriching children, institutionalizing childhood? Geographies of play, extracurricular activities, and parenting in England. *Ann. Am. Assoc. Geogr.* **2014**, *104*, 613–627. [CrossRef]
46. Bennett, P.R.; Lutz, A.C.; Jayaram, L. Beyond the schoolyard: The role of parenting logics, financial resources, and social institutions in the social class gap in structured activity participation. *Sociol. Educ.* **2012**, *85*, 131–157. [CrossRef]
47. UK Government. The Shared Parental Leave Regulations 2014. London: UK Government. 2014. Available online: https://www.legislation.gov.uk/uksi/2014/3050/contents/made (accessed on 5 September 2022).
48. Michie, S.; van Stralen, M.M.; West, R. The behaviour change wheel: A new method for characterising and designing behaviour change interventions. *Implement. Sci.* **2011**, *6*, 42. [CrossRef] [PubMed]
49. Marshall, T.; Rees, A.; Roberts, C. Active beyond the School Day. A Research Review. Cardiff: Sport Wales. 2022. Available online: https://www.sport.wales/research-and-insight/report-active-education-beyond-the-school-day/ (accessed on 5 September 2022).
50. Marmot, M. *Fair Society, Healthy Lives: The Marmot Review: Strategic Review of Health Inequalities in England Post-2010*; Parliament: London, UK, 2010; ISBN 9780956487001. Available online: https://www.parliament.uk/globalassets/documents/fair-society-healthy-lives-full-report.pdf (accessed on 5 September 2022).
51. United Nations Economic and Social Council. World Economic Situation and Prospects 2022. Statistical Annex. Geneva: United Nations Economic and Social Council. 2022. Available online: https://www.un.org/development/desa/dpad/wp-content/uploads/sites/45/WESP2022_ANNEX.pdf (accessed on 11 May 2022).
52. Khanom, A.; Evans, B.A.; Lynch, R.; Marchant, E.; Hill, R.A.; Morgan, K.; Rapport, F.; Lyons, R.A.; Brophy, S. Parent recommendations to support physical activity for families with young children: Results of interviews in deprived and affluent communities in South Wales (United Kingdom). *Health Expect.* **2020**, *23*, 284–295. [CrossRef] [PubMed]
53. Department of Health, Physical Activity, Health Improvement and Protection. Start Active, Stay Active: A Report on Physical Activity from the Four Home Countries' Chief Medical Officers. London: UK Government. 2011. Available online: https://assets.publishing.service.gov.uk/government/uploads/system/uploads/attachment_data/file/830943/withdrawn_dh_128210.pdf (accessed on 11 September 2022).

Article

Improving Body Mass Index in Students with Excess Weight through a Physical Activity Programme

Mădălina Doinița Scurt [1], Lorand Balint [2] and Raluca Mijaică [2,*]

[1] Sports Science and Physical Education, IOSUD—Transilvania University of Brasov, Resident Physician-Specialization: Radiology and Medical Imaging, Hospital CLINICO Valencia, 46010 Valencia, Spain

[2] Department of Physical Education and Special Motricity, Faculty of Physical Education and Mountain Sports, Transilvania University of Brasov, 500036 Brasov, Romania

* Correspondence: raluca_mijaica@unitbv.ro

Abstract: The obesity epidemic among young people can be tackled through regular physical activity. For this purpose, we developed and implemented a physical activity programme (PAP) that we carried out in students' free time during the school year 2018–2019. The target group consisted of 79 students with excess weight, aged between 12 and 15 years, selected from an initial sample of 495 students from 5 pre-university education units located in an urban area. That group followed a differentiated PAP for 26 weeks. The impact of the programme highlighted the following points: the average physical activity/week for the entire sample of subjects was 3.67 physical activities, with an allocated time/week ranging from 1 h 30 min to 3 h; in terms of effort intensity, 7.70% of the activities were performed at low intensity, 75.07% at medium intensity and 17.23% at submaximal intensity. At the end of the programme, out of 79 subjects who were overweight/obese at the initial testing, 37 improved their body composition at the final testing, with a healthy BMI. It was also found that there is a negative correlation coefficient (r = −0.23) between the time spent performing physical activities and the BMI of the subjects.

Keywords: physical activity; programme; children and adolescents; excess weight; urban environment

Citation: Scurt, M.D.; Balint, L.; Mijaică, R. Improving Body Mass Index in Students with Excess Weight through a Physical Activity Programme. *Children* **2022**, *9*, 1638. https://doi.org/10.3390/children9111638

Academic Editor: Luisa de Sanctis

Received: 14 August 2022
Accepted: 20 October 2022
Published: 27 October 2022

Publisher's Note: MDPI stays neutral with regard to jurisdictional claims in published maps and institutional affiliations.

Copyright: © 2022 by the authors. Licensee MDPI, Basel, Switzerland. This article is an open access article distributed under the terms and conditions of the Creative Commons Attribution (CC BY) license (https://creativecommons.org/licenses/by/4.0/).

1. Introduction

Physical activity habits have changed over the years, and today, increasing numbers of young people are tending to become less active from a motor perspective, a fact that is reflected in their lifestyles. The reality that current research reveals is that the evolution of physical activity has stagnated in recent decades among all categories of citizens, and the percentage of those who practiced but have stopped has increased alarmingly [1]. A study on current trends in physical inactivity in adolescents was published at the end of 2019 [2]. This study pointed out that in 2016, overall, 81% of the respondents (adolescents aged 11–17 years) were not physically active enough (77.60% of boys and 84.70% of girls), and that by territorial socioeconomic level, the prevalence of physical inactivity in 2016 was 84.90% in underdeveloped countries and 79.40% in economically developed countries.

The reality of the situation, particularly in adolescence, applies to virtually any other citizen, regardless of age or country of origin. About all of the above, by a relatively simple differentiation, it can be said that some are more physically active, and others are less likely to participate or to devote time consistently and significantly to this kind of biological effort (physical, mental, emotional). Explanations along this line of selective human behaviour are quite complex as the behaviours are multifactorially determined and, to simplify things, most researchers, rightly perhaps, choose one or another of the possible and relevant variables that determine physical activity or inactivity as the topic of study. Therefore, there are studies that generally address the following variables: age [3–11], gender identity [1,10,12–22], socioeconomic and education level [23–31], personality characteristics/motivation [12,15,32], social environment/family and friends [12,26,31,33], as well as environmental setting [4,34–36], all of which can influence the level of physical activity as a means of maintaining and/or improving health. Some of the variables listed above are not changeable (such as age, gender), but the others can be altered to a certain extent (level of education, subjects' motivations, social and proximal environment in which each individual lives).

Closely related to health, as a direct cause–effect relationship, is the concept of lifestyle [37,38], presented as a subjective term, encompassing different aspects. One of these is physical activity, which if performed with an appropriate frequency, intensity and duration, is an integrated factor in the so-called healthy lifestyle, which decisively contributes to the maintenance of health and quality of life [39]. In the same vein, any physical exercise performed sporadically does not become part of a stable lifestyle, and its beneficial influences cannot make their presence known.

Paradoxically perhaps, in this area of interest are taken as subjects of studies, to a large extent, only young people, namely "normal" people, and those who practice should be in the forefront of research attention; under these circumstances, it would be justified to be young people/adolescents who for one reason or another have deviated from the state of normality. Our study is also part of this category, which aims to support, in a proactive manner, adolescents with excess weight (of nonendocrine nature), considering that they are the first to either self-exclude or are excluded by others from the enjoyment of physical activities and as a result are not the beneficiaries of this kind of habit, becoming, in the not too distant future, those who will suffer from serious health conditions such as: cardiovascular diseases, diabetes, immune deficit liver diseases.

Specialized studies approach the category of overweight subjects, in most cases, through the prism of the subjects' physical activity/inactivity and only through questionnaires. These are more of an observational nature, being "passive" studies that only point out the problem in question. For example, there are international studies showing an inverse relationship between physical activity and body fat [40–42] and the degree to which physical inactivity contributes/predisposes to increased levels of overweight among the infantile population compared with those who declare themselves physically active [43]. Along the same lines, other studies using objective methods to assess physical activity levels have shown negative associations between physical activity and the development of excess weight, obesity and central/abdominal adipose tissue among children and adolescents [44–46]. Similarly, in a cross-sectional study conducted with adolescent schoolchildren [47], an inverse relationship was found between physical activity and body weight. The results indicated that those who suffered from overweight and obesity spent less time engaged in physical activity than adolescents with a normal body weight. In the same observational manner, a Norwegian study [48] revealed that a substantial decrease in the level of physical activity among adolescents led to a significant increase in body mass index over the same period. Based on this argument, the authors of the study emphasize the importance of maintaining physical activity levels in order to reduce negative trends in body mass index growth. In Romania, research in this area has been largely of a constative nature and, moreover, without any connection with physical activity, presenting only statistical data either from interested institutions or researchers in fields related to physical education and school sport. For example, in an urban study conducted in 2008 on a group of 7904 children and adolescents in grades I-XII from 20 schools and high schools, prevalences were shown of overweight of 12.84% and obesity of 8.29%. Adolescents in this study had lower-than-average prevalences of, respectively, 7.66% for overweight and 3.81% for obesity [49]. In Timiș County (located in western Romania), according to another study [50] conducted between 2010 and 2011 on a total of 3731 subjects aged 7 to 19 years, the overall prevalence of overweight was 18.20% (boys, 20.70% and girls, 16.30%), and that of obesity was 7.20% (boys, 9% and girls, 5.80%). Overweight was more common in the urban areas, a percentage of 18.20%, compared with 17.90% in the rural area. Another observational study carried out over a period of 10 years (2006–2015) on 25,060 subjects aged between 6 and 19 years coming from 8 counties in Romania [51] showed that 28.30% of them had excess weight (obesity and overweight). The same study also finds that the prevalence of excess weight in children and adolescents in urban areas (29.5% = 18.30% overweight and 11.20% obesity; n = 20,137 subjects) is higher than that of children and adolescents in rural areas (22.90% = 14% overweight + 9.90% obesity; n = 4923 subjects). The examples are endless, but the important point is that in Romania, there is still little emphasis on the health-generating value of physical exercise practiced in one form or another continuously throughout life.

All this research, as well as other studies not mentioned here, is indeed a wake-up call for responsible entities (institutional and nongovernmental alike), who must take appropriate measures to prevent this globalized phenomenon, currently called the obesity epidemic. For this goal, however, proactive actions/practices are needed, not only to identify but, importantly, also to mobilize young people with excess weight in the constant practice of physical activities, especially in their free time, following a proper methodology. This is also the major goal of our study, which aimed to design and implement a physical activity programme (PAP) exclusively for adolescents aged 12 to 15 years with excess weight (overweight or obesity). Our working hypothesis was that the body mass index in adolescents with excess weight would correlate with the total amount of physical activity that the subjects performed in their free time in completing an average-duration physical activity program.

Following the implementation of this programme, some results were obtained that we consider encouraging. They offer real/objective perspectives on improving the body mass index for the study subjects according to their proactive attitudes toward these kinds of activities. The data highlighted for each subject how much time they dedicated to physical activities (sequentially averaged/day/week/month and overall for the 6 months) and what type of activities they enjoyed, as well as with what effort intensity they chose to perform their preferred activities. These data were finally correlated with the body mass index to demonstrate the existence or nonexistence of relationships there between.

2. Materials and Methods

2.1. Methodology of the Physical Activity Programme

The physical activity programme (PAP) was created at our initiative as an operational model that could be disseminated among students with excess weight according to BMI and benefited, for the implementation of its contents, from the guidance of several specialists and volunteers in the field of sports science and physical education from the Transilvania University of Brasov Faculty of Physical Education and Mountain Sports. In order to be able to cover the entire range of activities related to our study, starting with the initial and final anthropometric measurements, the presentation and implementation of the PAP programme, as well as the monitoring/evidence and centralization of activities, we turned to the help of 5 physical education teachers from the subjects' schools of origin, as well as to a university professor (teacher), and last but not least, to 5 students from our institution, from the study program Kinetic Therapy and Special Motricity.

In addition, we followed the recommendations of the World Health Organization (WHO) regarding the characteristics of our target group (young people with excess weight), namely at least 60 min/day of moderate to vigorous physical activity. When designing the PAP, it was also taken into account the idea that physical activity should be mostly aerobic but include other types of activities that contribute to the strengthening of the musculoskeletal system [52].

To select the subjects for our research, we performed anthropometric measurements on a larger group of students (Section 2.3) from among whom we extracted those with excess weight. Specifically, we measured height with a thaliometer and weight with an electronic scale, respecting the recommended methodologies in both measurement procedures, including regarding the summary clothing of subjects and lack of footwear.

We next used these measurements to calculate the body mass index (BMI) of each subject (according to the formula: BMI = weight/height^2), thus obtaining the sample of subjects targeted in our approach.

In developing the methodological guidelines of the PAP, we started from the assumption/recommendation that, in general, children and adolescents (6–17 years) with excess weight and/or who are usually physically inactive may not be able to practice 60 min of moderate (according to WHO) let alone increased physical activity. In this context and for our subjects, we felt that they should start in the PAP with moderate intensity physical activity and gradually increase the frequency and time allocated to physical activity to reach the target of 60 min of physical activity/day after a reasonable period of time. Therefore, it was suggested that for the activities carried out with higher intensity, the students should progressively incorporate them into their individual activity plans until they can perform high-intensity activities 2–3 times/week.

In the model PAP, which was designed for 26 weeks and distributed to students to guide their choices, 5 weekly sessions of physical activity were proposed, with an intensity generally between 60 and 70% of the maximum heart rate/subject. During these sessions, we suggested for the weekend physical activities carried out in nature (hiking, cycling, sports orientation, etc.), all with a marked aerobic energy substrate and usually recommended for subjects with excess weight.

Additionally through the designed programme, we wanted to ensure that physical activity/exercise was not abandoned during the study, thus combating the sedentary lifestyle so common today at these ages. At the same time, the programme was developed in such a way as to allow the participants to modify its contents according to their personal preferences.

The physical activity volume, in terms of frequency, the intensity of physical effort and the duration and type of physical activity chosen, was monitored through the physical activity journals that we designed and distributed to the students. Periodically, from these journals, the different quantitative indicators needed for the processing of the executive/practical components of the study were extracted. The students recorded their frequency of physical activities, duration and activity type after each execution in tables in their journals.

In the case of identifying the zones of physical effort (low, aerobic 1, moderate, aerobic 2 and submaximal, aerobic–anaerobic), recorded in the same format of the journal, the subjects were asked to self-determine their heart rate (either by measuring the heart rate by palpating it at the level of the radial artery or by downloading the Heart Rate Monitor-Pulse App from any type of smartphone from the Play Store menu) at 6–8 min after starting an activity, as well as at 15, 20 and/or 30 min (depending on the type of activity), finally averaging the 2 or 3 values obtained. As indicative values for framing the effort intensity zones for ages 12–15 years, we chose the intervals: 130–150 rpm (low); 155–170 rpm (moderate); >175 (submaximal).

Due to the variety, attractiveness and fun that the physical activities of PAP can generate, we encouraged family and friends, even if they were not overweight, to participate together with the subjects to support, encourage and motivate them in their efforts to complete their activities.

Regarding the locations for practicing physical activities that make up the PAP (these being extracurricular activities), the following were suggested to students/their parents: spaces in one's own school, neighbourhood or home, a park or any other accessible public space that offers physical safety and has a purpose that allows the practice of these types of activities.

All this information, as well as the benefits of completing the PAP, were presented at the beginning of the school year (September 2018, after the target group was created), within the pre-university education units involved in the study: physical education teachers, students and their parents, from which the acceptance agreement was obtained. As a result, the study was conducted in accordance with the Declaration of Helsinki and approved by the Institutional Examination Board of the Faculty of Physical Education and Mountain Sports of the Transilvania University of Brasov (No. 243 from 26 September 2018), and informal consent was obtained from all subjects (the parents of the 79 students) involved in the study.

As a final clarification regarding the methodological aspects of the application of the PAP, we would like to mention that all the proposed activities were chosen in such a way as to have a non-competitive character, thus encouraging the development of confidence in one's own strengths, without favouring the manifestation of inferiority complexes and with the contents adapted to individual motor/physical capabilities.

2.2. Presentation of the Contents and Structure of the Physical Activity Programme (PAP) for Schoolchildren with Excess Weight

The PAP was designed on 4 modules of physical activities: aerobic endurance; strength and muscular endurance; flexibility and weekend activities. Please note that these modules were not randomly selected. Their development was based on the recommendations of several studies carried out internationally that indicate that these types of activities are the most effective for adolescents. It is considered that from a physiological point of view, children and adolescents (6–17 years) adapt easily to physical activities practiced to develop muscle strength and endurance [53,54], aerobic endurance [55,56] and flexibility/mobility [57,58]. However, because the musculoskeletal system is still immature at this age, children and adolescents should not engage in excessive amounts of high-intensity physical activity.

Within each programme out of the 4 proposed, for each physical activity, we estimated the total caloric consumption generated by its execution. These values were obtained experimentally by measuring the MET (the metabolic equivalent of a task/activity) in an obese 13-year-old girl weighing 70 kg and an overweight 15-year-old boy weighing 67 kg. Please note that MET is the unit of measurement for oxygen consumption, and for calculating the total calories consumed/physical activity, the oxygen consumption and the body weight were used, in addition to other parameters such as age, stature and total time dedicated to physical activity. The calculator used to obtain these data is available online at https://healthyeater.com/calories-burned (accessed on 3 October 2018). Regarding the weekend activities module, in order to estimate the caloric burns occurring in the body, we took into account constants such as: body weight, distance covered (in kilometres), type of terrain (presence of uphill and/or downhill sequences) and level difference. The calculator used is available online at: https://caloriesburnedhq.com/calories-burned-hiking/ (accessed on 5 October 2018).

Table 1 present the main indicators of each module.

These modules of the PAP represent diversified offers for physical effort. Due to this fact, they are not of interest in terms of their contents (which are obviously different) but more for the relevant indicators they develop, such as: the number of activities carried out/subject, their average duration/week and the intensity zones in which they fall.

As a way of presenting the PAP document to the students, physical education teachers and the students' parents, each module included: preamble, objectives, application requirements/recommendations and methodological conditions (as appropriate) and examples of activities in different intensity steps with the equivalent in caloric burn.

From a longitudinal perspective, the model PAP, designed over an effective duration of 6 months (26 weeks effectively), was divided into 4 sections with different steps/purposes of action as follows: weeks 1 to 4 (the settling-in period); weeks 5 to 8 (the development period); weeks 9 to 12 (the optimization period); weeks 13 to 26 (the completion period).

Each physical activity session (regardless of the module chosen) consisted of 3 parts that were recommended to students, namely:

1. A body preparation phase (warm-up) lasting 5–10 min, consisting of 4–5 min of brisk walking/running/cycling, etc., followed by dynamic joint mobility exercises, 5–6 min.

2. A main part, lasting 20–30 min, in which exercises on a suitable energy substrate/activity are combined to develop fitness (depending on the module chosen).
3. A body recovery phase after exercise (winding down), lasting 5–10 min, which is important for lowering the heart and breathing rates and is achieved by performing a active physical mobility exercises (stretching), static and/or dynamic, to avoid muscle pain.

Table 1. PAP program features.

PAP—Aerobic Endurance Module			
Session Frequency	**Intensity**	**Dosing**	**Type of Activity**
5–7 times/week	Moderate-high 130–150 heartbeats/min. 155–170 heartbeats/min.	15–60 min/session (With an upward dynamic in volume)	Running, skating, cycling, swimming, cross-country skiing or any other cyclical physical activity
PAP—Strength and Muscular Endurance module			
Session frequency	Intensity	Dosing	Type of activity
at least 3 times/week	Moderate 130–150 heartbeats/min.	8–10 repetitions; 8 to 10 exercises/set of exercises	A wide range of exercises for harmonious physical development, toning and muscle trophicity.
PAP—Flexibility module			
Session frequency	Intensity/Amplitude	Dosing	Type of activity
at least ≥3 times/week	There must be no pain	Dynamic exercises—Position must be held for 4–5 s, 3–5 repetitions Static exercises—Position must be held for 10–20 s.	Pilates stretching (Active and combined)
PAP—Weekend activities module			
Session frequency	Intensity	Dosing	Type of activity
1 time/week	Moderate 130–150 heartbeats/min.	Depending on the chosen route (2 to 4.2 km)	Hiking, Theme parks

2.3. Research Subjects, Inclusion/Exclusion Criteria

The experiment was conducted from September 2018 to June 2019. The target group for the study consisted of 79 subjects with excess weight (38 girls and 41 boys) from General Schools No. 11, No. 13, No. 15, No. 19 and No. 30 of the Municipality of Brasov, aged between 12 and 15 years. In the 5 educational units, we had access through the collaboration protocol concluded between our academic institution and the Brașov County School Inspectorate (partnership agreement no. 298/18 October 2017). The selection of subjects was targeted on multilayered sampling to cover a varied population in terms of age and gender. From an initial group of 494 pupils, the eligible population was selected: 79 subjects with excess weight (52 overweight subjects, group SP, and 27 obese subjects, group O). We divided the final study sample into 4 subgroups for analysis and comparison: obese girls; obese boys; overweight girls; overweight boys).

- Inclusion criteria: pupils from secondary school (grades V–VIII); obesity: BMI ≥ 95th percentile (+2SD) for gender and age; overweight: 85th percentile ≤ BMI < 95th percentile (+2SD) for gender and age; possibility of monitoring and evaluating the results by completing the proposed physical activity programme.
- Exclusion criteria: impossibility to monitor subjects; refusal of parents and/or students to participate in the study.

2.4. Statistical Methods Used in Research

The processing and interpretation of the collected data was carried out using the Statistical Package for the Social Sciences Program (SPSS version 25.0; IBM Corp., Armonk, NY, USA).

3. Results

3.1. Identification of Subjects with Excess Weight at the Initial Measurement

Tables 2 and 3 present the distributions of the students' height and weight at the initial measurement, respectively.

Table 2. Distribution of height in the target group–initial testing.

Statural Classification		Frequency	Percentage	Percentage of Measured Subjects	Cumulative Percentage
Height (cm)	Hyperstatural	1	1.30	1.30	1.30
	Normostatural	74	93.70	93.70	94.90
	Hypostatural	4	5.10	5.10	100
	Total	79	100	100	

Table 3. Distribution of weight in the target group–initial testing.

Body Weight Classification		Frequency	Percentage	Percentage of Measured Subjects	Cumulative Percentage
Weight (kg)	Extra Heavy Weight	23	29.10	29.10	29.10
	Healthy weight	56	70.90	70.90	100
	Very low weight	0	0.00	0.00	
	Total	79	100	100	

Table 3 indicates that most of the subjects had a heathy body weight and only about one-third had a very high body weight. However, calculating the body mass index showed that all subjects had excess weight (Table 4).

Table 4. Distribution of initial body mass indices by gender.

				BMI (kg/m^2)		Total
				Overweight	Obesity	
Gender		Girls	Frequency	25	13	38
			Percentage (%)	65.80%	34.20%	100%
		Boys	Frequency	27	14	41
			Percentage (%)	65.90%	34.10%	100%
	Total		Frequency	52	27	79
			Percentage (%)	65.80%	34.20%	100%

The subjects' distribution by grade is as follows: 15 students (18.99%) were in the 5th grade, 16 students (20.25%) in the 6th grade, 23 students (29.11%) in the 7th grade and 25 students (31.65%) in the 8th grade. In terms of age, the oldest students were the 8th-grade students, with an average age of 15 years; followed by the 7th-grade students, with an average age of 14 years; 6th grade, average age 13 years; and 5th grade, average age, 12 years.

At the end the physical activities programme (April 2019), we collected a significant amount of data in the following areas: volume of activity performed/subject, expressed as the number of activities performed; time allocated to physical activities (expressed in minutes and hours); the intensity of the physical activities; and the subjects' actual choices of the activities provided in the PAP. Below, we present the findings according to the four student subgroups: overweight boys and girls and obese boys and girls.

3.2. Average Volume/Frequency of Physical Activities Performed According to the Body Mass Index and Gender of Subjects, Post-Impact PAP

Over the 26 weeks, overall, the subjects in our study (the 79 subjects—38 girls and 41 boys) performed an average of 95.36 physical activities/subject, the fewest being 64 and the most, 124 (Table 5), for an average for the entire sample of subjects of 3.67 physical activities/week. This result was obtained by dividing the average physical activity/subject (95.36) by the duration of the PAP (26 weeks).

Table 5. Total physical activity by body mass index and gender.

BMI	Gender	N	Frequency Average	Standard Deviation	Average Standard Error	95% Conf. Range Min.	Max.	Minimum	Maximum
O	Girls	13	96.38	21.02	5.83	83.68	109.09	73	123
	Boys	14	85.07	13.30	3.55	77.39	92.75	69	123
	Total	27	90.72	18.04	3.47	83.38	97.66	69	123
SP	Girls	25	100.60	13.92	2.78	94.85	106.35	75	124
	Boys	27	95.37	18.87	3.63	87.91	102.83	64	124
	Total	52	97.98	16.72	2.32	93.23	102.54	64	124

O = obesity; SP = overweight.

It can also be observed that based on the body mass index, over the entire duration of the programme, obese girls performed more activities on average than obese boys, a fact found also in the case of overweight girls in relation to overweight boys (Table 5).

Both overweight boys and girls performed more physical activities than the obese girls and boys by an average difference of 7.36 activities. Over the 6 months of the programme, the 79 students performed a total of 7534 activities, an overall average of 3.67 activities/week/subject. Figure 1 presents their activities by month. It can be observed that most physical activities were carried out in April, while the fewest were carried out in November.

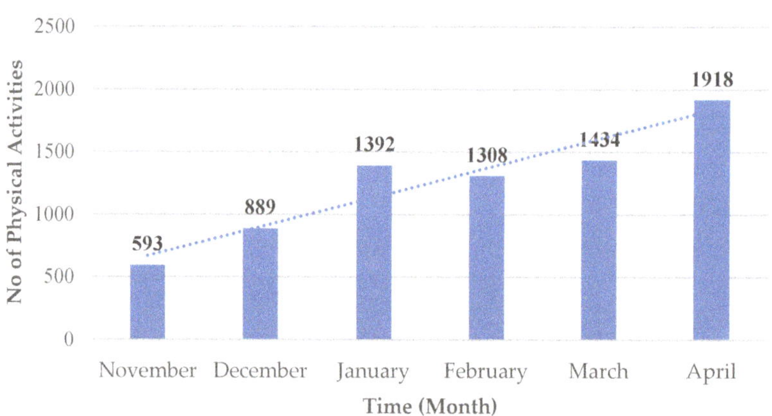

Figure 1. Distribution of the total number of physical activities /month.

3.3. Averages/Subgroups of the Time Allocated to the PAP Sessions

The total amount of time dedicated to the PAP was calculated for the entire period of the programme (1 November 2018–30 April 2019), according to the body mass index and gender of the subjects concerned. On an overall level (79 subjects), the time dedicated to the physical activity programme shows an increasing trend from month to month: from a total of 916 h 8 min in November 2018 to a total of 1101 h 10 min in April 2019. In obese subjects, girls spent more time on physical activity (983 h 24 min) than boys (883 h 57 min). In the case of overweight subjects, girls (1990 h 29 min) and boys (1976 h 23 min), allocated approximately the same total amount of time to the program. Table 6 presents average times spent on the physical activities according to weight group (obese or overweight) and time interval (between 1 h 30 min–2 h 29 min and $\geq$3 h), as well as their number/activity time interval/week (Table 6 and Figure 2).

Table 6. Distribution of total weekly time spent on physical activity by weight group and time interval.

BMI (kg/m^2) Quantitative Indicators			Average Time Volume/Week Dedicated to the Physical Activity Programme by Subject/Group					Total
			$\geq$1 h	$\geq$1 h 29 min	1 h 30 min–2 h 29 min	2 h 30 min and 2 h 59 min	$\geq$3 h	
Groups	Overweight	Frequency (N)	0	0	10	21	21	52
		Percentage (%)	0.00%	0.00%	19.20%	40.40%	40.40%	100%
	Obese	Frequency (N)	0	0	17	3	7	27
		Percentage (%)	0.00%	0.00%	63.00%	11.10%	25.90%	100%
Total		Frequency (N)	0	0	27	24	28	79
		Percentage (%)	0.00%	0.00%	34.20%	30.40%	35.40%	100%

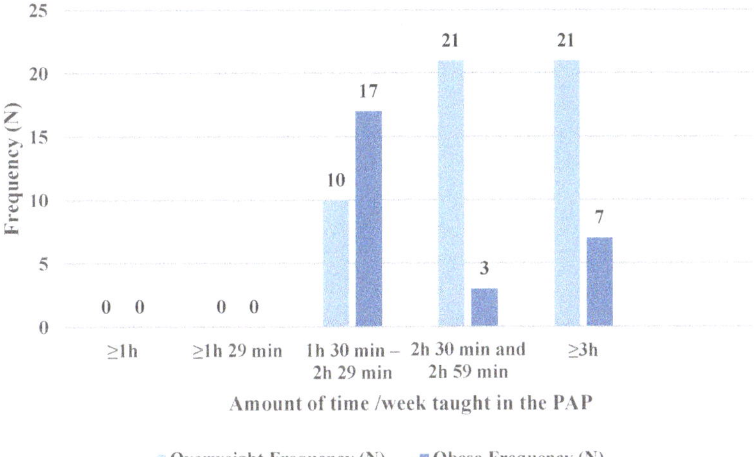

Figure 2. Distribution of total weekly time spent on physical activity.

3.4. Averages/Subgroup Effort Intensity of the Physical Activities

To determine the average of the physical activities/subject according to the intensity of the physical effort, we cumulated the frequency of the physical activities performed (percentages) by each subject, and subsequently, divided the obtained result by the total number of subjects corresponding to each of the 4 subgroups.

The collected data highlight the fact that during the 26 weeks of activity, in the group of obese girls (13 girls), out of the total of 1253 physical activities performed, 98 were performed with a low intensity of physical effort, 964 were performed with an average intensity and 191 were at submaximal intensity. The percentages physical activities/effort zones for obese girls are shown in Figure 3a.

In overweight girls, out of the total of 25 subjects, we found that out of the total of 2515 physical activities carried out, 158 were performed with a low intensity of physical effort, 1917 with an average intensity and 440 at submaximal intensity. In Figure 3b, these data are presented as percentages.

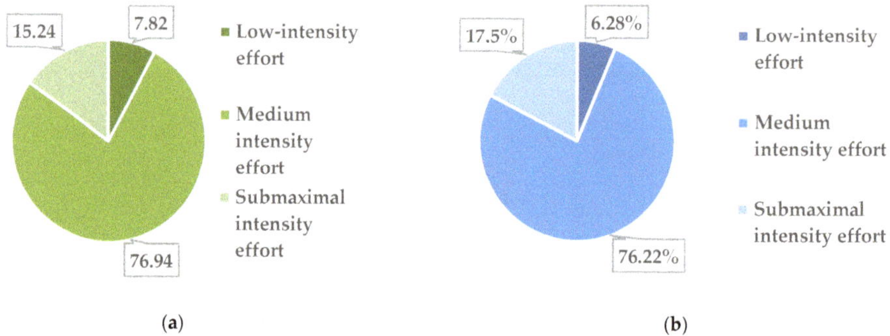

Figure 3. Distribution of physical activity by intensity of the physical effort in obese girls (**a**) and overweight girls (**b**).

In the subgroup of obese boys (14 subjects), out of the total of 1191 physical activities, 129 were performed with a low intensity of physical effort, 886 with an average intensity and 176 with a submaximal intensity. The corresponding percentages are presented in Figure 4a.

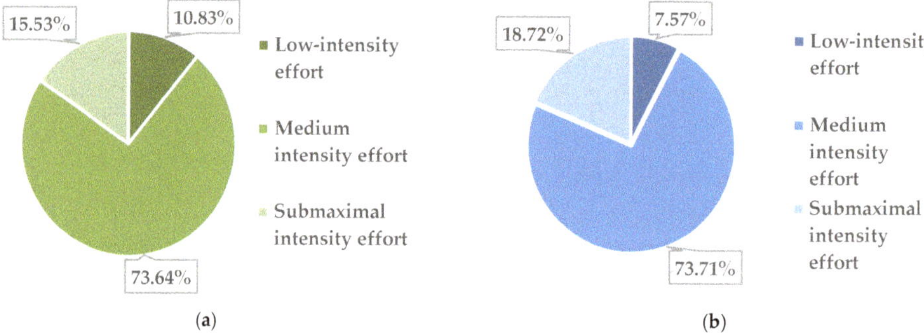

Figure 4. Distribution of physical activity by intensity of the physical effort in obese boys (**a**) and overweight boys (**b**).

In overweight boys, out of the total of 27 subjects, out of the total of 2575 physical activities, 195 were performed with a low intensity of physical effort, 1898 physical activities with an average intensity and 482 at submaximal intensity. The corresponding percentages can be found in Figure 4b.

Table 7 presents the 52 overweight and 27 obese (at initial measurement) study subjects' numbers of physical activities (AF), their percentages (%) and the intensity zones of the activities (A, B, C).

Table 7. Total number of performed physical activities by gender, weight group, and intensity of the physical effort.

BMI (kg/m^2)	Gender	No. of Subjects	Freq. N/AF	Total		
				A	B	C
overweight	Girls	25	No. AF	158	1917	440
			%	6.28%	76.22%	17.50%
	Boys	27	No. AF	195	1898	482
			%	7.57%	73.71%	18.72%

Table 7. *Cont.*

BMI (kg/m²)	Gender	No. of Subjects	Freq. N/AF	Total		
				A	B	C
Total		52	No. AF	353	3815	922
			%	6.94%	74.95%	18.11%
obesity	Girls	13	No. AF	98	964	191
			%	7.82%	76.94%	15.24%
	Boys	14	No. AF	129	877	185
			%	10.83%	73.64%	15.53%
Total		27	No. AF	227	1841	376
			%	9.29%	75.33%	15.38%
Total		79	No. AF	580	5656	1298
			%	7.70%	75.07%	17.23%

A = low intensity, B = medium intensity, C = submaximal intensity, AF = physical activity.

3.5. Types of Activities Chosen by the Study Subjects

Figures 5a,b and 6a,b present the percentages of participants in each activity module for the obese and overweight girls and boys, respectively, for all 79 subjects.

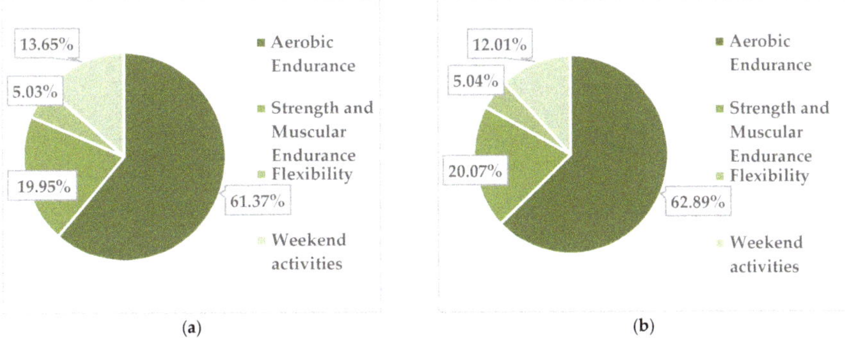

Figure 5. Frequency of preferred activities/module in the physical activity programme in obese girls (**a**), obese boys (**b**).

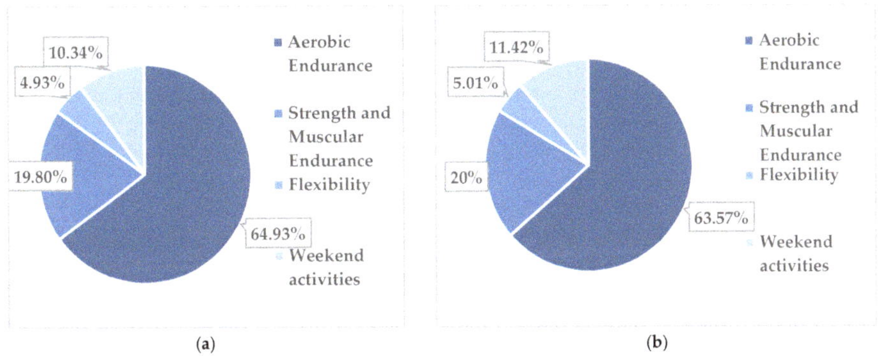

Figure 6. Frequency of preferred activities/module in the physical activity programme in overweight girls (**a**), overweight boys (**b**).

The data reflect the fact that all students, regardless of their subgroup, liked the physical activity modules in the following order: 1. aerobic endurance; 2. strength and muscular endurance; 3. flexibility; 4. weekend activities.

3.6. Interpretation of Final Results

At the end of the 6-month programme, we again measured the students' height and weight and again calculated body mass index for the subjects in the four subgroups in order to ascertain if there are, especially in the latter parameter (BMI), significant changes resulting from the effect of the continued practice of physical exercise.

- In the final testing, the majority of the subjects measured, namely 75 subjects, had a normal stature index (Table 8).
- In the final measurement of weight, the students showed an average body weight of 66.57 kg, and the results deviate from the average by 12.13 kg. The majority of the subjects measured (77 subjects) had a normal body weight, and only 2 subjects still had a high body weight relative to age (Table 9).

Table 8. Distribution of height in the target group–final testing.

Statural Classification		Frequency	Percentage	Percentage of Measured Subjects	Cumulative Percentage
Height (cm)	Hyperstatural	1	1.30	1.30	1.30
	Normostatural	75	94.90	94.90	96.20
	Hypostatural	3	3.80	3.80	100
	Total	79	100	100	

Table 9. Distribution of weight in the target group–final testing.

Body Weight Classification		Frequency	Percentage	Percentage of Measured Subjects	Cumulative Percentage
Weight (kg)	Extra Heavy Weight	2	2.50	2.50	2.50
	Normal weight	77	97.50	97.50	100
	Very low weight	0	0.00	0.00	
	Total	79	100	100	

There were no significant differences between boys and girls in either height or weight at both the initial and final testings. The majority of the subjects out of the total of 79 (38 girls and 41 boys) were of normal height and of weight within normal limits.

- The analysis of the body mass index at final testing reveals the BMI at the time of final testing by gender. Basically, in order to centralize the results, we defined the categories of subjects according to the new body mass indices obtained after participating in the PAP. Out of the total of 38 girls identified at the initial testing as having excess weight (25 overweight girls and 13 obese girls; see Table 4), at the final testing (Table 10) after completing the PAP, 23 girls acquired a healthy BMI, 9 girls were classified as overweight and 6 girls were obese. In boys, out of the total of 41, initially 27 overweight and 14 obese (Table 4) in the final testing (after completing the PAP), 14 boys reached a healthy BMI, 17 boys were classified as overweight and 10 boys as obese.

Regarding BMI, determined at the initial and final testings, it can be noted that at the initial test, 52 overweight subjects were recorded, and at the final test only 26 students remained classified as overweight. At the same time, of the 27 subjects identified with obesity at initial testing, only 16 subjects still showed obesity on final testing (Figure 7).

With regard to body mass index by gender, as determined in the initial and final testings (Figure 8), it can be observed that:

Table 10. Distribution of body mass indices at the final testing.

BMI Classification		Frequency	Percentage	Percentage of Evaluated Subjects	Cumulative Percentage
BMI (kg/m^2)	healthy	37	46.80	46.80	46.80
	overweight	26	32.90	32.90	79.70
	obesity	16	20.30	20.30	100
	Total	79	100	100	

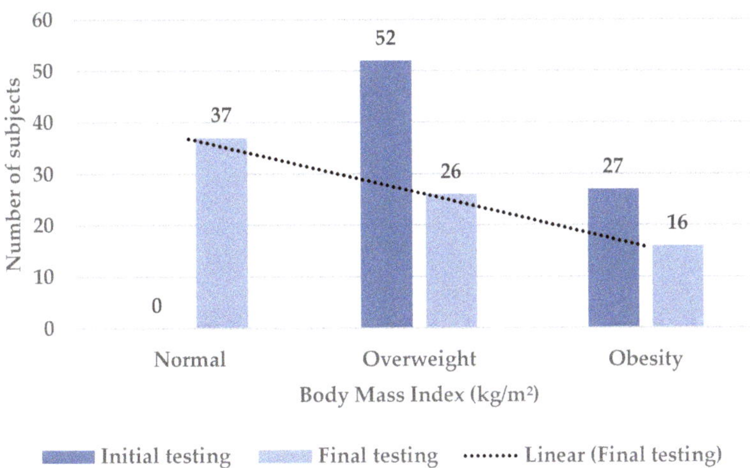

Figure 7. Body mass index distribution in initial and final testing for the target sample.

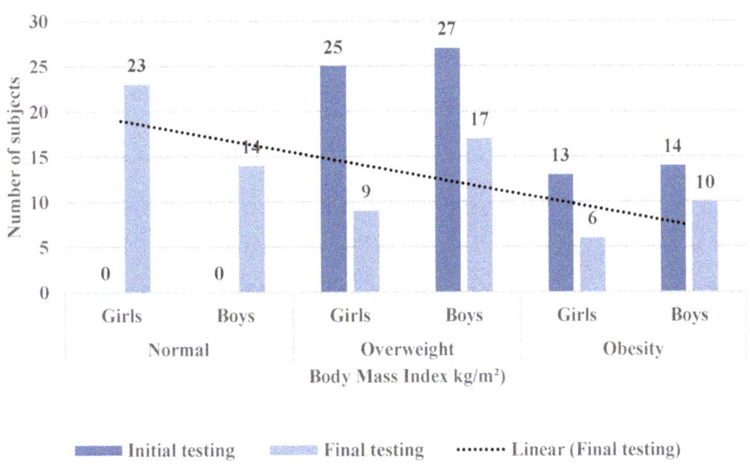

Figure 8. Distribution of initial and final body mass indices for the target sample by gender.

- In the initial testing, 25 girls were identified as overweight, while in the final testing, only 9 girls were still overweight;
- In the case of overweight boys, in the initial testing, 27 boys were overweight, and in the final testing, 17 boys with overweight were identified;
- In the case of obese girls, 13 girls were recorded with obesity at the initial testing, and 6 girls were obese at the final testing;

- For obese boys, 14 were identified in the initial testing, while in the final testing, only 10 boys had obesity;
- Out of the total number of subjects (79), at the final testing, 37 subjects with a healthy body mass index were reported (23 girls and 14 boys).
- In concluding this study, we wanted to determine whether there was a relationship between body mass index (acquired at the time of final testing) and the total amount of time allocated by the subjects with excess weight to the physical activity programme (PAP). Thus, the materiality threshold ($p = 0.04$) obtained in the bivariate analysis between the BMI determined at the final testing on the one hand and the total amount of time devoted to PAP on the other hand shows that there is a statistically significant correlation. As we can see, this correlation is reversed ($r = -0.23$), which allows us to state that in the entire sample (79 subjects with overweight and obesity), a higher body mass index was associated with less time dedicated to the physical activity program (PAP) (Table 11 and Figure 9).

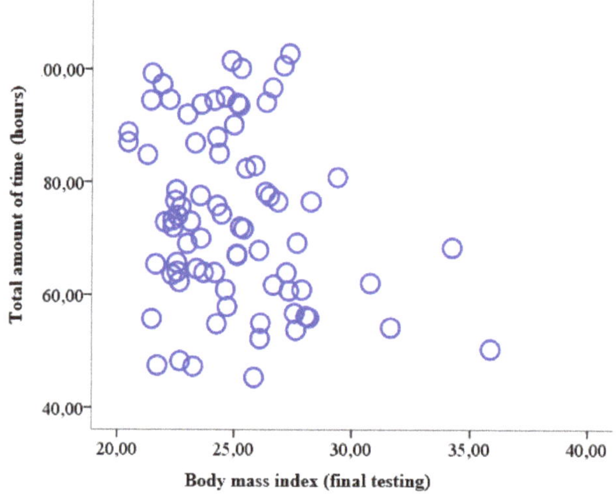

Figure 9. Scatter diagram for the relationship between the body mass index determined at the final testing and the total amount of time devoted to physical activity (hours).

Table 11. Results of the Pearson correlation analysis between the body mass index determined at final testing and the total amount of time devoted to the physical activity programme (PAP).

		BMI-TF	VT. PAP
BMI-TF	Correlation coefficient	1	−0.23
	Materiality threshold (p)		0.04
	N	79	79
VT. PAP	Correlation coefficient	−0.23	1
	Materiality threshold (p)	0.04	
	df	79	79

The correlation is significant at a threshold of 0.05; BMI-TF = body mass index determined at final testing; VT. PAP = total amount of time dedicated to the physical activity programme.

Furthermore, the statistical analysis using the chi-squared test shows that there is a significant difference in the sample selected to participate in the PAP based on body mass index in terms of the total amount of time spent weekly on physical activities ($\chi^2(9) = 19.33$, $p = 0.02$). Along the same lines, the analysis of the data with the help of Fisher's exact test leads us to the conclusion that there is a significant difference (Fischer's exact probability $p \leq 0.001$) in the total amount of time spent per week on physical activity between obese subjects (27 subjects, 13 girls and 14 boys) and overweight subjects (52 subjects, 25 girls and 27 boys) (Table 12). In other words, our study shows that overweight subjects (girls and boys) devoted more time to physical activity than obese students.

Table 12. Statistical significance of the total amount of time devoted weekly to the physical activity programme based on excess weight category and gender.

	Value	Degrees of Freedom	p
Pearson Chi-Square	16.01	2	0.001
Fisher's Exact Test	15.53		0.001
Valid responses	79		

The data presented confirm, from a statistical and mathematical point of view, our working hypothesis, which we mentioned in the introductory part of the study.

4. Discussion

In Romania, the National Center for Evaluation and Promotion reported on the state of health (CNEPSS) for the 2016–2017 school year (study published in 2018), centralizing data regarding the prevalence of chronic diseases found in preschoolers and students, and highlighted the fact that in the second place is obesity of nonendocrine causes in both urban and rural areas [59]. For the urban environment, the following values are presented: 66.95% of children and young people are healthy weight, 21.75% are overweight and 11.3% are obese. Analysing the preliminary sample from our research, the data show us that of the total number of subjects measured (n = 494), a percentage of 84.01% are healthy weight and 15.99% are overweight, of whom 10.54% are overweight and 5.46% obese. As can be seen, our sample falls into lower values of excess weight compared with the national averages for Romania. We believe that this difference is due to the geographic area of the city where the measurements were made. Brașov is one of the biggest mountain tourist towns in Romania, with a multi-ethnic population, where there is a certain culture of practicing physical activities. This fact can be an argument for the mentioned differences. On the other hand, also for Romania, the World Obesity Federation estimated in 2016 that 4% of girls and 9.20% of boys aged 10–19 overweight /obese, thus resulting in a sum of 13.2%. Comparing this value with that of our sample (15.99%), which fits into a narrower age range (12–15 years), we can consider that the two means are quite close. Also, the same source [60] estimates that in the year 2030, in Romania, approximately 14.30% of the population aged 10–19 will be obese, and our initial value, related to the excess weight of the subjects, was 15.99%. At the end of the PAP, this percentage dropped significantly, to 8.50% (42 subjects with excess weight out of n = 494), with a difference of almost half of the initial percentage, 7.49%. In this context, we can say that the independent variable through which we intervened and which we monitored was given only by the continuous practice of physical activities (through PAP), and this fact determined an improvement in the body mass index at the level of the majority of the targeted subjects.

Regarding the number/frequency of physical activities recommended for children and adolescents, it is true that the World Health Organization [52] recommends practicing at least 60 min/day of moderate to high intensity physical activity (vigorous). We did not succeed in reaching this level of participation with our subjects, our average of physical activities/week having a frequency of 3.66 (global value). However, the same source believes that currently there is no clear evidence of an association between physical activities and managing a healthy weight, and more research is needed to determine the level of association between them. Our study, based on the collected data, is also oriented towards this shortcoming, and the statistical correlation that emerged as a result of our calculations (r = −0.23) shows that the students in our interventions with higher body mass indices spent less time dedicated to the physical activity program (PAP). In this context, also in the direction of those indicated previously, we found through the obtained data that there was a statistically significant difference ($p \leq 0.001$) in terms of the total time dedicated weekly to the PAP between obese and overweight subjects. In other words, our study shows that overweight subjects (girls and boys), who in general allocated a higher proportion of time to PAP activities (80.80% in the intervals of 2 h 30 min and 2 h 59 min or $\geq$3 h), than the obese (37%), achieved a more significant improvement in body mass index in relation to the obese, a fact that can be considered as representing an association between physical activity and a healthy weight.

Regarding the contents of the four modules within the PAP (aerobic endurance; strength and muscular endurance; flexibility; weekend activities, mostly aerobic 1), the final results of our research show that they were correctly chosen, being effective for the intended purpose, and confirm the correctness of the recommendations offered by some specialized studies that consider that children and adolescents (6–17 years) can easily adapt to physical activities practiced in order to develop strength and muscular endurance [53,54], aerobic endurance [55,56] and flexibility/mobility [57,58].

Regarding the intensity zones of effort that we agreed on for the PAP, they are also supported by the recommendations of extensive specialist studies [61]. Thus, in these informative documents, for children and adolescents between 5 and 17 years of age, not only physical activities of moderate intensity but also physical activities of high intensity carried out at least 3 times a week including muscle strength exercises are recommended. We, through the PAP, offered to adolescents the opportunity to perform activities (regardless of their content) with the possibility of performing them in three intensity zones (low, moderate and submaximal). Monitoring this aspect during the 26 weeks, we found that overall, the intensity zone where the most activities were performed was the moderate one (75.07% of the total PAP activities), followed by the submaximal one (17.23%) and the one with low intensity (7.70%). Compared with the previously mentioned recommendations, our study does not fully comply with the provision for high-intensity physical activities, but taking into account the fact that the 79 subjects were all with excess weight at the beginning of the intervention and the source reference is given for those of healthy weight, we still consider that the indication in question was respected to a certain extent, and it follows that in the future, one should insist on performing physical activities that require a more sustained effort.

One last aspect we would like to point out is adolescents' physical activity according to gender. Summarized data [21] referring to 20 studies from different countries found that in children and adolescents, boys in all countries were more active than girls. Along the same lines, [12] it was found that girls practice less physical activity regardless of the age analysed (from 11 to 16 years). According to the authors, for a long time, the socialization process has caused sports inhibition among girls because sports have been labelled a "male activity".

Compared with the mentioned considerations, through the data of our research, we did not notice these differences in attitude between the 38 girls and 41 boys (79 subjects) included in the PAP. Our claims are based on the cumulative data obtained on the number of physical activities performed within the PAP, throughout the monitoring period (26 weeks). Thus, from the total of 7534 physical activities completed, the boys performed 3766 activities and the girls 3768. This equality in the data is given, we believe, by the way we constantly promoted and motivated the subjects (especially through the five physical education teachers who centralized the weekly records of the activities) to practice physical activities in view of our common goal, to improve the excess weight. We believe that the most important factor to consider when it comes to changing physical activity behavior is motivation, which is the inner force that determines the completion of a task for which a commitment has been agreed on.

Our study, admittedly conducted on a more specific sample, namely monitored adolescents with a certain level of excess weight, showed, we believe, that there are no major differences between the levels of participation of girls and boys in PAP activities and even in the types of agreed-on/accessed modules according to the centralized self-reports.

As a general educational aspect, we are also not sure whether it is a good idea, from a pedagogical perspective, to group students in physical education lessons as a constant organizational form of directed motor activity according to normal and overweight categories and compare the manifested attitudes. After all, all children and adolescents are ours (from a didactic perspective), and we are therefore bound to support them in their own training. Perhaps even based on this idea, we decided that the activities in our study should be carried out in the students' free time and differentiated according to their interests and enjoyment. We believe that when those responsible (we refer primarily to physical education teachers) adopt a positive pedagogy in their training approach (whether formal or non-formal), encourage the students, motivate them, show persuasive power through relevant arguments formulated according to the student's level of understanding, constantly highlight each student's progress, support the student in their most difficult moments, keep in touch with the students' families, adopt social learning techniques through movement, etc., the results and attitudes towards physical activities can be surprising. Thus, first in our minds, then in the minds of teachers, and from there through the education provided to all young people and their families, the ideas that discriminate between the two sexes with regard to the practice of physical exercise in any form will fade over time, and even more, the psychological boundary given by the criteria of body mass index will be eliminated, everything being redirected towards the real pleasure that movement can generate for all regardless of age, body weight and any other criteria of human segregation.

5. Research Limitations

The present study is only one section of a larger research project that also includes dietary aspects of students with excess weight, as well as measuring the levels of physical fitness in this target group before and after the impact of the physical activity programme. Due to lack of space, we have not been able to synthesize all the data we have available. The study is also focused on a selective sample in terms of inclusion criteria and limited in terms of number of participants, which is why the data presented need to be confirmed by other more comprehensive studies from a quantitative point of view and possibly extend to the rural school population.

Author Contributions: Conceptualization, M.D.S., L.B. and R.M.; methodology, M.D.S. and L.B.; validation, L.B. and R.M.; literature review: M.D.S., L.B. and R.M.; writing—original draft preparation, M.D.S. and L.B.; writing—review and editing, L.B. and R.M.; supervision, M.D.S., L.B. and R.M. All authors have read and agreed to the published version of the manuscript.

Funding: This research received no external funding.

Institutional Review Board Statement: The study was conducted in accordance with the Declaration of Helsinki, and approved by the Ethics Committee of the Faculty of Physical education and mountain sports at Transilvania University of Brașov no. 243 from 26 September 2018.

Informed Consent Statement: Informed consent was obtained from all subjects involved in the study, concurrent with the agreement of the ethics commissions of the Brasov County School Inspectorate and the Faculty of Physical Education and Mountain Sports, with no. 298 from 18 October 2017.

Data Availability Statement: Not applicable.

Conflicts of Interest: The authors declare no conflict of interest.

References

1. Valverde, J. Valoración de la Asignatura de Educación Física y su Relación con los Niveles de Actividad Física Habitual en Adolescentes Escolarizados de la Región de Murcia. Ph.D. Thesis, Universidad de Murcia, Murcia, Spain, 20 May 2008.
2. Guthold, R.; Stevens, G.A.; Riley, L.M.; Bull, F.C. Global trends in insufficient physical activity among adolescents: A pooled analysis of 298 population-based surveys with 1·6 million participants. *Lancet Child Adolesc. Health* **2020**, *4*, 23–35. [CrossRef]
3. López-Sánchez, G.; Emeljanovas, A.; Miežienė, B.; Díaz-Suárez, A.; Sánchez-Castillo, S.; Yang, L.; Roberts, J.; Smith, L. Levels of Physical Activity in Lithuanian Adolescents. *Medicina* **2018**, *54*, 84. [CrossRef] [PubMed]

4. Aznar, S.; Webster, T. Actividad física y salud en la infancia y la adolescencia. In *Guía para Todas las Personas que Participan en su Educación*; Grafo, S.A.: Madrid, Spain, 2006; pp. 42–47. Available online: https://www.sanidad.gob.es/ciudadanos/proteccionSalud/adultos/actiFisica/docs/ActividadFisicaSaludEspanol.pdf (accessed on 11 December 2018).
5. Ingram, D.K. Age-related decline in physical activity: Generalization to nonhumans. *Med. Sci. Sport. Exerc.* **2000**, *32*, 1623–1629. [CrossRef]
6. Sallis, J.F.; Prochaska, J.J.; Taylor, W.C. A review of correlates of physical activity of children and adolescents. *Med. Sci. Sport. Exerc.* **2000**, *32*, 963–975. [CrossRef] [PubMed]
7. Lasheras, L.; Aznar, S.; Merino, B.; López, E.G. Factors Associated with Physical Activity among Spanish Youth through the National Health Survey. *Prev. Med.* **2001**, *32*, 455–464. [CrossRef]
8. Corder, K.; Winpenny, E.; Love, R.; Brown, H.E.; White, M.; Sluijs, E.V. Change in physical activity from adolescence to early adulthood: A systematic review and meta-analysis of longitudinal cohort studies. *Br. J. Sport. Med.* **2019**, *53*, 496–503. [CrossRef] [PubMed]
9. Nader, P.R.; Bradley, R.H.; Houts, R.M.; McRitchie, S.L.; O'Brien, M. Moderate-to-vigorous physical activity from ages 9 to 15 years. *J. Am. Med. Asoc.* **2008**, *300*, 295–305. [CrossRef] [PubMed]
10. Lau, E.Y.; Dowda, M.; McIver, K.L.; Pate, R.R. Changes in physical activity in the school, afterschool, and evening periods during the transition from elementary to middle school. *J. Sch. Health* **2017**, *87*, 531–537. [CrossRef] [PubMed]
11. Katzmarzyk, P.T.; Lee, I.M.; Martin, C.K.; Blair, S.N. Epidemiology of Physical Activity and Exercise Training in the United States. *Prog. Cardiovasc. Dis.* **2017**, *60*, 3–10. [CrossRef]
12. Castillo-Fernández, I.; Balaguer-Solá, I. Dimensiones de los motivos de práctica deportiva de los adolescentes valencianos escolarizados. *Apunt. Educ. Física Esports* **2001**, *63*, 22–29.
13. Rowlands, A.V.; Pilgrim, E.L.; Eston, R.G. Patterns of habitual activity across weekdays and weekend days in 9–11-year-old children. *Prev. Med.* **2007**, *46*, 317–324. [CrossRef] [PubMed]
14. Sallis, J.F. Age-related decline in physical activity: A synthesis of human and animal studies. *Med. Sci. Sport. Exerc.* **2000**, *32*, 1598–1600. [CrossRef] [PubMed]
15. Riddoch, C.J.; Andersen, L.B.; Wedderkopp, N.; Harro, M.; Klasson-Heggebo, L.; Sardinha, L.B.; Cooper, A.R.; Ekelund, U.L.F. Physical activity levels and patterns of 9-and 15-yr-old European children. *Med. Sci. Sport. Exerc.* **2004**, *36*, 86–92. [CrossRef] [PubMed]
16. Whitt-Glover, M.C.; Taylor, W.C.; Floyd, M.F.; Yore, M.M.; Yancey, A.K.; Matthews, C.E. Disparities in Physical Activity and Sedentary Behaviors Among US Children and Adolescents: Prevalence, Correlates, and Intervention Implications. *J. Public Health Policy* **2009**, *30*, S309–S334. [CrossRef]
17. Slater, A.; Tiggemann, M. "Uncool to do sport": A focus group study of adolescent girls' reasons for withdrawing from physical activity. *Psychol. Sport Exerc.* **2010**, *11*, 619–626. [CrossRef]
18. Butt, J.; Weinberg, R.S.; Breckon, J.D.; Claytor, R.P. Adolescent physical activity participation and motivational determinants across gender, age, and race. *J. Phys. Act. Health* **2011**, *8*, 1074–1083. [CrossRef] [PubMed]
19. Bailey, D.; Fairclough, S.; Savory, L.; Denton, S.; Pang, D.; Deane, C.; Kerr, C. Accelerometry-assessed sedentary behaviour and physical activity levels during the segmented school day in 10–14-year-old children: The HAPPY study. *Eur. J. Pediatr.* **2012**, *171*, 1805–1813. [CrossRef] [PubMed]
20. Dumith, S.C.; Gigante, D.P.; Domingues, M.R.; Hallal, P.C.; Menezes, A.M.B.; Kohl, H.W. Predictors of physical activity change during adolescence: A 3.5-year follow-up. *Public Health Nutr.* **2012**, *15*, 2237–2245. [CrossRef] [PubMed]
21. Cooper, A.R.; Goodman, A.; Page, A.S.; Sherar, L.B.; Esliger, D.W.; van Sluijs, E.; Andersen, L.B.; Anderssen, S.; Cardon, G.; Davey, R.; et al. Objectively measured physical activity and sedentary time in youth: The international children's accelerometry database (ICAD). *Int. J. Behav. Nutr. Phys. Act.* **2015**, *12*, 113. [CrossRef]
22. Van Hecke, L.; Loyen, A.; Verloigne, M.; Van der Ploeg, H.P.; Lakerveld, J.; Brug, J.; de Bourdeaudhuij, I.; Ekelund, U.; Donnelly, A.; Hendriksen, I.; et al. Variation in population levels of physical activity in European children and adolescents according to cross-European studies: A systematic literature review within DEDIPAC. *Int. J. Behav. Nutr. Phys. Act.* **2016**, *13*, 70. [CrossRef]
23. Varo, J.J.; Martínez-González, M.A.; de Irala-Estévez, J.; Kearney, J.; Gibney, M.; Martínez, J.A. Distribution and determinants of sedentary lifestyles in the European Union. *Int. J. Epidemiol.* **2003**, *32*, 138–146. [CrossRef] [PubMed]
24. Janssen, I.; Boyce, W.F.; Simpson, K.; Pickett, W. Influence of individual-and area-level measures of socioeconomic status on obesity, unhealthy eating, and physical inactivity in Canadian adolescents. *Am. J. Clin. Nutr.* **2006**, *83*, 139–145. [CrossRef] [PubMed]
25. Rodenburg, G.; Oenema, A.; Pasma, M.; Kremers, S.P.J.; van de Mheen, D. Clustering of food and activity preferences in primary school children. *Appetite* **2013**, *60*, 123–132. [CrossRef] [PubMed]
26. Gubbels, J.S.; van Assema, P.; Kremers, S.P.J. Physical Activity, Sedentary Behavior, and Dietary Patterns among Children. *Curr. Nutr. Rep.* **2013**, *2*, 105–112. [CrossRef]
27. Perez-Villagrán, S.; Novalbos-Ruiz, J.P.; Rodríguez-Martín, A.; Martínez-Nieto, J.M.; Lechuga-Sancho, A.M. Implications of family socioeconomic level on risk behaviors in child-youth obesity. *Nutr. Hosp.* **2013**, *28*, 1951–1960. [CrossRef]
28. Baquet, G.; Ridgers, N.D.; Blaes, A.; Aucouturier, J.; Van Praagh, E.; Berthoin, S. Objectively assessed recess physical activity in girls and boys from high and low socioeconomic backgrounds. *BMC Public Health* **2014**, *14*, 192. [CrossRef]

29. Fernández, I.; Canet, O.; Giné-Garriga, M. Assessment of physical activity levels, fitness and perceived barriers to physical activity practice in adolescents: Cross-sectional study. *Eur. J. Pediatr.* **2016**, *176*, 57–65. [CrossRef]

30. Hankonen, N.; Heino, M.T.J.; Kujala, E.; Hynynen, S.; Absetz, P.; Araújo-Soares, V.; Haukkala, A. What explains the socioeconomic status gap in activity? educational differences in determinants of physical activity and screentime. *BMC Public Health* **2017**, *17*, 144. [CrossRef]

31. Al-Hazzaa, H.M. Physical inactivity in Saudi Arabia revisited: A systematic review of inactivity prevalence and perceived barriers to active living. *Int. J. Health Sci.* **2018**, *12*, 50–64.

32. Moreno-Murcia, J.A.; Cervelló, E.; Huéscar, E.; Llamas, L. Relación de los motivos de práctica deportiva en adolescentes con la percepción de competencia, imagen corporal y hábitos saludables. *Cult. Educ.* **2011**, *23*, 533–542. [CrossRef]

33. Molina, J. Un Estudio Sobre la Práctica de Actividad Física, la Adiposidad Corporal y el Bienestar Psicológico en Universitarios. Ph.D. Thesis, Universidad de Valencia, Valencia, Spain, 18 June 2007.

34. Garzón, P.C.; Fernández, M.D.; Sánchez, P.T.; Gross, M.G. Actividad físico-deportiva en escolares adolescentes. *Retos Nuevas Tend. Educ. Física Deporte Recreación* **2002**, *3*, 5–12. [CrossRef]

35. Casterad, J.Z.; Puyal, J.R.S.; Gurrola, O.C.; Lanaspa, E.G.; Ostariz, E.S.; Clemente, J.A.J. Los factores ambientales y su influencia en los patrones de actividad física en adolescentes. *RICYDE Rev. Int. Cienc. Deporte* **2006**, *2*, 1–14. [CrossRef]

36. Narciso, J.; Silva, A.J.; Rodrigues, V.; Monteiro, M.J.; Almeida, A.; Saavedra, R.; Costa, A.M. Behavioral, contextual and biological factors associated with obesity during adolescence: A systematic review. *PLoS ONE* **2019**, *14*, e0214941. [CrossRef] [PubMed]

37. Trost, S.G.; Loprinzi, P.D. Exercise—Promoting healthy lifestyles in children and adolescents. *J. Clin. Lipidol.* **2008**, *2*, 162–168. [CrossRef] [PubMed]

38. Melnyk, B.M.; Jacobson, D.; Kelly, S.; Belyea, M.; Shaibi, G.; Small, L.; Marsiglia, F.F. Promoting Healthy Lifestyles in High School Adolescents. *Am. J. Prev. Med.* **2013**, *45*, 407–415. [CrossRef] [PubMed]

39. Bădicu, G. Contribution to the Improvement of the Quality of Life in Adults through the Implementation of Leisure Sports Programs. Ph.D. Thesis, Transilvania University of Brașov, Brașov, Romania, 17 July 2015.

40. Jakicic, J.M.; Otto, A.D. Physical activity considerations for the treatment and prevention of obesity. *Am. J. Clin. Nutr.* **2005**, *82*, 226S–229S. [CrossRef]

41. Arocha Rodulfo, J.I. Sedentarism, a disease from XXI century. *Clínica Investig. Arterioscler.* **2019**, *31*, 233–240. [CrossRef]

42. Tur, J.A.; Martinez, J.A. Guide and advances on childhood obesity determinants: Setting the research agenda. *Obes. Rev.* **2021**, *23*, e13379. [CrossRef]

43. Garcia-Pastor, T.; Salinero, J.J.; Sanz-Frias, D.; Pertusa, G.; Del Coso, J. Body fat percentage is more associated with low physical fitness than with sedentarism and diet in male and female adolescents. *Physiol. Behav.* **2016**, *165*, 166–172. [CrossRef]

44. Gutin, B.; Yin, Z.; Humphries, M.C.; Barbeau, P. Relations of moderate and vigorous physical activity to fitness and fatness in adolescents. *Am. J. Clin. Nutr.* **2005**, *81*, 746–750. [CrossRef]

45. Ortega, F.B.; Tresaco, B.; Ruiz, J.R.; Moreno, L.A.; Martin-Matillas, M.; Mesa, J.L.; Warnberg, J.; Bueno, M.; Tercedor, P.; Gutiérrez, A.; et al. Cardiorespiratory fitness and sedentary activities are associated with adiposity in adolescents. *Obesity* **2007**, *15*, 1589–1599. [CrossRef] [PubMed]

46. An, R. Diet quality and physical activity in relation to childhood obesity. *Int. J. Adolesc. Med. Health* **2017**, *29*, 1–9. [CrossRef] [PubMed]

47. Ying-Xiu, Z.; Jin-Shan, Z.; Jing-Yang, Z.; Zun-Hua, C.; Guang-Jian, W. Comparison on physical activity among adolescents with different weight status in Shandong, China. *J. Trop. Pediatr.* **2013**, *59*, 226–230. [CrossRef] [PubMed]

48. Lagestad, P.; Van den Tillaar, R.; Mamen, A. Longitudinal changes in physical activity level, body mass index, and oxygen uptake among norwegian adolescents. *Front. Public Health* **2018**, *6*, 97. [CrossRef] [PubMed]

49. Vălean, C.; Tătar, S.; Nanulescu, M.; Leucuța, A.; Ichim, G. Prevalence of obesity and overweight among school schildren in Cluj-Napoca. *Acta Endocrinol.* **2009**, *5*, 213–219. [CrossRef]

50. Chiriță-Emandi, A.C.; Puiu, M.; Gafencu, M.; Pienar, C. Growth references for school aged children in western Romania. *Acta Endocrinol.* **2012**, *8*, 133–152. [CrossRef]

51. Chiriță-Emandi, A.; Barbu, C.G.; Cinteza, E.E.; Chesaru, B.I.; Gafencu, M.; Mocanu, V.; Pascanu, I.M.; Tatar, S.A.; Balgradean, M.; Dobre, M.; et al. Overweight and Underweight Prevalence Trends in Children from Romania—Pooled Analysis of Cross-Sectional Studies between 2006 and 2015. *Obes. Facts* **2016**, *9*, 206–220. [CrossRef]

52. World Health Organization. Guidelines on Physical Activity and Sedentary Behaviour. 2020. Available online: https://apps.who.int/iris/bitstream/handle/10665/336656/9789240015128-eng.pdf?sequence=1&isAllowed=y (accessed on 23 January 2019).

53. Gutin, B.; Barbeau, P.; Owens, S.; Lemmon, C.R.; Bauman, M.; Allison, J.; Kang, K.S.; Litaker, M.S. Effects of exercise intensity on cardiovascular fitness, total body composition, and visceral adiposity of obese adolescents. *Am. J. Clin. Nutr.* **2002**, *75*, 818–826. [CrossRef]

54. Obert, P.; Mandigouts, S.; Nottin, S.; Vinet, A.; N'Guyen, L.D.; Lecoq, A.M. Cardiovascular responses to endurance training in children: Effect of gender. *Eur. J. Clin. Investig.* **2003**, *33*, 199–208. [CrossRef]

55. Bar, O.; Rowland, W.T. *Paediatric Exercise Medicine from Physiological Principles to Health Care Application*; Human Kinetics: Champaign, IL, USA, 2004; pp. 500–501.

56. Malina, R.M. Weight Training in Youth-Growth, Maturation, and Safety: An Evidence-Based Review. *Clin. J. Sport Med.* **2006**, *16*, 478–487. [CrossRef]

57. MacKelvie, K.J.; Petit, M.A.; Khan, K.M.; Beck, T.J.; McKay, H.A. Bone mass and structure are enhanced following a 2-year randomized controlled trial of exercise in prepubertal boys. *Bone* **2004**, *34*, 755–764. [CrossRef] [PubMed]
58. MacDonald, H.M.; Kontulainen, S.A.; Khan, K.M.; McKay, H.A. Is a School-Based Physical Activity Intervention Effective for Increasing Tibial Bone Strength in Boys and Girls? *J. Bone Miner. Res.* **2006**, *22*, 434–446. [CrossRef] [PubMed]
59. România–Documente. Evaluarea Nivelului de Dezvoltare Fizică și a Stării de Sănătate. Available online: https://documente. net/document/evaluarea-nivelului-de-dezvoltare-fizica-si-a-starii-de-inspgovrositescnepsswp-contentuploads201805raport-national.html?page=13 (accessed on 16 June 2019).
60. World Obesity Federation. Childhood Obesity Atlas Report. Available online: https://s3-eu-west-1.amazonaws.com/wof-files/ WOF_Childhood_Obesity_Atlas_Report_Oct19_V2.pdf (accessed on 26 May 2019).
61. Tremblay, M.S.; Carson, V.; Chaput, J.P.; Gorber, S.C.; Dinh, T.; Duggan, M.; Faulkner, G.; Gray, C.E.; Gruber, R.; Janson, K.; et al. Canadian 24-Hour Movement Guidelines for Children and Youth: An Integration of Physical Activity, Sedentary Behaviour, and Sleep. *Appl. Physiol. Nutr. Metabolism.* **2016**, *416* (Suppl. 3), S311–S327. [CrossRef] [PubMed]

Article

Balance Rehabilitation Approach by Bobath and Vojta Methods in Cerebral Palsy: A Pilot Study

Andreea Ungureanu [1,†]**, Ligia Rusu** [1,*,†]**, Mihai Robert Rusu** [1,†] **and Mihnea Ion Marin** [2,†]

[1]   Faculty of Physical Education and Sport, University of Craiova, 200585 Craiova, Romania
[2]   Faculty of Mechanics, University of Craiova, 200585 Craiova, Romania
*   Correspondence: ligia.rusu@edu.ucv.ro
†   These authors contributed equally to this work.

Citation: Ungureanu, A.; Rusu, L.; Rusu, M.R.; Marin, M.I. Balance Rehabilitation Approach by Bobath and Vojta Methods in Cerebral Palsy: A Pilot Study. *Children* **2022**, *9*, 1481. https://doi.org/10.3390/children9101481

Academic Editors: Georgian Badicu, Ana Filipa Silva and Hugo Miguel Borges Sarmento

Received: 21 August 2022
Accepted: 24 September 2022
Published: 28 September 2022

Publisher's Note: MDPI stays neutral with regard to jurisdictional claims in published maps and institutional affiliations.

Copyright: © 2022 by the authors. Licensee MDPI, Basel, Switzerland. This article is an open access article distributed under the terms and conditions of the Creative Commons Attribution (CC BY) license (https://creativecommons.org/licenses/by/4.0/).

Abstract: In cerebral palsy (CP) the basis for rehabilitation comes from neuroplasticity. One of the leading therapeutic approaches used in the management of CP is the NDT Bobath therapy and Vojta therapy consists in trying to program the ideal movement patterns for the age. The aim of our research was to analyze, from a functional point of view, the evolution of the biomechanical parameters characterizing the balance, in children with CP. The group of 12 subjects average age of 7 ± 3.28 years. The subject's evaluation included a functional clinical evaluation by Berg pediatric scale and a biomechanical evaluation performed using the "Stabilometry footboard PoData 2.00" for evaluation the body weight distribution on the foot level. The rehabilitation program was developed based on two methods, NDT Bobath and Vojta. A 90-min physiotherapy session starts with a Vojta therapy activation, for 20 min. Between the two therapies there is a 10-min break, then the session continues with NDT Bobath exercises within the 3 physical exercises proposed for 60 min. 5 days per week, 6 months. The analysis of the data collected before and after the application of the rehabilitation program, regarding the using the Berg scale indicates a progress of 32.35%, ($p = 0.0001 < 0.05$) and the effect size is large. The evolution of the data that indicate the distribution of body weight at the level of the two lower limbs, at the two moments pre/post, evaluation. For left side a progress of 8.39%, ($p = 0.027 < 0.05$) but a small effect size of 0.86. For right side a progress of 10.36% ($p = 0.027 < 0.05$) and also a small effect size of 0.86. Analyzing the results, we find that there is a left-right rebalancing in most patients. The favorable results that were obtained by drawing up a physiotherapy program composed of the combination of the two Vojta and NDT Bobath methods are proof of the fact that both methods are based on the creation of a stimulating peripheral pressure, which, if maintained, generates an extended stereotyped motor response. A pattern of symmetrical muscle contraction is thus created and thus balance and postural control can be achieved. The left-right rebalancing, proven by the percentage distribution analysis of the weight at the lower segmental level, demonstrated that the body alignment approach through the Vojta method on the one hand and the inhibitory facilitating postures/exercises promoted by the NDT Bobath method, allows obtaining a symmetry.

Keywords: cerebral palsy; balance; therapy activation; rehabilitation

1. Introduction

Neuroplasticity refers to the central nervous system's ability to change, and this change is not singular for central nervous system (CNS) injuries or in response to rehabilitation protocols. Neurons have the ability to change their structure and function depending on the inputs generated by activity and learning; in fact, neural change is the basis for memory and behavioral change that results from experience [1]. Plasticity takes place constantly, regardless of whether we are subjected to intense training or not. In addition, plasticity can be positive (adaptive) or negative (maladaptive). The central nervous system (CNS) has an innovative capacity to recover and to adapt to the compensatory mechanisms following an injury [2]. The basis for rehabilitation comes from neuroplasticity, defined as the ability of neural network to make adaptive changes at both structural and functional levels, ranging from molecular, synaptic and cellular changes to global network changes [3]. Human balance and gait are the most complex tasks that require the coordination of different neural network in the brain, brainstem, cerebellum and spinal cord [4] and they can recover function after injury/insult to any of these structures by applying methods that indirectly use neuroplasticity. Moreover, the activity of muscles and joints involved in walking and balance will be controlled via the descending motor tracts [5], while and important intervention for gaining balance is achieved through the pyramidal tracts (corticospinal tract) and the extrapyramidal tracts (vestibulospinal tract) [6,7].

Cerebral palsy (CP) is a pathology that may involve an interruption of the descending tracts. It is caused by a non-progressive disorder of the brain that occurred during the fetal or developing infant period. This disorder can cause certain permanent limitations of posture and movement [8]. One of the major challenges for children with infantile cerebral palsy is spasticity which causes a deficit of balance and gait [9]. In these cases, neuroplasticity is the key to functional recovery. Physiotherapists therefore need to understand the impact of rehabilitation on neural reorganization during recovery. Many of the neurosciences researched over the past 3 decades and more recently have been focused on understanding the physiological mechanisms underlying neuromotor recovery, as well as its determinants, so that we can maximize the effects of applied treatments [10]. However, there are fewer studies that have reported the results of objectifying the application of recovery methods that activate neuroplasticity.

There is no unique way to action for curing infantile cerebral palsy (CP) or for eliminating brain lesions [11], but there are therapeutic methods such as NDT Bobath therapy [12] and Vojta therapy [13] that, through reflex mechanisms of neural stimulation, as a result of reflex postures or specific mobilizations, contribute to improving balance.

Neuroplasticity is stimulated by intense repetitive practice, but it should not exhaust the nervous system. It is necessary for the treatment not to focus exclusively on repetition, but to take into account the patient's mobility and activity levels. Studies of the brain response of a brain-injured child have provided important information and diverse opinions concerning the effects that age has on recovery. The Kennard Principle, first advanced by Margaret Kennard, has demonstrated the fact that the developing brain is capable of significant reorganization and recovery after trauma [14]. In addition, the younger brain, in contrast with the older one, is less likely to develop progressive cognitive decline [15].

The greatest potential to obtain a good evolution in children with CP, based on neuroplasticity, is found in children whose rehabilitation intervention started early. This is because the gross motor skills in children with CP stabilize at the age of 4–7 years, while after this period the motor skills are relatively stable. Up to this age, we are dealing with a critical period in which the effects of neuroplasticity are exerted in relation to the motor control centers in the brain [15].

Analyzing the scientific literature, in order to document the main therapeutic interventions in children with CP, shows that one of the leading therapeutic approaches used in the management of brain neuroplasticity underdevelopment is the NDT Bobath therapy [12] (Neurodevelopmental Therapy), which is based on the fact that the atypical development concerning postural control and reflexes is responsible for the observed motor anomalies, due to the basic dysfunction of the central nervous system. This approach aims to facilitate standard motor development and functioning and to prevent the development of secondary disorders caused by muscle contractures and joint deformities of the limbs. The basic principle is that any stimulus of supraliminal intensity can "awaken" a clinical sign of CNS [16]. All therapeutic interventions based on the concept of neurodevelopment were built with the aim of improving movements and the active participation in maintaining body alignment and stabilizing each part of the body. The exercise program proposed by most authors was designed to lengthen the extensor muscles of the lower limbs with a greater focus on the hip muscles [17].

Despite the widespread use of NDT Bobath therapy there is no rigorous research regarding its clinical effectiveness. NDT aims at the normalization of muscle tone, its primitive and abnormal inhibition and the facilitation of normal movements. The goal of this approach is to correct postural tone abnormalities and facilitate more regular movement patterns for performing daily activities. Despite its widespread use in the clinic, limited studies have demonstrated evidence for its effectiveness in children with CP. Thus, a valid and reliable assessment tools are needed to measure the effectiveness of therapy in CP rehabilitation [18].

Following the four main patterns present at the level of the lower limbs of children with infantile CP, spastic form, a method was developed aiming to overcome spasticity with the help of spinal reflexes, by using "reflex unlocking" techniques, so that through peripheral stimulation the higher nerve centers are activated, as their activity is modeled on the basis of neuroplasticity. The techniques were adopted by Doman, and later by V. Vojta, whose approach was based on combining the techniques of Kabat and Fay [12]. According to Vojta, the role of therapy consists in trying to program the ideal movement patterns for the age of the newborn and of the infant with an affected central nervous system, as much as it is possible. This means that neurophysiological programming attempts to introduce an automatic coordination of the body's position, with well-defined angles of the upper and lower limbs relative to the trunk and vice versa, and of different body parts to each other, in a regular and reciprocal manner (alternatively on both sides of the body, left and right), with a change in the center of gravity's position, as it is common with each movement [19].

At a neurophysiological level, stimulation via Vojta therapy [13] induces psychomotor changes [20]. Vojta therapy [13] is a method that suppresses abnormal movement and induces normal motor development by promoting postural control and suppressing the patient's compensatory movements that were wrongfully acquired [21].

Vykuntaraju Kammasandra Nanjundagowda (2014) notes that medical recovery in the pathology of cerebral palsy aims to promote normal movement patterns and inhibit abnormal ones to maximize the independence of motor functionality. He mentions that several physical therapy techniques can be used in the rehabilitation plan: "Bobath Neurodevelopmental technique, the Vojta method, the Kabat, Philps and Denver method, but there is no clear evidence to support the superiority of one over the other" [22].

In the recovery process of children with infantile CP, there have been several major therapeutic practices in recent years, including the NDT Bobath concept and Vojta therapy; these treatment models have been adopted as best practices and accepted as conventional treatment approaches. However, additional, well-regulated, randomized studies are needed in order to determine the efficiency and the most appropriate roles for new technologies in physical rehabilitation interventions [13].

The aim of our research study was to analyze, from a functional point of view, the evolution of the biomechanical parameters characterizing postural control and balance, in children suffering from CP, aged between 3 and 11 years old and who were included in physical therapy programs consisting of exercises based on the principles and means of the NDT Bobath and Vojta methods. We chose this approach because currently there is no research to provide actual results regarding the quantification of the combined application of the two methods in the recovery of children with CP.

2. Materials and Methods

2.1. Subjects

We studied a number of 12 subjects diagnosed with CP, tetrapharesis, aged between 3–11 years. The small number of subjects is due to the fact that the consistent participation of these children in a rehabilitation program is limited. Therefore, we cannot cover a complete application of rehabilitation program. The group of subjects included 4 boys and 8 girls, with an average age of 7 ± 3.28 years, average height of 121.7 ± 22.48 cm, average weight of 28.68 ± 15.95 Kg (mean value $\pm$ SD). Score I GMFCS for 3 children, score II for 6 children, score III for 3 children. The research has been approved by Etich Committee of Research

Center-Human Body Movement Research Center (approved number 1540/1.10.2021) and respect the rules of Helsinki Declaration about the research that included human subjects. In the same time we have the informed consent for all children that have been included in the research. The informed consent has been signed by their parents.

The values of the anthropometric measurements for the group of subjects are those recorded in the table below (Table 1):

Table 1. The values of the main anthropometric parameters of the subjects in the experimental group.

Subject Code (n = 12)	Age [Years]	Weight G [kg]	Height I [cm]	BMI = G/I*I [Kg/m^2]
P1	7	28.20	121.00	19.31
P2	3	12.63	96.00	13.71
P3	10	44.50	131.00	25.93
P4	3	14.07	85.33	19.32
P5	6	17.43	103.75	16.19
P6	3	23.10	105.00	20.95
P7	11	56.23	151.67	24.45
P8	11	60.27	158.33	24.04
P9	10	27.90	136.00	15.08
P10	6	23.93	127.00	14.84
P11	4	20.70	108.33	17.64
P12	10	43.20	137.00	23.02

The selection of the 12 subjects was made based on inclusion and exclusion criteria as follows:

Inclusion criteria:

- children diagnosed with CP;
- children aged between 3–11 years;
- children capable of understanding and executing commands;
- children without other associated diseases;
- children who did not participate in recovery programs that were based on the two methods, NDT Bobath and Vojta;
- children with GMFCS level I-IV;
- children who can adopt the orthostatic position necessary to assess balance.

Exclusion criteria:

- children with mental retardation;
- children who cannot participate constantly in physical therapy sessions;
- children with deformities of the locomotor system;
- children with visual or hearing impairments;
- children with spasticity more then 2 on Ashworth scale.

2.2. Evaluation of Subjects with CP

The subject's evaluation included a functional clinical evaluation and a biomechanical evaluation. The clinical functional evaluation of the 12 subjects included the balance assessment using the Berg scale [23]. The biomechanical evaluation was performed using the "Stabilometry footboard PoData 2.00" bipodalic platform with a built-in podoscope that can be directly connected to a computer via USB ports. The device is represented by six load cells that can be positioned to detect the distribution of body weight at the points corresponding to the 1st metatarsal, the 5th metatarsal and the heel of each foot. It is also used to measure the average position of the body's center of gravity and its small movements around this position. This equipment works based on the principle of gathering information from the plantar level, and the data provided comes from the stimulation of the platform's sensors [24].

The patients were prepared before the start of the assessment, being minimally dressed, as the assessment was carried out in basal conditions. At the time of evaluation, we must take into accounted the fact that the CNS system, through its extero-proprioceptive receptors, is able to identify the best postural strategies, for each moment, adapting them to the contingent situation. Regarding the vertical position, its efficiency is determined by the distribution of body weight on both legs. For this reason, three measurements were made.

The patient must be without shoes, the skin must be bare. Alone he/she steps on the platform. The platform is divided into two equal parts, the child puts one foot on the right side and one foot on the left side, and must remain still for 20 s (Figure 1). The result is the footprint obtained as shown in Figure 2.

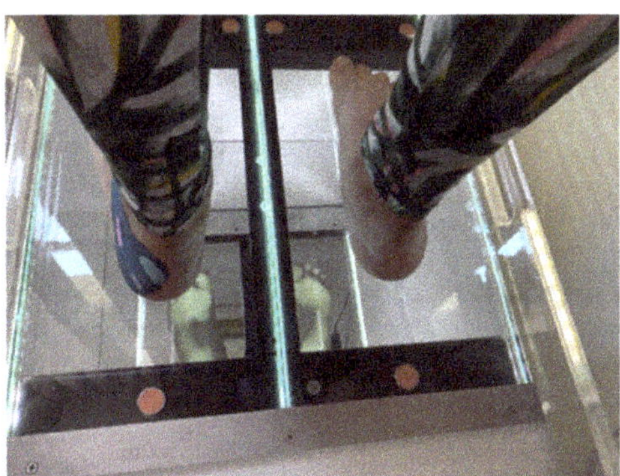

Figure 1. Patient preparation and positioning.

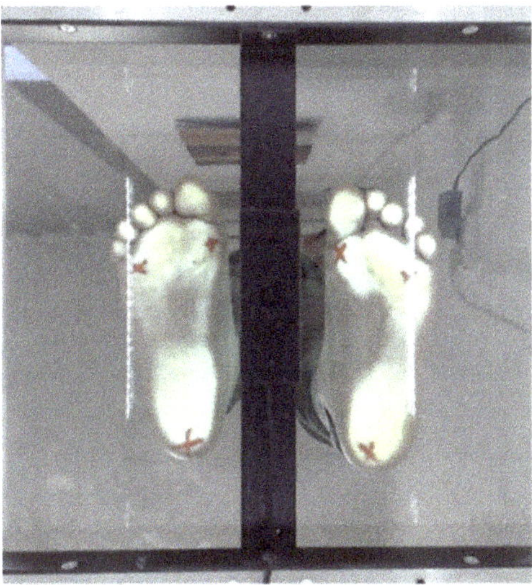

Figure 2. Plantogram picture (left spastic hemiparesis patient).

This platform supported the analysis of some parameters such as the left-right weight distribution, a feature that we believe is relevant for assessing the evolution of balance depending on the regional loading at the plantar level. The subjects' evaluation was carried out at two times—evaluation 1 (EV1) and evaluation 2 (EV2), 6 months apart.

Exercise 2. First position

The objective of the exercise: Lifting against gravity, extension of the spine, targeted and precise movements of the arms and legs.

The execution method: the child is positioned in a crouching position, on the edge of the bed, with the seat on the heel, and the legs are positioned outside the physiotherapy table in order to monitor the kinesiological response and not to be restrained—position represented in Figure 4. The head is placed on the surface work and turned to one side, with the occipital part towards the physiotherapist, being called the occipital part, and the facial part looks forward. The facial upper limb is in 180° flexion, and the occipital upper limb is in extension.

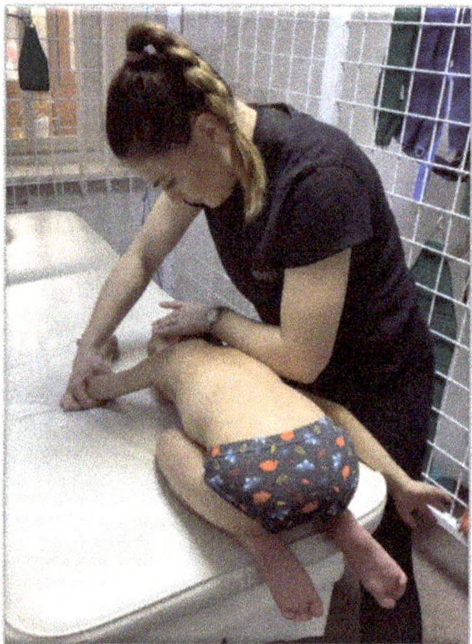

Figure 4. "First position".

The movement takes place in a so-called cross scheme, in which the right leg and the left arm move at the same time, as well as vice versa. One leg and arm on the opposite side supports the body and moves the trunk forward.

In therapy, when the child begins to rotate the head, the therapist applies appropriate resistance. This increases the activation of the entire body's musculature, thereby achieving the prerequisites for the verticalization process.

The execution technique: will activate from this position 3 times on each side, for a 15–50 s activation. We start with the activation on the right facial side, the child is in the prone position, with the right upper limb (UL) in 180° flexion at the shoulder level and a slight flexion at the elbow level. From this position we press on the "calcaneus" area at the LL level on the occipital side and the "epicondyle" area at the elbow level on the facial UL pressing gently, it remains in position pressing continuously and slowly for 15–50 s, until the expected kinesiological response is obtained: at the level of UL on the facial side, an extension of the fingers is obtained, UL on the occipital side initiates the flexion movement, the leg on the facial side performs an eversion movement with extension and abduction of the fingers, the leg on the occipital side performs an inversion with flexion of the fingers. After the kinesiological response is obtained, the activation side is changed, now the facial side will be the child's left side. These activations are repeated 3 times for each part.

Dosage: 2 activations per each part;

Sets: 3 sets, no rest between sets.

2.4.2. NDT Bobath Therapy

NDT Bobath therapy had as its objective the re-education of balance, coordination and balance obtained within the 3 physical exercises proposed and applied.

Exercise 1. Quadruped imbalances (Figure 5)

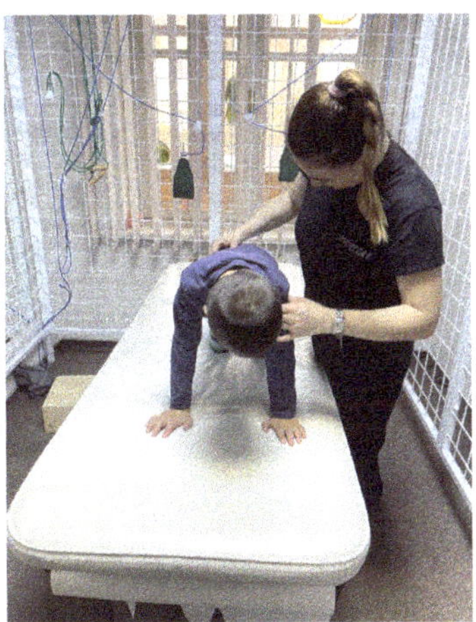

Figure 5. The exercise "Quadruped imbalances".

The execution method: The subject is positioned on all fours, the therapist imbalances him by pushing him sideways and backwards from the shoulders and sideways and forwards from the pelvis. The amplitude and force that the physical therapist applies to imbalance the subject is dosed according to the subject's ability to rebalance.

Dosage: 10 repetitions;

Sets: 2, with 30 s rest between sets.

Exercise 2—Imbalances from the "kneeling" position (Figure 6)

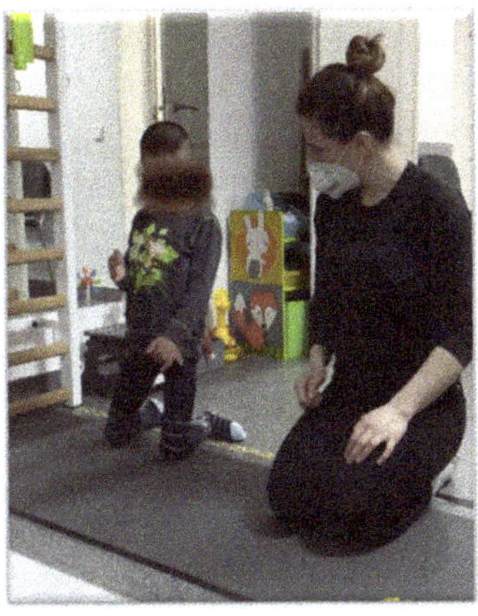

Figure 6. Imbalances from the "kneeling" position.

The execution method: With the subject supported on his knees, the therapist imbalances him from all directions. The force and amplitude of the imbalance is dosed according to the subject's ability to rebalance.

Dosage: 5 repetitions;

Sets: 2, with a 30 s break between repetitions.

Exercise 3—The Cervant Knight (Figure 7)

Figure 7. "The Servant Knight" position.

The execution method: The subject loads on one knee, the contralateral lower limb performs a triple flexion (flexion at the hip level, flexion at the knee level, and flexion at the ankle joint level)—the position of the serving knight (pelvis in retroversion on the side without load), inhibits spasticity of the adductors and extensors of the hip on the non-loading side, and the flexors of the hip on the loaded side, facilitating stabilization of the pelvis.

Dosage: 5 repetitions per each leg;

Sets: 2, with a 1-min break between sets.

3. Results

3.1. Functional Clinical Evaluation

The analysis of the data collected before and after the application of the rehabilitation program, regarding the assessment of balance using the Berg scale for pediatrics, indicates an improvement in the Berg scale score, the results being reproduced in Figures 8 and 9 for the 12 subjects.

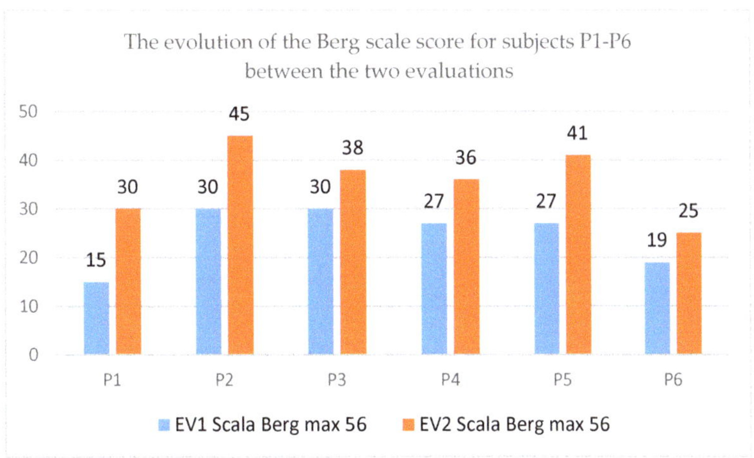

Figure 8. The evolution of the Berg scale score for subjects P1–P6.

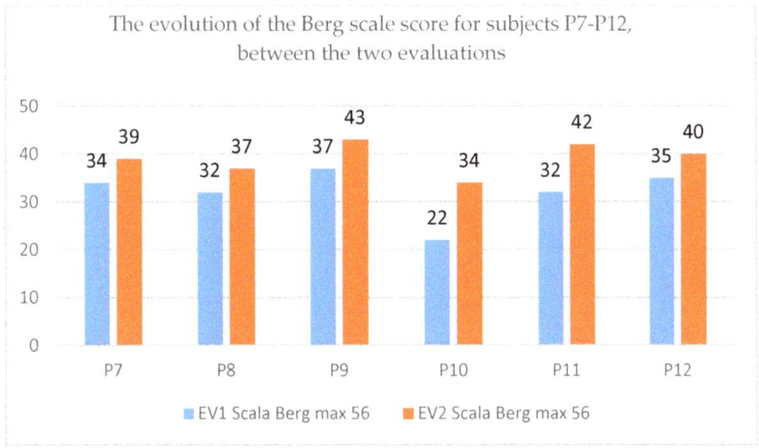

Figure 9. The evolution of the Berg scale score for subjects P7–P12.

The evolution from a clinical functional point of view is in the sense of an improvement in the balance evaluated by the Berg scale, more accentuated in patient P3, P5 and P10.

The results of the statistical analysis of the clinical functional evaluation by means of the Berg scale are presented in Tables 2–4.

Table 2. The mean value of the Berg scale.

Test	Average	SD	Minimum	Maximum	CV
EV1	28.33	6.69	15	37	23.62
EV2	37.5	5.68	25	45	15.15

Table 3. The table of statistical indicators of the difference of the Berg scale.

Statistical Indicators of the Resulting Difference (EV2-EV1)				Bilateral Dependent Student *t*-Test			
Average	Standard Deviation	95% Confidence Interval	Effect Size (Cohen D'test)	t Obs	df	*p* *	
9.16	3.97	(6.64; 11.69)	1.47	7.99	11	0.0001	

* The significance threshold is *p* = 0.05.

Table 4. Table of average differences of the Berg scale.

Average Difference (EV1-EV2)	Progress	Difference Size	The Progress Is	Null Hypotesis (Averages Are Equal)
9.16	32.35%	Very high	Statistically significant	It is rejected

The statistical analysis indicates a progress of 32.35%, with an average increase in the Berg scale value of 9.16. Two-sided *t*-test reveals a statistically significant difference in averages $p = 0.0001 < 0.05$, for $t_{obs} = 7.99$ and df = 11. It is observed that the effect size is large.

Regarding evolution of GMFCS score we present the results in the next figure (Figure 10 for P1–P6 and Figure 11 for P7–P12).

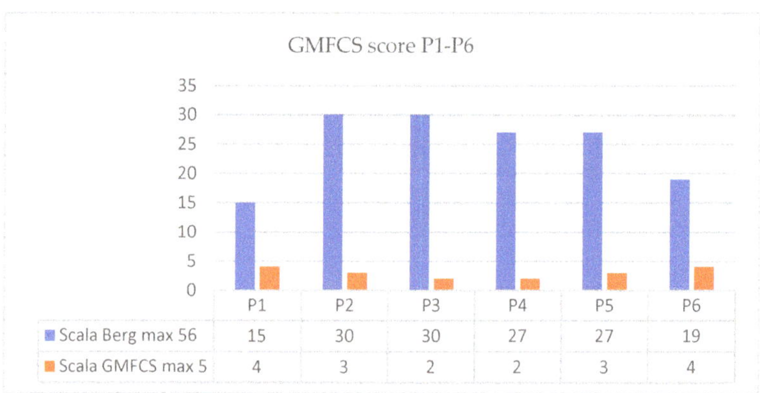

	P1	P2	P3	P4	P5	P6
Scala Berg max 56	15	30	30	27	27	19
Scala GMFCS max 5	4	3	2	2	3	4

Figure 10. Evolution of GMFCS score patient P1–P6.

3.2. The Results of the Biomechanical Evaluation

In this paper, we present the evolution of the data that indicate the distribution of body weight at the level of the two lower limbs, at the two moments of the EV1 and EV2 evaluation.

In order to evaluate the evolution as objectively as possible from the recovery of balance and implicitly the efficiency of the combined therapy point of view, we consider that it is necessary to focus our attention on how the patients manage to balance left and right.

In this sense, Table 5 shows the body weight distributions (expressed in kg and %) at the level of each leg, at the two moments of the evaluation (EV1 and EV2).

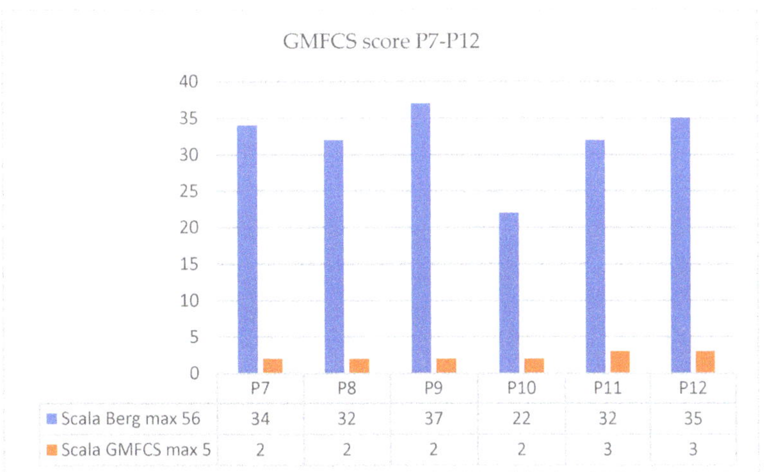

Figure 11. Evolution of GMFCS score patient P7–P12.

Table 5. Weight values and their distributions on each leg, at the two evaluations.

Patient Code	Total Weight (Kg)		Weight Distribution on Right Leg of Total Weight [%]		Weight Distribution on Right Leg of Total Weight [Kg]		Weight Distribution on Left Leg of Total Weight [%]		Weight Distribution on Left Leg of Total Weight [Kg]	
	EV1	EV2	EV1	EV2	EV1	EV2	EV1	EV2	EV1	EV2
P1	28.44	28.36	45	48	12.92	13.61	55	52	15.52	14.75
P2	11.72	12.93	41	47	4.84	6.08	59	53	6.88	6.85
P3	42.78	45.81	44	51	18.67	23.23	56	49	24.12	22.58
P4	12.76	15.33	52	51	6.59	7.82	48	49	6.18	7.51
P5	15.77	17.46	53	52	8.29	9.08	47	48	7.48	8.38
P6	19.37	23.65	42	48	8.05	11.35	58	52	11.31	12.30
P7	51.51	58.83	49	50	25.30	29.26	51	50	26.22	29.57
P8	57.35	64.37	44	47	25.50	30.26	56	53	31.84	34.12
P9	25.07	30.68	32	45	8.13	13.85	68	55	16.94	16.83
P10	22.67	24.07	57	53	12.87	12.75	43	47	9.80	11.31
P11	13.98	21.04	35	54	4.88	11.35	65	46	9.10	9.68
P12	42.74	43.80	43	47	18.41	20.54	57	53	24.32	23.26
Min	11.72	12.93	32.42	45.15	4.84	6.08	43.24	46.03	6.18	6.85
Max	57.35	64.37	56.76	53.97	25.50	30.26	67.58	54.85	31.84	34.12
AV *	28.68	32.19	44.74	49.37	12.87	15.77	55.26	50.63	15.81	16.43
SD **	15.95	17.11	7.06	2.78	7.46	8.14	7.06	2.78	8.80	9.04
CV ***	55.62	53.13	15.78	5.62	58.00	51.62	12.77	5.48	55.64	55.00
JB ****	0.521	0.504	0.908	0.685	0.532	0.502	0.908	0.685	0.553	0.510

* Average value. ** Standard deviation. *** CV is the coefficient of variation. **** Jarque-Bera test, calculated at the significance threshold of $p = 0.05$.

Analyzing Table 5, it is observed that all measured values have a normal distribution according to the Jarque-Bera test ($p \geq 0.05$, so H0 is confirmed).

In order to carry out a deeper analysis, we considered it useful to follow how the weight is distributed at the level of each leg, expressed as a percentage of the body weight.

3.2.1. Weight Distribution on the Left Leg

The distribution of the weight on the left leg shown in percentage in Figure 12, under the aspect of evolution between the two moments of the evaluation.

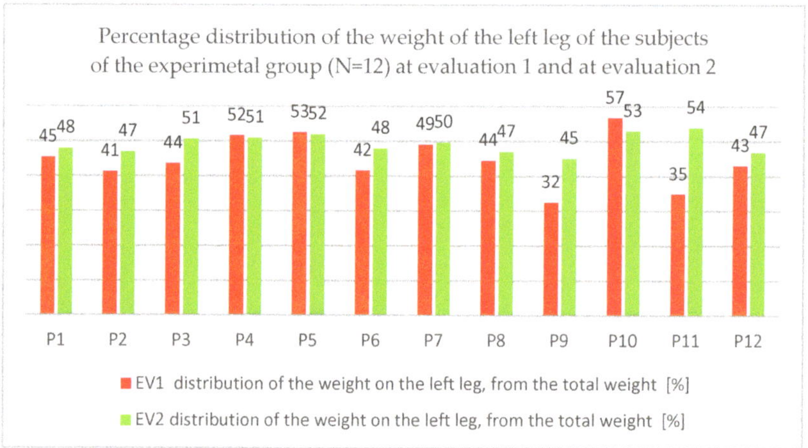

Figure 12. Percentage distribution of the weight on the left leg, from the total weight at the first and second evaluation of the subjects of the experimental group (N = 12).

An evolution of the load on the left leg is observed, towards 50% of the body weight.

Tables 6–8 show the statistical indicators that highlight the degree of percentage distribution at the level of the left leg, percentage of weight.

Table 6. The average value of the percentage distribution of the weight on the left leg.

Test	Average	SD	Minimum	Maximum	CV
EV1	55.26	7.06	43.24	67.58	12.77
EV2	50.63	2.78	46.03	54.85	5.48

Table 7. The table of statistical indicators of the difference in percentage distribution of weight on the left leg.

Statistical Indicators of the Resulting Difference (EV2-EV1)				Bilateral Dependent Student *t*-Test		
Average	Standard Deviation	95% Confidence Interval	Effect Size (Cohen D'test)	t obs	df	*p* *
4.64	6.29	(0.63; 8.63)	0.86	2.552	11	0.027

* The significance threshold is $p = 0.05$.

Table 8. Table of average differences of percentage weight distribution on the left leg.

Average Difference (EV1-EV2)	Progress	Difference Size	The Progress Is	Null Hypotesis (Averages Are Equal)
4.64	8.39%	Average	Statistically significant	It is rejected

The statistical analysis indicates a progress of 8.39%, with an average decrease of 4.64. The bilateral t-test reveals a statistically significant difference in averages, $p = 0.027 < 0.05$, for t_{obs} 2.552 and df = 11. A small effect size of 0.86 is observed.

3.2.2. Weight Distribution on the Right Leg

The same aspect of the percentage distribution of body weight at the level of the right leg can be found in Figure 13.

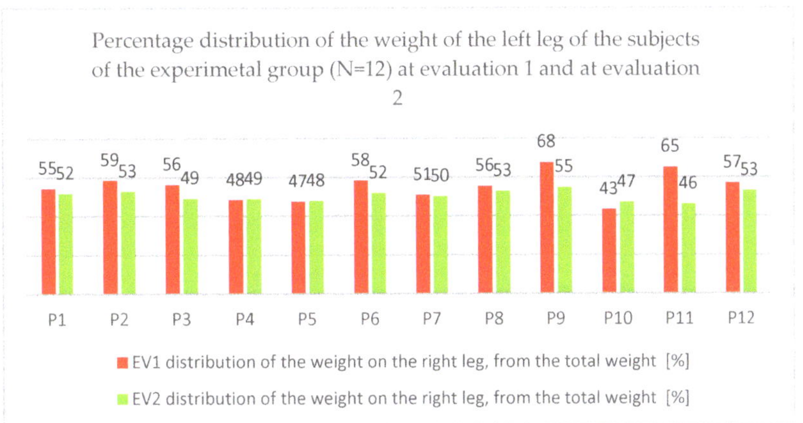

Figure 13. Percentage distribution of the weight on the right leg, from the total weight at the first and second evaluation of the subjects of the experimental group (N = 12).

The same evolution is observed in the sense of decreasing or increasing the distribution so that one can talk about an attempt to rebalance.

Tables 9–11 highlight the statistical indicators.

Table 9. The average value of the percentage distribution of the weight on the right leg.

Test	Average	SD	Minimum	Maximum	CV
EV1	44.74	7.06	32.42	56.76	15.78
EV2	49.37	2.78	45.15	53.97	5.62

Table 10. The table of statistical indicators of the difference in percentage distribution of weight on the right leg.

Statistical Indicators of the Resulting Difference (EV2-EV1)			Bilateral Dependent Student *t*-Test			
Average	Standard Deviation	95% Confidence Interval	Effect Size (Cohen D'test)	t Obs	df	*p* *
−4.64	6.29	(8.63; 6.37)	0.86	2.552	11	0.027

* The significance threshold is *p* = 0.05.

Table 11. Table of average differences of percentage weight distribution on the right leg.

Average Difference (EV1-EV2)	Progress	Difference Size	The Progress Is	Null Hypotesis (Averages Are Equal)
−4.64	10.36%	Big	Statistically significant	It is rejected

The statistical analysis indicates a progress of 10.36%, with an average decrease of 4.64. The bilateral t-test reveals a statistically significant difference in means, *p* = 0.027 < 0.05, for t_{obs} = 2.552 and df = 11. A small effect size of 0.86 is observed.

Analyzing the previous results, we find that there is a left-right rebalancing in most patients.

4. Discussion

The favorable results that were obtained by drawing up a physiotherapy program composed of the combination of the two Vojta and NDTBobath methods are proof of the fact that both methods are based on the creation of a stimulating peripheral pressure, which, if maintained, generates an extended stereotyped motor response. A pattern of symmetrical muscle contraction is thus created, at the level of the neck, trunk and limbs, as a result of the summation, and thus postural control can be achieved.

We found the clinical and functional expression of this aspect in the assessment of balance with the help of the Berg scale, which proves to be particularly useful in establishing the rehabilitation program, being an objective means of monitoring the evolution. This aspect of the importance of the Berg scale in the preparation of the balance rehabilitation program was also highlighted by Louie and colab [27] who used this scale in the admission stage of the stroke patient in the recovery program, considering that it is a predictive element of the evolution of balance and gait, which registered an important improvement 6–7 weeks after the initiation of the rehabilitation program. Thus, the authors consider that a patient who has a Berg score of at least 29 has a chance to regain his balance in 6–7 weeks of therapy. In the case of our study, we note that all subjects registered scores that exceed the value of 29 of the Berg scale score. Consequently, we can consider that the evolution of the subjects, objectified by this scale, confirms the fact that the complex therapeutic approach is effective and can be supported by concrete data.

The left-right rebalancing, proven by the percentage distribution analysis of the weight at the lower segmental level, demonstrated that the body alignment approach through the Vojta method on the one hand and the inhibitory facilitating postures/exercises promoted by the NDT Bobath method, allows obtaining a symmetry of the base and control of the trunk. It is known that there is a correlation between the asymmetry of the pelvis and the control of the trunk in children with PCI. The objectification of the effects of Vojta therapy was found in studies carried out in patients with stroke, in which it was observed that the Trunk Control Test (TCT) recorded an average value of 25 (Normal 0–43), compared to the result obtained in a group of patients who followed classical therapy and in which the average score was 46 [28].

We found in our research that there is a favorable evolution of weight distribution at the level of the two legs, with the achievement of a rebalancing with the increase of the load at the level of the right leg and the decrease at the level of the left leg, but what is more important is a closeness to the value of 50% within the rating from time EV2 as opposed to time EV1.

The effectiveness of the Vojta therapy was also reported by Hyungwon and Kim who noted positive results in the changes in each joint angle in the sagittal plane after the Vojta therapy. The conclusions of the study indicate that Vojta therapy can play an important role in improving the spatiotemporal parameters of the gait of children with spastic diplegia [22], this being a study that proves that through this approach an improvement in balance is obtained.

Proximo-distal stabilization in supine, prone, sitting and standing positions favors proprioceptive and vestibular stimulation.

Regarding the NDTBobath therapy, as I mentioned before, the objectification of the effectiveness of this therapy is minimal, the results of our study are supported by the study of Erdogan Kavlak (2018) who investigated the effects of this therapy after 8 weeks of application for the re-education of balance for children with CP [29]. The result was a significant improvement in balance, gross motor function and the level of functional independence.

NDTA study by Moazma Jamil compared effects of convention and NDT Bobath therapy to improve GMF among 24 children with CP. Children with CP received 3 months of intervention, 40 min/day, 5 days/week All children were tested with the Modified Ashworth Scale before starting treatment and 16 weeks after treatment. The result of the study showed that NDT Bobath therapy is more effective compared to conventional treatment [30]. The reduction of spasticity as well as the promotion of inhibitory reflex postures cause cortical activation, which, based on neuroplasticity, can trigger postural control and rehabilitation with recovery postural balance. This aspect was proven in our study by adjusting the left/right weight distribution, objectified by the percentage values of the loaNDT d.

The recovery process after brain injury is long, but with emerging evidence for neuroplasticity, the outlook for recovery is no longer so bleak. The exact mechanism remains unknown; however, many hypotheses are currently being investigated.

Zanon et.al. [31] in their paper consider that effect of NDT therapy is still uncertain and need more studies for assess the evolution of children with CP. Much more does not exist studies that reflects the results of NDT in balance rehabilitation.

Regarding Vojta therapy we found the paper of Mengibar et.al [32] that spoke about the role of Vojta therapy in accelerate the acquisition of GMFM-88-items and seems to activate the postural control. Furthermore, in this study this method seems to improve the balance in CP, but without biomechanical evaluation.

Our results demonstrate that there is a need to have a holistic approach of children with CP, means that also in clinical field the physiotherapist could apply combination of method for improve the movement and balance patterns. In the same time use of this evaluation, periodically could help the physician to monitoring the evolution and together with the physiotherapist design the rehabilitation goals.

We consider that these findings demonstrate the importance to use combination of the two techniques, but of course this combination could be applied to children with severed CP. For this reason, the physical therapist must have good knowledge about NDTBobath and Vojta methods. They have to be habilitated in field of these two methods and they will be able to design the interventions.

5. Conclusions

The training strategies promoted by the combination of the two methods are shown to be useful in the recovery program, which was reflected in the improvement of the Berg scale scores.

The results of the biomechanical evaluations complement the functional clinical evaluation and highlight as the main aspect the right recovery, which is in agreement with the Bobath method's objectives, namely the fact that the efficiency of the method is to make the patient maintain his balance and distribute his weight bilaterally by reducing right/left asymmetry.

At the level of the foot, this selective control of the movement is very strong, a fact also demonstrated by the loading mode of the foot in the studied patients. We can talk about an element of prediction in terms of estimating the effectiveness of the therapeutic program.

The effect of the combined therapy, at the level of the foot that presents a behavior that seems to prepare the dynamic balance.

Left-right weight transfer must be carried out while maintaining postural control.

6. Limitations

The study has limitations due to the small number of research subjects. In addition, during the moment of measurements some of children have not full cooperation. The same could be say regarding the participation to the physical therapy program and how the children answer to the physical therapy tasks.

Author Contributions: A.U.: data collection, and methodology; L.R.: conceptualization, methodology, writing—original draft preparation, analysis the results; M.R.R. data collect, translation, writing the draft; M.I.M.: methodology, software, and visualization. All authors have read and agreed to the published version of the manuscript.

Funding: This research received no external funding.

Institutional Review Board Statement: The study was conducted in accordance with the Declaration of Helsinki and approved by the Ethics Committee of the University of Craiova Research Centre for Human body motricity, study nr. 1540/1.10/2021.

Informed Consent Statement: Informed consent was obtained from all subjects involved in the study.

Data Availability Statement: Not applicable.

Conflicts of Interest: The authors declare no conflict of interest.

References

1. Puderbaugh, M.; Emmady, P.D. Neuroplasticity. In *StatPearls*; StatPearls Publishing: Treasure Island, FL, USA, 2022.
2. Nettle, D.; Bateson, M. Adaptive developmental plasticity: What is it, how can we recognize it and when can it evolve? *Proc. R. Soc.* **2015**, *282*, 20151005. [CrossRef] [PubMed]
3. Faingold, C.L. Chapter 7-Network Control Mechanisms: Cellular Inputs, Neuroactive Substances, and Synaptic Changes. In *Hal Blumenfeld, Neuronal Networks in Brain Function, CNS Disorders, and Therapeutics*; Faingold, C.L., Ed.; Academic Press: Cambridge, MA, USA, 2014; pp. 91–101; ISBN 9780124158047. [CrossRef]
4. Yang, J.F.; Livingstone, D.; Brunton, K.; Kim, D.; Lopetinsky, B.; Roy, F.; Gorassini, M. Training to enhance walking in children with cerebral palsy: Are we missing the window of opportunity? *Semin. Pediatr. Neurol.* **2013**, *20*, 106–115. [CrossRef]
5. Nielsen, J.B. How we walk: Central control of muscle activity during human walking. *Neuroscientist* **2003**, *9*, 195–204. [CrossRef] [PubMed]
6. Jang, S.H.; Kwon, H.G. Delayed gait recovery with recovery of an injured corticoreticulospinal tract in a chronic hemiparetic patient: A case report. *Medicine* **2016**, *95*, e5277. [CrossRef] [PubMed]
7. Trompetto, C.; Marinelli, L.; Mori, L.; Pelosin, E.; Currà, A.; Molfetta, L.; Abbruzzese, G. Pathophysiology of Spasticity: Implications for Neurorehabilitatio. *BioMed Res. Int.* **2014**, *8*, 354906. [CrossRef] [PubMed]
8. Johnston, M.V.; Hoon, A.H., Jr. Cerebral palsy. *Neuromolecular Med.* **2006**, *8*, 435–450. [CrossRef]
9. Abo, M.; Kakuda, W. *Rehabilitation with Rtms*; Springer: New York, NY, USA, 2015; 208p.
10. Bhinde, S.; Patel, K.; Kori, V.; Rajagopala, S. Management of spastic cerebral palsy through multiple Ayurveda treatment modalities. *AYU Int. Q. J. Res. Ayurveda* **2014**, *35*, 462. [CrossRef]
11. Levin, H.S.; Song, J.; Ewing-Cobbs, L.; Chapman, S.B.; Mendelsohn, D. Word fluency in relation to severity of closed head injury, associated frontal brain lesions, and age at injury in children. *Neuropsihologia* **2001**, *39*, 122–131. [CrossRef]
12. Bobath, K.; Bobath, B. The Neuro-Developmental Treatment of Cerebral Palsy. *Phys. Ther.* **1967**, *47*, 1039–1043. [CrossRef]
13. Hyungwon Lim, T.K. Effects of Vojta Therapy on Gait of Children with Spastic Diplegia. *J. Phys. Ther. Sci.* **2013**, *25*, 1605–1608. [CrossRef]
14. Marquez de la Plata, C.D.; Hart, T.; Hammond, F.M.; Frol, A.B.; Hudak, A.; Harper, C.R.; O'Neil-Pirozzi, T.M.; Whyte, J.; Carlile, M.; Diaz-Arrastia, R. Impact of age on long-term recovery from traumatic brain injury. *Arch. Phys. Med. Rehabil.* **2008**, *89*, 896–903. [CrossRef] [PubMed]
15. Beckung, E.; Carlsson, G.; Carlsdotter, S.; Uvebrant, P. The natural history of gross motor development in children with cerebral palsy aged 1 to 15 years. *Dev. Med. Child. Neurol.* **2007**, *49*, 751–756. [CrossRef] [PubMed]
16. Berger, M.S.; Pitts, L.H.; Lovely, M.; Edwards, M.S.B.; Bartkowski, H.M. Outcome from severe head injury in children and adolescents. *J. Neurosurg.* **1985**, *62*, 194–199. Available online: https://thejns.org/view/journals/j-neurosurg/62/2/article-p194.xml (accessed on 4 September 2022). [CrossRef]
17. Rain, S. The Current theoretical assumptions of the bobath concept as determined by the members of BBTA. *Psysiotherapy Theory Pract.* **2007**, *23*, 137–157. [CrossRef] [PubMed]
18. Dos Santos, G. Humeral external rotation handling by using the Bobath concept approach affects trunk extensor muscles electromyography in children with cerebral palsy. *Reserch Dev. Disabil.* **2014**, *20*, 134–141. [CrossRef]
19. Rogers, B.; Msall, M.; Owens, T.; Guernsey, K.; Brody, A.; Buck, G.; Hudak, M. Cystic periventricular leukomalacia and type of cerebral palsy in preterm infants. *J. Pediatr.* **1994**, *125*, S1–S8. [CrossRef]
20. Al Tawil, K.I.; El Mahdy, H.S.; Al Rifai, M.T.; Tamim, H.M.; Ahmed, I.A.; Al Saif, S.A. Risk Factors for Isolated Periventricular Leukomalacia. *Pediatr. Neurol.* **2012**, *46*, 149–153. [CrossRef]
21. Ment, L.R.; Hirtz, D.; Hüppi, P.S. Imaging biomarkers of outcome in the developing preterm brain. *Lancet Neurol.* **2009**, *8*, 1042–1055. [CrossRef]
22. Nanjundagowda, V.K. *Cerebral Palsy Early Stimulation*; Jaypee Brothers Medical Publishers (P) Ltd.: New Delhi, India, 2014; pp. 90–145. ISBN 10: 9350903016/13: 9789350903018.
23. Berg, K.; Wood-Dauphinėe, S.; Williams, J.I.; Gayton, D. Measuring balance in the elderly: Preliminary development of an instrument. *Physiother. Can.* **1989**, *41*, 304–311. [CrossRef]
24. Caṭan, L.; Cerbu, S.; Amaricai, E.; Suciu, O.; Horhat, D.I.; Popoiu, C.M.; Adam, O.; Boia, E. Assessment of Static Plantar Pressure, Stabilometry, Vitamin D and Bone Mineral Density in Female Adolescents with Moderate Idiopathic Scoliosis. *Int. J. Environ. Res. Public Health* **2020**, *17*, 2167. [CrossRef]

25. Vincent, W.; Weir, J. Statistics in Kinesiology. In *Human Kinetics*, 4th ed.; NRC Research Press: Champaign, IL, USA, 2012; ISBN 13 978-1-4504-0254-5. Available online: www.HumanKinetics.com (accessed on 4 September 2022).
26. Addinsoft. *XLSTAT Statistical and Data Analysis Solution*; Addinsoft: New York, NY, USA, 2022; Available online: https://www.xlstat.com/en (accessed on 4 September 2022).
27. Louie, D.R.; Eng, J.J. Berg balance scale score at admission can predict walking suitable for community ambulation at discharge from inpatient stroke rehabilitation. *J. Rehabil. Med.* **2018**, *50*, 37–44. [CrossRef] [PubMed]
28. Epple, C.; Maurer-Burkhard, B.; Lichti, M.C.; Steiner, T. Vojta therapy improves postural control in very early stroke rehabilitation: A randomised controlled pilot trial. *Neurol. Res. Pract.* **2020**, *2*, 23. [CrossRef] [PubMed]
29. Kavlak, E. Effectiveness of Bobath therapy on balance in cerebral palsy. *Cukurova Med. J.* **2018**, *43*, 975–981. [CrossRef]
30. Moazma, J.S. Effectiveness of Bobath and conventional tratment in cerebral palsy children. *Rawal. Med. J.* **2020**, *45*, 974–976. Available online: https://www.rmj.org.pk/fulltext/27-1563944706.pdf (accessed on 4 September 2022).
31. Zanon, M.A.; Pacheco, R.L.; Latorraca, C.O.C.; Martimbianco, A.L.C.; Pachito, D.V.; Riera, R. Neurodevelopmental Treatment (Bobath) for Children with Cerebral Palsy: A Systematic Review. *J. Child. Neurol.* **2019**, *34*, 679–686. [CrossRef] [PubMed]
32. Sanz-Mengibar, J.M.; Menendez-Pardiñas, M.; Santonja-Medina, F. Is the implementation of Vojta therapy associated with faster gross motor development in children with cerebral palsy? *Clin. Neurosci.* **2021**, *74*, 329–336. [CrossRef]

Article

Physical Activity among U.S. Preschool-Aged Children: Application of Machine Learning Physical Activity Classification to the 2012 National Health and Nutrition Examination Survey National Youth Fitness Survey

Soyang Kwon [1,*], Megan K. O'Brien [2], Sarah B. Welch [3] and Kyle Honegger [4]

1   Department of Pediatrics, Northwestern University, 225 E Chicago Ave. Box 157, Chicago, IL 60611, USA
2   Shirley Ryan Ability Lab, Department of Physical Medicine and Rehabilitation, Northwestern University, 710 N. Lake Shore Drive, Chicago, IL 60611, USA
3   Buehler Center for Health Policy and Economics, Northwestern University, 420 E Superior St., Chicago, IL 60611, USA
4   Ann & Robert H Lurie Children's Hospital of Chicago, 225 E Chicago Ave. Box 157, Chicago, IL 60611, USA
*   Correspondence: skwon@luriechildrens.org

Abstract: Early childhood is an important development period for establishing healthy physical activity (PA) habits. The objective of this study was to evaluate PA levels in a representative sample of U.S. preschool-aged children. The study sample included 301 participants (149 girls, 3–5 years of age) in the 2012 U.S. National Health and Examination Survey National Youth Fitness Survey. Participants were asked to wear an ActiGraph accelerometer on their wrist for 7 days. A machine learning random forest classification algorithm was applied to accelerometer data to estimate daily time spent in moderate- and vigorous-intensity PA (MVPA; the sum of minutes spent in running, walking, and other moderate- and vigorous-intensity PA) and total PA (the sum of MVPA and light-intensity PA). We estimated that U.S. preschool-aged children engaged in 28 min/day of MVPA and 361 min/day of total PA, on average. MVPA and total PA levels were not significantly different between males and females. This study revealed that U.S. preschool-aged children engage in lower levels of MVPA and higher levels of total PA than the minimum recommended by the World Health Organization.

Keywords: ActiGraph accelerometer; machine learning; walking; early childhood

Citation: Kwon, S.; O'Brien, M.K.; Welch, S.B.; Honegger, K. Physical Activity among U.S. Preschool-Aged Children: Application of Machine Learning Physical Activity Classification to the 2012 National Health and Nutrition Examination Survey National Youth Fitness Survey. *Children* **2022**, *9*, 1433. https://doi.org/10.3390/children9101433

Academic Editors: Georgian Badicu, Ana Filipa Silva and Hugo Miguel Borges Sarmento

Received: 19 August 2022
Accepted: 19 September 2022
Published: 21 September 2022

Publisher's Note: MDPI stays neutral with regard to jurisdictional claims in published maps and institutional affiliations.

Copyright: © 2022 by the authors. Licensee MDPI, Basel, Switzerland. This article is an open access article distributed under the terms and conditions of the Creative Commons Attribution (CC BY) license (https://creativecommons.org/licenses/by/4.0/).

1. Introduction

Early childhood is an important developmental period for establishing healthy physical activity (PA) habits [1]. In early childhood, PA is particularly important to develop gross motor (GM) skills and acquire movement proficiency [2]. Studies showed that moderate- and vigorous-intensity PA (MVPA) was associated with GM competency among preschool-aged children aged 3–5 years [3,4]. However, according to previous studies [5–9], a large proportion of U.S. preschool-aged children do not engage in recommended levels of PA. Further, a review of publications from 1980 to 2007 [8] and some later studies [5–7], but not all [10], have reported that preschool-aged females in particular are less active than preschool-aged males. The World Health Organization (WHO) recently established new PA guidelines for children under 5 years of age [11]. For children 3–4 years of age, the PA guidelines recommend at least 180 min/day of PA at any intensity, of which at least 60 min should be MVPA. To evaluate whether PA health behaviors of U.S. preschool-aged children meet these guidelines, it is critical to collect PA data among preschool-aged children at a national level.

The 2011 U.S. National Health and Examination Survey National Youth Fitness Survey (NNYFS) provides an opportunity to evaluate PA and GM development among U.S. preschool-aged children. The NNYFS conducted accelerometer assessments and the Test of GM Development-2 (TGMD-2) in a representative sample of U.S. children including preschool-aged children. Processed accelerometer data from the NNYFS became publicly available in Monitor-Independent Movement Summary (MIMS) units in 2020. From a previous analysis of the MIMS data from the 2011 NNYFS, we found that daily accumulated MIMS were not different between males and females in U.S. preschool-aged children [12]. However, no PA classification algorithms were available for analyzing MIMS data. Thus, we were unable to evaluate how much time preschool-aged children spent in MVPA and total PA (the sum of light-intensity PA [LPA] and MVPA), which limited our ability to use the MIMS data from the NNYFS to inform public health messages and actions. More recently, NNYFS raw accelerometer data have become publicly available, allowing researchers to apply validated PA classification algorithms.

Using the NNYFS data, we also reported that one in three preschool-aged children (33.9%) scored below average on the overall GM level [13]. Of the two GM subsets (locomotion and object control), locomotor competency was shown to be associated with self-reported participation in specific types of PA (i.e., bike riding, scooter riding, trampoline, soccer, and swimming) [13,14]. However, our prior study [12] that used an overall PA measure (MIMS) that encompassed LPA and MVPA failed to show its positive association with locomotor competency.

This present paper aims to expand on the findings from our prior studies [12,13], by utilizing a machine learning PA classification algorithm to analyze NNYFS raw accelerometer data. Machine learning is a type of artificial intelligence (AI) that provides the ability to automatically learn/find patterns of complex data from a large amount of data. The machine learning algorithm was shown to have higher accuracy than traditional cut-point methods [15]. The primary aim of this study was to report MVPA and total PA levels in a representative sample of U.S. preschool-aged children, using NNYFS raw accelerometer data. The secondary aim was to examine the associations of MVPA and LPA with GM skills among preschool-aged child participants.

2. Materials and Methods

2.1. Study Sample

The study sample included U.S. children 3–5 years of age who participated in the 2012 NNYFS. The 2012 NNYFS was a cross-sectional survey that collected data on PA and fitness levels to provide an evaluation of the health and fitness of children 3–15 years of age living in the U.S. The NNYFS used complex staged and stratified sampling methods to select a representative sample of U.S. children 3–5 years of age (n = 368). Accelerometer data for NNYFS participants were collected using wrist-worn accelerometers; thus, 16 participants with missing arms bilaterally were excluded from accelerometer assessment. Ethical approval and consent were obtained from participants' parents and guardians. Of 352 eligible participants, 301 (149 females) had accelerometer data for at least 1 day and were included in this analysis. Our prior study [12] reported no difference in participant characteristics between those included and excluded, except that the excluded participants tended to be younger.

2.2. Accelerometer Assessment

ActiGraph GT3X+ accelerometers (Pensacola, FL, USA) were used to assess PA at 80 Hz. During a mobile examination center visit, participants were asked to wear an accelerometer wristband (dorsal orientation) on the non-dominant hand for 9 consecutive days (including 7 full days). Participants returned the accelerometer wristband via mail using pre-paid padded envelopes.

Publicly available raw accelerometer data files were downloaded, from which data for participants 3–5 years of age were extracted. The data cleaning process was identical to that described in our prior study [12]. Briefly, we excluded data collected on day 1 (when the accelerometer was placed on the participant) and day 9 (when the accelerometer was removed from the participant, which occurred in the morning). We further excluded data collected between 10 PM and 6 PM (nighttime sleep period), which resulted in a maximum of 16 h of accelerometer data per full day of wear [16]. To be considered as a valid wear day, each day should have ≥600 wear minutes. Detailed information about the wear/non-wear detection algorithm can be found in the NNYFS Analytic Notes [17]. We included any participants who had ≥1 valid wear day for data analysis [12].

To process and analyze the wrist accelerometer data, we applied a validated random forest PA classification algorithm developed for preschool-aged children by Trost and the colleagues [15,18]. The algorithm development and validation have been described in prior publications [15,18]. Briefly, recognizing the low performance of laboratory-trained PA classifiers for free-living data [19], random forest PA classifiers were trained using free-living accelerometer data among preschool-aged children. Tri-axial accelerometer signals were transformed into a single-dimension vector magnitude, and time and frequency domain (base) features as well as temporal features were used to train random forest classifiers with various window sizes: 1, 5, 10, and 15 s [18]. Among the wrist random forest classifiers trained, a 15-s window classifier with both base and temporal (base + temporal) features was reported to have the highest performance for predicting five activity classes: "run", "walk", "other MVPA", 'LPA', and "sedentary" (weight average F-score = 81%) [18]. Therefore, we selected the 15-s base + temporal feature classifier to analyze NNYFS wrist accelerometer data.

The NNYFS wrist accelerometer data were segmented into 15-s non-overlapping windows. Feature extraction was performed using the R script obtained from the model developer [18]. Using the R script for the 15-s base + temporal feature classifier (https://github.com/QUTcparg/PS_PAClassification; accessed on 20 September 2022), each of the 15-s windows was predicted as "run", "walk", "other MVPA", 'LPA', or "sedentary". We calculated the number of windows for each activity class a day. The number of windows for each activity class was divided by 4 to express the time spent in each activity in minutes a day. For example, 40 15-s windows predicted as "walk" a day were converted to 10 min of "walk" a day. Multiple days of estimated time spent in each activity were averaged per participant. MVPA (minutes/day) was calculated by summing minutes spent in "run", "walk", and "other MVPA". Total PA (minutes/day) was calculated by summing time spent in MVPA and LPA. Participants who engaged in MVPA ≥ 60 min/day were considered to have met the WHO MVPA recommendation. Similarly, participants with total PA ≥ 180 min/day were considered to have met the WHO total PA recommendation.

2.3. Gross Motor Assessment

GM skills were evaluated using the TGMD-2 [20]. The TGMD-2 is a widely accepted tool to evaluate GM among young children that includes two GM subtests: locomotor and object control. The detailed TGMD-2 assessment protocol is described in a prior publication [13] as well as on the NYFS website [21]. In accordance with the TGMD-2 manual [20], a locomotor standard score (range: 1 to 20) and an object control standard score (range: 1 to 20) were calculated.

2.4. Statistical Analyses

We used SAS 9.4 (Cary, NC, USA) for data analyses. We incorporated the complex sampling design of the NNYFS in analyses. To achieve the primary aim, we calculated the means and 95% confidence intervals (CIs) of the PA metrics. We repeated this analysis, separately, for participant characteristics, such as biological sex, age (3, 4, or 5 years), race and ethnicity (Hispanic, non-Hispanic White, non-Hispanic Black, or Other), language spoken at home (English only or at least some non-English), family income levels (ratio of family income to poverty, reported as <1.0 [below the poverty level], 1.0 to <3.0, or ≥3.0) [12]. A correlation coefficient (r) between MVPA and LPA was calculated. Frequency analyses were conducted to identify the proportion of children meeting the WHO MVPA and total PA recommendations.

To achieve the secondary aim, we conducted multivariable linear regression analyses. Independent variables of interest included MVPA and LPA. Dependent variables included locomotor and object control standard scores. Covariates were selected based on our prior analysis [13]: sex, age, family income, and living with a child(ren) $\leq$5 years of age for the locomotor outcome; and sex, age, and living with a child(ren) 6–17 years of age for the object control outcome.

3. Results

Among 301 participants, accelerometer wear time was on average 915 min (standard error: 7 min). Wear time did not differ by sex, age, or racial/ethnic group. As shown in Table 1, U.S. preschool-aged children were estimated to run for 4 min/day (95% CI = 3, 5) and walk for 14 min/day (95% CI = 13, 15) on average. MVPA was estimated at 28 min/day (95% CI = 25, 30) and LPA was estimated at 361 min/day (95% CI = 343, 378). MVPA and total PA tended to be higher in older preschool-aged children. MVPA and total PA levels did not differ by sex, racial and ethnic group (Table 1), language spoken at home, or family income (Supplementary Table S1). The correlation coefficient r between MVPA and LPA was 0.53 ($p < 0.01$). Only 2% of participants met the WHO MVPA recommendation, while 95% met the WHO total PA recommendation.

Table 1. Estimated daily minutes spent in physical activity in U.S. preschool-aged children. The 2012 NNYFS.

	Run	Walk	Other MVPA	LPA	MVPA [a]	Total PA [b]
	Weighted mean (95% CI), minutes/day					
All (n = 301)	4 (3, 5)	14 (13, 15)	10 (8, 11)	361 (343, 378)	28 (25, 30)	388 (368, 408)
	Sex					
Male (n = 152)	5 (4, 6)	14 (12, 16)	10 (9, 12)	361 (334, 387)	29 (25, 33)	389 (359, 419)
Female (n = 149)	3 (2, 4)	15 (13, 16)	9 (7, 10)	360 (342, 379)	27 (24, 29)	387 (367, 407)
	Age					
3 years (n = 96)	3 (2, 4)	13 (10, 16)	8 (6, 10)	333 (300, 366)	24 (19, 29)	357 (320, 393)
4 years (n = 101)	4 (3, 5)	14 (12, 16)	10 (6, 12)	370 (346, 394)	28 (25, 31)	398 (372, 424)
5 years (n = 104)	4 (3, 5)	16 (14, 18)	11 (9, 12)	376 (361, 392)	31 (27, 34)	407 (389, 425)
	Race/ethnicity					
Non-Hispanic White (n = 121)	4 (2, 5)	13 (12, 14)	9 (8, 11)	362 (336, 388)	26 (23, 29)	388 (359, 416)
Hispanic (n = 102)	4 (3, 5)	16 (13, 20)	10 (8, 12)	364 (340, 387)	31 (26, 35)	394 (368, 421)
Non-Hispanic Black (n = 61)	4 (3, 5)	15 (13, 17)	11 (9, 12)	368 (342, 395)	29 (26, 33)	398 (368, 427)
Other (n = 17)	4 (2, 5)	13 (7, 19)	8 (5, 12)	320 (275, 364)	25 (15, 35)	345 (292, 398)

Physical activity types were classified based on the random forest activity classification algorithm by Ahmed and Trost [15]. [a] Sum of run, walk, other MVPA. [b] Sum of MVPA and LPA. CI, confidence interval; LPA, light-intensity physical activity; MVPA, moderate- and vigorous-intensity physical activity; PA, physical activity.

Table 2 presents the multivariable linear regression analysis results. An additional 10 min spent in MVPA was significantly associated with a 0.7-point higher locomotor standard score (95% CI = 0.4, 1.0). However, LPA was not statistically significantly associated with a locomotor standard score. Similarly, an additional 10 min spent in MVPA was significantly associated with a 0.4-point higher object control standard score (95% CI = 0.04, 0.7). However, LPA was not statistically significantly associated with an object control standard score.

Table 2. Multivariable linear regression models for locomotor and object control standard scores in U.S. preschool-aged children. The 2012 NNYFS.

Predictors	Locomotor Standard Score	Object Control Standard Score
	Estimate (95% CI)	Estimate (95% CI)
Intercept	7.6 (6.1, 9.1)	8.2 (6.6, 9.9)
Sex: female vs. male	1.3 (0.7, 2.0)	0.1 (−0.6, 0.8)
Age: 4 vs. 3 years	0.8 (0.1, 1.5)	−0.3 (−1.1, 0.4)
Age: 5 vs. 3 years	0.4 (−0.4, 1.2)	−0.01 (−0.9, 0.9)
Ratio of family income to poverty: 1.0 to <3.0 vs. ≥3.0	0.7 (−0.3, 1.7)	NA [a]
Ratio of family income to poverty: <1.0 vs. ≥3.0	0.3 (−0.6, 1.1)	NA [a]
Living with child(ren) ≤5 years of age: yes vs. no	0.3 (−1.1, 1.6)	NA [a]
Living with child(ren) 6–17 years of age: yes vs. no	NA [a]	0.5 (−0.3, 1.2)
Additional 1 h/day of LPA	−0.2 (−0.5, 0.1)	−0.1 (−0.5, 0.2)
Additional 10 min/day of MVPA	0.7 (0.4, 1.0)	0.4 (0.04, 0.7)

[a] Not applicable. CI, confidence interval; LPA, light-intensity physical activity; MVPA, moderate- and vigorous-intensity physical activity; NA, not applicable; OR, odds ratio.

4. Discussion

This study utilized a recently validated machine learning algorithm to estimate free-living PA levels among preschool-aged children. Applying the algorithm, this study identified an estimated national MVPA level among preschool-aged children of 28 min/day, which is much lower than the WHO recommendation of 60 min/day, and an estimated total PA level of 361 min/day, which is much higher than the WHO recommendation of 180 min/day. This study also confirmed that MVPA and LPA levels were not different between preschool-aged males and females. MVPA, but not LPA, was associated with better scores on locomotor and object control skills.

This study was innovative in the application of an accelerometer-based machine learning PA classification algorithm that is more accurate (weighted kappa = 0.72) than available cut-point algorithms (weighted kappa = 0.31–0.44) [15]. Accelerometer activity monitors have vastly facilitated the evolution of child PA research over the past three decades, allowing researchers to objectively assess free-living PA levels. Using limited raw data, researchers have established accelerometer count cut-points to define PA intensities/classes (i.e., sedentary, light, moderate, and vigorous). However, significant limitations of the cut-point approach, including low accuracy and the inability to detect activity type, have also been recognized. To address these limitations, a new analytic approach, machine learning, has been adopted in PA measurement research. Machine learning can process abundant raw tri-axial acceleration signal data for pattern recognition. To date, the machine learning PA classifier by Trost and the colleagues [15,18] is the most rigorously validated accelerometer-based algorithm for preschool-aged children, which was utilized in this study to estimate PA levels among preschool-aged children.

This is one of the first studies to report key PA indicators among U.S. preschool-aged children at the national level. Our national estimation of 28 min/day MVPA indicates that many U.S. preschool-aged children do not sufficiently engage in MVPA. However, in interpreting the study results, caution should be taken to consider the direction (i.e., over- or under-estimation) and the size of potential misclassification errors. In a study by Ahmadi and Trost [15], the confusion matrix of the PA classifier indicated that the machine learning algorithm correctly classified 87.5% of LPA and misclassified only 4.7% of LPA as sedentary behavior and only 7.8% of LPA as MVPA. However, a large proportion of MVPA (31.4%) was misclassified as LPA [15]. Altogether, the algorithm slightly over-estimated LPA (140 min by prediction vs. 132 min by the ground truth) and slightly under-estimated MVPA (42 min by prediction vs. 47 min by the ground truth). Therefore, we assume that our MVPA estimate may be underestimated, and LPA estimate may be overestimated.

It is an important finding that MVPA, but not LPA, was positively associated with GM skills. For GM development and other health benefits, MVPA should be encouraged among preschool-aged children. However, it is concerning that many preschool-aged children do not engage in sufficient levels of MVPA. Our evaluation revealed that preschool-aged children on average walked or ran for 18 min a day and engaged in MVPA for 28 min a day. We suggest that US preschool-aged children should be offered greater opportunities to engage in specific PA types that help GM development, such as bike riding, scooter riding, trampoline, soccer, and swimming [13,14]. In addition, although there are no PA guidelines specific to walking and running activities, we suggest that US preschool-aged children should be encouraged to engage in more walking and running activities throughout the day. Spending more time outdoors [12] and increasing opportunities for walking/running, as opposed to being restrained in a stroller or a car [11], would likely increase levels of MVPA to align with WHO recommendations.

The limitation of this study includes that the cross-sectional examination for a relationship between MVPA and GM skills cannot establish a temporal relationship; thus, we are unable to determine whether limited engagement in MVPA hinders GM development or limited GM competency restricts participation in MVPA. Second, PA levels among preschool-aged children may have changed over the past decade since 2012. In particular, child PA may have decreased during the COVID-19 pandemic [22]. Therefore, our PA estimation based in 2012 data may not accurately reflect the current status of PA among U.S. preschool-aged children. Nonetheless, this study used the most recent available accelerometer data to evaluate key PA indicators among U.S. preschool-aged children at a national level.

5. Conclusions

Our evaluation of PA among preschool-aged children using the most accurate PA classification approach available showed that on average, U.S. preschool-aged children engaged in 28 min/day of MVPA and 361 min/day of total PA. We also found that time spent in MVPA, but not in LPA, was positively associated with GM skills.

Supplementary Materials: The following supporting information can be downloaded at: https://www.mdpi.com/article/10.3390/children9101433/s1, Table S1: Estimated time spent in various types of physical activity in a representative sample of U.S. preschool-aged children.

Author Contributions: S.K. conceived the study, analyzed and interpreted the data, and drafted the manuscript. K.H. analyzed data and critically reviewed the manuscript. M.K.O. contributed to data analysis and critically reviewed the manuscript. S.B.W. critically reviewed the manuscript. All authors have read and agreed to the published version of the manuscript.

Funding: This work was partly funded by National Institutes of Health (grant number: R01HL155113).

Institutional Review Board Statement: Ethical review and approval were waived for this study due to use of de-identified data for this secondary analysis.

Informed Consent Statement: Informed consent was obtained from all subjects involved in the study.

Data Availability Statement: The data that support the findings of this study are available from https://www.cdc.gov/nchs/nnyfs/index.htm (accessed on 20 September 2022).

Conflicts of Interest: The authors declare no conflict of interest.

References

1. Goldfield, G.S.; Harvey, A.; Grattan, K.; Adamo, K.B. Physical activity promotion in the preschool years: A critical period to intervene. *Int. J. Environ. Res. Public Health* **2012**, *9*, 1326–1342. [CrossRef] [PubMed]
2. Gallahue, D.; Ozmun, J.; Goodway, J. *Understanding Motor Development: Infants, Children, Adolescents, Adults*, 7th ed.; McGraw-Hill: Boston, MA, USA, 2012.
3. Fisher, A.; Reilly, J.J.; Kelly, L.A.; Montgomery, C.; Williamson, A.; Paton, J.Y.; Grant, S. Fundamental movement skills and habitual physical activity in young children. *Med. Sci. Sports Exerc.* **2005**, *37*, 684–688. [CrossRef] [PubMed]
4. Foweather, L.; Knowles, Z.; Ridgers, N.D.; O'Dwyer, M.V.; Foulkes, J.D.; Stratton, G. Fundamental movement skills in relation to weekday and weekend physical activity in preschool children. *J. Sci. Med. Sport* **2015**, *18*, 691–696. [CrossRef] [PubMed]
5. Hinkley, T.; Salmon, J.; Okely, A.D.; Crawford, D.; Hesketh, K. Preschoolers' physical activity, screen time, and compliance with recommendations. *Med. Sci. Sports Exerc.* **2012**, *44*, 458–465. [CrossRef] [PubMed]
6. Beets, M.W.; Bornstein, D.; Dowda, M.; Pate, R.R. Compliance with national guidelines for physical activity in U.S. preschoolers: Measurement and interpretation. *Pediatrics* **2011**, *127*, 658–664. [CrossRef] [PubMed]
7. Pate, R.R.; O'Neill, J.R.; Brown, W.H.; Pfeiffer, K.A.; Dowda, M.; Addy, C.L. Prevalence of compliance with a new physical activity guideline for preschool-age children. *Child Obes.* **2015**, *11*, 415–420. [CrossRef] [PubMed]
8. Hinkley, T.; Crawford, D.; Salmon, J.; Okely, A.D.; Hesketh, K. Preschool children and physical activity: A review of correlates. *Am. J. Prev. Med.* **2008**, *34*, 435–441. [CrossRef] [PubMed]
9. Kwon, S.; Mason, M.; Becker, A. Environmental factors associated with child physical activity at childcare. *Health Behav. Policy Rev.* **2015**, *2*, 260–267. [CrossRef]
10. Bergqvist-Norén, L.; Hagman, E.; Xiu, L.; Marcus, C.; Hagströmer, M. Physical activity in early childhood: A five-year longitudinal analysis of patterns and correlates. *Int. J. Behav. Nutr. Phys. Act.* **2022**, *19*, 47. [CrossRef] [PubMed]
11. World Health Organization. *Guidelines on Physical Activity, Sedentary Behaviour and Sleep for Children under 5 Years of Age*; World Health Organization: Geneva, Switzerland, 2019.
12. Kwon, S.; Tandon, P.; O'Neill, M.; Becker, A. Cross-sectional association of light sensor-measured time outdoors with physical activity and gross motor competency among U.S. preschool-aged children: The 2012 NHANES National Youth Fitness Survey. *BMC Public Health* **2022**, *22*, 833. [CrossRef] [PubMed]
13. Kwon, S.; O'Neill, M. Socioeconomic and familial factors associated with gross motor skills among US children aged 3–5 years: The 2012 NHANES National Youth Fitness Survey. *Int. J. Environ. Res. Public Health* **2020**, *17*, 4491. [CrossRef] [PubMed]
14. Wood, A.P.; Imai, S.; McMillan, A.G.; Swift, D.; DuBose, K.D. Physical activity types and motor skills in 3–5-year old children: National Youth Fitness Survey. *J. Sci. Med. Sport* **2020**, *23*, 390–395. [CrossRef] [PubMed]
15. Ahmadi, M.N.; Trost, S.G. Device-based measurement of physical activity in pre-schoolers: Comparison of machine learning and cut point methods. *PLoS ONE* **2022**, *17*, e0266970. [CrossRef] [PubMed]
16. Dias, K.I.; White, J.; Jago, R.; Cardon, G.; Davey, R.; Janz, K.F.; Pate, R.R.; Puder, J.J.; Reilly, J.J.; Kipping, R. International comparison of the levels and potential correlates of objectively measured sedentary time and physical activity among three- to four-year-old children. *Int. J. Environ. Res. Public Health* **2019**, *16*, 1929. [CrossRef] [PubMed]
17. National Center for Health Statistics. NHANES National Youth Fitness Survey 2012 Data Documentation, Codebook, and Frequencies 2020. Available online: https://wwwn.cdc.gov/Nchs/Nnyfs/Y_PAXMIN.htm#Analytic_Notes (accessed on 20 September 2022).
18. Ahmadi, M.N.; Pavey, T.G.; Trost, S.G. Machine learning models for classifying physical activity in free-living preschool children. *Sensors* **2020**, *20*, 4364. [CrossRef] [PubMed]
19. Ahmadi, M.N.; Brookes, D.; Chowdhury, A.; Pavey, T.; Trost, S.G. Free-living evaluation of laboratory-based activity classifiers in preschoolers. *Med. Sci. Sports Exerc.* **2020**, *52*, 1227–1234. [CrossRef] [PubMed]
20. Ulrich, D. *Examiner's Manual Test of Gross Motor Development (TGMD-2)*, 2nd ed.; PRO-ED: Austin, TX, USA, 2000.
21. Centers for Disease Prevention and Control. National Youth Fitness Survey (NYFS) Test of Gross Motor Development (TGMD-2) Procedures Manual. 2012. Available online: https://www.cdc.gov/nchs/data/nnyfs/tgmd.pdf (accessed on 15 September 2022).
22. Tulchin-Francis, K.; Stevens, W.; Gu, X.; Zhang, T.; Roberts, H.; Keller, J.; Dempsey, D.; Borchard, J.; Jeans, K.; VanPelt, J. The impact of the coronavirus disease 2019 pandemic on physical activity in U.S. children. *J. Sport Health Sci.* **2021**, *10*, 323–332. [CrossRef] [PubMed]

Article

The Level and Factors Differentiating the Physical Fitness of Adolescents Passively and Actively Resting in South-Eastern Poland—A Pilot Study

Anna Puchalska-Sarna [1,2], Rafał Baran [1,3], Magdalena Kustra [1], Teresa Pop [1], Jarosław Herbert [4] and Joanna Baran [1,5,*]

1   Institute of Health Sciences, Medical College, University of Rzeszów, 35-310 Rzeszów, Poland
2   Regional Clinical Rehabilitation and Education Center for Children and Youth in Rzeszów, 35-301 Rzeszów, Poland
3   Solution-Statistical Analysis, 35-120 Rzeszów, Poland
4   Institute of Physical Culture Sciences, Medical College, University of Rzeszów, 35-326 Rzeszów, Poland
5   Natural and Medical Center for Innovative Research, 35-310 Rzeszów, Poland
*   Correspondence: jbaran@ur.edu.pl

Abstract: Physical activity (PA) is defined as any bodily movement produced by skeletal muscles that requires energy expenditure. Due to civilization's development, we can observe a global decline in physical activity which negatively affects the state of physical and mental health. The physical activity of children and adolescents is a counterpart to their physical fitness. There is also more frequent spending of free time in a passive way rather than actively. The aim of the study was to determine whether there are differences in the physical fitness of young people who rest passively in relation to those who rest actively. In addition, it was checked whether factors, such as age, weight, body height and BMI differentiate the level of fitness in adolescents. Study group: 25 boys and 25 girls declaring active leisure activities. Control group: 25 boys and 25 girls declaring passive leisure activities. Age of the respondents ranged from 11 to 15 years (Me = 13; SD = 1.23). The research used: the author's questionnaire and the Index of Physical Fitness of K. Zuchora. The results were statistically developed. The youth who spend their free time actively were characterised by a higher level of physical fitness than their peers who choose passive recreation. The students with a higher BMI obtained worse results than the children with a lower body mass index. In both groups, slightly better results were obtained by girls. A significant relationship between age and results has been observed in the control group—the results increased with increasing age. The level of physical fitness is higher in active forms of recreation than in passive rest. Physical fitness tends to increase with age but decreases with increasing BMI. Girls are characterised by a higher level of physical fitness than boys.

Keywords: active free time; adolescents; factors affecting physical fitness; passive free time; physical fitness; Zuchora test

Citation: Puchalska-Sarna, A.; Baran, R.; Kustra, M.; Pop, T.; Herbert, J.; Baran, J. The Level and Factors Differentiating the Physical Fitness of Adolescents Passively and Actively Resting in South-Eastern Poland—A Pilot Study. *Children* **2022**, *9*, 1341. https://doi.org/10.3390/children9091341

Academic Editors: Niels Wedderkopp, Georgian Badicu, Ana Filipa Silva and Hugo Miguel Borges Sarmento

Received: 13 July 2022
Accepted: 26 August 2022
Published: 1 September 2022

Publisher's Note: MDPI stays neutral with regard to jurisdictional claims in published maps and institutional affiliations.

Copyright: © 2022 by the authors. Licensee MDPI, Basel, Switzerland. This article is an open access article distributed under the terms and conditions of the Creative Commons Attribution (CC BY) license (https://creativecommons.org/licenses/by/4.0/).

1. Introduction

Physical fitness (PF) is defined as the current ability to perform motor activities that require the involvement of strength, speed, endurance, motor coordination (agility) and suppleness. It is not only a function of our locomotor system but also the basis for the proper biological functioning of the whole body. PF consists of motor features, such as muscle strength, speed of movement, endurance and flexibility of the body, which is sometimes also called suppleness or agility. These features have an impact and depend on our health. In order to assess the level of PF (not taking movement tests into account), the psychophysical properties and the features of the body structure are usually assessed [1,2].

The level of PF is also determined by sex and age, along with which human motor skills change. A baby starts to sit up in the fourth month of life, at the end of the first year—to walk. From that moment on, motor coordination develops intensively, reaching a relatively high level at the age of 4–5 (the so-called golden age of motor skills) [3]. Generally, boys at this age are more mobile than girls, but all children prefer active play rather than spending time, for example, watching TV [4]. This situation changes with age. Adolescents are capable of acquiring complex motor systems, and at the age of 12–14, they begin the stage of sexual maturation. This period often causes changes in PF, as well as regression of some motor skills. It is related to the action of hormones, modification of body proportions (mainly in the case of girls in whom hormonal fluctuations also contribute to a decrease in self-esteem, lack of acceptance of their own body, and thus—aversion to physical activity (PA)) and changes in BMI [3]. The research of the last decade shows that adolescents with a high degree of body fat and with a high BMI show a lower level of fitness, as well as worse results in endurance tests [5,6].

Special PF means the ability to perform movements that are mainly related to equipment specific to a given type of sport (apparatus, accessories). All types of fitness have an impact on the overall physical development of a person. Therefore, the largest pool of exercises is focused on general PF [7,8]. The level of PF is determined by simple tests that have been developed for specific populations, e.g., pupils, students, manual or white-collar workers, considering age categories and sex [9–11]. For people who practice competitive sports, separate tests with slightly higher criteria have been developed [12].

PF is closely related to PA. which is any body movement generated by skeletal muscles that requires energy consumption. Physical activity includes exercise and other activities that are performed in the form of play, housework, work and recreational activities. This activity has a large impact on shaping, maintaining and increasing fitness. In general, the more physically active we are, the more often and regularly we exercise, and the more physically fit we become. However, the effectiveness of the exercises performed is subject to the individual circumstances of each person's body and is associated with genetic predisposition. The PA of children and adolescents is a counterpart of their physical fitness. It also has a significant impact on other spheres of life. From a biological point of view, appropriate physical activity shapes a healthy and physically fit individual, who reacts to environmental threats [13–15]. Exercise as a subcategory of PA is planned, structured, purposeful and repetitive [16].

There is scientific evidence that PA provides essential health benefits for children and adolescents [17–19]. This conclusion is based on observations in which greater PA was associated with an increase in beneficial health parameters. The documented benefits include increased physical fitness (both cardio-respiratory and muscle strength), increased bone density, reduced body fat and a reduced risk of metabolic and cardiovascular diseases. There is also a reduction in the occurrence of symptoms of depression. Maintaining regular PA that begins in childhood and continues into adulthood reduces the risk of death from cardiovascular and metabolic diseases in later years [20–22].

According to WHO recommendations, the PA of children and adolescents aged 5–17 should include play, games, sports, transport, recreation, physical education or planned exercises in the context of family, school and community activities [23] to improve cardiorespiratory and muscular fitness, bone health, cardiovascular and metabolic biomarkers, and to reduce symptoms of anxiety and depression:

1. Children and adolescents 5–17 years of age should spend at least 60 min a day in Moderate to Vigorous Physical Activity (MVPA).
2. Physical activity for more than 60 min a day will provide additional health benefits.
3. Most of the daily physical activity should be aerobic. Vigorous Physical Activity (VPA) activities, including those that strengthen muscles and bones should be practiced at least three times a week.

According to WHO, in 2010 81% of adolescents between 11 and 17 years of age worldwide were not sufficiently physically active. Adolescent girls were less active than their peers: 84% of girls did not meet WHO recommendations, and among boys—78% of adolescents [15].

Comparing the data obtained in two stages of the Polish research, a team from HBSC found statistically significant differences from 2014 to 2018 in the level of adolescents' PA. Recommendations for MVPA are met by only 17.2% of adolescents, which is less than one-fifth of the studied population. There was also a clear negative trend in this period, i.e., a decrease in the percentage of adolescents meeting the WHO recommendations for moderate PA, from 24.2% in 2014 to 17.2% in 2018 [24].

Researchers emphasise that parents play an important role in creating appropriate habits of movement. However, only 44 percent of children engage in PA after school (e.g., cycling, walking) with their parents. On the other hand, 48 percent of children are driven to school by their parents by car [25].

Importantly, the levels of fitness appropriate to sex and age in children and adolescents tend to remain in adulthood, and their determination and strengthening are the basis for a physically active lifestyle throughout childhood, adolescence and adulthood [26,27]. Studies have shown that physically fit children were willing to engage in physical activity and maintain their behaviours during adolescence, while children who are less physically capable tended to be physically inactive during adolescence [28–30]. Children who are more physically fit have the fundamental skills required to participate successfully at different levels of PA. Therefore, they are more likely to keep PA within their field of interest [31].

The aim of the study was to determine whether there are differences in the physical fitness of young people who rest passively in relation to those who rest actively. In addition, it was checked whether factors, such as age, weight, body height and BMI differentiate the level of fitness in adolescents.

2. Materials and Methods

2.1. Participants

The study covered 100 pupils of the Municipal School Complex No. 3 in Krosno (Poland). After obtaining the written consent of the parents and children, the project began. The study group consisted of 25 girls and 25 boys. The control group also consisted of 25 girls and 25 boys. All children were examined in the spring period at the turn of April and May. All the subjects lived in the town.

The subjects were divided according to the way they spend their free time into two groups: people declaring active leisure (study group) and people declaring passive leisure (control group). By completing the questionnaire, the child answered the question about how they like to spend their free time: passively or actively. It was a subjective self-assessment of each child. Examples of passive activity (e.g., sitting in front of a TV set, in front of a computer) and active activity (e.g., cycling, walking, playing football) are given so that the child can easily notice the difference between the concept of passive and active leisure.

The criteria for inclusion in the study group are pupils aged 11–15 years, declared active way of spending free time, consent of the legal guardian to the child's participation in the study, and informed consent of the pupil to participate in the study.

The criteria for exclusion from the research are disability or injuries in the lower limbs that make it impossible to take up PA, health contraindications for participation in the fitness test, the pupil feeling unwell during the test, withdrawal of the consent of the child or his/her guardian to participate in the study, even during the tests.

2.2. Research Tools

The study used the assessment of physical fitness with the use of K. Zuchora's Physical Fitness Index, which consists of a six-step assessment of the pupil's fitness [32]. This assessment checks:

1. Speed (sprint run on the spot with simultaneous clapping of hands under the knees—the result is determined by the number of claps within 10 s),

2. Jumping (long jump from the spot—distance measured in the pupil's feet),
3. Shoulder strength (hanging on gymnastic wall bars: with both hands, one hand, both hands with a pull-up, for boys, additionally hanging with a pull-up and lowering, once with the right hand, once with the left hand, with holding for every 10 s—time measured in seconds),
4. Suppleness (bending the torso forward—the result is determined by the ability to reach as low as possible with the hands and touch the head to the knees),
5. General endurance (endurance run on the spot—time measured in minutes),
6. Strength of the abdominal muscles (performing transverse "scissors" while lying on the back—time measured in seconds/minutes).

Depending on their sex and the results of the individual attempts, pupils were assigned points (from 0 to 6) which were then added up and compared with the age-appropriate norms. In this way, the level of the children's physical fitness was determined (Tables 1 and 2).

Table 1. The level of the test execution and the score in points.

Sex	Minimum 1	Sufficient 2	Good 3	Very good 4	High 5	Excellent 6
			Test 1.			
F	12 claps	16 claps	20 claps	25 claps	30 claps	35 claps
M	15 claps	20 claps	25 claps	30 claps	35 claps	40 claps
			Test 2.			
F	5 feet	6 feet	7 feet	8 feet	9 feet	10 feet
M	5 feet	6 feet	7 feet	8 feet	9 feet	10 feet
			Test 3.			
F	Overhang AA 3 s	Overhang AA 10 s	Overhang A 3 s overhang and pull up 3 s	Overhang A 10 s overhang and pull up 10 s	overhang and pull up 3 s pull up A 10 s	overhang and pull up 10 s LA/RA
M	overhang AA 10 s	Overhang A 10 s				
			Test 4.			
F	Ankle grip	fingers-toes	fingers-ground	fingers wide-ground/	hands-ground	head-knees
M						
			Test 5.			
F	1 min	3 min	6 min	10 min	15 min	20 min
M	2 min	5 min	10 min	15 min	20 min	30 min
			Test 6.			
F	10 s	30 s	1 min	1.5 min	2 min	3 min
M	20 s	1 min	1.5 min	2 min	3 min	4 min

A—one arm; AA—double arms; F—female; M—male; LA—left arm; min—minutes; RA—right arm; s—seconds.

Table 2. Standards for individual age categories.

Marks	11–12 Years	13–15 Years
Minimum	6	6
Sufficient	11	12
Good	16	17
Very good	20	22
High	25	27
Excellent	29	31

2.3. Statistical Analysis

The Statistica 10.0 program (TIBCO Software Inc., Palo Alto, CA, USA) was used for statistical analysis. In the case of variables expressed on qualitative scales, the results were presented in the form of frequency distributions with percentage values, while for variables on quantitative scales, the basic measures of descriptive statistics were calculated: arithmetic mean, median and standard deviation. Compliance with normal distribution was tested using the Shapiro–Wilk test. Due to the fact that the variables did not meet the assumptions regarding the use of parametric methods (the distributions significantly deviated from the normal distribution), non-parametric methods were used to verify the hypotheses.

The Mann–Whitney U test was used for comparisons between the two groups, while Spearman's rank correlation coefficient was used to analyse the relationships between the variables. Comparative analyses for the qualitative variables were performed using the Chi-square test. The significance level was assumed to be $\alpha = 0.05$. The results were considered statistically significant when the calculated test probability p satisfied $p < 0.05$.

3. Results

The age of all subjects in the study ranged from 11 to 15 years of age. The average BMI of the subjects was 20.68. The shortest pupil was 1.40 m and the tallest was 1.80 m. The children's body weight ranged from 33 to 90 kg (Table 3).

Table 3. Characteristics of the studied children.

Variables	N	$\bar{x}$	Me	Min.	Max.	SD
Age [years]	100	13.02	13.00	11.00	15.00	1.23
Body height [m]	100	1.61	1.60	1.40	1.80	0.08
Body weight [kg]	100	53.79	50.00	33.00	90.00	12.52
BMI [kg/m^2]	100	20.68	20.01	14.82	31.16	3.73

N—numbers of participants, Max.—maximum value, Me—median, Min—minimum value, SD—standard deviation, $\bar{x}$—average value.

The analysis carried out using the Mann–Whitney U test showed statistically significant differences between the study group and the control group with regard to the final result of Zuchora's test ($p < 0.001$). The comparison of arithmetic means and medians clearly shows that adolescents who actively spend their free time are characterised by a higher level of physical fitness than their peers choosing passive forms of recreation.

The analysis showed no statistically significant differences between girls and boys in the study group with regard to the obtained results, but the test probability p result is at a level tending towards significance. On the other hand, in the control group, there is a statistically significant difference ($p = 0.003$) between girls and boys in relation to the obtained results. On the basis of the arithmetic means and medians, it can be seen that in both groups girls obtained slightly better results (Table 4).

The comparison of the distribution of scores using the Chi-square test confirms the earlier observations that the study group is characterised by higher fitness ($p < 0.001$). In the group actively spending free time, the majority (58%) are adolescents with a good level of fitness, 28% are pupils with very good fitness, 10%—high, and only 4% obtained a satisfactory score (with no minimum scores). In the group preferring passive forms, as many as two out of three pupils (66%) are only sufficiently fit, 32% obtained the minimum score, and only 2% are good (with no very good or high scores) (Table 5).

Table 4. Comparative analysis of the results obtained in the Zuchora test in both groups.

				Zuchora Test Results			
Group	Sex	N	$\bar{x}$	Me	SD	Z	p
Study	All	50	20.38	20	3.25	8.47	<0.001
Control	All	50	12.34	13	2.57		
Study	Female	25	21.16	21	3.22	1.79	0.073
	Male	25	19.60	19	3.15		
Control	Female	25	13.44	14	2.12	2.94	0.003
	Male	25	11.24	11	2.54		

N—numbers of participants, Max.—maximum value, Me—median, Min—minimum value, SD—standard deviation, $\bar{x}$—average value, p—test probability, Z—U Mann–Whitney test value.

Table 5. Distribution of grades awarded on the basis of the Zuchora test result and the applicable standards.

	Zuchora Test Results					
	Study		Control		All	
Mark	N	%	N	%	N	%
Minimum	0	0.0%	16	32.0%	16	16.0%
Sufficient	2	4.0%	33	66.0%	35	35.0%
Good	29	58.0%	1	2.0%	30	30.0%
Very good	14	28.0%	0	0.0%	14	14.0%
High	5	10.0%	0	0.0%	5	5.0%
All	50	100.0%	50	100.0%	100	100.0%
Chi-sq. test	$\chi^2 = 88.6$; df = 4; $p < 0.001$					

N—numbers of participants, %—percent of participants, χ^2—chi-squared test.

The statistical analysis showed a very clear and statistically significant difference ($p < 0.001$) between the groups in the scope of the performed attempts. Pupils in the test group achieved significantly higher results in each of the individual attempts of Zuchora's test than their peers in the control group (Table 6).

Table 6. Comparisons of the results obtained in individual attempts of the Zuchora test between the study and control groups.

	Zuchora Test Results						
	Study			Control			p
Test	$\bar{x}$	Me	SD	$\bar{x}$	Me	SD	
Test 1	3.66	4	0.85	2.58	3	0.84	<0.001
Test 2	3.90	4	0.89	2.84	3	1.00	<0.001
Test 3	3.20	3	1.28	1.62	1.5	0.78	<0.001
Test 4	3.78	3.5	1.39	2.22	2	1.37	<0.001
Test 5	4.10	4	1.34	2.18	2	0.90	<0.001
Test 6	1.74	2	0.94	0.90	1	0.65	<0.001

Me—median, SD—standard deviation, $\bar{x}$—average value, p—test probability.

The analysis showed the existence of statistically significant differences ($p < 0.001$) between the study and control group in terms of body weight and BMI value. However, the differences between the groups with regard to age and body height were not statistically significant ($p > 0.05$). The analysis carried out with the use of the Mann–Whitney U test proved the existence of statistically significant differences ($p = 0.042$) between girls and boys in the study group in relation to the BMI value. However, the differences between sexes in relation to age, body height and body weight turned out to be statistically insignificant in this group ($p > 0.05$). In the control group, however, there were no statistically significant differences ($p > 0.05$) between girls and boys in terms of age, height and body weight. In the case of BMI, the value of the test probability p tends towards significance ($p = 0.051$) (Table 7).

Table 7. Comparisons of age, weight, height and BMI between girls and boys in the study and control group.

Variables	Sex	N	$\bar{x}$	Me	SD	Z	p
All							
Age	Study	50	12.98	13.00	1.29	−0.36	0.717
	Control	50	13.06	13.00	1.19		
Body height	Study	50	1.60	1.60	0.08	−1.12	0.263
	Control	50	1.62	1.61	0.08		
Body weight	Study	50	48.04	47.00	7.67	−4.45	<0.001
	Control	50	59.54	59.50	13.79		
BMI	Study	50	18.79	18.36	2.05	−4.89	<0.001
	Control	50	22.56	21.93	4.09		
Study group							
Age	Female	25	13.00	13.00	1.22	0.12	0.907
	Male	25	12.96	13.00	1.37		
Body height	Female	25	1.60	1.60	0.08	0.42	0.677
	Male	25	1.60	1.58	0.09		
Body weight	Female	25	46.32	46.00	5.16	−1.48	0.138
	Male	25	49.76	48.00	9.34		
BMI	Female	25	18.19	18.07	1.46	−2.04	0.042
	Male	25	19.40	19.48	2.38		
Control group							
Age	Female	25	13.12	13.00	1.17	0.23	0.816
	Male	25	13.00	13.00	1.22		
Body height	Female	25	1.63	1.61	0.07	0.76	0.449
	Male	25	1.61	1.59	0.10		
Body weight	Female	25	56.96	54.00	12.75	−1.58	0.114
	Male	25	62.12	63.00	14.56		
BMI	Female	25	21.37	20.72	3.61	−1.95	0.051
	Male	25	23.74	23.88	4.26		

N—numbers of participants, Max.—maximum value, Me—median, Min—minimum value, SD—standard deviation, $\bar{x}$—average value, p—test probability, Z—U Manna-Whitney test value.

The calculations made for all the subjects revealed statistically significant correlations between the final result of Zuchora's test and body weight ($R = -0.46$; $p < 0.001$) and BMI ($R = -0.49$; $p < 0.001$). Children with a higher body weight or a higher BMI had lower results in Zuchora's test.

Subgroup analysis allows for the discovery of differences in the relationships between the variables. In the group of active pupils, the final test result is correlated with the BMI value ($R = -0.38$; $p = 0.006$) as well as its components, body height ($R = -0.30$; $p = 0.035$) and body weight ($R = -0 49$; $p < 0.001$). As with the calculations for the entire sample, we are dealing with negative correlations of average strength. In the control group, only the relationship between age and results is significant ($R = 0.34$; $p = 0.017$), and its direction is positive, which means that older children had higher results in Zuchora's test (Table 8).

Table 8. Correlations of the results in the Zuchora test with the parameters of children.

Group	A Pair of Variables	N	R Spearman	T (N − 2)	p
All					
	Zuchora & Age	100	0.01	0.12	0.901
	Zuchora & Body height	100	−0.13	−1.31	0.192
	Zuchora & Body weight	100	−0.46	−5.17	<0.001
	Zuchora & BMI	100	−0.49	−5.58	<0.001
Study					
	Zuchora & Age	50	−0.15	−1.06	0.293
	Zuchora & Body height	50	−0.30	−2.17	0.035
	Zuchora & Body weight	50	−0.49	−3.86	<0.001
	Zuchora & BMI	50	−0.38	−2.86	0.006
Control					
	Zuchora & Age	50	0.34	2.47	0.017
	Zuchora & Body height	50	0.15	1.03	0.310
	Zuchora & Body weight	50	0.06	0.39	0.699
	Zuchora & BMI	50	0.02	0.11	0.916

4. Discussion

The aim of the study is to check to what extent the physical fitness of pupils spending leisure time actively differs from that of pupils spending their free time passively.

According to ECOG, it is necessary to explain the differences and similarities between physical activity and physical fitness. Of course, they are highly dependent and interrelated; however, they are two separate concepts that also need to be considered separately. Fitness depends on the level of physical ability (e.g., being able to participate in open-access physical education classes would require walking or running skills) [33].

Several studies have investigated physical activity and its effects on obesity and health, showing that regular physical activity combined with improved physical fitness reduces the risk of obesity and several metabolic problems (e.g., diabetes mellitus, metabolic syndrome, heart disease) and also improves overall health [34].

Vincent et al., Used the daily step counting method to assess physical fitness. The study covered three countries: Sweden, Australia and the United States. The authors found that students from Sweden had the highest level of physical fitness, bearing in mind the impact of different conditions in which students live. It also found that in Sweden, about 70 percent of students participated in extracurricular sports activities, and in the United States, only 20 percent of students were physically active in extracurricular activities [35].

The research shows that pupils who spend their free time actively are characterised by a higher level of physical fitness than their peers who spend their free time passively. There are statistically significant differences between the study group (active adolescents) and the control group (passive adolescents) in relation to the final result of K. Zuchora's Physical Fitness Index. Most of the study group (58%) completed the test with a good result, 28% with a very good result, 10% with a high result, and only 4% of the group obtained a satisfactory result (with no minimum scores). In the control group (preferring passive forms of recreation) as many as two out of three pupils (66%) showed only sufficient fitness, minimum—as many as 32%, and only 2% were able to perform at a good level. In this group, no one received a very good or high score. Moreover, pupils in the test group obtained significantly better results in each individual attempt at Zuchora's test than their peers in the control group.

These results are consistent with those of other authors. Children actively spending their free time dancing obtained much higher results in individual attempts of the test (the final results were very good and high, only two boys received a good score) than the group of boys who did not engage in any PA (the final scores were sufficient, with a smaller percentage minimum and good). The level of PF was assessed in both groups using K. Zuchora's Physical Fitness Index [36].

Frömel et al., found that the most common forms of physical activity chosen by boys were team games, swimming and skateboarding, while in girls the most common forms of activity were swimming, dancing and skateboarding [37].

The authors' own research also checked whether the test results depended on age, height, body weight and BMI. The calculations carried out for all the subjects showed statistically significant correlations between the final results of the physical fitness test and body weight and BMI. This means that body weight and BMI have an influence on fitness level. The analysis carried out separately in the study and control groups allowed for the observation of differences in the relationships between the variables. In the study group, the final result of the test was influenced by the BMI value, as well as its components, i.e., body height and body weight. Here too, the increasing values of the parameters determining body structure were accompanied by a decrease in the final values of Zuchora's test. In the control group, there is only a significant relationship between age and results, and in this case, the test results increase with age. These results are consistent with the results of other researchers who found a statistically significant decrease in the level of children's fitness in the study group with their height and BMI. Higher efficiency positively correlated with declarations of active forms of recreation. The authors assessed physical fitness with the use of Zuchora's test [3].

Moreover, in another study 1/5 of pupils with excess body weight assessed themselves as not very active or physically inactive, 40% hardly or never attended physical education classes, and 1/3 did not systematically engage in any physical activity except PE classes. Comparing this group with peers with a normal BMI, obese junior high school pupils turned out to be significantly less physically active, and much less often participated in PE and sports outside school. However, no statistically significant differences were found between the groups in terms of the amount of time spent passively [38].

Based on the analysis of the results of Zuchora's test, it turned out that there are no statistically significant differences between the girls and boys in the study group, but the test probability score is at a level tending towards significance. This means that the level of fitness of both sexes is at a similar level (although we can observe a slightly higher level in the girls surveyed). In the control group, however, there is a statistically significant difference between the sexes in relation to the results of Zuchora's test—girls obtained better results than boys. Hoos et al., used other studies, claiming that there are no significant differences between the level of physical fitness of girls and boys; the level of energy spent on physical activity was the only difference between the sexes [39].

These results are consistent with the results of other researchers who, on the basis of PF tests carried out with children aged 11–15 years, found that the physical fitness of the pupils was at a satisfactory level. There were no clear differences in the obtained results between girls and boys, which indicated a similar development of the skills of all the pupils. The assessment of the pupils' free time activity showed that they preferred spending free time actively [32].

As shown by Riddoch et al., European studies of children's physical fitness show a higher level of fitness in boys than in girls [40].

Another test to assess your physical fitness is the Fitnessgram Battery Test. Studies carried out with its use show that men presented significantly higher results in the test of upper strength ($p < 0.001$) and aerobic capacity ($p < 0.001$), while women showed higher results in the sit and reach test ($p < 0.001$) in the torso lifting test ($p < 0.005$) and in the value of fat mass ($p < 0.001$) [41].

Standardised studies carried out using the EUROFIT test showed that Boys performed significantly (standardised differences > 0.2) better than girls in the tests of muscle strength, muscle power, muscle endurance, speed and CRF, but worse in the flexibility test. Physical function generally improved faster in boys than in girls, especially in adolescence [42].

In another study which used Zuchora's PF Index more than half of the examined pupils undertook physical activity two to three times a week; 12% of the subjects declared low physical activity. Only 9% of the adolescents undertook physical activity five times a week or more. Pupils in this group showed better fitness in the test than those who practiced sports less frequently. Moreover, decreased physical fitness was found in the group of pupils with an excessive BMI index, and the examined girls showed higher fitness than boys. The analysis of the results proved that age does not significantly affect the level of physical fitness of the pupils [43].

Analysing our own research, it was found that in the study group there were statistically significant differences between girls and boys ($p < 0.05$) in relation to the BMI value. However, in relation to age, height and weight, the differences between the sexes in this group turned out to be statistically insignificant ($p > 0.05$).

In the control group, there were no statistically significant differences ($p > 0.05$) with respect to age, height, weight and BMI. However, in relation to BMI, the value of the test probability p tended towards significance ($p = 0.051$)—boys were characterised by a slightly higher body mass index. The better results obtained by girls may be a consequence of lower body mass indexes of girls as compared to boys.

In a large Greek cohort of 424,328 girls and boys aged 6–18 years, boys typically outperformed girls for cardiovascular endurance, muscular strength, muscular endurance, and speed/agility, but lower flexibility (all $p < 0.001$). Older boys and girls achieved better results than younger ones ($p < 0.001$). Physical fitness test scores peaked around age 15 for both sexes [44].

The boys' fitness was better than that of the girls, with the exception of the sit down and press the back test, in which the girls performed better. Except for the sit down and back press test and the 10×5 m pendulum run test, physical fitness was significantly related to age. These results were obtained from a sample of 11,186 children and adolescents (5546 boys and 5640 girls) aged 10 to 15 years old and were assessed in the French national BOUGE study. Participants were tested for cardiorespiratory fitness, muscular endurance, speed, flexibility and agility with the following tests: the 20 m Pendulum Test, Rollback Test, 50 m Sprint Test, Sit and Bench Test, and the $10\times$ Test 5-m shuttle test. Percentile values were estimated for French youth as a function of age stratified by sex using the Generalised Additive Location, Scale and Shape (GAMLSS) model [45].

As for results from our own research, the correlations between BMI and the results of the fitness test turned out to be very significant; the score decreased with an increase in BMI. Moreover, girls were shown to be better than boys in Zuchora's test. This may be related to the average lower BMI of female pupils than male pupils. Free time physical activity significantly influenced the level of physical fitness in the children. Pupils who declared active recreation obtained significantly higher results in individual attempts of Zuchora's test, compared to their peers whose way of spending their free time was almost completely passive.

Many authors also compare the level of physical fitness of students living in different regions. Loucaides et al., studied the physical fitness of students from rural and urban areas living in Cyprus, taking into account the summer and winter periods. The authors found that the children showed much greater physical activity in the summer, taking into account the place of residence of the children. Students from urban regions were slightly more active in winter than in rural areas, where they were more active in summer [46].

Children's physical fitness is an extremely important aspect of their future health. The conducted research proved that physical fitness depends on whether we spend our free time passively or actively. That is why it is so important for parents to encourage their children to spend their free time actively while setting an example themselves.

It is worth emphasising that the literature does not contain many items of literature comparing the assessment of physical fitness using the Zuchora index. Many researchers describe PA under the concept of PF, which is a mistake because they are two different concepts.

Limitations

The data are cross-sectional and do not allow for an analysis of cause-and-effect relationships. Consequently, we do not know the direction of the relationships found.

Another limitation is the small study group and the lack of a normal distribution of results, which prevents more advanced analyses.

The results may be limited to some extent because they were conducted in one school and the research sample was not an ideal representation of the child population.

5. Conclusions

The level of physical fitness is higher in the case of those undertaking active forms of recreation than in the case of those engaged in passive leisure. Physical fitness tends to increase with age but decreases with increasing BMI. Girls are characterised by a higher level of physical fitness than boys.

Future research should focus on more people and should be repeated annually. It would also be worth comparing the above results at different times of the year, or with a sports intervention for a certain period of time.

Author Contributions: Conceptualization, J.B., M.K., A.P.-S.; methodology, J.B., R.B., M.K., formal analysis, R.B., J.H., resources, M.K., A.P.-S.; writing—original draft preparation, A.P.-S., J.B., J.H.; writing—review and editing, A.P.-S., R.B., M.K., J.B., T.P., J.H.; supervision, T.P. All authors have read and agreed to the published version of the manuscript.

Funding: This research received no external funding.

Institutional Review Board Statement: The study was conducted in accordance with the Declaration of Helsinki and approved by the Ethics Committee of University in Rzeszów (No. 18/12/2015 from 2 December 2015).

Informed Consent Statement: Informed consent was obtained from all subjects involved in the study.

Data Availability Statement: Data available from the authors of this publication.

Conflicts of Interest: The authors declare no conflict of interest.

References

1. Klimczyk, M.; Goździk, A. Physical fitness of children aged 12 years attending classes of sports and overall profile. *J. Health Sci.* **2014**, *4*, 198–206.
2. Nazar, K. Fizjologiczne podstawy wysiłków fizycznych. Zmiany treningowe w organizmie człowieka. In *Wybrane Zagadnienia Medycyny Sportowej*; Jegier, A., Krawczyk, J., Eds.; Wydawnictwo Lekarskie PZWL: Warszawa, Poland, 2012; pp. 11–24.
3. Nieczuja-Dwojacka, J.; Grzelak, J.; Siniarska, A.; Chwaścińska, K. Sprawność fizyczna warszawskiej młodzieży gimnazjalnej. *Studia Ecol. Bioethicae* **2015**, *13*, 67–84. [CrossRef]
4. Hinkley, T.; Crawford, D.; Salmon, J.; Okely, A.D.; Hesketh, K. Preschool children and physical activity: A review of correlates. *Am. J. Prev. Med.* **2008**, *34*, 435–441. [CrossRef]
5. Barańska, E.; Gajewska, E. Ocena sprawności motorycznej występującej u dzieci z nadwagą i otyłością. *Now. Lek.* **2009**, *78*, 182–185.
6. Silva-Santos, S.; Santos, A.; Vale, S.; Mota, J. Motor fitness and preschooler children obesity status. *J. Sports Sci.* **2017**, *35*, 1704–1708. [CrossRef] [PubMed]
7. Dudkowski, A.; Majorowski, M.; Rokita, A.; Naglak, K. Związki sprawności fizycznej ogólnej, ukierunkowanej i specjalnej u młodych piłkarzy ręcznych w rocznym cyklu szkolenia. *Rozpr. Nauk. AWF We Wrocławiu* **2011**, *35*, 131–136.
8. Kochański, Ł: Sprawność ogólna młodych piłkarzy nożnych na tle reprezentacji Województwa Mazowieckiego U-16. *Zesz. Nauk. WSKFiT* **2014**, *9*, 43–48.
9. Soubra, R.; Chkeir, A.; Novella, J.L. A Systematic Review of Thirty-One Assessment Tests to Evaluate Mobility in Older Adults. *BioMed Res. Int.* **2019**, *2019*, 1354362. [CrossRef]
10. Williams, H.G.; Pfeiffer, K.A.; Dowda, M.; Jeter, C.; Jones, S.; Pate, R.R. A Field-Based Testing Protocol for Assessing Gross Motor Skills in Preschool Children: The CHAMPS Motor Skills Protocol (CMSP). *Meas. Phys. Educ. Exerc. Sci.* **2009**, *13*, 151–165. [CrossRef]
11. Platvoet, S.; Faber, I.R.; De Niet, M.; Kannekens, R.; Pion, J.; Elferink-Gemser, M.T.; Visscher, C. Development of a Tool to Assess Fundamental Movement Skills in Applied Settings. *Front. Educ.* **2018**, *3*, 75. [CrossRef]
12. Talaga, J. *A-Z Sprawności Fizycznej, Atlas Ćwiczeń. Wyd*; Ypsylon Warszawa: Warszawa, Poland, 1995; p. 9.

13. Sjøgaard, G.; Christensen, J.R.; Justesen, J.B.; Murray, M.; Dalager, T.; Fredslund, G.H.; Søgaard, K. Exercise is more than medicine: The working age population's well-being and productivity. *J. Sport Health Sci.* **2016**, *5*, 159–165. [CrossRef] [PubMed]
14. Kubusiak-Słonina, A.; Grzegorczyk, J.; Mazur, A. Ocena sprawności i aktywności fizycznej dzieci szkolnych z nadmierną i prawidłową masą ciała. *Endokrynol. Otył. Zab. Przem. Mat.* **2012**, *8*, 16–23.
15. World Health Organization. Physical Actvity. Available online: http://www.who.int/news-room/fact-sheets/detail/physical-activity (accessed on 15 May 2022).
16. Caspersen, C.J.; Powell, K.E.; Christenson, G.M. Aktywność fizyczna, ćwiczenia i definicje fizyczne: I rozróżnienie w badaniach ze zdrowiem. *Rap. Zdrowiu* **1985**, *100*, 126.
17. Warburton, D.E.; Nicol, C.W.; Bredin, S.S. Health benefits of physical activity: The evidence. *CMAJ* **2006**, *174*, 801–809. [CrossRef]
18. Ruegsegger, G.N.; Booth, F.W. Health Benefits of Exercise. *Cold Spring Harb. Perspect. Med.* **2018**, *8*, a029694. [CrossRef]
19. Reiner, M.; Niermann, C.; Jekauc, D.; Woll, A. Long-term health benefits of physical activity–a systematic review of longitudinal studies. *BMC Public Health* **2013**, *13*, 813. [CrossRef]
20. Janssen, I. Physical activity guidelines for children and youth. *Appl. Physiol. Nutr. Metab.* **2007**, *32*, 109–121. [CrossRef]
21. Janssen, I.; LeBlanc, A. Systematic Review of the Health Benefits of Physical Activity in School-Aged Children and Youth. *Appl. Physiol. Nutr. Metab.* **2010**, *7*, 40–55. [CrossRef]
22. US Department of Health and Human Services. *Physical Activity Guidelines Advisory Committee Report*; US Department of Health and Human Services: Washington, DC, USA, 2008; pp. 22–23.
23. World Health Organization. *Recommended Population Levels of Physical Activity for Health*; World Health Organization: Geneva, Switzerland, 2010; pp. 17–21.
24. Kleszczewska, D.; Dzielska, A. Aktywność fizyczna młodzieży. In *Zdrowie Uczniów w 2018 Roku na Tle Nowego Modelu Badań HBSC*; Mazur, J., Małkowskiej-Szkutnik, A., Eds.; Instytut Matki i Dziecka: Warszawa, Poland, 2018; pp. 87–92. ISBN 978-83-951033-3-9.
25. Sprawność Fizyczna Dzieci w Polsce Systematycznie Spada Medexpress. Available online: https://www.medexpress.pl/sprawnosc-fizyczna-dzieci-w-polsce-systematycznie-spada/72962 (accessed on 15 May 2022).
26. Chen, W.; Hammond-Bennett, A.; Hypnar, A.; Mason, S. Health-related physical fitness and physical activity in elementary school students. *BMC Public Health* **2018**, *18*, 195–206. [CrossRef]
27. Blair, S.N.; Cheng, Y.; Holder, J.S. Is physical activity or physical fitness more important in defining health benefits? *Med. Sci. Sports Exerc.* **2001**, *33*, 379–399. [CrossRef]
28. Barnett, L.M.; Morgan, P.J.; van Beurden, E. Perceived sports competence mediates the relationship between child hood motor skill proficiency and adolescent physical activity and fitness; a longitudinal assessment. *J. Behav. Nutr. Phys. Act.* **2008**, *5*, 40–52. [CrossRef] [PubMed]
29. Welk, G.J. The youth physical activity promotion model: A conceptual bridge between theory and practice. *Quest* **1999**, *51*, 5–23. [CrossRef]
30. Ewrin, H.E.; Castelli, D.M. National physical education standards: A summary of student performance and its correlates. *Res. Q. Exerc. Sport* **2008**, *79*, 495–505.
31. Stodden, D.F.; Goodway, J.D.; Langendorfer, S.J. A developmental perspective on the role of motor skill competence in physical activity: An emergen relationship. *Quest* **2008**, *60*, 290–306. [CrossRef]
32. Dąbrowska, K.; Cybulski, M.; Konopka, A.; Krajewska-Kułak, E. Ocena sprawności fizycznej dzieci i młodzieży z terenów wiejskich na przykładzie uczniów Zespołu Szkół w Ołdakach. *Pediatr. Med. Rodz.* **2017**, *13*, 527–539. [CrossRef]
33. O'Malley, G.; Ring-Dimitriou, S.; Nowicka, P.; Vania, A.; Frelut, M.L.; Farpour-Lambert, N.; Weghuber, D.; Thivel, D. Physical Activity and Physical Fitness in Pediatric Obesity: What are the First Steps for Clinicians? Expert Conclusion from the 2016 ECOG Workshop. *Int. J. Exerc. Sci.* **2017**, *10*, 487–496.
34. Kyröläinen, H.; Santtila, M.; Nindl, B.C.; Vasankari, T. Physical fitness profiles of young men: Associations between physical fitness, obesity and health. *Sports Med.* **2010**, *40*, 907–920. [CrossRef]
35. Vincent, S.; Pangrazi, R.; Raustorp, A.; Tomson, M.; Cuddihy, T. Activity levels and body mass index of children in the United States, Sweden, and Australia. *Med. Sci. Sports Exerc.* **2003**, *35*, 1367–1373. [CrossRef]
36. Byzdra, K.; Skrzypczyńska, A.; Piątek, M.; Stępniak, R. Physical activity and development of physical fitness at boys at age between 13 and 15 years old. *J. Health Sci.* **2013**, *3*, 261–274.
37. Frömel, K.; Formánková, S.; Sallis, J.F. Physical activity and sport preferences of 10 to 14-year-old children: A 5-year prospective study. *Acta Univ. Palacki Olomuc. Gymn.* **2002**, *32*, 11–16.
38. Jodkowska, M.; Tabak, I.; Oblacińska, A. Aktywność fizyczna i zachowania sedenteryjne gimnazjalistów z nadwagą i otyłością w Polsce w 2005r. *Probl. Hig. Epidemiol.* **2007**, *88*, 149–156.
39. Hoos, M.B.; Gerver, W.J.M.; Kester, A.D.; Westerterp, K.R. Physical activity levels in children and adolescents. *Int. J. Obes. Relat. Metab. Disord.* **2003**, *27*, 605–609. [CrossRef] [PubMed]
40. Riddoch, C.J.; Andersen, L.B.; Wedderkopp, N.; Harro, M.; Klasson-Heggebø, L.; Sardinha, L.B.; Cooper, A.R.; Ekelund, U. Physical activity levels and patterns of 9- and 15-yr-old European children. *Med. Sci. Sports Exerc.* **2004**, *36*, 86–92. [CrossRef] [PubMed]
41. Gouveia, J.P.; Forte, P.; Coelho, E. Youth's Physical Activity and Fitness from a Rural Environment of an Azores Island. *Soc. Sci.* **2021**, *10*, 96. [CrossRef]

42. Tomkinson, G.R.; Carver, K.D.; Atkinson, F.; Daniell, N.D.; Lewis, L.K.; Fitzgerald, J.S.; Lang, J.J.; Ortega, F.B. European normative values for physical fitness in children and adolescents aged 9–17 years: Results from 2 779 165 Eurofit performances representing 30 countries. *Br. J. Sports Med.* **2018**, *52*, 1445–14563. [CrossRef]
43. Leszczyńska, K.; Golec, S.; Maciejewska-Paszek, I.; Paszek, P.; Irzyniec, T.; Szostak-Trybuś, M. Ocena sprawności fizycznej i zachowań zdrowotnych u młodzieży gimnazjalnej w Ozimku. *Zagroż. Życia Zdrowia Człowieka* **2017**, *16*, 217–229.
44. Tambalis, K.D.; Panagiotakos, D.B.; Psarra, G.; Daskalakis, S.; Kavouras, S.A.; Geladas, N.; Tokmakidis, S.; Sidossis, L.S. Physical fitness normative values for 6–18-year-old Greek boys and girls, using the empirical distribution and the lambda, mu, and sigma statistical method. *Eur. J. Sport Sci.* **2016**, *16*, 736–746. [CrossRef]
45. Vanhelst, J.; Labreuche, J.; Béghin, L.; Drumez, E.; Fardy, P.S.; Chapelot, D.; Mikulovic, J.; Ulmer, Z. Physical Fitness Reference Standards in French Youth: The BOUGE Program. *J. Strength Cond. Res.* **2017**, *31*, 1709–1718. [CrossRef]
46. Loucaides, C.A.; Chedzoy, S.M.; Bennett, N. Differences in physical activity levels between urban and rural school children in Cyprus. *Health Educ. Res.* **2004**, *19*, 138–147. [CrossRef]

Systematic Review

School-Based Exercise Programs for Promoting Cardiorespiratory Fitness in Overweight and Obese Children Aged 6 to 10

Stefan Mijalković [1], Dušan Stanković [1,*], Mario Tomljanović [2], Maja Batez [3], Maki Grle [4], Ivana Grle [5], Ivan Brkljačić [6], Josip Jularić [6], Goran Sporiš [6] and Suzana Žilič Fišer [7]

[1] Faculty of Sport and Physical Education, University of Niš, 18000 Niš, Serbia
[2] Faculty of Kinesiology, University of Split, Ruđera Boškovića 31, 21000 Split, Croatia
[3] Faculty of Sport and Physical Education, University of Novi Sad, 21000 Novi Sad, Serbia
[4] Orthopedic Clinical Department, University Clinical Hospital in Mostar, 88000 Mostar, Bosnia and Herzegovina
[5] Clinical Department for Physical Medicine and Rehabilitation, University Clinical Hospital in Mostar, 88000 Mostar, Bosnia and Herzegovina
[6] Faculty of Kinesiology, University of Zagreb, 10000 Zagreb, Croatia
[7] Institute of Media Communications, Faculty of Electrical Engineering and Computer Science, University of Maribor, Koroška Cesta 46, 2000 Maribor, Slovenia
* Correspondence: dukislavujac@gmail.com

Abstract: The aim of this study was to conduct a systematic review of the school-based exercise programs for promoting cardiorespiratory fitness in overweight and obese children aged 6 to 10. Electronic databases (Web of Science and PubMed) were used as searching tools for collecting adequate studies published in the past 20 years. A total of 13 studies met the criteria for inclusion in this review, with a total of 2810 participants, both male and female. According to the results of this systematic review, overweight and obese children aged 6 to 10 who underwent certain interventions had their CRF improved. Furthermore, evidence suggested that interventions carried out during a longer period of time suggested led to greater improvement of cardiorespiratory fitness than a shorter one, but the level of cardiorespiratory fitness gradually decreases after the intervention.

Keywords: physical activity; physical fitness; motor competence; children; health; monitoring and promoting; sport; sedentary behavior; obesity; well-being

Citation: Mijalković, S.; Stanković, D.; Tomljanović, M.; Batez, M.; Grle, M.; Grle, I.; Brkljačić, I.; Jularić, J.; Sporiš, G.; Fišer, S.Ž. School-Based Exercise Programs for Promoting Cardiorespiratory Fitness in Overweight and Obese Children Aged 6 to 10. *Children* **2022**, *9*, 1323. https://doi.org/10.3390/children9091323

Academic Editor: Georgian Badicu

Received: 7 July 2022
Accepted: 27 August 2022
Published: 30 August 2022

Publisher's Note: MDPI stays neutral with regard to jurisdictional claims in published maps and institutional affiliations.

Copyright: © 2022 by the authors. Licensee MDPI, Basel, Switzerland. This article is an open access article distributed under the terms and conditions of the Creative Commons Attribution (CC BY) license (https://creativecommons.org/licenses/by/4.0/).

1. Introduction

Cardiorespiratory fitness (CRF) is one of the most important health components of physical fitness [1], which is mainly expressed in maximal oxygen intake (VO2max) or in metabolic equivalents (MET) [2]. Cardiorespiratory fitness in children has a well-established link to overall health in youth and can lower the risk of cardiovascular disease (CVD) in later life [3–5]. Furthermore, it has been shown that there is a link between low CRF in childhood and early mortality in adulthood [6]. Therefore, in order to minimize the effects of CVD later in life and to prevent early mortality, it is strongly advised to concentrate on improving CRF from a young age [7,8]. It is well documented that overweight and obese children have lower CRF, and they are not able to train as hard and intensively as children with normal weight [9]. Furthermore, being overweight or obese as a child raises the risk of CVD in adulthood [10,11]. The negative effects of CRF and obesity from early age may be affected by being physically inactive [12,13]. On the other hand, it is proven that being physically active greatly influences the improvement of CRF in overweight and obese children aged 6 to 10 [14–17].

There is an increase in school interventions aiming to improve and promote CRF in early childhood [18–22]. Castro-Piñero et al. [23] state in their study that a CRF is a reliable indicator of CVD risk and should be tracked to identify children who may be at CVD risk. This study also suggests that, between baseline and follow-up, VO2max considerably decreased in both boys and girls ($p < 0.001$). They also concluded that the CRF should be a monitored system in order to prevent the potential occurrence of CVD. Regarding the program frequency, in order for overweight and obese children to have positive results, the CRF exercise program should be conducted three to four times a week for at least 6 weeks [24]. Recently, high-intensity circuit training (HIIT) has shown to be an effective exercise intervention that led to significant improvements and, therefore, could be included in regular classes [25]. Studies suggest that the HIIT method leads to a large improvement in CRF in children and affects the parameters related to neuromuscular and aerobic performance [26]. Furthermore, Stanly and Dharuman [27] state, in their study, that tai-chi, pilates, and yoga have proven to be methods that greatly influence the improvement of CRF.

It is essential to increase people's understanding of how low CRF may have a range of negative effects throughout life [28], and its development should start in childhood. Therefore, improving CRF should be an integral part of physical education programs in all lower grades of primary schools. The aim of this work was to conduct a systematic review of the school-based exercise programs for promoting CRF in overweight and obese children aged 6 to 10.

2. Materials and Methods

2.1. Literature Identification

According to the PRISMA (Preferred Reporting Items for Systematic Reviews and Meta-Analyses) standards, studies were searched and analyzed [29]. The following databases were searched to collect relevant literature for this study: PubMed and Web of Science. The following terms were used during the search: ((school-based OR school-program OR intervention OR preschool OR primary school OR elementary school) AND (cardio-respiratory fitness OR CRF OR cardio fitness OR VO2max OR maximal oxygen consumption OR heart rate)) AND (overweight OR obese) NOT disease. Child: 6–12 years filter was turned on (Table 1). Studies are selected on the basis of titles, keywords, and abstracts, but primarily on the basis of the content of the study published in its entirety.

Table 1. Search strategy to identify articles.

Search 1	Search 2	Search 3	Filters
school-based	cardio-respiratory fitness		
school program	CRF		
Intervention	cardio fitness	overweight	child: 6–12 years
preschool	VO2max	obese	
primary school	maximal oxygen consumption		
elementary school	heart rate		

The data were analyzed using the descriptive approach, and the titles and abstracts evaluating CRF in overweight and obese children were used to determine whether or not a particular study was included. Studies were carefully identified, and they were only deemed pertinent if they fit the inclusion criteria. Two authors (D.S. and S.M.) carried out the research search, value evaluation, and data extraction. Each author then carried out cross-identification of studies, after which the study was either accepted or rejected for further analysis.

2.2. Inclusion Criteria

To be taken into account for the final analysis, the study had to meet the following criteria: The first requirement was that the study examines the relationship between school exercise programs and CRF among overweight and obese children aged 6 to 10. This selection criterion was used to rule out studies that included children who were not of this age and studies whose goal was not to determine how the school exercise program affected cardiorespiratory fitness. The second requirement was that the study's participants had to be overweight or obese. The research had to have been published within the last 20 years, which was the third requirement. The fourth criterion was that the studies were published in English. The fifth criterion was that studies were original research (Figure 1).

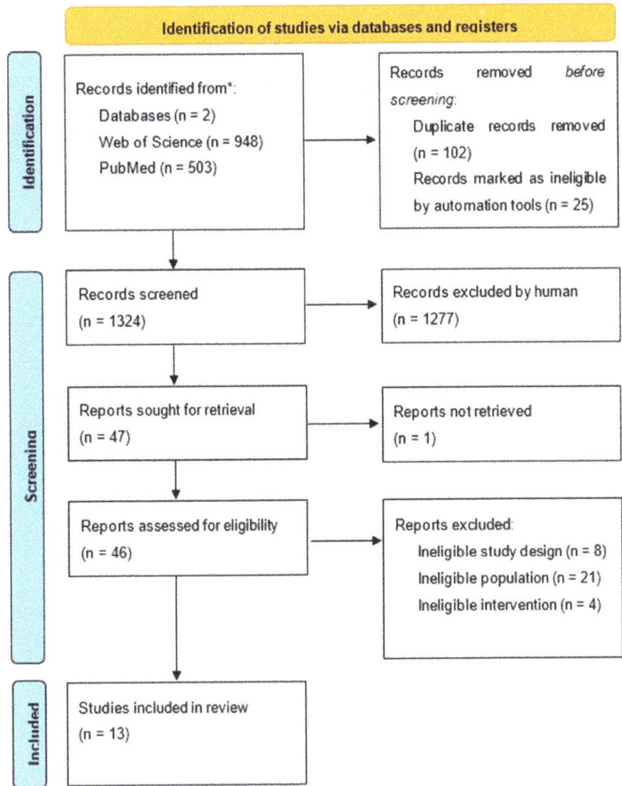

Figure 1. PRISMA flow diagram.

2.3. Risk of Bias Assessment

The study's quality and viability for inclusion, in the final analysis, were evaluated by two separate authors (S.M. and D.S.). The "Rayyan" web tool was used to do blind reviewing. A third reviewer (M.T.), who made the ultimate determination in cases of dispute on the findings on the assessment of the risk of bias, evaluated the collected data.

3. Results

3.1. Quality of the Studies

Pedro scale results were shown in Table 2. The total number of studies included in the quantitative synthesis and the points each study obtained on the PEDro scale were used to generate the study assessment scores [30]. The first criterion, which determines eligibility, is concerned with external validity but is not factored into the final result.

Table 2. PEDro scale results.

Study						Criterion						
	1	2	3	4	5	6	7	8	9	10	11	Σ
Thivel et al. (2011) [31]	Y	Y	Y	Y	N	N	N	Y	Y	Y	Y	7
Resaland et al. (2011) [32]	Y	N	N	Y	N	N	N	N	Y	Y	Y	4
Yin et al. (2012) [33]	Y	Y	Y	Y	N	N	N	N	Y	Y	Y	6
Krustrup et al. (2014) [34]	Y	Y	Y	Y	N	N	Y	Y	Y	Y	Y	8
Khan et al. (2014) [35]	Y	Y	Y	Y	Y	N	N	Y	Y	Y	Y	8
Tan et al. (2015) [36]	Y	Y	Y	Y	N	N	N	Y	Y	Y	Y	7
Martinez et al. (2016) [19]	Y	Y	Y	Y	N	N	N	Y	Y	Y	Y	7
Leeuwen et al. (2018) [17]	Y	N	N	N	N	N	N	Y	Y	N	Y	3
Ye et al. (2019) [22]	Y	Y	Y	N	N	N	N	Y	N	Y	Y	5
Davis et al. (2019) [16]	Y	Y	Y	Y	Y	N	N	Y	Y	Y	Y	8
Espinoza-Silva et al. (2019) [37]	Y	N	N	Y	N	N	N	Y	Y	Y	Y	5
Leandro et al. (2021) [14]	Y	Y	Y	Y	N	N	N	Y	Y	Y	Y	7
Martinez-Viscaiano et al. (2022) [15]	Y	Y	Y	Y	Y	Y	Y	Y	N	Y	Y	9

Legend: 1—eligibility criteria; 2—random allocation; 3—concealed allocation; 4—baseline comparability; 5—blind subject; 6—blind clinician; 7—blind assessor; 8—adequate follow-up; 9—intention-to-treat analysis; 10—between-group analysis; 11—point estimates and variability; Y—criterion is satisfied; N—criterion is not satisfied; Σ—total awarded points.

3.2. Selection and Characteristics of Studies

The electronic databases were searched, and 1451 studies were located. Following the elimination of duplicate research, systematic reviews, and meta-analyses, 1324 studies were left. After 1277 research were disqualified owing to inclusion requirements, 42 studies were evaluated for eligibility. A total of 13 papers were included in the final analysis after the remaining studies were reviewed and thoroughly read (Table 3).

Table 3. Participants, variables, interventions and results of included studies.

First Author and Year of Publication	Sample of Participants		PF	Type of Intervention	Duration of Intervention	Results
	Number	Age			Weeks	
Thivel et al. (2011) [31]	N—457	6–10	SRT, CPP, RHR, HRR	AL, 2/week (60 min)	26	HRR↑ RHR↑ (E&C)
Resaland et al. (2011) [32]	N—256 M—125 F—131	9–10	TRE, VO_{2peak}, HR_{peak}	Daily, MVPA (60 min)	104	VO_{2peak}↑ (E)
Yin et al. (2012) [33]	N—574	8.7 ± 0.5	ST, HR	Daily, Kids4Fit (120 min)	156	HR↑ (E&C)
Krustrup et al. (2014) [34]	N—51 M—21 F—30	9–10	CTE, IVRTglobal, LVSEF, LVPWD, RHR, RBP, HR, HR_{max}	SSF, 3x/week (40 min)	10	LVSEF↑ IVRTglobal↑(E)
Khan et al. (2014) [35]	N—220 M—117 F—103	8–9	TRE, HR, VO_{2max}	Daily MVPA (70 min)	39	VO_{2max}↑ (E)
Tan et al. (2015) [36]	M—46	8–10	SRT, VO_{2max}, HR, FAT_{max}	Daily PA (40 min)	10	VO_{2max}↑ (E)
Martinez et al. (2016) [19]	N—94 M—52 F-42	7–9	SRT, TRE, VO_{2max}, EPOC, HR_{max}	HIIT, 2x/week (90 min)	13	VO_{2max}↑ EPOC↑ (E)
Leeuwen et al. (2018) [17]	N—154 M—66 F—88	8.5 ± 1.8	SRT, VO_{2max}, BP	Kids4Fit, 2xweek; 1xweek (20 min)	13	VO_{2max}↑ BP↑ (E)
Ye et al. (2019) [22]	N—81 M—42 F—39	9.23 ± 0.62	HMR, VO_{2max}	EXG, 1/week (50 min)	35	/

Table 3. *Cont.*

First Author and Year of Publication	Sample of Participants		PF	Type of Intervention	Duration of Intervention	Results
	Number	Age			Weeks	
Davis et al. (2019) [16]	N—75 M—29 F—46	9.5–9.8	TRE, PWV, BP, VO$_{2peak}$	Daily, ASAE (40 min)	35	VO$_{2peak}$↑ (E)
Espinoza-Silva et al. (2019) [37]	N—274 M—120 F—154	7–9	6MWT, BP, VO$_{2max}$	HIIT, 2x/week (40–50 min)	30	VO$_{2max}$↑ (E)
Leandro et al. (2021) [14]	M—41	7–9	ABPM, RHR, BP	PLT 3x/week (20 min)	13	BP↑ RHR↑ (E)
Martinez-Viscaiano et al. (2022) [15]	N—487 M—233 F—254	9.89 ± 0.71	SRT, BP, VO$_{2max}$	HIIT, 4x/week (60 min)	39	BP↑ VO$_{2max}$↑ (E/onlyF)

Legend: ↑ significant improvement; N—number of respondents; M—male participants; F—female participanrts; E—experimental group; C—control group; PF—physical fitness test; PA—physical activity; SSF—small-sided football; AL—additional lessons; PLT—plyometric training; HIIT—high intensity interval training; VO$_{2max}$—maximal oxygen consumption; VO$_{2peak}$—peak oxygen uptake; HR—heart rate; RHR—rest heart rate; HRR—heart rate reserve; HR$_{peak}$—peak heart rate; HR$_{max}$—maximum heart rate; LVSEF—left ventricular systolic ejection fraction; IVRTglobal—global isovolumetric relaxation time; LVPWD—left ventricular posterior wall diameter; PWV—Carotid-femoral pulse wave velocity; BP—blood pressure; RBP—rest blood pressure; EPOC—excess post-exercise oxygen consumption; ST—step test; 6MWT—6 min walk test; HMR—half-mile run; Kids4Fit—multidisciplinary weight reduction program; CTE—comprehensive transthoracic echocardiography; ABPM—automatic arterial blood pressure monitor; TRE—treadmill protocol; CPP—cycle peak power; SRT—shuttle run test; MVPA—moderate-to-vigorous intensity physical activity; EXG—exergaming; ASAE—after school aerobic exercise; FATmax—the intensity of maximal fat oxidation rate.

Thirteen studies met the inclusion criteria for inclusion in this review. The oldest study was published in 2011 [31], and the most recent one is from 2022 [15]. The total number of participants was 2810. The highest number of participants was 574 [33] (Yin), and the lowest number of participants was 41 [14]. In almost all studies, the participants were both sexes. However, in two studies, the participants were only male [14,36], while no study was performed with females only. The longest intervention (36 months) was by [33], and the shortest interventions (two and a half months) were by Krustrup et al. [34] and Tan et al. [36]. All studies aimed to improve CRF, and post-intervention CRF improvement was found in all studies except in one [22]. The interventions most used in the studies were high-intensity interval training [15,19,37] and daily physical activity [16,32,33,35,36].

There were eight studies that increased VO$_{2max}$/VO$_{2peak}$ with physical activity sessions that lasted between 20 to 90 min; session frequency varied between daily, four times a week, two times a week, and once a week [15–17,19,32,35–37]. Heart rate and RHR were improved in three studies [14,31,33] by physical activity sessions that lasted either 20, 60, or 120 min; session frequencies varied between daily, two times a week, and three times a week. Blood pressure was improved in three studies [14,15,17] by physical activity sessions that lasted 20 or 60 min, and session frequencies varied between two, three, and four times a week.

4. Discussion

The current study aimed to conduct a systematic review of the school-based exercise programs for promoting cardiorespiratory fitness (CRF) in overweight and obese children aged 6 to 10. CRF has an important role in children's health status. According to the results of the reviewed studies, twelve school-based programs have shown to affect the improvement of CRF in overweight and obese children aged 6 to 10 to some extent. Consequently, interventions such as high-intensity interval training, plyometric training, multidisciplinary weight reduction program (Kids4Fit), football, and active video gaming have a positive influence on CRF in children and reduce the risk of CVD. Therefore, the school's physical education program should include exercises for promoting CRF in order to increase their aerobic capacity.

A key goal of lowering cardiovascular complications is to increase exercise capacity and CRF [38]. The main parameters that indicate CRF improvements tend to be VO_{2max}, VO_{2peak}, excess post-exercise oxygen consumption (EPOC), and resting HR [14,15,19,33,35–37]. Except for one study [22], all studies that have been reviewed achieved improvements in CRF with specific intervention. However, the question is whether these are safe and appropriate interventions for children aged 6 to 10 in terms of individualization and specificity [39]. Yin et al. [33], in their study, showed a positive trend in improving CRF, which is shown by heart rate (bpm) ($p < 0.001$), in favor of the intervention group. Additionally, plyometric training has proved to be another beneficial tool for lowering heart rate in a resting position [14]. Tan et al. [36] tend to improve CRF through various 40 min physical activities (walking, running, and ball games). A positive method in this study was that assessors had constant control over the subjects' HRs which they wore during every training session. In another study, Thivel et al. [31] showed similar results according to CRF in the control and intervention groups. Actually, both control and intervention groups had significant increases in CRF, which was defined by the number of fully completed stages in the shuttle run.

There is strong evidence that high-intensity interval training (HIIT) can be a feasible and powerful tool for improving CRF [40–43]. The literature review reveals that HIIT training is popular among older populations [44–47]; however, there are papers that integrate school-based programs for younger groups aged 6 to 10. Martinez-Visciano et al. [15] concluded that HIIT training improved the girls' CRF throughout one school year. On the other hand, Martinez et al. [19] proved that VO_{2max} in overweight children was enhanced in only three months while conducting two HIIT training sessions per week and using high-intensity intermittent exercises and sports activities such as: half-squats, sprints, jumps, and horizontal shot puts. In addition, strategies with exercise machines such as bicycles and treadmills, as well as basic motor skills (running, jumping, throwing), were applied in high-intensity programs [37]. Evidence suggests that school-based HIIT training program leads to improvement of aerobic capacity of overweight and obese children.

School-based exercise programs for increasing CRF in overweight and obese children aged 6 to 10 included after-school aerobic workouts, moderate to vigorous physical activity (MVPA), the multidisciplinary weight reduction program (Kids4Fit), and the intensity of maximal fat oxidation rate (FATmax) [16,32,33,35,36]. It can be said that CRF and aerobic capacity of children aged 6 to 10 have significantly improved as a result of everyday participation in these physical activities. Additionally, Leeuwen et al. [17] used Kids4Fit as an intervention to enhance CRF, with the intervention taking place twice a week during the initial six weeks while being carried out the intervention once a week during the final six weeks. While conducting this study, a significant positive effect on CRF was also noticed in overweight and obese children, but after the intervention, CRF gradually declined. These findings would suggest that Kids4Fit was a good school-based intervention program for promoting CRF in children aged 6 to 10, but it is necessary to do the intervention daily and for a longer period of time for the CRF to continue improving. The regular classes of physical education are not enough in order to promote the children's CRF. However, if two additional workouts, which include exercises to improve coordination, strength, endurance, speed, and flexibility, are added to regular classes, the improvement of the CRF in overweight and obese children can be greatly influenced [31]. Furthermore, exergaming and small-sided football are two interesting and fun ways to include additional classes of vigorous intensity. In addition to children's enjoyment, they also regulate their CRF by using these interventions [22,45]. It is necessary for the realization of children's regular classes to be fun and playable in order to improve their CRF.

The main limitation of this review was the small number of articles engaging CRF in children aged 6 to 10. In addition, the majority of studies included children with average body weight. Secondly, the study covered only overweight and obese children who are more likely to increase their CRF due to them being less active than children of average weight. The third limitation of the study is the variety of ethnicities among participants since the mentalities of different cultures regarding physical activity and motivation differences.

5. Conclusions

The results of this systematic review showed that there were interventions that led to improvement in CRF in overweight and obese children aged 6 to 10. Long-lasting interventions led to greater improvement of CRF than a shorter intervention. Our findings provide evidence that school-based exercise programs greatly influence the CRF parameters such as maximal oxygen intake, peak oxygen uptake, heart rate, and resting heart rate, but it gradually declines after the intervention.

Author Contributions: Conceptualization, S.M. and D.S.; methodology, S.M. and D.S.; software, J.J.; validation, M.T., G.S. and S.Ž.-F.; formal analysis, M.B. and I.G.; investigation, M.G.; resources, S.M. and I.B.; data curation, M.G.; writing—original draft preparation, D.S.; writing—review and editing, S.Ž.-F. and J.J.; visualization, M.T.; supervision, G.S. and S.Ž.-F.; project administration, S.M.; funding acquisition, M.B. All authors have read and agreed to the published version of the manuscript.

Funding: This research received no external funding.

Institutional Review Board Statement: Not applicable.

Informed Consent Statement: Not applicable.

Data Availability Statement: Not applicable.

Conflicts of Interest: The authors declare no conflict of interest.

References

1. Veijalainen, A.; Tompuri, T.; Haapala, E.A.; Viitasalo, A.; Lintu, N.; Väistö, J.; Laitinen, T.; Lindi, V.; Lakka, T.A. Associations of cardiorespiratory fitness, physical activity, and adiposity with arterial stiffness in children. *Scand. J. Med. Sci. Sport* **2016**, *26*, 943–950. [CrossRef] [PubMed]
2. Lee, D.-c.; Artero, E.G.; Sui, X.; Blair, S.N. Mortality trends in the general population: The importance of cardiorespiratory fitness. *J. Psychopharmacol.* **2010**, *24*, 27–35. [CrossRef] [PubMed]
3. Ortega, F.B.; Ruiz, J.R.; Castillo, M.J.; Moreno, L.A.; Urzanqui, A.; González-Gross, M.; Sjöström, M.; Gutiérrez, A. Health-related physical fitness according to chronological and biological age in adolescents. The AVENA study. *J. Sports Med. Phys. Fit.* **2008**, *48*, 371–379.
4. Ruiz, J.R.; Castro-Piñero, J.; Artero, E.G.; Ortega, F.B.; Sjöström, M.; Suni, J.; Castillo, M.J. Predictive validity of health-related fitness in youth: A systematic review. *Br. J. Sports Med.* **2009**, *43*, 909–923. [CrossRef]
5. Ruiz, J.R.; Cavero-Redondo, I.; Ortega, F.B.; Welk, G.J.; Andersen, L.B.; Martinez-Vizcaino, V. Cardiorespiratory fitness cut points to avoid cardiovascular disease risk in children and adolescents; what level of fitness should raise a red flag? A systematic review and meta-analysis. *Br. J. Sports Med.* **2016**, *50*, 1451–1458. [CrossRef]
6. Sui, X.; LaMonte, M.J.; Laditka, J.N.; Hardin, J.W.; Chase, N.; Hooker, S.P.; Blair, S.N. Cardiorespiratory fitness and adiposity as mortality predictors in older adults. *JAMA* **2007**, *298*, 2507–2516. [CrossRef]
7. Rodrigues, A.N.; Perez, A.J.; Carletti, L.; Bissoli, N.S.; Abreu, G.R. The association between cardiorespiratory fitness and cardiovascular risk in adolescents. *J. Pediatr.* **2007**, *83*, 429–435. [CrossRef]
8. Cao, C.; Yang, L.; Cade, W.T.; Racette, S.B.; Park, Y.; Cao, Y.; Friedenreich, C.M.; Hamer, M.; Stamatakis, E.; Smith, L. Cardiorespiratory Fitness Is Associated with Early Death among Healthy Young and Middle-Aged Baby Boomers and Generation Xers. *Am. J. Med.* **2020**, *133*, 961–968.e3. [CrossRef]
9. Setty, P.; Padmanabha, B.V.; Doddamani, B.R. Correlation between obesity and cardio respiratory fitness. *Int. J. Med. Sci. Public Health* **2013**, *2*, 300–304. [CrossRef]
10. Jastreboff, A.M.; Kotz, C.M.; Kelly, A.S.; Heymsfield, S.B. Obesity as a disease: The obesity society 2018 position statement. *Obesity* **2019**, *27*, 7–9. [CrossRef]
11. Saha, A.K.; Sarkar, N.; Chatterjee, T. Health consequences of childhood obesity. *Indian J. Pediatrics* **2011**, *78*, 1349–1355. [CrossRef] [PubMed]
12. Katzmarzyk, P.T.; Lear, S.A. Physical activity for obese individuals: A systematic review of effects on chronic disease risk factors. *Obes. Rev.* **2012**, *13*, 95–105. [CrossRef] [PubMed]

13. Vasconcellos, F.; Seabra, A.; Katzmarzyk, P.T.; Kraemer-Aguiar, L.G.; Bouskela, E.; Farinatti, P. Physical activity in overweight and obese adolescents: Systematic review of the effects on physical fitness components and cardiovascular risk factors. *Sports Med.* **2014**, *44*, 1139–1152. [CrossRef]

14. Góis Leandro, C.; Arnaut Brinco, R.; Góes Nobre, G.; Góes Nobre, I.; Silva-Santiago, L.C.; Aires-Dos-Santos, B.R.; Marinho-Dos-Santos, R.; Rodrigues-Ribeiro, M.; Marinho-Barros, M.R.; Alves-Macedo, F.; et al. Post-exercise hypotension effects in response to plyometric training of 7- to 9-year-old boys with overweight/obesity: A randomized controlled study. *J. Sports Med. Phys. Fit.* **2021**, *61*, 1281–1289. [CrossRef] [PubMed]

15. Martínez-Vizcaíno, V.; Soriano-Cano, A.; Garrido-Miguel, M.; Cavero-Redondo, I.; de Medio, E.P.; Madrid, V.M.; Martínez-Hortelano, J.A.; Berlanga-Macías, C.; Sánchez-López, M. The effectiveness of a high-intensity interval games intervention in schoolchildren: A cluster-randomized trial. *Scand. J. Med. Sci. Sports* **2022**, *32*, 765–781. [CrossRef] [PubMed]

16. Davis, C.L.; Litwin, S.E.; Pollock, N.K.; Waller, J.L.; Zhu, H.D.; Dong, Y.B.; Kapuku, G.; Bhagatwala, J.; Harris, R.A.; Looney, J.; et al. Exercise effects on arterial stiffness and heart health in children with excess weight: The SMART RCT. *Int. J. Obes.* **2020**, *44*, 1152–1163. [CrossRef] [PubMed]

17. Van Leeuwen, J.; Andrinopoulou, E.R.; Hamoen, M.; Paulis, W.D.; Van Teeffelen, J.; Kornelisse, K.; Van Der Wijst-Ligthart, K.; Koes, B.W.; Van Middelkoop, M. The effect of a multidisciplinary intervention program for overweight and obese children on cardiorespiratory fitness and blood pressure. *Fam. Pract.* **2018**, *36*, 147–153. [CrossRef]

18. Jankowski, M.; Niedzielska, A.; Brzezinski, M.; Drabik, J. Cardiorespiratory fitness in children: A simple screening test for population studies. *Pediatr. Cardiol.* **2015**, *36*, 27–32. [CrossRef]

19. Martínez, S.R.; Ríos, L.J.C.; Tamayo, I.M.; Almeida, L.G.; López-Gomez, M.A.; Jara, C.C. An After-School, high-intensity, interval physical activity programme improves health-related fitness in children. *Motriz Rev. Educ. Fis.* **2016**, *22*, 359–367. [CrossRef]

20. Meng, C.; Yucheng, T.; Shu, L.; Yu, Z. Effects of school-based high-intensity interval training on body composition, cardiorespiratory fitness and cardiometabolic markers in adolescent boys with obesity: A randomized controlled trial. *BMC Pediatr.* **2022**, *22*, 112. [CrossRef]

21. Larsen, M.N.; Nielsen, C.M.; Helge, E.W.; Madsen, M.; Manniche, V.; Hansen, L.; Hansen, P.R.; Bangsbo, J.; Krustrup, P. Positive effects on bone mineralisation and muscular fitness after 10 months of intense school-based physical training for children aged 8-10 years: The FIT FIRST randomised controlled trial. *Br. J. Sports Med.* **2018**, *52*, 254–260. [CrossRef] [PubMed]

22. Ye, S.; Pope, Z.C.; Lee, J.E.; Gao, Z. Effects of school-based exergaming on urban children's physical activity and cardiorespiratory fitness: A quasi-experimental study. *Int. J. Environ. Res. Public Health* **2019**, *16*, 4080. [CrossRef] [PubMed]

23. Castro-Piñero, J.; Perez-Bey, A.; Segura-Jiménez, V.; Aparicio, V.A.; Gómez-Martínez, S.; Izquierdo-Gomez, R.; Marcos, A.; Ruiz, J.R.; Veiga, O.L.; Bandres, F.; et al. Cardiorespiratory Fitness Cutoff Points for Early Detection of Present and Future Cardiovascular Risk in Children: A 2-Year Follow-up Study. *Mayo Clin. Proc.* **2017**, *92*, 1753–1762. [CrossRef] [PubMed]

24. Braaksma, P.; Stuive, I.; Garst, R.; Wesselink, C.F.; van der Sluis, C.K.; Dekker, R.; Schoemaker, M.M. Characteristics of physical activity interventions and effects on cardiorespiratory fitness in children aged 6–12 years-A systematic review. *J. Sci. Med. Sport* **2018**, *21*, 296–306. [CrossRef] [PubMed]

25. Engel, F.A.; Wagner, M.O.; Schelhorn, F.; Deubert, F.; Leutzsch, S.; Stolz, A.; Sperlich, B. Classroom-Based Micro-Sessions of Functional High-Intensity Circuit Training Enhances Functional Strength but Not Cardiorespiratory Fitness in School Children—A Feasibility Study. *Front. Public Health* **2019**, *7*, 291. [CrossRef] [PubMed]

26. Bauer, N.; Sperlich, B.; Holmberg, H.C.; Engel, F.A. Effects of High-Intensity Interval Training in School on the Physical Performance and Health of Children and Adolescents: A Systematic Review with Meta-Analysis. *Sport Med. Open* **2022**, *8*, 50. [CrossRef]

27. Stanly, S.L.; Maniazhagu, D. Individual and combined interventions of tai chi, pilates and yogic practices on cardio respiratory endurance of b.ed. trainees. *Int. J. Phys. Educ. Sport Manag. Yogic Sci.* **2020**, *10*, 25–31. [CrossRef]

28. Kodama, S.; Saito, K.; Tanaka, S.; Maki, M.; Yachi, Y.; Asumi, M.; Sugawara, A.; Totsuka, K.; Shimano, H.; Ohashi, Y.; et al. Cardiorespiratory fitness as a quantitative predictor of all-cause mortality and cardiovascular events in healthy men and women: A meta-analysis. *JAMA* **2009**, *301*, 2024–2035. [CrossRef]

29. Page, M.J.; McKenzie, J.E.; Bossuyt, P.M.; Boutron, I.; Hoffmann, T.C.; Mulrow, C.D.; Shamseer, L.; Tetzlaff, J.M.; Akl, E.A.; Brennan, S.E.; et al. The PRISMA 2020 statement: An updated guideline for reporting systematic reviews. *BMJ* **2021**, *372*, n71. [CrossRef]

30. Maher, C.G.; Sherrington, C.; Herbert, R.D.; Moseley, A.M.; Elkins, M. Reliability of the PEDro scale for rating quality of randomized controlled trials. *Phys. Ther.* **2003**, *83*, 713–721. [CrossRef]

31. Thivel, D.; Isacco, L.; Lazaar, N.; Aucouturier, J.; Ratel, S.; Doré, E.; Meyer, M.; Duché, P. Effect of a 6-month school-based physical activity program on body composition and physical fitness in lean and obese schoolchildren. *Eur. J. Pediatr.* **2011**, *170*, 1435–1443. [CrossRef]

32. Resaland, G.K.; Andersen, L.B.; Mamen, A.; Anderssen, S.A. Effects of a 2-year school-based daily physical activity intervention on cardiorespiratory fitness: The Sogndal school-intervention study. *Scand. J. Med. Sci. Sport* **2011**, *21*, 302–309. [CrossRef] [PubMed]

33. Yin, Z.; Moore, J.B.; Johnson, M.H.; Vernon, M.M.; Gutin, B. The impact of a 3-year after-school obesity prevention program in elementary school children. *Child. Obes.* **2012**, *8*, 60–70. [CrossRef] [PubMed]

34. Krustrup, P.; Hansen, P.R.; Nielsen, C.M.; Larsen, M.N.; Randers, M.B.; Manniche, V.; Hansen, L.; Dvorak, J.; Bangsbo, J. Structural and functional cardiac adaptations to a 10-week school-based football intervention for 9-10-year-old children. *Scand. J. Med. Sci. Sport* **2014**, *24*, 4–9. [CrossRef] [PubMed]

35. Khan, N.A.; Raine, L.B.; Drollette, E.S.; Scudder, M.R.; Pontifex, M.B.; Castelli, D.M.; Donovan, S.M.; Evans, E.M.; Hillman, C.H. Impact of the fitkids physical activity intervention on adiposity in prepubertal children. *Pediatrics* **2014**, *133*, e875–e883. [CrossRef]

36. Tan, S.; Wang, J.; Cao, L. Exercise training at the intensity of maximal fat oxidation in obese boys. *Appl. Physiol. Nutr. Metab.* **2015**, *41*, 49–54. [CrossRef]

37. Espinoza-Silva, M.; Latorre-Román, P.; Párraga-Montilla, J.; Caamaño-Navarrete, F.; Jerez-Mayorga, D.; Delgado-Floody, P. Response of obese schoolchildren to high-intensity interval training applied in the school context. *Endocrinol. Diabetes Nutr.* **2019**, *66*, 611–619. [CrossRef]

38. Al-Mallah, M.H.; Sakr, S.; Al-Qunaibet, A. Cardiorespiratory Fitness and Cardiovascular Disease Prevention: An Update. *Curr. Atheroscler. Rep.* **2018**, *20*, 1. [CrossRef]

39. Ferguson, B. ACSM's Guidelines for Exercise Testing and Prescription 9th Ed. 2014. *J. Can. Chiropr. Assoc.* **2014**, *58*, 328.

40. Burgomaster, K.A.; Heigenhauser, G.J.F.; Gibala, M.J. Effect of short-term sprint interval training on human skeletal muscle carbohydrate metabolism during exercise and time-trial performance. *J. Appl. Physiol.* **2006**, *100*, 2041–2047. [CrossRef]

41. Gibala, M.J.; Little, J.P.; Macdonald, M.J.; Hawley, J.A. Physiological adaptations to low-volume, high-intensity interval training in health and disease. *J. Physiol.* **2012**, *590*, 1077–1084. [CrossRef] [PubMed]

42. Whyte, L.J.; Gill, J.M.R.; Cathcart, A.J. Effect of 2 weeks of sprint interval training on health-related outcomes in sedentary overweight/obese men. *Metabolism* **2010**, *59*, 1421–1428. [CrossRef] [PubMed]

43. Bogataj, Š.; Trajković, N.; Cadenas-Sanchez, C.; Sember, V. Effects of School-Based Exercise and Nutrition Intervention on Body Composition and Physical Fitness in Overweight Adolescent Girls. *Nutrients* **2021**, *13*, 238. [CrossRef] [PubMed]

44. Eather, N.; Riley, N.; Miller, A.; Smith, V.; Poole, A.; Vincze, L.; Morgan, P.J.; Lubans, D.R. Efficacy and feasibility of HIIT training for university students: The Uni-HIIT RCT. *J. Sci. Med. Sport* **2019**, *22*, 596–601. [CrossRef]

45. Berge, J.; Hjelmesæth, J.; Hertel, J.K.; Gjevestad, E.; Småstuen, M.C.; Johnson, L.K.; Martins, C.; Andersen, E.; Helgerud, J.; Støren, Ø. Effect of Aerobic Exercise Intensity on Energy Expenditure and Weight Loss in Severe Obesity-A Randomized Controlled Trial. *Obesity* **2021**, *29*, 359–369. [CrossRef]

46. Jiménez-García, J.D.; Hita-Contreras, F.; de la Torre-Cruz, M.J.; Aibar-Almazán, A.; Achalandabaso-Ochoa, A.; Fábrega-Cuadros, R.; Martínez-Amat, A. Effects of HIIT and MIIT Suspension Training Programs on Sleep Quality and Fatigue in Older Adults: Randomized Controlled Clinical Trial. *Int. J. Environ. Res. Public Health* **2021**, *18*, 1211. [CrossRef]

47. Jiménez-García, J.D.; Martínez-Amat, A.; De La Torre-Cruz, M.J.; Fábrega-Cuadros, R.; Cruz-Díaz, D.; Aibar-Almazán, A.; Achalandabaso-Ochoa, A.; Hita-Contreras, F. Suspension Training HIIT Improves Gait Speed, Strength and Quality of Life in Older Adults. *Int. J. Sports Med.* **2019**, *40*, 116–124. [CrossRef]

children

MDPI

Article

Relationships between Functional Movement Quality and Sprint and Jump Performance in Female Youth Soccer Athletes of Team China

Junjie Zhang [1], Junlei Lin [2,*], Hongwen Wei [2] and Haiyuan Liu [1]

[1] Graduate School, Capital University of Physical Education and Sports, Beijing 100091, China
[2] School of Strength and Conditioning Training, Beijing Sport University, Beijing 100084, China
* Correspondence: linjunlei@bsu.edu.cn

Abstract: This study aimed to determine the optimal functional movement screen (FMS) cut score for assessing the risk of sport injury, and to investigate the correlations between functional movement quality and sprint and jump performance. Twenty-four (N = 24) athletes performed all tests in one day at 10–30 min intervals, and the FMS test was performed first, without a warm-up session. After a standard warm-up, athletes then completed the Y-balance Test (YBT), sprint, counter-movement jump (CMJ), and standing long jump (SLJ), in turn. For each test, the best of three attempts was recorded for further analysis. A receiver operating characteristic (ROC) curve and area-under-the-curve (AUC) were used to determine the optimal FMS cut score for assessing the risk of sport injuries, and Spearman's rank correlation analysis was used to quantify associations between functional movement scores and athletic performance. The average FMS score was 16.2 and the optimal FMS cut score for assessing the risk of sport injuries was 14.5. There were moderate relationships between total FMS score and 10–20 m sprint time (r = −0.46, $p < 0.05$), between In-line Lunge and 0–20 m sprint time (r = −0.47, $p < 0.05$), between Shoulder Mobility and 0–10 m sprint time (r = −0.48, $p < 0.05$), and between Trunk-stability Push-up and 10–20 m sprint time (r = −0.47, $p < 0.05$). Moreover, Hurdle Step score was largely correlated with 0–10 m time (r = −0.51, $p < 0.05$). For Y-balance, moderate correlations were observed between CMJ height and anterior asymmetry score (r = −0.47, $p < 0.05$) and posteromedial asymmetry score (r = −0.44, $p < 0.05$). However, there were no significant associations between YBT performance (asymmetric in three directions and composite score) and sprint performance ($p > 0.05$). Taken together, the results indicate that a FMS score of 14 is not a gold standard for assessing the risk of injury in all populations; we recommend that the FMS cut score of 14.5 should be the optimal score for assessing risk of injury in young female elite soccer players. Moreover, the FMS and YBT were introduced to assess the quality of functional movements, and they cannot be used to assess sprint and jump performance. Practitioners can use components of the FMS that have similar characteristics to specific sports to assess athletic performance.

Keywords: youth; sprint; jump; YBT; FMS

Citation: Zhang, J.; Lin, J.; Wei, H.; Liu, H. Relationships between Functional Movement Quality and Sprint and Jump Performance in Female Youth Soccer Athletes of Team China. *Children* **2022**, *9*, 1312. https://doi.org/10.3390/children9091312

Academic Editors: Georgian Badicu, Ana Filipa Silva and Hugo Miguel Borges Sarmento

Received: 2 August 2022
Accepted: 16 August 2022
Published: 29 August 2022

Publisher's Note: MDPI stays neutral with regard to jurisdictional claims in published maps and institutional affiliations.

Copyright: © 2022 by the authors. Licensee MDPI, Basel, Switzerland. This article is an open access article distributed under the terms and conditions of the Creative Commons Attribution (CC BY) license (https://creativecommons.org/licenses/by/4.0/).

1. Introduction

The game of professional soccer demands a high level of physical performance from players, all of whom need to have excellent strength, speed, endurance, etc. For soccer athletes, several tests have been conducted to assess performance. The most common measurement tools include 1-repetition maximum (1RM) [1], rate of force development [2], 505 change of direction [3], and the yo-yo test [4]. Notably, these tests only provide information on musculoskeletal and cardiopulmonary fitness, and lack functional insight. Insufficient functional ability can reduce athletic performance and increase the risk of injury; it is thus essential to identify functional deficits using functional screening tools [5].

Presently available functional screening tools include functional movement screening (e.g., functional movement screen, FMS) and functional asymmetric screening (e.g., Y-balance Test, YBT). FMS was introduced to evaluate functional limitations, and was essential in determining mobility and stability when performing fundamental movement skills [6]. It can also assess maximal strength, postural control, and range of motion [7]. The capability of functional movement refers to the ability to perform essential motor skills (locomotor, stabilization, and manipulation) under controlled conditions [6]. The FMS test comprises seven distinct tests, which are scored from 0 to 3, and a composite score ranges from 0 to 21. Furthermore, FMS is a reliable assessment tool for interrater (ICC: 0.81; 95% CI: 0.74–0.8) and interrater reliability (ICC: 0.81, 95% CI: 0.70–0.92) [8]. YBT was derived from the Star Excursion Balance Test (SEBT) [9] and can be used to assess dynamic balance, functional asymmetric, proprioception, and strength [10]. Numerous studies have proved that YBT is a reliable assessment of dynamic balance (ICCS: 0.85–1.00) [9,11].

Due to their role in identifying functional deficits, these functional screening tools can be a useful predictor of injury risk. Massive studies have examined the validity of the FMS as a tool for predicting injury risk. A study conducted by Kiesel et al. [12] found that football athletes with FMS score of ≤ 14 showed greater risk of sports injury than those with higher scores. Elsewhere, this finding was also proved by other studies [13,14]. However, FMS cut score is influenced by a variety of factors, including training state, sport participating, gender, age, etc. It is therefore necessary to determine the optimal FMS cut score for different populations, a number which can then provide useful information for athletes and practitioners. However, to the author's knowledge, no studies have been conducted on the optimal FMS cut score in populations of young female football athletes.

These functional screening tools were introduced to assess movement quality, which also can be indicative of sports performance [15]. Okada et al. [16] found that the FMS score is relatively strongly associated with lower-limb strength performance (r = $-0.38 \sim -0.46$) in recreational mixed-gender athletes. Similarly, significant correlations between FMS and vertical jump and agility have been observed in youth soccer athletes [17]. Furthermore, Liang et al. [18] explored whether FMS score was associated with athletic performance in collegiate baseball players. The authors reported that greater FMS scores indicated better speed and agility performance. However, conflicting results have been reported in other studies. Parchmann and McBride [19] reported no significant correlations between FMS scores and sprint time, jumping, or agility performance in college golfers. Similarly, the FMS score was not associated with selected physical performance in active subjects [20]. It is worth noting that the authors of these studies creatively combined athletes of different genders or studied male players only.

Therefore, this study aims to: (1) determine the optimal FMS cut score for assessing risk of sports injury in female youth football players; and (2) investigate the correlations between functional movement quality (FMS and YBT) and sprint and jump performance. It is hypothesized that the optimal FMS cut score is higher than 14, and that functional movement quality is strongly correlated with sprint time and jump performance.

2. Materials and Methods

A cross-sectional design was used to determine the optimal FMS cut score and whether the functional movement quality (FMS and YBT) was related to sprint and jump performance in young female soccer athletes. Subjects performed all tests in one day at 10–30 min intervals. The FMS test was conducted without warm-up. After the FMS test, all athletes underwent 20 min standard warm-up sessions, including 5 min cycling and 15 min static and dynamic stretch [21]. Afterward, the athletes completed Y-balance, sprint, and jump tests in turn. For all tests, all athletes took three attempts at 2–3 min intervals, and the best performance was recorded for further analysis [22].

2.1. Participants

Twenty-four female youth soccer athletes (age: 14.79 ± 0.40 years, height: 166.63 ± 5.29 cm, mass: 53.33 ± 6.19, BMI: 19.17 ± 1.71 kg/m^2) from Team China participated in this study. Before this study, all athletes were provided information about the procedures and study objectives. Participants were excluded if they were in their menstrual cycle. This study was approved by the Ethics Committee of Beijing Sport University (14 January 2022), and all subjects voluntarily participated in this study and signed informed consent forms.

2.2. 20 m Sprint

The 20 m sprint was measured by the light timing system (Swift EZE Jump, Version 2.5.28, Brisbane, Australia). Three pairs of light gates were placed at 0, 10 and 20 m with a height of 60 cm. To avoid triggering the beam of the timing gates before the start of the test, subjects started 30 cm behind the starting line in a 2-point stance.

2.3. Vertical and Horizontal Jump

The countermovement jump (CMJ) was measured using a force plate form (Kistler, Version 2822A1-1, Winterthur, Switzerland). Subjects were instructed to rapidly squat down to a predetermined degree (approximately 60° knee flexion angle) and then jump as high as possible. The distance of the standing long jump (SLJ) was measured using a standard tape.

2.4. Functional Movement Screen

All athletes completed a functional movement assessment through the FMS test, which was performed according to the standard guidelines [23]. Athletes were not allowed to have any warm-up prior to the test. Each component of the FMS was scored from 0 to 3. The score of 0 indicated that participants felt pain during the test, and the score of 3 indicated perfect performance. Each of the seven movements was repeated no more than three times, and only the best of the three repeats was recorded.

2.5. Y-Balance Test

In YBT, the dynamic balance was examined in three directions: anterior, posteromedial, and posterolateral. The athlete stood on one leg in the stationary center of the Y-balance kit with hands on the hips, and was instructed to stretch the contralateral lower limb as far forward as possible. After three formal trials on one limb, the athlete stood on the opposite lower limb and stretched in the same direction. Each direction was repeated two times with 1 min interval, and the best repetition was recorded. In addition, the lower-limb length was measured from the athlete's anterior superior spine to Epicondylar medial ankle [24,25]. YBT composite score was calculated using the following formula: (anterior + posteromedial + posterolateral performances in cm)/3 × lower-limb length in cm) × 100 [24].

2.6. Statistical Analysis

Data are presented as mean ± standard deviation (SD). The FMS cut score was determined using the receiver operating characteristic (ROC) curve and area-under-the-curve (AUC). Spearman's rank correlation analysis was used to quantify associations between selected variables. The strength of correlation was determined as small (<0.29), moderate (0.3~0.49), large (0.5–0.69), or very large (>0.7) [26]. The statistical analysis was performed using SPSS (Version 22 for Windows, Armonk, NY, USA. IBM Corp). Statistical significance was set at $p < 0.05$.

3. Results

3.1. Descriptive Characteristics

Table 1 demonstrates the descriptive statistics for characteristics of participants, functional screen, and physical performance. All athletes completed all five tests. The players' positions were striker ($n = 7$), midfielder ($n = 6$), fullback ($n = 8$) and goalkeeper ($n = 3$). The

FMS score ranged from 13.90 to 18.50, with a mean of 16.20 ± 2.30. Moreover, the mean YBT composite score was 96.65 ± 4.14, ranging from 92.51 to 100.79.

Table 1. Descriptive statistics for characteristics of participants, functional screening, and physical performance.

Variables (Mean ± SD)	Total (n = 24)	Striker (n = 7)	Midfielder (n = 6)	Fullback (n = 8)	Goalkeeper (n = 3)
Age (years)	14.79 ± 0.40	14.42 ± 0.49	15.00 ± 0.00	15.00 ± 0.00	14.67 ± 0.47
Height (cm)	166.63 ± 5.29	162.57 ± 3.89	165.50 ± 3.54	167.50 ± 3.35	176.00 ± 1.63
Mass (kg)	53.33 ± 6.19	50.00 ± 5.63	55.33 ± 8.13	54.00 ± 4.30	55.33 ± 3.39
BMI (kg/m^2)	19.17 ± 1.71	16.86 ± 1.43	20.11 ± 2.11	19.22 ± 1.13	17.87 ± 1.21
FMS score	16.20 ± 2.30	16.14 ± 1.25	16.50 ± 2.36	15.87 ± 2.80	16.70 ± 2.49
YBT score	96.91 ± 4.00	98.15 ± 3.58	96.66 ± 4.18	95.99 ± 4.23	95.55 ± 4.7
0–10 m (s)	2.06 ± 0.05	2.08 ± 0.05	2.07 ± 0.04	2.05 ± 0.06	2.04 ± 0.02
10–20 m (s)	1.45 ± 0.05	1.47 ± 0.07	1.44 ± 0.03	1.44 ± 0.03	1.47 ± 0.04
0–20 m (s)	3.52 ± 0.89	3.55 ± 0.12	3.53 ± 0.7	3.49 ± 0.07	3.51 ± 0.06
CMJ (cm)	29.62 ± 3.62	30.07 ± 2.80	28.32 ± 4.38	30.79 ± 3.88	28.00 ± 1.24
SLJ (cm)	208.17 ± 10.60	208.14 ± 7.47	199.67 ± 8.05	210.63 ± 10.96	218.67 ± 6.94

BMI, Body Mass Index; FMS, Functional Movements Screen; YBT, Y-Balance Test; CMJ, Countermovement Jump; SLJ, Standing Long Jump; n, number; years, years; cm, centimeter; m, meter; s, second; kg, kilogram.

3.2. Receiver Operating Characteristic Curve and Area under the Curve Analyses

For all athletes, the FMS cut score was 14.5. A relatively high AUC of 0.90 was observed (p = 0.004), and the sensitivity and specificity were 0.889 and 0.833, respectively. Moreover, 17 of the 24 athletes (70.8%) scored higher than 14.5.

3.3. Associations between FMS, Y-Balance and Sprint and Jumping Performance

Moderate correlations were observed between FMS score and 10–20 m sprint time (r = −0.46, p < 0.05), between In-line Lunge score and 0–20 m sprint time (r = −0.47, p < 0.05), between Shoulder Mobility score and 0–10 m sprint time (r = −0.48, p < 0.05), and between Trunk-stability Push-up score and 10–20 m sprint time (r = −0.47, p < 0.05). Hurdle Step score was largely correlated with 0–10 m time (r = −0.51, p < 0.05). Moreover, SLJ and CMJ performance did not correlate with FMS score and its seven movements (p > 0.05). Moderate correlations were observed between CMJ height and Anterior Asymmetry score (r = −0.47, p < 0.05) and Posteromedial Asymmetry score (r = −0.44, p < 0.05). Nevertheless, there were no significant correlations (p > 0.05) between YBT performance (asymmetric in three directions and composite score) and sprint performance (0–10 m, 10–20 m and 0–20 m sprint time) (Table 2).

Table 2. Associations between FMS and athletic performance (r, 95% CI).

Tests	0–10 m	10–20 m	0–20 m	SLJ	CMJ
FMS Score	−0.29 (−0.65, 0.13)	−0.46 * (−0.76, −0.04)	−0.35 (−0.66, 0.06)	0.20 (−0.29, 0.64)	−0.13 (−0.56, 0.36)
Deep Squat	0.10 (−0.36, 0.57)	−0.24 (−0.68, 0.22)	0.06 (−0.39, 0.49)	0.17 (−0.29, 0.58)	−0.29 (−0.70, 0.25)
Hurdle Step	−0.51 * (−0.75, −0.17)	−0.01 (−0.44, −0.38)	−0.40 (−0.68, −0.02)	0.32 (−0.18, 0.70)	0.19 (−0.26, 0.63)
In-line Lunge	−0.38 (−0.75, 0.07)	−0.37 (−0.71, 0.03)	−0.47 * (−0.80, −0.06)	0.27 (−0.23, 0.71)	0.03 (−0.45, 0.51)
Shoulder Mobility	−0.48 * (−0.80, −0.08)	−0.13 (−0.53, 0.27)	−0.37 (−0.70, 0.02)	0.10 (−0.34, 0.56)	0.14 (−0.29, 0.54)
ASLR	0.14 (0.01, 0.35)	0.21 (0.11, 0.44)	0.14 (0.00, 0.33)	−0.14 (−0.35, 0.00)	−0.15 (−0.39, −0.02)

Table 2. *Cont.*

Tests	0–10 m	10–20 m	0–20 m	SLJ	CMJ
TSPU	−0.27 (−0.66, 0.17)	−0.47 * (−0.79, −0.08)	−0.33 (−0.71, 0.12)	0.35 (−0.06, 0.67)	−0.07 (−0.50, 0.35)
t Rotary Stability	−0.02 (−0.21, 0.17)	−0.12 (−0.33, 0.02)	−0.06 (−0.26, 0.09)	−0.20 (−0.45, −0.08)	0.23 (0.11, 0.46)
Anterior (%)	0.35 (−0.11, 0.69)	−0.10 (−0.49, 0.30)	0.10 (−0.36, 0.49)	−0.29 (−0.67, 0.15)	−0.47 * (−0.72, −0.10)
Posteromedial (%)	0.13 (−0.32, 0.56)	−0.05 (−0.44, 0.30)	0.15 (−0.26, 0.51)	−0.02 (−0.47, 0.41)	−0.44 * (−0.72, −0.06)
Posterolateral (%)	0.40 (−0.06, 0.69)	0.02 (−0.41, 0.44)	0.28 (−0.17, 0.62)	−0.16 (−0.60, 0.38)	−0.13 (−0.57, 0.33)
YBT Score	−0.22 (−0.58, 0.27)	−0.19 (−0.60, 0.28)	−0.29 (−0.64, 0.18)	0.11 (−0.38, 0.57)	−0.30 (−0.68, 0.18)

*, $p \leq 0.05$; ASLR, Active Straight Leg Raise; TSPU, Trunk Stability Push-up; CMJ, Countermovement Jump; SLJ, Standing Long Jump; YBT, Y-balance test; FMS, Fundamental Movement Screen; CI, confidence interval.

4. Discussion

To the best of our knowledge, this is the first study to determine the optimal FMS cut score for female youth football players, and the first to investigate the correlations between the quality of functional movement (FMS and YBT) and sprint and jump performance. This study found that the optimal FMS cut score was 14.5, with a sensitivity of 0.889 and Youden index of 0.722, which indicates a higher sensitivity and a lower false-positive rate of FMS in elite young female football players. Moreover, FMS score was moderately correlated with 10–20 m sprint time, but not with SLJ and CMJ performance. Moreover, significant relationships were also observed between In-line Lunge score and 0–20 m sprint time, between Shoulder Mobility score and 0–10 m sprint time, and between Trunk-stability Push-up score and 10–20 m sprint time, and between Hurdle Step score and 0–10 m time. There were no significant associations between YBT composite score and jump and sprint performance.

Massive studies have proved that FMS score has been significantly correlated with the risk of sport injuries in elite athletes [27]. A study conducted by Kiesel et al. [12] found that for elite American football athletes, the risk of non-contact injury was increased when the FMS score was below 14. However, elite youth soccer players with an FMS total score of >14 had less groin and hip dysfunction [28]. Notably, for different populations, an FMS score of 14 is not necessarily a gold standard. In the present study, using a score of 14.5 as the FMS cut score, the sensitivity is high, and the misclassification rate is low (16.7%). If the FMS cut score is 15.0, although the misclassification rate is 0, the Youden index is 0.611, which is less than 0.722, while the sensitivity of 0.611 is less than 0.889. Players with an FMS score of 14.5 have a more significant risk of injury than elite youth female football with an FMS > 15.0. Similarly, Zhao and Zhou [29] examined the FMS cut score in Chinese national team swimmers, and the authors reported an optimal FMS cut score of 16.5. Moreover, for elite fencers and shooters, the optimal FMS cut score was 15 [30]. These differences in optimal FMS cut score can be explained by several factors, including different training characteristics, competition requirements, sports of athletes participating, the presence of contact injuries, the sensitivity of the observation, the specificity, the chi-square values, the pre- and post-test values, etc. Therefore, the characteristics of the specific sport must be considered when using FMS score to assess the risk of sport injuries.

Much attention has been focused on whether the quality of functional movements can be a valid and effective method of assessing the athletic performance in elite players. Bennett et al. [31] examined the relationship between movement quality and physical performance and explored whether improved movement ability positively transferred to sport performance. The authors reported that FMS score was significantly and slightly associated with 5 m sprint time and agility performance. Moreover, significant increases in FMS score and speed performance were observed after one year, but there were no significant relationships between these improved variables. Similarly, no significant correlation was observed between FMS score and athletic performance in another previous study [32]. These findings were in line with the results of the present study, which found only a significant association observed between FMS score and 10–20 m sprint time. These results can be explained by several reasons. First, for each specific sport, the movement quality plays a small role compared to other factors that contribute to improved athletic performance. For example, speed and power output are the most critical factors for success in the 100 m sprint race [33,34]. Second, the FMS only reflects the level of selected isolated fundamental movement patterns, which compromises its predictive effects on capabilities such as execution speed, agility, etc. Third, the specific technical movement skill of force application is a determinant factor of sprint performance, rather than is fundamental movement [35]. The quality of specific technical skills affects the efficiency of force generation during sprinting. However, there are significant differences between fundamental movements skills and specific technical skills. Though the ability to perform the fundamental movement can influence the development of specific techniques skills to some extent, for elite athletes, specific technical skills are also influenced by cognitive, habitual, and other aspects. Therefore, FMS score cannot be used to predict performance in youth soccer players.

Specifically, in the present study, significant relationships were observed between sprint performance and In-line Lunge score, Trunk-stability Push-up score, Shoulder Mobility score, and Hurdle Step score. There are several similar factors in the development of these tasks (sprint and movements of FMS mentioned above) that could explain these findings. Both In-line Lunge and Hurdle Step tasks reflect the capability of total-body coordination and integration in horizontal orientation [16], these abilities are also required in sprint performance. Moreover, the Hurdle Step task assesses the abilities of knee and hip flexion, as well as single-leg postural control. These abilities also play a crucial role in sprint performance. Trunk-stability can also be referred to as core stability. Core stability helps to efficiently transfer the force through the kinetic chain and avoids "energy leak" during the sprint [36]. Shoulder Mobility impacts the arm movement during the sprint. The role of arm-leg coupling is to reduce the rotation of the body, and arm action may be essential to optimize the body position for ground force application during the sprint. When arm movement is restricted by other factors, for example, limited shoulder mobility, sprint performance will be compromised [37]. Therefore, this study suggests that it is considered reasonable to use components of FMS that have characteristics similar to a specific sport to assess specific performance, rather than looking to the FMS composite score.

In the present study, we found that there were no significant correlations between Y-balance composite score and sprint and jump performance. This finding was consistent with the result of a previous study [38], which found no significant relationships between YBT performance and vertical and horizontal jump. Similarly, the Y-balance test was introduced to assess the stability of the lower or upper limbs, and it plays a relatively small role in athletic performance. It is thus not appropriate for predicting athletic performance. In contrast, Bennett et al. [39] found significant correlations between YBT performance and physical performance variables, but the strength of correlation was small (r: 0.21~0.36). Moreover, many studies have reported that YBT performance was related to knee extensor strength [40], hip abduction strength [41], hip extension [41], and external rotation strength [41]. However, there is no direct evidence to suggest that improving hip and knee strength by increasing YBT score can transfer to athletic performance. Therefore, YBT composite score is also not suitable for assessing physical performance.

The strengths of this study included: (1) the subjects of this study were lite athletes who were familiar with valid running and jumping movement skills, which can avoid compromising the results of this study due to invalid movement skills; (2) excessive fatigue induced by high-intensity competition that would negatively affect the results of this study. But this study was conducted during pre-season, athletes without excessive fatigue can perform better, which can improve the accuracy of the results of this study. However, there are some limitations in this study. First, this study did not record the participants' sports injuries, so we could not determine whether the FMS score can be used to predict the risk of sport injuries. Second, this study only included 24 players, the sample size is small, which may compromise the accuracy of the results. Therefore, future studies should include more participants. Third, this study only included female youth soccer players, and did not study male youth athletes. Therefore, the results of this study may be limited to male youth athletes.

5. Conclusions

In conclusion, a FMS score of 14 is not the gold standard for assessing risks of injury in all populations; this study has suggested that the optimal FMS cut score for assessing the risk of sport injury is 14.5 in female youth elite football players. Moreover, the FMS and YBT were introduced to screen for assessing injury risk, and neither test's composite score is appropriate for assessing physical performance. We suggest that practitioners can instead use components of the FMS that have characteristics similar to specific sports to assess athletic performance.

Author Contributions: Conceptualization, J.Z. and J.L.; methodology, J.Z.; software, J.L.; validation, J.Z. and J.L.; formal analysis, H.W.; investigation, J.Z.; resources, J.Z.; data curation, H.W.; writing—original draft preparation, J.Z.; writing—review and editing, J.L.; visualization, J.L.; supervision, H.W. and H.L.; project administration, J.L.; funding acquisition, H.L. All authors have read and agreed to the published version of the manuscript.

Funding: This work was supported by the National Philosophy Social Science Foundation of China (Grant No.: 14BTY045).

Institutional Review Board Statement: This study was approved by the Sports Science Experiment Ethics Committee of Beijing Sport University (No. 2021200H).

Informed Consent Statement: Before the study, all athletes and parents were informed the purpose of this study. Written informed consent was obtained from all athletes and parents. This investigation was conducted as an anonymous survey.

Data Availability Statement: The data are available from Junjie Zhang but restrictions apply to the availability of these data. Therefore, the data are not publicly available.

Acknowledgments: We would like to acknowledge all athletes who participated in the test.

Conflicts of Interest: The authors declare no conflict of interest.

References

1. Faigenbaum, A.D.; Milliken, L.A.; Westcott, W.L. Maximal strength testing in healthy children. *J. Strength Cond. Res.* **2003**, *17*, 162–166. [CrossRef] [PubMed]
2. Maffiuletti, N.A.; Aagaard, P.; Blazevich, A.J.; Folland, J.; Tillin, N.; Duchateau, J. Rate of force development: Physiological and methodological considerations. *Eur. J. Appl. Physiol.* **2016**, *116*, 1091–1116. [CrossRef] [PubMed]
3. Negra, Y.; Chaabene, H.; Amara, S.; Jaric, S.; Hammami, M.; Hachana, Y. Evaluation of the Illinois Change of Direction Test in Youth Elite Soccer Players of Different Age. *J. Hum. Kinet.* **2017**, *58*, 215–224. [CrossRef] [PubMed]
4. Bangsbo, J.; Iaia, F.M.; Krustrup, P. The Yo-Yo intermittent recovery test: A useful tool for evaluation of physical performance in intermittent sports. *Sports Med.* **2008**, *38*, 37–51. [CrossRef]
5. Lockie, R.G.; Schultz, A.B.; Jordan, C.A.; Callaghan, S.J.; Jeffriess, M.D.; Luczo, T.M. Can selected functional movement screen assessments be used to identify movement deficiencies that could affect multidirectional speed and jump performance? *J. Strength Cond. Res.* **2015**, *29*, 195–205. [CrossRef]
6. Cook, G.; Burton, L.; Hoogenboom, B. Pre-participation screening: The use of fundamental movements as an assessment of function—Part 2. *N. Am. J. Sports Phys. Ther.* **2006**, *1*, 132–139.
7. Cook, G.; Burton, L.; Hoogenboom, B.J.; Voight, M. Functional movement screening: The use of fundamental movements as an assessment of function—Part 1. *Int. J. Sports Phys. Ther.* **2014**, *9*, 396–409.
8. Bonazza, N.A.; Smuin, D.; Onks, C.A.; Silvis, M.L.; Dhawan, A. Reliability, Validity, and Injury Predictive Value of the Functional Movement Screen: A Systematic Review and Meta-analysis. *Am. J. Sports Med.* **2017**, *45*, 725–732. [CrossRef]
9. Plisky, P.J.; Gorman, P.P.; Butler, R.J.; Kiesel, K.B.; Underwood, F.B.; Elkins, B. The reliability of an instrumented device for measuring components of the star excursion balance test. *N. Am. J. Sports Phys. Ther.* **2009**, *4*, 92–99.

10. Chimera, N.J.; Smith, C.A.; Warren, M. Injury history, sex, and performance on the functional movement screen and Y balance test. *J. Athl. Train* **2015**, *50*, 475–485. [CrossRef]

11. Powden, C.J.; Dodds, T.K.; Gabriel, E.H. The Reliability of the Star Excursion Balance Test and Lower Quarter Y-Balance Test in Healthy Adults: A Systematic Review. *Int. J. Sports Phys. Ther.* **2019**, *14*, 683–694. [CrossRef] [PubMed]

12. Kiesel, K.; Plisky, P.J.; Voight, M.L. Can Serious Injury in Professional Football be Predicted by a Preseason Functional Movement Screen? *N. Am. J. Sports Phys. Ther.* **2007**, *2*, 147–158.

13. O'Connor, F.G.; Deuster, P.A.; Davis, J.; Pappas, C.G.; Knapik, J.J. Functional movement screening: Predicting injuries in officer candidates. *Med. Sci. Sports Exerc.* **2011**, *43*, 2224–2230. [CrossRef] [PubMed]

14. Lisman, P.; O'Connor, F.G.; Deuster, P.A.; Knapik, J.J. Functional movement screen and aerobic fitness predict injuries in military training. *Med. Sci. Sports Exerc.* **2013**, *45*, 636–643. [CrossRef] [PubMed]

15. Lockie, R.; Schultz, A.; Callaghan, S.; Jordan, C.; Luczo, T.; Jeffriess, M. A preliminary investigation into the relationship between functional movement screen scores and athletic physical performance in female team sport athletes. *Biol. Sport* **2015**, *32*, 41–51. [CrossRef] [PubMed]

16. Okada, T.; Huxel, K.C.; Nesser, T.W. Relationship between core stability, functional movement, and performance. *J. Strength Cond. Res.* **2011**, *25*, 252–261. [CrossRef]

17. Lloyd, R.S.; Oliver, J.L.; Radnor, J.M.; Rhodes, B.C.; Faigenbaum, A.D.; Myer, G.D. Relationships between functional movement screen scores, maturation and physical performance in young soccer players. *J. Sports Sci.* **2015**, *33*, 11–19. [CrossRef]

18. Liang, Y.P.; Kuo, Y.L.; Hsu, H.C.; Hsia, Y.Y.; Hsu, Y.W.; Tsai, Y.J. Collegiate baseball players with more optimal functional movement patterns demonstrate better athletic performance in speed and agility. *J. Sports Sci.* **2019**, *37*, 544–552. [CrossRef]

19. Parchmann, C.J.; McBride, J.M. Relationship between functional movement screen and athletic performance. *J. Strength Cond. Res.* **2011**, *25*, 3378–3384. [CrossRef]

20. Hartigan, E.H.; Lawrence, M.; Bisson, B.M.; Torgerson, E.; Knight, R.C. Relationship of the functional movement screen in-line lunge to power, speed, and balance measures. *Sports Health* **2014**, *6*, 197–202. [CrossRef]

21. Contreras, B.; Vigotsky, A.D.; Schoenfeld, B.J.; Beardsley, C.; McMaster, D.T.; Reyneke, J.H.; Cronin, J.B. Effects of a Six-Week Hip Thrust vs. Front Squat Resistance Training Program on Performance in Adolescent Males: A Randomized Controlled Trial. *J. Strength Cond. Res.* **2017**, *31*, 999–1008. [CrossRef] [PubMed]

22. Gillett, J.; De Witt, J.; Stahl, C.A.; Martinez, D.; Dawes, J.J. Descriptive and Kinetic Analysis of Two Different Vertical Jump Tests Among Youth and Adolescent Male Basketball Athletes Using a Supervised Machine Learning Approach. *J. Strength Cond. Res.* **2021**, *35*, 2762–2768. [CrossRef] [PubMed]

23. Cook, G.; Burton, L.; Hoogenboom, B. Pre-participation screening: The use of fundamental movements as an assessment of function—Part 1. *N. Am. J. Sports Phys. Ther.* **2006**, *1*, 62–72. [PubMed]

24. Plisky, P.J.; Rauh, M.J.; Kaminski, T.W.; Underwood, F.B. Star Excursion Balance Test as a predictor of lower extremity injury in high school basketball players. *J. Orthop. Sports Phys. Ther.* **2006**, *36*, 911–919. [CrossRef]

25. Lockie, R.G.; Schultz, A.B.; Callaghan, S.J.; Jeffriess, M.D. The effects of traditional and enforced stopping speed and agility training on multidirectional speed and athletic function. *J. Strength Cond. Res.* **2014**, *28*, 1538–1551. [CrossRef]

26. Hopkins, W.G.; Marshall, S.W.; Batterham, A.M.; Hanin, J. Progressive statistics for studies in sports medicine and exercise science. *Med. Sci. Sports Exerc.* **2009**, *41*, 3–13. [CrossRef]

27. Moore, E.; Chalmers, S.; Milanese, S.; Fuller, J.T. Factors Influencing the Relationship Between the Functional Movement Screen and Injury Risk in Sporting Populations: A Systematic Review and Meta-analysis. *Sports Med.* **2019**, *49*, 1449–1463. [CrossRef]

28. Linek, P.; Booysen, N.; Sikora, D.; Stokes, M. Functional movement screen and Y balance tests in adolescent footballers with hip/groin symptoms. *Phys. Ther. Sport* **2019**, *39*, 99–106. [CrossRef]

29. Zhao, H.; Zhou, A. Correlation between core stability, lower extremity explosiveness and functional movements in swimmers. *J. Henan Norm. Univ. Nat. Sci. Ed.* **2019**, *47*, 8.

30. Wang, J. Correlation analysis of injuries and injuries of athletes of the National Shooting Team and functional action screening. *J. Cap. Ins. Phys. Educ.* **2016**, *28*, 5.

31. Bennett, H.; Fuller, J.; Milanese, S.; Jones, S.; Moore, E.; Chalmers, S. Relationship Between Movement Quality and Physical Performance in Elite Adolescent Australian Football Players. *J. Strength Cond. Res.* **2021**, 1–6. [CrossRef] [PubMed]

32. Stapleton, D.T.; Boergers, R.J.; Rodriguez, J.; Green, G.; Johnson, K.; Williams, P.; Leelum, N.; Jackson, L.; Vallorosi, J. The Relationship Between Functional Movement, Dynamic Stability, and Athletic Performance Assessments in Baseball and Softball Athletes. *J. Strength Cond. Res.* **2021**, *35*, S42–S50. [CrossRef]

33. Morin, J.B.; Bourdin, M.; Edouard, P.; Peyrot, N.; Samozino, P.; Lacour, J.R. Mechanical determinants of 100-m sprint running performance. *Eur. J. Appl. Physiol.* **2012**, *112*, 3921–3930. [CrossRef]

34. Delecluse, C. Influence of strength training on sprint running performance. Current findings and implications for training. *Sports Med.* **1997**, *24*, 147–156. [CrossRef] [PubMed]

35. Morin, J.B.; Edouard, P.; Samozino, P. Technical ability of force application as a determinant factor of sprint performance. *Med. Sci. Sports Exerc.* **2011**, *43*, 1680–1688. [CrossRef]

36. Stockbrugger, B.A.; Haennel, R.G. Validity and reliability of a medicine ball explosive power test. *J. Strength Cond. Res.* **2001**, *15*, 431–438. [PubMed]

37. Brooks, L.C.; Weyand, P.G.; Clark, K.P. Does restricting arm motion compromise short sprint running performance? *Gait Posture* **2022**, *94*, 114–118. [CrossRef]

38. Kramer, T.A.; Sacko, R.S.; Pfeifer, C.E.; Gatens, D.R.; Goins, J.M.; Stodden, D.F. The Association between the Functional Movement Screen(Tm), Y-Balance Test, and Physical Performance Tests in Male and Female High School Athletes. *Int. J. Sports Phys. Ther.* **2019**, *14*, 911–919. [CrossRef]

39. Bennett, H.; Chalmers, S.; Milanese, S.; Fuller, J. The association between Y-balance test scores, injury, and physical performance in elite adolescent Australian footballers. *J. Sci. Med. Sport* **2022**, *25*, 306–311. [CrossRef]

40. Guirelli, A.R.; Carvalho, C.A.; Dos Santos, J.M.; Ramiro Felicio, L. Relationship between the strength of the hip and knee stabilizer muscles and the Y balance test performance in adolescent volleyball athletes. *J. Sports Med. Phys. Fit.* **2021**, *61*, 1326–1332. [CrossRef]

41. Wilson, B.R.; Robertson, K.E.; Burnham, J.M.; Yonz, M.C.; Ireland, M.L.; Noehren, B. The Relationship Between Hip Strength and the Y Balance Test. *J. Sport Rehabil.* **2018**, *27*, 445–450. [CrossRef] [PubMed]

Article

Optimizing the Development of Space-Temporal Orientation in Physical Education and Sports Lessons for Students Aged 8–11 Years

Denisa-Mădălina Bălănean, Cristian Negrea, Eugen Bota, Simona Petracovschi * and Bogdan Almăjan-Guță

Faculty of Physical Education and Sports, West University of Timișoara, 300223 Timișoara, Romania
* Correspondence: simona.petracovschi@e-uvt.ro

Abstract: The purpose of this research was to analyze how we can improve the space–temporal orientation ability with the help of physical exercises in physical education and sports lessons. In total,148 children between the ages of 8 and 11 participated in this study (M = 9.70; SD = 0.79). They were subjected to three tests, which measured general intelligence (Raven Progressive Matrices) and space–temporal orientation skills (Piaget-Head test and Bender–Santucci test). The tests were carried out both in the pre-test and in the post-test period. In the case of participants in the experimental group, a specific program was applied for a period of 12 weeks. The results showed that general intelligence level was identified as a predictor of spatial–temporal orientation (beta = 0.17, t = 2.08, $p = 0.03$) but only for the Piaget-Head test. Similarly, no differences between children's age groups were identified in any of the spatial–temporal orientation test scores. However, children in the "+9" age category had higher scores on the intelligence test compared to younger children (77.31 vs. 35.70). In conclusion, the intervention program had a positive effect on spatial orientation skills.

Keywords: space–temporal orientation; general intelligence; psychomotricity; specific intervention

Citation: Bălănean, D.-M.; Negrea, C.; Bota, E.; Petracovschi, S.; Almăjan-Guță, B. Optimizing the Development of Space-Temporal Orientation in Physical Education and Sports Lessons for Students Aged 8–11 Years. *Children* **2022**, *9*, 1299. https://doi.org/10.3390/children9091299

Academic Editors: Georgian Badicu, Ana Filipa Silva and Hugo Miguel Borges Sarmento

Received: 8 August 2022
Accepted: 25 August 2022
Published: 27 August 2022

Publisher's Note: MDPI stays neutral with regard to jurisdictional claims in published maps and institutional affiliations.

Copyright: © 2022 by the authors. Licensee MDPI, Basel, Switzerland. This article is an open access article distributed under the terms and conditions of the Creative Commons Attribution (CC BY) license (https://creativecommons.org/licenses/by/4.0/).

1. Introduction

Conceptual Foundation

Psychomotricity is the ability of the human being to coordinate thought (analysis) and reaction (movement) in an optimal time, before a certain stimulus [1]. Until the beginning of the 20th century, psychomotricity was included in the field of psychology. As time passed, a new conception was born regarding the integral formation of the human being [2]. Thus, Suasnabas and his collaborators [3] refer to psychomotricity through its two factors: the psychic and the motor parts, both making possible the physical interaction with the mental one, forming a whole that develops emotions and knowledge [4]. This function plays a primary role in the stimulation and development of bodily capacities [5], positively influencing self-esteem and independence in a developmental process [6,7]. Therefore, adequate psychomotor development allows children to improve their balance, coordination [6], and space–time orientation, which is of great value for increasing intelligence and reasoning when performing motor actions [8].

Space–temporal structuring is the ability to perceive, to relate, to move, and to orientate, with everything that exists, with its education being the basis for obtaining adequate motor and affective development [9]. The development of space–temporal orientation is a process that takes place starting from an early age. It contributes at the same time to the cognitive, physical, and psychomotor growth of the child. It has an important contribution to the educational system as well [10–12]. A total or partial lack of space–temporal orientation can have repercussions not only in the teaching–learning process [13,14] but also in the social context, influencing cognitive development and the quality of life [15].

The development of spatiality is considered by some authors to be an evolutionary process which is progressively acquired until realization, throughout psychomotor development [8,9]. The importance of space–time perception is essential in human development in the sense that one can observe the shapes, structure, or compositions of objects as well as their location in relation to one's own person or to those around him or her [16,17]. Thus, certain authors believe that [18] certain aspects must be considered for the development of spatial perception and structuring: orientation in space, appreciation of distances, appreciation of trajectories, and the space–time relationship. The authors of the previously mentioned study also believe that time is, in principle, closely related to space, being the duration that separates two successive spatial perceptions [18], constituting fundamental concepts in learning and cognitive development [19].

All children use spatial concepts in various areas of their lives as they are useful for reading, writing, running, playing, etc. However, this capacity is often not developed adequately [20]. There are cases when the child perceives space in relation to his/her body, it is clear to him/her that he/she is surrounded by peers and objects, but it is difficult for him/her to differentiate between them, to classify them in order or to identify a given distance [21]. Psychology states that this confusion occurs due to the inability to distinguish and perceive what they see, so they mainly use what they think [22]. Thus, there is a need for small children to establish a spatial order centered on their own person and, only then, on the connections they develop with the environment [23,24]. Simultaneously they develop the understanding of temporal notions, which helps them identify the past time and the future as well as moments of the day [25]. As a central aspect of development and evolution, an improvement in these skills helps children identify the spatial relationships between real objects and imaginary objects [26], recognizing space as much as they come to perceive and master it [27].

The general objective of the experimental study was to improve the spatial-temporal orientation ability of primary school students through specific exercise intervention. Given the general objective, the novel element is the testing of a possible predictor of spatial-temporal ability, namely the level of general intelligence. Currently, there are few studies that focus on these variables. Thus, obtaining results oriented in this direction could give value to this research.

2. Materials and Methods

2.1. Study Participants

The studied sample was composed of 148 children, students in the 3rd and 4th grades, of which 70 were boys and 78 were girls. The students were placed in groups using convenience sampling. The age range varied between 8.1 years and 11.9 years (M = 9.70; SD = 0.79). The children included in the study underwent an assessment for the level of intelligence. As a result, none of the students were excluded. Therefore, the number of participants in each group was equal. The experimental group included students from the 3rd grade and 4th grade (N = 74), aged between 8 years and 1 month, respectively 11 years and 9 months (M = 9.62, SD = 0.81). The control group included students from the 3rd grade and 4th grade (N = 74), aged between 8 years and 2 months and 11 years and 9 months, respectively (M = 9.78, SD = 0.77).

2.2. Objectives and Hypotheses of the Research

An experimental study was carried out, with the general objective of improving the ability of space–temporal orientation, with the help of intervention through specific exercises for primary school students.

Hypotheses H1. *It was assumed that the level of general intelligence influenced the ability of space–temporal orientation.*

Hypotheses H2. *There was a statistically significant difference between the age categories of the children in terms of the results on the assigned tests.*

2.3. Research Tools

The students were subjected to 3 psychological tests that determined general intelligence on the one hand and space–temporal orientation skills on the other.

2.3.1. Raven Progressive Matrices

Raven's Progressive Matrices Test is a test used in the field of psychology and psychopedagogy, which measures the general intelligence level of the subject. The main feature of this test is to encourage analytical reasoning, perception, and the ability to abstract. The students were given the first 4 series (from A to D), out of a total of 5, with 12 questions for each series (from A1 to D12), with these being organized according to progressive difficulty. The intellectual performance thus "measured" allowed the inclusion of the subject in one of the 5 different degrees (levels) of intelligence. For an explicit understanding of how to interpret results and convert test scores to percentiles, see Măłkiński and Măńdziuk [28]. It was decided to use this sample for 2 reasons:

1. One of the purposes of the study was to demonstrate that the sample consists of children of age-appropriate intelligence and the results will not be affected in any way by a possible low IQ of the students.

2. Children's general intelligence was one of the dependent variables used to determine the degree of its association or influence on space–temporal orientation skills.

2.3.2. Piaget-Head Space–Temporal Orientation Test

This test assesses the right–left spatial orientation of children between the ages of six and eleven [29]. The number of items the child must get right is 3 out of 3 at the age of 8, 6 out of 8 at the ages of 9 and 10, and 5 out of 6 at the age of 11. Successful trials are marked with +, unsuccessful ones with −, and spontaneously corrected trials with − +. The rating is done by adding up the successfully completed items.

2.3.3. The Perceptual–Motor Test of Bender–Santucci Spatial Configuration

This test targeted the perceptual–motor function of spatial configuration by testing the ability of children between the ages of 5 and 15 to perceive spatial configurations and to make comparisons between the respective configurations, thus making a rendering of space and shape on paper [30]. The testing took place individually, with the necessary materials being represented by 8 cards. The rating was made considering certain criteria: the construction of the angles, the orientation of the figures or the component elements, and the position of the figures or the elements that make up the model.

2.4. Research Method

In the first stage, an agreement was made between Secondary School No. 24 from Timișoara, the Faculty of Physical Education and Sport from Timișoara, and the Timiș School Inspectorate, which allowed the study to continue. Then, a collaboration contract was drawn up between the author of the study and a specialized psychologist under whose guidance the psychological tests were applied and interpreted.

In the second stage, the consent of the parents or legal guardians was requested and obtained, and then the children participating in the study were tested with the "Raven Progressive Matrices" test, to determine that the included sample consisted of children with age-appropriate intelligence.

In the third stage, students were randomly assigned to one of the experimental or control groups. The investigated variable "space–temporal orientation" was measured in the pre-test phase, so that in the case of the participants in the experimental group, the intervention would be applied to this independent variable. The control group did not benefit from any kind of intervention or manipulation of the mentioned variable.

In the post-test phase, the variable "space–temporal orientation" was measured again, to determine if the intervention generated changes in the test results.

The Intervention Plan

The actual research took place between January and June 2022. With a frequency of 2 physical education and sports lessons per week, the students from the experimental group were given the intervention plan, which had as general themes and objectives the elements detailed in Table 1.

Table 1. General themes and objectives.

Lesson	General Theme and Objective
Space Orientation	T: Body References (left side, right side): Knowing and identifying the right side and left side of the body
Space–Temporal Orientation	T: Recognition and operation with spatial and space–temporal notions G.O: Formation, recognition, and operation with space and space–temporal notions of one's own body (left side, right side). Interval duration perception
Space Orientation	T: Recognition and operation with space notions of one's own body, in relation to surrounding objects G.O: Formation, recognition, and operation with space notions of one's own body, in relation to surrounding objects (left side, right side)
Space–Temporal Orientation	T: Recognition and estimation of distances G.O: Recognition and estimation of distances, in due time. Left side/right side reminder
Space Orientation	T: Establishing the direction/position of objects G.O: Establishing the direction/position of objects in relation to one's own person but also to each other, as well as imitating certain actions
Space–Temporal Orientation	T: Knowledge of direction and axes G.O: Establishing the direction/position of objects in relation to one's own person, as well as orientation in space and time
Space Orientation	T: Working with space notions G.O: Recognizing and operating with space notions located in space near and far from one's own body. Linear orientation
Space–Temporal Orientation	T: Knowledge of two-dimensional space G.O: Recognition and estimation of distances (quantities) in two-dimensional space
Space Orientation	T: Knowledge of spatial concepts G.O: Identifying the notions of: "outside, inside/full, empty"
Space–Temporal Orientation	T: Strengthening the estimation of distances G.O: Recognition and estimation of distances. Interval duration perception
Space Orientation	T: Knowledge of spatial concepts G.O: Identifying the notions of "above and below"
Space–Temporal Orientation	T: Strengthening the estimation of distances G.O: Recognition and estimation of distances (quantities) in two-dimensional space and time
Space Orientation	T: Knowledge of spatial concepts G.O: Recognition and estimation of distances (length-width, near-far)
Space–Temporal Orientation	T: Relating to the environment G.O: Identifying the position of objects in relation to one's own person and orientation in space and time
Space Orientation	T: Knowing dimensions G.O: Knowledge of the notion of size and sensory-motor practice: organizing objects of the same nature according to size criteria: big-small, long-short, tall-short, etc., presented or not in the perceptual field
Space–Temporal Orientation	T: Strengthening the estimation of distances, in relation to time G.O: Recognition and estimation of distances, correlated with time

Table 1. *Cont.*

Lesson	General Theme and Objective
Space Orientation	T: Consolidation of spatial concepts G.O: Formation, recognition and use of spatial notions located in near and far space in relation to one's own body and in relation to others. Practicing mathematical calculations
Space–Temporal Orientation	T: Knowledge of direction and axes G.O: Establishing the direction/position of objects in relation to one's own person, but also to each other
Space Orientation	T: Consolidation of spatial concepts G.O: Recognizing and estimating some distances (quantities) in two-dimensional space, as well as operating with the notions of "up-down, forward and backward"
Space–Temporal Orientation	T: Relating to the environment G.O: Identifying one's own body and other objects in space and time
Space Orientation	T: Strengthening the estimation of distances and direction of movement G.O: Recognizing and estimating distances, as well as determining direction
Space–Temporal Orientation	T: Consolidation of spatial concepts G.O: Formation, recognition and use of spatial notions of one's own body and of other people (left-right, above-below, forward-backward, up-down, etc.).

2.5. Statistical Analysis

The collected data were entered, processed, and analyzed using the IBM SPSS Statistics 20 program (IBM Corp, Armonk, NY, USA). Each hypothesis was tested using appropriate statistical techniques. To identify the predictor factor on the space–temporal orientation, linear regression was analyzed. Correlations between variables were checked using the Pearson test.

3. Results

3.1. Intelligence Test Results, with the Purpose of Including Children in the Sample

The score with the highest frequency was 75 (17), followed by 85 (frequency 11). This aspect showed us that the average of the students included in the sample had an above average intelligence level, taking into account the age, measured in years and months. With the help of this result, they intended to keep the level of intelligence under control, and the scores obtained were suitable for the inclusion of the children in the study.

3.2. Hypothesis Testing

Hypothesis 1. *It was assumed that the level of general intelligence influenced the ability of space–temporal orientation.*

To test Hypothesis 1, a linear regression was used; however, before applying it, the correlation between the variables was verified by the Pearson test at a level of statistical significance of p less than 0.05. Thus, in Table 2 it can be seen that there is a statistically significant direct but weak relationship between IQ and spatial orientation for the Piaget-Head test ($r = 0.17, p = 0.03$).

After the application of the Pearson correlation we can expect IQ to be a predictive factor for spatial orientation measured by the Piaget-Head test.

a.   Linear regression with the dependent variable spatial orientation, measured with the Piaget-Head test

According to the results in Tables 3 and 4, we could conclude that the model was successful ($F(1,146) = 4.36, p = 0.03$) and that the variance explained by IQ was 2.9% for spatial orientation measured by the Piaget-Head test. Furthermore, IQ could be considered a predictive factor for spatial orientation by the Piaget-Head test (beta = 0.17, t = 2.08, p = 0.03).

Table 2. Correlations between variables.

		Correlations		
		Piaget-Head Test after the Intervention	**Bender–Santucci Test after the Intervention**	**Raven Spm Classical Progressive Matrices**
Piaget-Head test after the intervention	Pearson Correlation Sig. (2-tailed) N	1 148	0.051 0.540 148	0.170 * 0.039 148
Bender–Santucci test after the intervention	Pearson Correlation Sig. (2-tailed) N	0.051 0.540 148	1 148	0.095 0.250 148
Raven Spm Classical Progressive Matrices	Pearson Correlation Sig. (2-tailed) N	0.170 * 0.039 148	0.095 0.250 148	1 148

* Correlation is significant at the 0.05 level (2-tailed).

Table 3. Variance explained in Piaget-Head spatial orientation by IQ.

Model	R	R Square	Adjusted R Square	Std. Error of the Estimate
1	0.170 [a]	0.029	0.022	5.53416

[a] Predictors: (constant), Raven Spm Classical Progressive Matrices.

Table 4. Significant coefficients of the model with the dependent variable, spatial orientation, for the Piaget-Head test.

	Model	Unstandardized Coefficients		Standardized Coefficients			95.0% Confidence Interval for B	
		B	Std. Error	Beta	T	Sig.	Lower Bound	Upper Bound
1	(Constant)	16.921	1.444		11.716	0.000	14.067	19.776
	Raven Spm Classical Progressive Matrices	0.044	0.021	0.170	2.088	0.039	0.002	0.086

Linear regression with the dependent variable, spatial orientation, measured with the Bender–Santucci test.

According to Table 5, we could see that the model was not successful (F(1.146) = 1.33, p = 0.25), and the rest of the interpretations did not make sense.

Table 5. Explained variance of Bender–Santucci spatial orientation by IQ.

Model	R	R Square	Adjusted R Square
1	0.095 [a]	0.009	0.002

[a] Predictors: (constant), Bender-Santucci test.

Hypothesis 2. *There was a statistically significant difference between the age categories of the children in terms of the results on the assigned tests. Given that the variable "age" was continuous and the average age was 9.70, we transformed it into a categorical variable with the value 9. The new variable "age" can be seen in Table 6 and has the following composition:*

Table 6. Age range.

	Age per Range		
		Frequency	Percent
	Age < 9 years	10	6.8
Valid	Age > 9 years	138	93.2
	Total	148	100.0

Regarding the difference between the two age categories, the results of the spatial orientation tests did not reveal any statistically significant index according to the month and year of birth, but a statistically significant difference was recorded at the IQ level (M − W = 302.00, p = 0.003). Thus, children older than 9 years had a higher IQ (77.31 vs. 35.70), which can be seen in Figure 1.

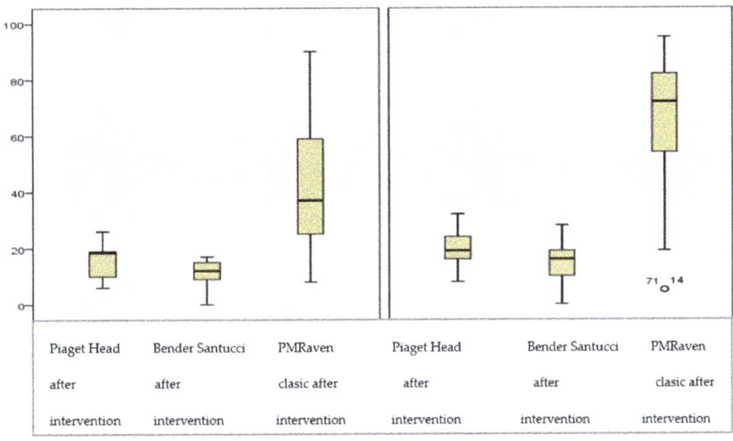

Figure 1. Age range of participants (<9 years/>9 years).

In conclusion, Table 7 illustrates the situation of the hypotheses.

Table 7. Situation of the hypotheses.

Hypothesis No.	Status
I1.	Partially Accepted
I2.	Partially Accepted

4. Discussion

The present study had as its general objective the improvement of the ability of spatial orientation in children between the ages of 8 and 11. Being an experimental study in which the proposed sample was divided into two groups (experiment and control) by convenience, one of the objectives was to keep under control the intelligence of the children included, before any other approach. Thus, the results of the Raven Progressive Matrices testing showed us that most of the included students had an above average level of intelligence, taking into account the age, measured in years and months (between 8 and 11 years). At the same time, general intelligence was identified as a predictive factor for spatial orientation, but only for the Piaget-Head test. In the same context, spatial ability, which refers to "the location of objects, their shapes, the relationship between them and the paths they take as they move" [31], has long been recognized as an ability partially independent of general intelligence [32,33]. In addition to being distinct from other cognitive abilities, spatial thinking itself has often been conceptualized in a multidimensional manner, consisting of several separate but related abilities. Thus, a result that supports the partial confirmation of our first hypothesis is a relatively recent study which discovered a relationship between certain spatial components involved in a puzzle game and the results of a preschool and primary intelligence test (WPPSI) [34]. Even though a direct relationship between general intelligence and space–temporal orientation has not been found in the studies conducted in recent years, certain studies did identify a relationship between executive function and spatial abilities [35,36]. However, we cannot affirm following these researches that the two components can be predictive factors. Thus, the relationship between the two is more assumed, studies being directed to the spatial components related to performance in mathematics and geometry [37–39]. Similarly, we do not know whether these assumed relationships are variable with age or remain stable across the lifespan. Although most children of this age are developing many of the cognitive skills necessary for successful spatial orientation [40,41], there is increased neural activity in areas of the brain associated with visual–spatial processing compared to young adults [42]. Thus, the ability to orient and navigate is a cognitive process that undergoes a maturation with the progression of skills and strategies during a large period of the childhood. These findings support that as children mature, they increase and refine their proficiency in visual and spatial skills, increasing network connectivity and enabling the successful use of spatial orientation strategies.

Verification of Hypothesis 2 confirmed the fact that students in the age category >9 years (10/11) recorded higher scores in terms of IQ, compared to the category ≤9 (8/9 years).

This result was consistent with two other studies in the field of psychology, in which the younger age groups showed a lack of maturation of the functionality of the prefrontal regions, which is involved in these types of tasks [43,44]. However, studies of brain development have shown that around 8 to 9 years of age, significant structural and functional changes [45] affect the whole brain and gray matter volume [46], synaptic pruning processes [47], and functional connectivity [48–50]. Furthermore, studies have shown that around the age of 10, there is a significant increase in cortical thickness in the parietal and frontal areas as well as in higher-order cortical areas such as the dorsolateral prefrontal cortex and the cingulate cortex [51,52]. Therefore, it is possible that such changes in brain development support a greater development of cognitive potential in the 8- to 10-year-old groups than in the younger ones, which is not consistent with the result of our study.

This study might have some limitations. First, the study sample consisted of children between the ages of 8 and 11, so the results may not be suitable for the entire general population. Additionally, the findings of the present study apply to subjects attending academic institutions in the urban region, which may be inappropriate for children in rural regions or who have dropped out of school. Second, although the variable "general intelligence" was kept under control, the results may have been influenced by other factors, such as the extracurricular activities that many of the subjects attended during the intervention period. At the same time, the large number of participating students could add value to the study.

5. Conclusions

The intervention program through specific exercises had the expected effect. Moreover, it was determined that the level of intelligence, as measured by percentiles, was a predictive factor only for the first spatial orientation test but not for the second. This reason might have been due to the content of the Santucci test, which relied to a greater or lesser extent on graphic qualities and imitability, aspects for which we do not know whether they are associated with children's IQ. This finding can lead to a new direction of research in which a possible connection between graphic qualities and the ability of space–temporal orientation and the level of general intelligence can be highlighted.

That is precisely why, in order to further explore the influence of some intervention programs on space–temporal orientation, it is essential to investigate its impact on several components, keeping under control factors related to age, gender, social environment, and level of general intelligence.

Author Contributions: D.-M.B.: data collection, study design, measurements, investigation, and methodology; S.P.: conceptualization, methodology, writing—original draft preparation; E.B.: methodology; B.A.-G.: data curation; C.N.: resources, writing—original draft preparation, and writing—review and editing. All authors have read and agreed to the published version of the manuscript.

Funding: This research received no external funding.

Institutional Review Board Statement: The manuscript has the Faculty of Physical Education and Sports Ethics Committee's approval (No. 21/5 September 2021).

Informed Consent Statement: Informed consent was obtained from all subjects involved in the study.

Data Availability Statement: Data can be available for consultation when requested from the corresponding author.

Conflicts of Interest: The authors declare no conflict of interest.

References

1. Garófano, V.V.; Guirado, L.C. Importancia de la motricidad para el desarrollo integral del niño en la etapa de educación infantil. *EmásF Rev. Digit. Educ. Fís.* **2017**, *47*, 89–105.
2. Flores Preciado, F.D.C. *Talleres Grafiplasti en la Motricidad Fina en Niños de Cuatro Años de Institucion Educativa Inicial N° 074 "las Ardillitas" Tumbes, 2020*; Universidad Nacional de Tumbes: Tumbes, Peru, 2020.
3. Pacheco, S.R.S.; Delgado, K.L.C.; Parra, M.J.S.; Pacheco, L.S.S. Influencia de la estimulación temprana en el desarrollo psicomotor en los niños y niñas de 1 y 2 años. *Reciamuc* **2017**, *1*, 105–127.
4. Coello Herrera, Y.N.; Quiroz Calderón, G.L. *Desarrollo de la Psicomotricidad a Través del Juego en los Niños de 3 a 5 Años, en la Unidad Educativa Adolfo María Astudillo, Ciudad Babahoyo, Provincia Los Ríos*; UTB: Babahoyo, Ecuador, 2022.
5. Vega Gaspar, J.O. *Psicomotricidad Gruesa en los Infantes de 5 años en una Institución Educativa del Distrito de San Marcos, Provincia de Huari-Ancash*; USCC: Lima, Perú, 2022.
6. Pacheco, R.J.P.; Bravo, N.O.; Viteri, B.P.; Ñacato, J.M.; Arufe-Giráldez, V. Análisis de la influencia de un programa estructurado de educación física sobre la coordinación motriz y autoestima en niños de 6 y 7 años. *J. Sport Health Res.* **2022**, *14*, 9.
7. Fresneda Ramos, D.C.; Pinzón Daza, M.A. *Aumentando la Autoestima por Medio de la Imagen Corporal*; Universidad Pedagógica Nacional: Bogotá, Colombia, 2015.
8. Villavicencio, J.L.G.; Caymayo, D.M.V.; Brito, D.Y.T.; Oviedo, P.B.S. El juego simbólico como estrategia para el desarrollo psicomotriz de los niños. Polo del Conocimiento. *Rev. Cient.-Prof.* **2022**, *7*, 81.
9. Peñuela Ladino, C.A. *Propuesta Didáctica Para el Desarrollo y Estimulación de la Ubicación Espacial en los Estudiantes de Grado Primero de Primaria del Colegio Nueva Constitución IED Mediado por las TIC's*; Universidad Libre: Colombia, Bogotá, 2021.
10. Cui, X.; Guo, K. Supporting mathematics learning: A review of spatial abilities from research to practice. *Curr. Opin. Behav. Sci.* **2022**, *46*, 101176. [CrossRef]
11. Fanari, R.; Meloni, D.; Massidda, D. Visual and spatial working memory abilities predict early math skills: A longitudinal study. *Front. Psychol.* **2019**, *10*, 2460. [CrossRef]
12. Oqueso Huanaco, N. *Los Juegos Corporales y su Incidencia en el Desarrollo de las Nociones Espaciales y Temporales en los Niños y Niñas de 4 Años de la Institución Educativa Inicial N° 571 Pumaorcco del Distrito de Sicuani Provincia de Canchis Región Cusco-2017*; Universidad Nacional Daniel Alcides Carrión: Pasco, Peru, 2019.
13. Zapateiro-Segura, J.C.; Poloche Arango, S.K.; Camargo Uribe, L. *Orientación Espacial: Una Ruta de Enseñanza y Aprendizaje Centrada en Ubicaciones y Trayectorias*; Facultad de Ciencia y Tecnología: Bilbao, Spain, 2018.
14. Polinsky, N.; Flynn, R.; Wartella, E.A.; Uttal, D.H. The role of spatial abilities in young children's spatially-focused touchscreen game play. *Cogn. Dev.* **2021**, *57*, 100970. [CrossRef]
15. Fernandez-Baizan, C.; Arias, J.L.; Mendez, M. Spatial orientation assessment in preschool children: Egocentric and allocentric frameworks. *Appl. Neuropsychol. Child* **2021**, *10*, 171–193. [CrossRef]
16. Canales-Vasquez, M.I.; Reyes Ferrel, R.R. *La Percepción Espacio-Temporal de los Estudiantes del Primer Grado de la Institución Educativa N° 36596-Castrovireyna Huancavelica*; Universidad Nacional de Huancavelica: Huancavelica, Peru, 2019.
17. Cordones-Tasigchana, J.M. *Recurso Didáctico Geoplano en el Desarrollo del Pensamiento Espacial en Niños y Niñas de 5 a 6 Años de la Unidad Educativa "Victoria Vásconez Cuvi-Simón Bolívar-Elvira Ortega*; Universidad Tècnica de Ambato; Facultad de Ciencias Humanas: Ambato, Ecuador, 2020.
18. Cortes Echeverri, S.M.; Muñoz González, L.; Giraldo Ortiz, Y.; Rendón Muñoz, L.; Agudelo Cárdenas, Y.; Gómez Jaramillo, Y.; Sierra Vásquez, L.; Agreda Mora, Y.; Yepes Aguirre, D.; López Betancur, A.; et al. *Favoreciendo los Componentes de la Psicomotricidad (Esquema Corporal, Respiración, Relajación, Tono, Postura, Orientación Espacial y Lateralidad) a Través de las Actividades Rectoras*; Tecnologico de Antioquia Institucion Universitaria: Medellín-Antioquia, Colombia, 2021.

19. Rodríguez, A.Y.; Becerra, G.; Quintero, L. *Evaluación del Factor Psicomotor de la Estructuración Espacio Temporal en Niños Pertenecientes a las Escuelas de la Ciudad de Pereira, con Edades Entre 4 a 14 años, Basados en la Batería de Vítor Da Fonseca*; Universidad Tecnológica de Pereira, Facultad de Ciencias de la Salud: Pereira, Colombia, 2013.

20. Tafur Castillejo, N.Y. *La Danza Infantil Como Estrategia Didáctica para Mejorar la Orientación Espacial en los Niños de la IEI N° 233, la Soledad–Huaraz*; Universidad Católica Los Ángeles de Chimbote: Chimbote, Peru, 2018.

21. Blakemore, S.J. The social brain in adolescence. *Nat. Rev. Neurosci.* **2008**, *9*, 267–277. [CrossRef]

22. Zamora-Rodríguez, V.; Barrantes Masot, M.C.; Barrantes López, M. Enseñanza y aprendizaje de la orientación espacial. *Números Rev. Didact. Mat.* **2021**, *107*, 129–146.

23. Saavedra-Cordova, P.M. *Las Nociones Espaciales de Niños y Niñas de 4 Años de Edad de la Cuna Jardín del Colegio Aplicación "José Antonio Encinas"-Provincia y Región TUMBES, Perú-2020*; Universidad Católica los Ángeles de Chimbote: Chimbote, Peru, 2020.

24. García-Chapoñan, Z.M. *Aplicación de un Programa de Juegos "me Ubico en el Espacio" Para Desarrollar las Nociones Espaciales en los Niños de 4 Años de la Institución Educativa Inicial N° 203 de Bagua Grande, provincia de Utcubamba, Región Amazonas*; Universidad Nacional Pedro Ruiz Gallo: Lambayeque, Peru, 2019.

25. Alyamani, A.H.; Jabali, S.M. The Effectiveness of an Educational Program Based on Pictures and Graphics in Developing Some Spatial and Temporal Concepts among Kindergarten Children. *Int. J. High. Educ.* **2021**, *10*, 319–328. [CrossRef]

26. Gifford, S.; Gripton, C.; Williams, H.J.; Lancaster, A.; Bates, K.E.; Williams, A.Y.; Farran, E.K. *Spatial Reasoning in Early Childhood*; Routledge: London, UK, 2022.

27. Chugá Benítez, S.P. *Estrategias Metodológicas Para Estimular las Nociones Temporo-Espaciales en los Niños de 3 a 4 Años del Centro Infantil La Primavera de la Ciudad de Ibarra en el año 2014–2015*; Universidad Técnica Del Norte Facultad De Educación, Ciencia Y Tecnología: Ibarra, Equador, 2019.

28. Małkiński, M.; Mańdziuk, J. Deep Learning Methods for Abstract Visual Reasoning: A Survey on Raven's Progressive Matrices. *arXiv* **2022**, arXiv:2201.12382.

29. Vrasmas, E.; Oprea, V. *Set de Instrumente, Probe și Teste Pentru Evaluarea Educațională a Copiilor cu Dizabilități*; Marlink: Bucuresti, Romania, 2003.

30. Radu, I.D.; Ulici, G. *Evaluarea și Educarea Psihomotricității*; Editura Fundației Humanitas: București, Romania, 2003.

31. Hodgkiss, A.; Gilligan, K.A.; Tolmie, A.K.; Thomas, M.S.; Farran, E.K. Spatial cognition and science achievement: The contribution of intrinsic and extrinsic spatial skills from 7 to 11 years. *Br. J. Educ. Psychol.* **2018**, *88*, 675–697. [CrossRef]

32. Hegarty, M. Spatial thinking in undergraduate science education. *Spat. Cogn. Comput.* **2014**, *14*, 142–167. [CrossRef]

33. Rimfeld, K.; Shakeshaft, N.G.; Malanchini, M.; Rodic, M.; Selzam, S.; Schofield, K.; Plomin, R. Phenotypic and genetic evidence for a unifactorial structure of spatial abilities. *Proc. Natl. Acad. Sci. USA* **2017**, *114*, 2777–2782. [CrossRef]

34. Jirout, J.J.; Newcombe, N.S. Building blocks for developing spatial skills: Evidence from a large, representative US sample. *Psychol. Sci.* **2015**, *26*, 302–310. [CrossRef] [PubMed]

35. Verdine, B.N.; Golinkoff, R.M.; Hirsh-Pasek, K.; Newcombe, N.S.; Filipowicz, A.T.; Chang, A. Deconstructing building blocks: Preschoolers' spatial assembly performance relates to early mathematical skills. *Child Dev.* **2014**, *85*, 1062–1076. [CrossRef]

36. Cragg, L.; Gilmore, C. Skills underlying mathematics: The role of executive function in the development of mathematics proficiency. *Trends Neurosci. Educ.* **2014**, *3*, 63–68. [CrossRef]

37. Cragg, L.; Keeble, S.; Richardson, S.; Roome, H.E.; Gilmore, C. Direct and indirect influences of executive functions on mathematics achievement. *Cognition* **2017**, *162*, 12–26. [CrossRef]

38. Hawes, Z.; Moss, J.; Caswell, B.; Seo, J.; Ansari, D. Relations between numerical, spatial, and executive function skills and mathematics achievement: A latent-variable approach. *Cogn. Psychol.* **2019**, *109*, 68–90. [CrossRef]

39. Mix, K.S.; Levine, S.C.; Cheng, Y.L.; Young, C.; Hambrick, D.Z.; Ping, R.; Konstantopoulos, S. Separate but correlated: The latent structure of space and mathematics across development. *J. Exp. Psychol. Gen.* **2016**, *145*, 1206. [CrossRef] [PubMed]

40. Bullens, J.; Iglói, K.; Berthoz, A.; Postma, A.; Rondi-Reig, L. Developmental time course of the acquisition of sequential egocentric and allocentric navigation strategies. *J. Exp. Child Psychol.* **2010**, *107*, 337–350. [CrossRef] [PubMed]

41. Negen, J.; Nardini, M. Four-year-olds use a mixture of spatial reference frames. *PLoS ONE* **2015**, *10*, e0131984.

42. Liu, I.; Levy, R.M.; Barton, J.J.; Iaria, G. Age and gender differences in various topographical orientation strategies. *Brain Res.* **2011**, *1410*, 112–119. [CrossRef] [PubMed]

43. Duncan, J.; Seitz, R.J.; Kolodny, J.; Bor, D.; Herzog, H.; Ahmed, A.; Newell, F.N.; Emslie, H. A neural basis for general intelligence. *Science* **2000**, *289*, 457–460. [CrossRef]

44. Smirni, D.; Precenzano, F.; Magliulo, R.M.; Romano, P.; Bonifacio, A.; Gison, G.; Bitetti, I.; Terracciano, M.; Ruberto, M.; Sorrentino, M. Inhibition, set-shifting and working memory in Global Developmental Delay preschool children. *Life Span Disabil.* **2018**, *21*, 191–206.

45. De Moor, M.H.; Costa, P.T.; Terracciano, A.; Krueger, R.F.; De Geus, E.J.C.; Toshiko, T.; Penninx, B.W.J.H.; Esko, T.; Madden, P.A.F.; Derringer, J. Meta-analysis of genome-wide association studies for personality. *Mol. Psychiatry* **2012**, *17*, 337–349. [CrossRef] [PubMed]

46. Lebel, C.; Beaulieu, C. Longitudinal development of human brain wiring continues from childhood into adulthood. *J. Neurosci.* **2011**, *31*, 10937–10947. [CrossRef]

47. Smirni, P.; Smirni, D. Current and Potential Cognitive Development in Healthy Children: A New Approach to Raven Coloured Progressive Matrices. *Children* **2022**, *9*, 446. [CrossRef]

48. Fair, D.A.; Dosenbach, N.U.F.; Church, J.A.; Cohen, A.L.; Brahmbhatt, S.; Miezin, F.M.; Barch, D.M.; Raichle, M.E.; Petersen, S.E.; Schlaggar, B.L. Development of distinct control networks through segregation and integration. *Proc. Natl. Acad. Sci. USA* **2007**, *104*, 13507–13512. [CrossRef]
49. Uda, S.; Matsui, M.; Tanaka, C.; Uematsu, A.; Miura, K.; Kawana, I.; Noguchi, K. Normal development of human brain white matter from infancy to early adulthood: A diffusion tensor imaging study. *Dev. Neurosci.* **2015**, *37*, 182–194. [CrossRef] [PubMed]
50. Steinberg, L.; Bornstein, M.H.; Vandell, D.L. *Life-Span Development: Infancy through Adulthood*; Cengage Learning: Belmont, CA, USA, 2010.
51. Lenroot, R.K.; Giedd, J.N. Brain development in children and adolescents: Insights from anatomical magnetic resonance imaging. *Neurosci. Biobehav. Rev.* **2006**, *30*, 718–729. [CrossRef] [PubMed]
52. Shaw, P.; Kabani, N.J.; Lerch, J.P.; Eckstrand, K.; Lenroot, R.; Gogtay, N.; Greenstein, D.; Clasen, L.; Evans, A.; Rapoport, J.L. Neurodevelopmental trajectories of the human cerebral cortex. *J. Neurosci.* **2008**, *28*, 3586–3594. [CrossRef] [PubMed]

children

MDPI

Article

Association between Physical Fitness Index and Psychological Symptoms in Chinese Children and Adolescents

Jinkui Lu [1], Hao Sun [2], Jianfeng Zhou [1] and Jianping Xiong [3,*]

1   School of Physical Education, Shangrao Normal University, Shangrao 334000, China
2   Physical Education College, Jiangxi Normal University, Nanchang 330022, China
3   School of Sports, Jiangxi University of Finance and Economics, Nanchang 330013, China
*   Correspondence: xiongjianpingjxcd@126.com

Abstract: The aim of this study was to determine the relationship between different physical fitness indices (PFIs) and psychological symptoms and each dimension (emotional symptoms, behavioral symptoms, social adaptation difficulties) of Chinese children and adolescents. Methods: A total of 7199 children and adolescents aged 13–18 in Jiangxi Province, China, were tested for grip strength, standing long jump, sit-ups, sit and reach, repeated straddling, 50 m run, 20 m shuttle run test (20 m SRT) items. The physical fitness indicators were standardized, converted to Z score and added up to obtain the PFI, and the self-assessment of the psychological section of the multidimensional sub-health questionnaire of adolescents (MSQA) to test the psychological symptoms, using the chi-square test to determine the psychological symptoms of different types of children and adolescents and binary logistic regression analysis to determine the association between psychological symptoms and different PFI grades. Results: The higher the PFI of Chinese children and adolescents, the lower the detection rate of psychological symptoms, emotional symptoms and social adaptation difficulties, from 25.0% to 18.4%, 31.3% to 25.7% and 20.1% to 14.4%, respectively. These results were statistically significant (χ^2 = 14.073, 9.332, 12.183, $p < 0.05$). Taking the high-grade PFI as a reference, binary logistic regression analysis was performed. Generally, compared with the high-grade PFI, children and adolescents with a low-grade PFI (OR = 1.476, 95% CI: 1.200–1.814) or medium-grade PFI (OR = 1.195, 95% CI: 1.010–1.413) had a higher risk of psychological symptoms ($p < 0.05$). Conclusions: The lower the PFI of Chinese children and adolescents, the higher the detection rate of psychological symptoms, showing a negative correlation. In the future, measures should be taken to improve the physical fitness level of children and adolescents in order to reduce the incidence of psychological symptoms and promote the healthy development of children and adolescents.

Keywords: mental health; PFI; children and adolescents; association analysis; cross-sectional survey

Citation: Lu, J.; Sun, H.; Zhou, J.; Xiong, J. Association between Physical Fitness Index and Psychological Symptoms in Chinese Children and Adolescents. *Children* **2022**, *9*, 1286. https://doi.org/10.3390/children9091286

Academic Editors: Georgian Badicu, Ana Filipa Silva and Hugo Miguel Borges Sarmento

Received: 26 July 2022
Accepted: 22 August 2022
Published: 26 August 2022

Publisher's Note: MDPI stays neutral with regard to jurisdictional claims in published maps and institutional affiliations.

Copyright: © 2022 by the authors. Licensee MDPI, Basel, Switzerland. This article is an open access article distributed under the terms and conditions of the Creative Commons Attribution (CC BY) license (https://creativecommons.org/licenses/by/4.0/).

1. Introduction

As the economy continues to grow, and the level of medical care and technology are improved, people's quality of life is greatly improved. However, this also leads to many adverse effects on human health, such as the increase in obesity [1], screen time [2] and the detection rate of psychological symptoms as well as decreased physical activity [3], resulting in a continuous increase in the proportion of people with psychological symptoms. Adolescence is a transitional stage from childhood to adulthood in which physiological development gradually matures, but psychological development relatively lags behind, causing a serious imbalance [4,5]. In addition, this stage is affected by complex changes such as interpersonal relationships and academic pressure, which can easily lead to various psychological symptoms [6–8]. Research shows that the proportion of psychological symptoms in children and adolescents is increasing year by year and is much higher than that of other age groups [9]. Epidemiological survey data show that 10%~20% of adolescents worldwide are affected by different psychological symptoms, accounting for a large part of the global disease burden [10], and 50% of psychological symptoms in adulthood are caused by childhood, indicating a trajectory effect [11]. Studies have confirmed [12] that the appearance of psychological symptoms in children and adolescents leads to the decline of learning and self-care ability and has a serious negative impact on the healthy growth of children, adolescents and their families. Other studies have shown that [13] psychological symptoms are the initial stage of mental disorders, which will develop without timely intervention.

Past research has shown that the world is facing a serious issue in the prevention and control of psychological symptoms. About 6 million people in the United States suffer from psychological symptoms every year, along with 35% of the population in Japan and 37% of the population in Australia [14]. Seventy percent of the population in China suffers from various psychological symptoms, with children and adolescents making up the largest proportion [15]. There are also studies showing that the proportion of Chinese children and adolescents who suffer from various psychological symptoms affecting their daily life and study is increasing [16,17]. In addition, the self-injury and suicide behaviors of children and adolescents due to various psychological symptoms are also increasing, which poses a serious threat to the healthy development of Chinese children and adolescents [18,19]. Another survey shows that about 30 million children and adolescents in China are experiencing psychological symptoms, and the detection rate is increasing year by year, with a higher incidence among younger children [19,20]. It can be seen from these findings that children and adolescents in China are facing a severe situation of continuous increase in psychological symptoms.

It is an indisputable fact that modern lifestyle changes have led to decreased physical fitness and physical activity and increased screen time among children and adolescents [21]. Studies have shown that among various factors leading to death, death due to reduced physical fitness and reduced exercise time has become the fourth most prevalent [22]. Research also shows that people with higher muscle strength have improved psychological symptoms, and the two are positively correlated [23,24]. A survey conducted by Tao et al. [25] of 5453 middle school students in China showed that low- to moderate-intensity physical activity was a protective factor for depression in children and adolescents. Research showed that there was a negative correlation between physical fitness levels and psychological symptom indicators of boys and girls [25,26]. An increase in the weekly physical exercise time of 1 h would reduce the proportion of the number of people suffering from depression by 8% [27]. A 4-year follow-up study on children and adolescents aged 14~24 showed that adequately ensuring physical exercise time and promoting physical fitness can reduce the occurrence of depression, anxiety and other adverse psychological problems [27]. There are also studies showing that the improvement of cardiorespiratory fitness level plays a positive role in reducing the occurrence of anxiety and depression, as active physical exercise can improve body shape, enhance self-confidence and thus reduce the occurrence of psychological symptoms [28]. Active physical exercise can also increase the secretion of endorphins and dopamine in the brain, which can promote physical and mental pleasure and improve psychological symptoms [27,29]. Regular physical exercise can effectively reduce the risk of various diseases, such as hypertension, type 2 diabetes, depression, anxiety and various types of cancer and coronary heart disease. The reduction in physical fitness is an important reason for the continued decline in psychological symptoms [29].

Combining the above studies on physical fitness, physical exercise and psychological symptoms by scholars from various countries, it can be seen that there is an association between them [30–32]. However, most of the previous studies on the psychological symptoms of children and adolescents have used the clinical diagnostic scale of "mental disease" to conduct research and evaluation, mainly with the dimensions of anxiety, emotional disorders, depression and hostility [33]. These scales are suitable for use in the clinical, psychiatric and psychological counseling fields but cannot effectively screen children and adolescents with psychological symptoms, so early intervention and prevention of psychological symptoms cannot be carried out. Furthermore, previous studies have mainly focused on the relationship between physical fitness programs and psychological symptoms, while the relationship between PFI and psychological symptoms of children and adolescents has been less studied [33,34]. In view of this, this study investigated the psychological symptoms of 7199 Chinese children and adolescents aged 13–18 using the MSQA scale and tested seven physical fitness indicators to determine the relationship between different PFIs and psychological symptoms.

2. Methods

2.1. Data Source and Participants

The data collection for this study was carried out with the help of the National Student Physical Health Standard (NSPHS) test. The NSPHS test is organized and implemented by the Chinese Ministry of Education to investigate the physical fitness of school students across the country [35].

This study investigated 6 age groups from 13 to 18 years old. Surveys were conducted in 6 cities in Jiangxi, and it was concluded that the sample size should be 6660 people. After the investigation, after removing 288 (3.85%) responses with missing basic demographic information, 7199 (96.15%) valid surveys were retained in this study. The specific sampling process of the participants is shown in Figure 1.

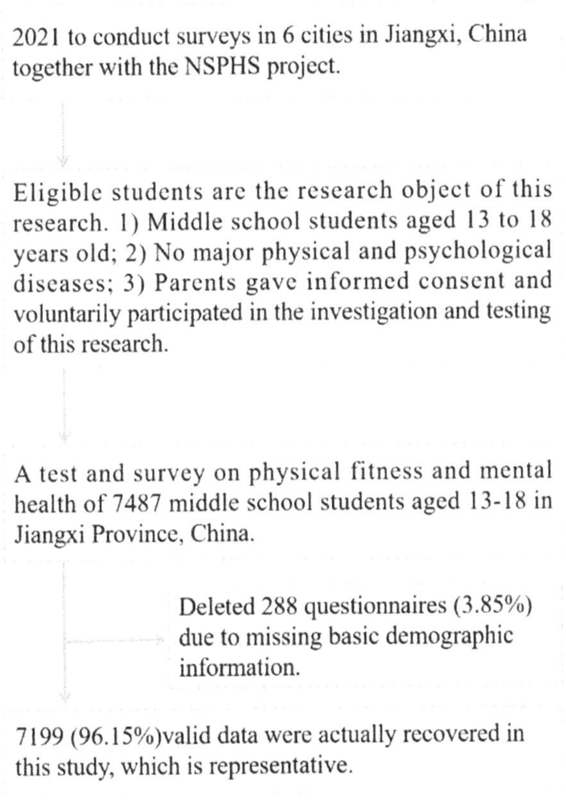

2021 to conduct surveys in 6 cities in Jiangxi, China together with the NSPHS project.

Eligible students are the research object of this research. 1) Middle school students aged 13 to 18 years old; 2) No major physical and psychological diseases; 3) Parents gave informed consent and voluntarily participated in the investigation and testing of this research.

A test and survey on physical fitness and mental health of 7487 middle school students aged 13-18 in Jiangxi Province, China.

Deleted 288 questionnaires (3.85%) due to missing basic demographic information.

7199 (96.15%)valid data were actually recovered in this study, which is representative.

Figure 1. The specific sampling process of the participants.

This study was approved by the Ethics Committee of Shangrao Normal University (2020R-0125). Written informed consent was obtained from the school, the students and their parents before the investigation. The questionnaire was coded to strictly protect the privacy of students.

2.2. Psychological Symptoms

Psychological symptoms were evaluated using the multidimensional sub-health questionnaire of adolescents (MSQA) [36–40]. The MSQA consists of 39 items, which are divided into 3 dimensions: emotional symptoms, behavioral symptoms and social adaptation difficulties. Emotional symptoms were tested with 17 items, behavioral symptoms were tested with 9 items and social adaptation difficulties were tested with 13 items. Each item is scored on a 6-level scale, and the subjects choose the corresponding duration according to their own conditions: none or less than 1 week = 1, more than 1 week = 2, more than 2 weeks = 3, more than 1 month = 4, more than 2 months = 5, more than 3 months = 6. The results were reclassified during the statistical analysis of the questionnaire, as the responses of more than 1 month, more than 2 months and more than 3 months were recorded as 1, and the others were recorded as 0, and psychological symptoms were calculated as the sum of the scores of all 39 items. The three dimension scores, namely emotional symptoms, behavioral symptoms and social adaptation difficulties, were calculated separately according to the items.

According to the provisions of the "National Norm Development of the Multidimensional Assessment Questionnaire for Adolescents' Sub-health" [37–40], the 90th percentile was used as the demarcation value of the psychological symptoms of adolescent students of all ages and as the criterion for judging the psychological symptoms of children and adolescents. That is, the values for emotional symptoms, behavioral symptoms, social adaptation difficulties and psychological symptoms were ≥ 3, ≥ 1, ≥ 4 and ≥ 8, respectively.

The test–retest correlation coefficient of the scale was 0.868, the Cronbach α coefficient was 0.957 and the split-half reliability was 0.942. The self-rating symptom scale (SCL-90) and the Cornell Medical Index (CMI) questionnaire were used as the criteria, and the criterion-related validity was 0.636 and 0.649, respectively [39].

2.3. Physical Fitness Index

The physical fitness index (PFI) can comprehensively reflect the development level of physical fitness of children and adolescents to a certain extent and has been applied and recognized in many studies [41,42]. The scores of the 7 physical fitness test indicators were standardized into groups according to gender and age, and the Z score was calculated: Z score = (measured value of each physical fitness indicator − average value of each physical fitness indicator in each group)/standard deviation of each physical fitness index in each group. The PFI is obtained by adding up the Z scores of each physical fitness index and taking the opposite number for the Z score of the 50 m run because the higher the Z score, the lower the subject's performance.

$$\mathrm{PFI} = Z_{\text{grip strength}} + Z_{\text{standing long jump}} + Z_{\text{sit-ups}} + Z_{\text{sit and reach}} + Z_{\text{repeated straddling}} + Z_{\text{20 m SRT}} - Z_{\text{50 m run}}.$$

According to the PFI percentiles of different gender and age groups, PFI was stratified [43] and divided into three grades: low, medium and high: low PFI < P15, medium P15 $\leq$ PFI < P85 and high PFI $\leq$ P85.

The physical fitness test includes 7 items, namely grip strength, standing long jump, sit-ups, sit and reach, repeated straddling, 50 m run and 20 m SRT.

Grip strength: The subjects were required to use their left or right hand to perform the electronic grip dynamometer test. During the test, the subjects were required to relax their hands naturally and perform two grip strength tests with maximum effort. The highest result, accurate to 0.1 kg, was recorded.

Standing long jump: The subjects stood naturally with their feet apart behind the jumping line. Participants jumped with both feet, with no step-by-step jumping. The distance from the subject to the closest touchdown point was measured. Each subject jumped twice, and the best score was recorded. The test results are accurate to 0.1 cm.

Sit-ups: Before the test, the staff prepared the mat and stopwatch. The subjects lay in the supine position on the mat, with relaxed bodies, hands are crossed on chest, knees bent at 60–90 degrees, and feet flat on the ground. The assistant pressed the subjects' ankles with both hands to immobilize their lower limbs. When the subjects heard the "start" command, they began to do sit-ups, with their elbows touching or exceeding their knees, and then quickly returned to the initial lying supine position with their back touching the mat. The above actions were repeated as much as possible. The number of sit-ups (where the elbows touched the knees or the thighs) was recorded.

Sit and reach: Before the test, the inspector prepared the instrument and cushion, selected a flat ground and checked the working state of the sit and reach instrument. The subject sat upright on the mat, with the head, back and buttocks close to the wall, and the feet were placed under the instrument, but the angle of the feet was not fixed. The subject extended their arms shoulder-width apart, placed their palms on the test instrument board, expanded their chest and kept their hands close to the instrument board with their elbows stretched forward and back straight. Initially in a sitting position, the subject flexed and stretched forward and slowly pushed the instrument in the forward direction. When the body flexed and stretched forward to the maximum, the data were recorded, and the subject left the test board using both hands to end the test. The movement distance of the instrument from the initial position to the maximum flexion and extension was recorded. Measurements were in centimeters (cm) with one decimal place. Each subject was tested twice, with the best result recorded.

Repeated straddling: Two parallel lines were drawn at a distance of 100 cm from the central line, for a total of 3 lines. The subjects stood with their feet open on the central line. When they heard the "start" command, they crossed the horizontal line in the order of right → middle → left middle. Participants were instructed not to jump with both feet. The number of completed actions within 20s was recorded [44].

50 m run: Before the test, several straight runways were drawn with a length of 50 m and a width of 1.22 m on the flat ground with clear runway lines. The timekeeper stood on the side of the finish line, started timing when the starting flag was waving and stopped when the subject's chest reached the vertical plane of the finish line. The time score was accurate to 0.1s.

20 m SRT: This test involves the subject running back and forth between 2 lines at a distance of 20 m. Each 20 m completed was recorded as 1 lap (time). The running speed was indicated by music, and the initial speed was 8.0 km/h, the speed in the second minute was 9.0 km/h and the speed was increased by one speed level every minute, that is, 0.5 km/h each lap. The subject endeavored to complete the running speed level. The test ended when they could not follow the rhythm to reach the 20 m end line for two consecutive times, and the number of laps completed was recorded [45,46].

2.4. Statistical Analysis

In the basic condition part, each physical fitness index of 7199 children and adolescents was expressed as mean ± standard deviation (M ± SD), and the detection rate of psychological symptoms and each dimension was expressed as a percentage of classification.

The chi-square test was used to compare the detection rates of different categories, namely psychological symptoms and dimensions (emotional symptoms, behavioral symptoms, and social adaptation difficulties) of children and adolescents with different PFIs. The relationship between different levels of PFI and various dimensions of psychological symptoms in children and adolescents was analyzed with binary logistic regression analysis and interaction effect analysis, and the OR value, 95% confidence interval and p value were obtained; $p < 0.05$ was regarded as a statistically significant difference. Our research hypothesis is that children and adolescents with higher PFI have a lower detection rate of psychological symptoms.

Data analysis was performed with SPSS 25.0 software, graphs were produced with GraphPad Prism and the statistical significance level was set at 0.05.

3. Results

The results showed that the average age of the 7199 participating Chinese children and adolescents was (15.50 ± 1.71) years old. All physical fitness indicators for boys were higher than those of girls except for the sit and reach test. Low, middle and high PFIs were segmented by percentiles, so the proportion of males and females was the same at each level. The detection rate of psychological symptoms and each dimension (emotional symptoms, behavioral symptoms, social adaptation difficulties) in males was higher than that in females (Table 1).

Table 1. Physical fitness indicators and psychological symptoms of Chinese children and adolescents.

Classification	Total	Boys	Girls
N	7199	3600	3599
Physical fitness indicators (M ± SD)			
Grip strength (kg)	31.45 ± 9.99	37.05 ± 9.83	25.85 ± 6.36
Standing long jump (cm)	188.24 ± 33.64	209.43 ± 30.15	167.06 ± 21.37
Sit-ups	23.01 ± 6.90	24.80 ± 6.98	21.21 ± 6.33
20 m SRT	36.76 ± 17.17	43.12 ± 18.61	30.40 ± 12.75
Sit and reach (cm)	38.77 ± 11.72	38.12 ± 11.74	39.42 ± 11.66
Repeated straddling	31.54 ± 9.19	33.42 ± 9.72	29.66 ± 8.21
50 m run (s)	8.53 ± 1.26	7.86 ± 1.01	9.21 ± 1.12
PFI, *n* (%)			
Low	1080 (15.0)	540 (15.0)	540 (15.0)
Middle	5040 (70.0)	2520 (70.0)	2520 (70.0)
High	1080 (15.0)	540 (15.0)	539 (15.0)
Psychological symptoms, *n* (%)			
Emotional symptoms	1995 (27.7)	1001 (27.8)	993 (27.6)
Behavioral symptoms	1981 (27.5)	1021 (28.4)	960 (26.7)
Social adaptation difficulties	1236 (17.2)	656 (18.2)	580 (16.1)
Psychological symptoms	1540 (21.4)	797 (22.1)	743 (20.6)

Note: N, sample size; M ± SD, mean ± standard deviation; PFI, physical fitness index; 20 m SRT, 20 m shuttle run test.

The results of our study showed that the higher the PFI of Chinese children and adolescent boys, the lower the detection rate of psychological symptoms and social adaptation difficulties, and there was statistical significance ($\chi 2$ = 8.106, 7.065, $p < 0.05$). The higher the PFI of girls, the lower the detection rate of psychological symptoms and social adaptation difficulties, and there was statistical significance ($\chi 2$ = 10.922, 6.172, $p < 0.05$). In general, the higher the PFI of Chinese children and adolescents, the lower the detection rate of psychological symptoms, emotional symptoms and social adaptation difficulties, from 25.0% to 18.4%, 31.3% to 25.7% and 20.1% to 14.4%, respectively, and there was statistical significance ($\chi 2$ = 14.073, 9.332, 12.183, $p < 0.05$) (Table 2).

Using the psychological symptoms of children and adolescents in the high-grade PFI group as reference, binary logistic regression analysis was performed. Boys with a low-grade (OR = 1.485, 95% CI: 1.104–1.998) or middle-grade (OR = 1.375, 95% CI: 1.082–1.749) PFI had a higher risk of psychological symptoms, with statistical significance ($p < 0.05$). Psychological symptoms were risk factors for low grades of PFI in girls (OR = 1.467, 95% CI: 1.100–1.957) compared with high grades of PFI ($p < 0.05$). Overall, compared with the high-grade PFI, children and adolescents with a low-grade (OR = 1.476, 95% CI: 1.200–1.814) or medium-grade PFI (OR = 1.195, 95% CI: 1.010–1.413) were more likely to develop psychological symptoms ($p < 0.05$) (Table 3).

Figure 2 shows that compared with the high-grade PFI group, the middle-grade PFI and especially the low-grade PFI group are more inclined to the right; that is, the risk of psychological symptoms is higher.

Table 2. Comparison of detection rates of psychological symptoms in different categories of Chinese children and adolescents (%).

Category		N	Emotional Symptoms	χ2-Value	p-Value	Behavioral Symptoms	χ2-Value	p-Value	Social Adaptation Difficulties	χ2-Value	p-Value	Psychological Symptoms	χ2-Value	p-Value
Boys	PFI (Low)	540	167 (30.9)			164 (30.4)			112 (20.7)			130 (24.1)		
	PFI (Middle)	2520	702 (27.9)	5.662	0.059	720 (28.6)	3.505	0.173	465 (18.5)	7.065	0.029	572 (22.7)	8.106	0.017
	PFI (High)	540	132 (24.4)			137 (25.4)			79 (14.6)			95 (17.6)		
Girls	PFI (Low)	540	171 (31.7)			156 (28.9)			105 (19.4)			140 (25.9)		
	PFI (Middle)	2520	677 (26.9)	5.234	0.073	661 (26.2)	1.921	0.383	395 (15.7)	6.172	0.046	499 (19.8)	10.922	0.004
	PFI (High)	539	146 (27.0)			138 (25.6)			77 (14.3)			104 (19.3)		
Total	PFI (Low)	1080	338 (31.3)			320 (29.6)			217 (20.1)			270 (25.0)		
	PFI (Middle)	5040	1379 (27.4)	9.332	0.009	1381 (27.4)	4.724	0.094	860 (17.1)	12.183	0.002	1071 (21.3)	14.073	0.001
	PFI (High)	1079	278 (25.7)			275 (25.5)			156 (14.4)			199 (18.4)		

Note: N, sample size; PFI, physical fitness index.

Table 3. Binary logistic regression analysis of psychological symptoms of Chinese children and adolescents in different categories.

Category		Emotional Symptoms		p-Value	Behavioral Symptoms		p-Value	Social adaptation Difficulties		p-Value	Psychological Symptoms		p-Value
		OR	(95% CI)		OR	(95% CI)		OR	(95% CI)		OR	(95% CI)	
Boys	PFI (Low)	1.384	1.059–1.809	0.017	1.283	0.983–1.675	0.067	1.527	1.113–2.096	0.009	1.485	1.104–1.998	0.009
	PFI (Middle)	1.194	0.963–1.479	0.106	1.177	0.952–1.455	0.133	1.320	1.019–1.711	0.035	1.375	1.082–1.749	0.009
	PFI (High)	1.000	1.000		1.000	1.000		1.000	1.000		1.000	1.000	
Girls	PFI (Low)	1.251	0.962–1.626	0.095	1.183	0.905–1.548	0.219	1.451	1.052–2.003	0.023	1.467	1.100–1.957	0.009
	PFI (Middle)	0.991	0.804–1.222	0.935	1.036	0.837–1.281	0.746	1.118	0.858–1.455	0.409	1.035	0.818–1.310	0.774
	PFI (High)	1.000	1.000		1.000	1.000		1.000	1.000		1.000	1.000	
Total	PFI (Low)	1.314	1.089–1.585	0.004	1.233	1.020–1.489	0.030	1.489	1.188–1.866	0.001	1.476	1.200–1.814	0.000
	PFI (Middle)	1.087	0.935–1.262	0.277	1.105	0.951–1.284	0.193	1.219	1.013–1.466	0.036	1.195	1.010–1.413	0.038
	PFI (High)	1.000	1.000		1.000	1.000		1.000	1.000		1.000	1.000	

Note: PFI, physical fitness index; OR (95% CI), odds ratio (95% confidence interval).

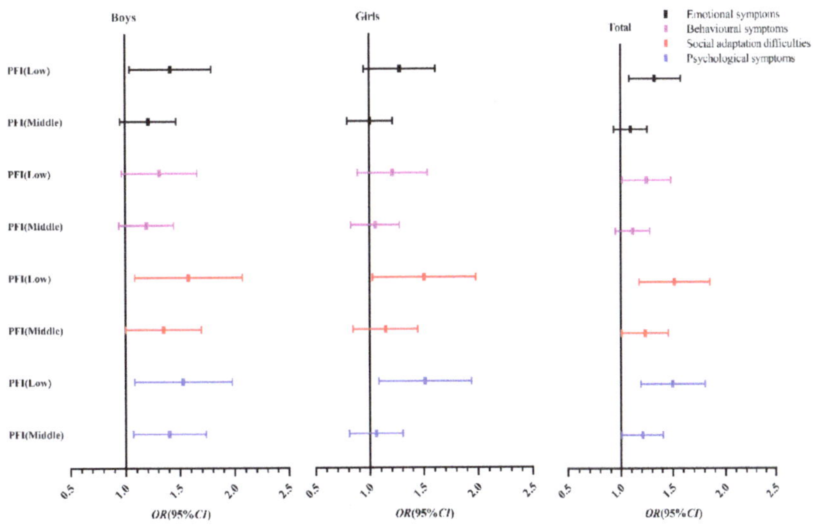

Figure 2. Binary logistic regression analysis *OR* value (95% CI) of psychological symptoms of Chinese children and adolescents. Note: The high grade of PFI is 1.0; OR (95% CI), odds ratio (95% confidence interval).

4. Discussion

The results of our study showed that the detection rate of psychological symptoms in Chinese children and adolescents was 21.4%, which was lower than the results of the psychological symptoms survey involving middle school students in nine cities in China (36.73%) [47]. A survey of middle school students showed that the detection rate of psychological symptoms was 64.1% for middle school students and 68.9% for high school students [48], lower than the results of this study. The reason for this difference is related to the differences in the selected regions, age groups and ethnic groups of the surveyed subjects and the differing research length. For example, in recent years, China has increased investment in the education of children and adolescents' mental health and strengthened teachers, which is also an important reason for the low proportion of psychological symptoms [49]. In addition, there are differences in the evaluation criteria or questionnaires of psychological symptoms used in different studies, which is also an important reason for the large differences in the results of different studies [47,48]. Although the detection rate of psychological symptoms in children and adolescents continues to decrease, it is still at a high level. Therefore, social attention is needed to continuously reduce the detection rate of psychological symptoms in children and adolescents. Although the results are different, it is still vital to conduct research on the influencing factors and interventions of psychological symptoms in children and adolescents, and the results also provide theoretical support for the intervention aimed at children and adolescents with psychological symptoms. Our research also shows that the detection rate of different dimensions of psychological symptoms also differs between genders. The detection rate of boys is higher than that of girls in emotional symptoms, behavioral symptoms, social adaptation difficulties and psychological symptoms. The reason may be related to the congenital personality differences between boys and girls. Society and the family endow boys with greater responsibilities and expectations, resulting in a higher detection rate of psychological symptoms.

In this study, the PFI was obtained by standardizing the Z score of nine physical fitness items and divided into high-, medium- and low-level groups according to the percentile. The higher the PFI, the lower the detection rate of each dimension of psychological symptoms in children and adolescents, and these findings were statistically significant ($p < 0.01$). Relevant studies have shown that physical activity plays a positive role in improving psychological symptoms [50–52]. Other studies have shown that active participation in physical exercise and a good level of physical fitness play a positive role in promoting the mental health of children and adolescents, and mentally healthy children and adolescents are also more willing to improve their athletic ability through physical exercise [53,54]. The Matute-Llorente [55] study showed that the longer the daily exercise time in children and adolescents, the higher the level of cardiorespiratory fitness. The Aires [56] study also confirmed that the longer children and adolescents participate in exercise every day, the higher their cardiorespiratory fitness level. Similarly, studies have confirmed that the higher the level of cardiorespiratory fitness, the lower the detection rate of psychological symptoms in children and adolescents, which further confirms the close relationship between physical fitness level and psychological symptoms [57–59]. They also confirmed the conclusion that children and adolescents with higher PFI had lower detection rate of psychological symptoms in this study. Our study also showed that the influence of PFI on psychological symptoms in boys was significantly higher than that in girls, especially when the PFI was at a middle level. The reason may be that boys prefer physical exercise, their physical fitness is better and the impact on psychological symptoms is more obvious. Conversely, girls' participation in exercise is lower, so the effect on psychological symptoms is not significant.

Our research also has certain advantages. The sample size of the study is large, leading to representative results. The use of PFI, which comprehensively reflects physical fitness, can also more accurately analyze the relationship between physical fitness and psychological symptoms. Our research also has some shortcomings. First, this study is a cross-sectional study, which can determine the association between physical fitness and psychological symptoms but not the causal relationship between the two. Prospective cohort studies should be conducted in the future on this topic. Second, the sample population of this study was located in Jiangxi, China, and the survey area was thus limited. The scope of the survey should be expanded in the future to support these findings. Third, the covariates in the investigation and analysis of this study were limited. The investigation of covariates, such as sleep quality, physical activity time and screen time, should be added in future studies to verify our results.

Based on the above studies, it can be seen that PFI is closely related to the detection rate of psychological symptoms in Chinese children and adolescents. Reasonable arrangements for active participation in physical exercise and improving physical fitness play a positive role in preventing the occurrence of psychological symptoms in children and adolescents. Of course, the reduction in psychological symptoms in children and adolescents cannot be achieved solely by the improvement of physical fitness level. Although a large number of theories and practices have confirmed that physical fitness improvement can cultivate good mentality and values in children and adolescents, thereby promoting a good life and habits and the healthy development of the body and mind, the development of children and adolescents' mental health is affected by multiple factors. We should also start from the aspects of children and adolescents' health education, family guidance and social environment and jointly promote the physical and mental health of children and adolescents.

5. Conclusions

By testing seven physical fitness items of 7199 children and adolescents aged 13–18 in China, the physical fitness indicators were standardized and converted into Z scores, and the PFI was obtained after adding them up. The higher the PFI, the lower the detection rate of psychological symptoms, emotional symptoms, behavioral symptoms and social adaptation difficulties, showing a negative correlation between PFI and psychological symptoms. In addition, compared with girls, boys' PFI had a more significant effect on psychological symptoms. In the future, the physical fitness level of Chinese children and adolescents should be improved, and the detection rate of psychological symptoms in children and adolescents should be reduced so as to jointly promote the physical and mental development of Chinese children and adolescents.

Author Contributions: Conceptualization, J.L.; data curation, J.L.; formal analysis, J.Z.; funding acquisition, J.X.; investigation, J.X.; methodology, J.L.; project administration, J.X.; resources, J.L.; software, H.S.; supervision, H.S.; validation, J.Z.; visualization, J.X.; writing—original draft, J.L., H.S., J.X.; writing—review and editing, J.X.; All authors have read and agreed to the published version of the manuscript.

Funding: The research is supported by the national general project of the 13th Five-Year Plan of National Education Science: Research on Rural School Development (no. bha170138).

Institutional Review Board Statement: Not applicable.

Informed Consent Statement: The study was in accordance with the requirement of the World Medical Association Declaration of Helsinki, and written consent was obtained from the students, schools and parents. The survey was approved by the Ethics Committee of the School of Physical Education, Shangrao Normal University, and was carried out after the approval (2020R-0125).

Data Availability Statement: To protect the privacy of participants, the questionnaire data will not be disclosed to the public. If necessary, you can contact the corresponding author.

Acknowledgments: We thank the students and parents who participated in this study as well as the staff who participated in the data testing of this study.

Conflicts of Interest: The authors declare no conflict of interest.

References

1. Pan, L.; Li, X.; Feng, Y.; Hong, L. Psychological assessment of children and adolescents with obesity. *J. Int. Med. Res.* **2018**, *46*, 89–97. [CrossRef] [PubMed]
2. Zhang, F.; Yin, X.; Bi, C.; Ji, L.; Wu, H.; Li, Y.; Sun, Y.; Ren, S.; Wang, G.; Yang, X.; et al. Psychological symptoms are associated with screen and exercise time: A cross-sectional study of Chinese adolescents. *BMC Public Health* **2020**, *20*, 1695. [CrossRef] [PubMed]
3. Violant-Holz, V.; Gallego-Jimenez, M.G.; Gonzalez-Gonzalez, C.S.; Munoz-Violant, S.; Rodriguez, M.J.; Sansano-Nadal, O.; Guerra-Balic, M. Psychological Health and Physical Activity Levels during the COVID-19 Pandemic: A Systematic Review. *Int. J. Environ. Res. Public Health* **2020**, *17*, 9419. [CrossRef] [PubMed]
4. Jones, E.; Mitra, A.K.; Bhuiyan, A.R. Impact of COVID-19 on Mental Health in Adolescents: A Systematic Review. *Int. J. Environ. Res. Public Health* **2021**, *18*, 2470. [CrossRef] [PubMed]
5. Harris, T.B.; Udoetuk, S.C.; Webb, S.; Tatem, A.; Nutile, L.M.; Al-Mateen, C.S. Achieving Mental Health Equity: Children and Adolescents. *Psychiatr. Clin. N. Am.* **2020**, *43*, 471–485. [CrossRef]
6. Nesi, J. The Impact of Social Media on Youth Mental Health: Challenges and Opportunities. *North Carol. Med. J.* **2020**, *81*, 116–121. [CrossRef]
7. O'Reilly, M. Social media and adolescent mental health: The good, the bad and the ugly. *J. Ment. Health* **2020**, *29*, 200–206. [CrossRef]
8. Ortuno-Sierra, J.; Lucas-Molina, B.; Inchausti, F.; Fonseca-Pedrero, E. Special Issue on Mental Health and Well-Being in Adolescence: Environment and Behavior. *Int. J. Environ. Res. Public Health* **2021**, *18*, 2975. [CrossRef]
9. Viejo, C.; Gomez-Lopez, M.; Ortega-Ruiz, R. Adolescents' Psychological Well-Being: A Multidimensional Measure. *Int. J. Environ. Res. Public Health* **2018**, *15*, 2325. [CrossRef]
10. Kieling, C.; Baker-Henningham, H.; Belfer, M.; Conti, G.; Ertem, I.; Omigbodun, O.; Rohde, L.A.; Srinath, S.; Ulkuer, N.; Rahman, A. Child and adolescent mental health worldwide: Evidence for action. *Lancet* **2011**, *378*, 1515–1525. [CrossRef]
11. Minh, A.; Bultmann, U.; Reijneveld, S.A.; van Zon, S.; McLeod, C.B. Childhood Socioeconomic Status and Depressive Symptom Trajectories in the Transition to Adulthood in the United States and Canada. *J. Adolesc. Health* **2021**, *68*, 161–168. [CrossRef] [PubMed]
12. Francis, A.; Pai, M.S.; Badagabettu, S. Psychological Well-being and Perceived Parenting Style among Adolescents. *Compr. Child Adolesc. Nurs.* **2021**, *44*, 134–143. [CrossRef] [PubMed]
13. Mayou, R. Are treatments for common mental disorders also effective for functional symptoms and disorder? *Psychosom. Med.* **2007**, *69*, 876–880. [CrossRef] [PubMed]
14. Belfer, M.L. Child and adolescent mental disorders: The magnitude of the problem across the globe. *J. Child Psychol. Psychiatry* **2008**, *49*, 226–236. [CrossRef]
15. Li, W.; Yang, Y.; Liu, Z.H.; Zhao, Y.J.; Zhang, Q.; Zhang, L.; Cheung, T.; Xiang, Y.T. Progression of Mental Health Services during the COVID-19 Outbreak in China. *Int. J. Biol. Sci.* **2020**, *16*, 1732–1738. [CrossRef]
16. Shah, K.; Mann, S.; Singh, R.; Bangar, R.; Kulkarni, R. Impact of COVID-19 on the Mental Health of Children and Adolescents. *Cureus* **2020**, *12*, e10051. [CrossRef]
17. Duan, L.; Shao, X.; Wang, Y.; Huang, Y.; Miao, J.; Yang, X.; Zhu, G. An investigation of mental health status of children and adolescents in china during the outbreak of COVID-19. *J. Affect. Disord.* **2020**, *275*, 112–118. [CrossRef]
18. Zhang, Y.; Wu, C.; Yuan, S.; Xiang, J.; Hao, W.; Yu, Y. Association of aggression and suicide behaviors: A school-based sample of rural Chinese adolescents. *J. Affect. Disord.* **2018**, *239*, 295–302. [CrossRef]
19. Yin, X.; Li, D.; Zhu, K.; Liang, X.; Peng, S.; Tan, A.; Du, Y. Comparison of Intentional and Unintentional Injuries Among Chinese Children and Adolescents. *J. Epidemiol.* **2020**, *30*, 529–536. [CrossRef]
20. Wang, Q.L.; Zhang, X.D.; Wu, X.Y.; Zhang, Q.; Zhang, Y.; Sun, J.; Zhang, S.C.; Wang, X.; Zong, Q.; Tao, S.M.; et al. Sleep status associated with psychological and behavioral problems in adolescents and children. *Zhonghua Liu Xing Bing Xue Za Zhi* **2021**, *42*, 859–865. [CrossRef]
21. Whiting, S.; Buoncristiano, M.; Gelius, P.; Abu-Omar, K.; Pattison, M.; Hyska, J.; Duleva, V.; Music, M.S.; Zamrazilova, H.; Hejgaard, T.; et al. Physical Activity, Screen Time, and Sleep Duration of Children Aged 6-9 Years in 25 Countries: An Analysis within the WHO European Childhood Obesity Surveillance Initiative (COSI) 2015-2017. *Obes. Facts* **2021**, *14*, 32–44. [CrossRef] [PubMed]
22. Eisenmann, J.C.; Bartee, R.T.; Smith, D.T.; Welk, G.J.; Fu, Q. Combined influence of physical activity and television viewing on the risk of overweight in US youth. *Int. J. Obes. (Lond.)* **2008**, *32*, 613–618. [CrossRef] [PubMed]
23. Hwang, I.C.; Ahn, H.Y.; Choi, S.J. Association between handgrip strength and mental health in Korean adolescents. *Fam. Pr.* **2021**, *38*, 826–829. [CrossRef] [PubMed]
24. Ludyga, S.; Puhse, U.; Gerber, M.; Mucke, M. Muscle strength and executive function in children and adolescents with autism spectrum disorder. *Autism Res.* **2021**, *14*, 2555–2563. [CrossRef] [PubMed]
25. Tao, F.B.; Xu, M.L.; Kim, S.D.; Sun, Y.; Su, P.Y.; Huang, K. Physical activity might not be the protective factor for health risk behaviours and psychopathological symptoms in adolescents. *J. Paediatr. Child Heal.* **2007**, *43*, 762–767. [CrossRef]
26. Rothon, C.; Edwards, P.; Bhui, K.; Viner, R.M.; Taylor, S.; Stansfeld, S.A. Physical activity and depressive symptoms in adolescents: A prospective study. *BMC Med.* **2010**, *8*, 32. [CrossRef] [PubMed]

27. Sampasa-Kanyinga, H.; Colman, I.; Goldfield, G.S.; Janssen, I.; Wang, J.; Podinic, I.; Tremblay, M.S.; Saunders, T.J.; Sampson, M.; Chaput, J.P. Combinations of physical activity, sedentary time, and sleep duration and their associations with depressive symptoms and other mental health problems in children and adolescents: A systematic review. *Int. J. Behav. Nutr. Phys. Act.* **2020**, *17*, 72. [CrossRef]
28. Strohle, A.; Hofler, M.; Pfister, H.; Muller, A.G.; Hoyer, J.; Wittchen, H.U.; Lieb, R. Physical activity and prevalence and incidence of mental disorders in adolescents and young adults. *Psychol. Med.* **2007**, *37*, 1657–1666. [CrossRef]
29. Jalene, S.; Pharr, J.; Shan, G.; Poston, B. Estimated Cardiorespiratory Fitness Is Associated With Reported Depression in College Students. *Front. Physiol.* **2019**, *10*, 1191. [CrossRef]
30. Alpsoy, S. Exercise and Hypertension. *Adv. Exp. Med. Biol.* **2020**, *1228*, 153–167. [CrossRef]
31. Bakker, E.A.; Sui, X.; Brellenthin, A.G.; Lee, D.C. Physical activity and fitness for the prevention of hypertension. *Curr. Opin. Cardiol.* **2018**, *33*, 394–401. [CrossRef] [PubMed]
32. Kanaley, J.A.; Colberg, S.R.; Corcoran, M.H.; Malin, S.K.; Rodriguez, N.R.; Crespo, C.J.; Kirwan, J.P.; Zierath, J.R. Exercise/Physical Activity in Individuals with Type 2 Diabetes: A Consensus Statement from the American College of Sports Medicine. *Med. Sci. Sports Exerc.* **2022**, *54*, 353–368. [CrossRef] [PubMed]
33. McGough, J.; Curry, J.F. Utility of the SCL-90-R with depressed and conduct-disordered adolescent inpatients. *J. Pers. Assess* **1992**, *59*, 552–563. [CrossRef] [PubMed]
34. Alves, D.A.; Waclawovsky, A.J.; Tonello, L.; Firth, J.; Smith, L.; Stubbs, B.; Schuch, F.B.; Boullosa, D. Association between cardiorespiratory fitness and depressive symptoms in children and adolescents: A systematic review and meta-analysis. *J. Affect. Disord.* **2021**, *282*, 1234–1240. [CrossRef]
35. Mao, Y.; He, Y.; Xia, T.; Xu, H.; Zhou, S.; Zhang, J. Examining the Dose-Response Relationship between Outdoor Jogging and Physical Health of Youths: A Long-Term Experimental Study in Campus Green Space. *Int. J. Environ. Res. Public Health* **2022**, *19*, 5648. [CrossRef]
36. Cheng, S. A study on the relationship between mental health literacy and anxiety and depression symptoms in college students. Master's Thesis, Shandong University, Jinan, China, 2021.
37. Tao, F.; Hu, C.; Sun, Y.; Hao, J. The development and application of multidimensional sub-health questionnaire of adolescents (MSQA). *Chin. J. Dis. Control Prev.* **2008**, *12*, 309–314.
38. Tao, S.; Wan, Y.; Wu, X.; Sun, Y.; Xu, S.; Zhang, S.; Hao, J.; Tao, F. Psychological evaluation and application of "Concise Questionnaire for Adolescent Mental Health Assessment". *Chin. Sch. Health* **2020**, *41*, 1331–1334.
39. Wan, Y.; Hu, C.; Tao, F. Research on responsiveness of multidimensional sub-health questionnaire of adolescents. *Chin. J. Public Health* **2008**, *24*, 1035–1036.
40. Xing, C.; Tao, F.; Yuan, C. Evaluation of reliability and validity of the multi-dimensional sub-health questionnaire of adolescents. *Chin. J. Public Health* **2008**, *24*, 1031–1033.
41. Li, Y.; Zhang, F.; Chen, Q.; Yin, X.; Bi, C.; Yang, X.; Sun, Y.; Li, M.; Zhang, T.; Liu, Y.; et al. Levels of Physical Fitness and Weight Status in Children and Adolescents: A Comparison between China and Japan. *Int. J. Environ. Res. Public Health* **2020**, *17*, 9569. [CrossRef]
42. Bi, C.; Zhang, F.; Gu, Y.; Song, Y.; Cai, X. Secular Trend in the Physical Fitness of Xinjiang Children and Adolescents between 1985 and 2014. *Int. J. Environ. Res. Public Health* **2020**, *17*, 2195. [CrossRef] [PubMed]
43. Bi, C.; Yang, J.; Sun, J.; Song, Y.; Wu, X.; Zhang, F. Benefits of normal body mass index on physical fitness: A cross-sectional study among children and adolescents in Xinjiang Uyghur Autonomous Region, China. *PLoS ONE* **2019**, *14*, e220863. [CrossRef] [PubMed]
44. Zhang, F.; Bi, C.; Yin, X.; Chen, Q.; Li, Y.; Liu, Y.; Zhang, T.; Li, M.; Sun, Y.; Yang, X. Physical fitness reference standards for Chinese children and adolescents. *Sci. Rep.* **2021**, *11*, 4991. [CrossRef] [PubMed]
45. Zou, Z.; Chen, P.; Yang, Y.; Xiao, M.; Wang, Z. Assessment of physical fitness and its correlates in Chinese children and adolescents in Shanghai using the multistage 20-M shuttle-run test. *Am. J. Hum. Biol.* **2019**, *31*, e23148. [CrossRef]
46. Zhang, F.; Yin, X.; Bi, C.; Li, Y.; Sun, Y.; Zhang, T.; Yang, X.; Li, M.; Liu, Y.; Cao, J.; et al. Normative Reference Values and International Comparisons for the 20-Metre Shuttle Run Test: Analysis of 69,960 Test Results among Chinese Children and Youth. *J. Sports Sci. Med.* **2020**, *19*, 478–488.
47. FB, T.; C, X.; CJ, Y. Development of national norm for multidimensional sub-health questionnaire of adolescents. *Chin. J. Sch. Health* **2009**, *30*, 292–295.
48. Xuan, G.; Hong, Y.; Jian, J.; Yu, H. Analysis of related factors between health literacy and mental sub-health symptoms of middle school students in Bengbu City, Anhui Province. *Chin. Health Educ.* **2017**, *33*, 8–11.
49. Li, X.; Krumholz, H.M.; Yip, W.; Cheng, K.K.; De Maeseneer, J.; Meng, Q.; Mossialos, E.; Li, C.; Lu, J.; Su, M.; et al. Quality of primary health care in China: Challenges and recommendations. *Lancet* **2020**, *395*, 1802–1812. [CrossRef]
50. Diamond, R.; Waite, F. Physical activity in a pandemic: A new treatment target for psychological therapy. *Psychol. Psychother. Theory Res. Pr.* **2021**, *94*, 357–364. [CrossRef]
51. Haegele, J.A.; Zhu, X.; Wilson, P.B.; Kirk, T.N.; Davis, S. Physical activity, nutrition, and psychological well-being among youth with visual impairments and their siblings. *Disabil. Rehabil.* **2021**, *43*, 1420–1428. [CrossRef]
52. Moral-García, J.E.; Román-Palmero, J.; García, S.L.; García-Cantó, E.; Pérez-Soto, J.J.; Rosa-Guillamón, A.; Urchaga-Litago, J.D. Self-esteem and sports practice in adolescents. *Int. J. Med. Sci. Phys. Act. Sport* **2021**, *21*, 157–174.

53. Moral-Garcia, J.E.; Lopez-Garcia, S.; Urchaga, J.D.; Maneiro, R.; Guevara, R.M. Relationship Between Motivation, Sex, Age, Body Composition and Physical Activity in Schoolchildren. *Notes Phys. Educ. Sports* **2021**, *144*, 1–9. [CrossRef]
54. Lema-Gomez, L.; Arango-Paternina, C.M.; Eusse-Lopez, C.; Petro, J.; Petro-Petro, J.; Lopez-Sanchez, M.; Watts-Fernandez, W.; Perea-Velasquez, F. Family aspects, physical fitness, and physical activity associated with mental-health indicators in adolescents. *BMC Public Health* **2021**, *21*, 2324. [CrossRef] [PubMed]
55. Matute-Llorente, A.; Gonzalez-Aguero, A.; Gomez-Cabello, A.; Vicente-Rodriguez, G.; Casajus, J.A. Physical activity and cardiorespiratory fitness in adolescents with Down syndrome. *Nutr. Hosp.* **2013**, *28*, 1151–1155. [CrossRef] [PubMed]
56. Aires, L.; Silva, P.; Silva, G.; Santos, M.P.; Ribeiro, J.C.; Mota, J. Intensity of physical activity, cardiorespiratory fitness, and body mass index in youth. *J. Phys. Act. Health* **2010**, *7*, 54–59. [CrossRef]
57. Chen, W.; Gu, X.; Chen, J.; Wang, X. Association of Cardiorespiratory Fitness and Cognitive Function with Psychological Well-Being in School-Aged Children. *Int. J. Environ. Res. Public Health* **2022**, *19*, 1434. [CrossRef]
58. Dimitri, P.; Joshi, K.; Jones, N. Moving more: Physical activity and its positive effects on long term conditions in children and young people. *Arch. Dis. Child.* **2020**, *105*, 1035–1040. [CrossRef]
59. Jimenez, B.R.; Arriscado, A.D.; Gargallo, I.E.; Dalmau, T.J. [Determinants of health in adolescence: Cardiorespiratory fitness and body composition]. *Nutr. Hosp.* **2021**, *38*, 697–703. [CrossRef]

Review

Effects of Participating in Martial Arts in Children: A Systematic Review

Aleksandar Stamenković [1], Mila Manić [1], Roberto Roklicer [2], Tatjana Trivić [2], Pavle Malović [3] and Patrik Drid [2,*]

1 Faculty of Sport and Physical Education, University of Niš, 18000 Niš, Serbia
2 Faculty of Sport and Physical Education, University of Novi Sad, 21000 Novi Sad, Serbia
3 Faculty for Sport and Physical Education, University of Montenegro, 81400 Nikšić, Montenegro
* Correspondence: patrikdrid@gmail.com

Abstract: Background: The application of various martial arts programs can greatly contribute to improving the of physical fitness of preschool and school children. The purpose of this review paper was to determine the effects and influence that martial arts program intervention has on children's physical fitness, which includes motor skills and the aerobic and anaerobic abilities of children. Method: We searched the following electronic scientific databases for articles published in English from January 2006 to April 2021 to gather data for this review paper: Google Scholar, Pub Med, and Web of Science. Results: After the search was completed, 162 studies were identified, of which 16 studies were selected and were systematically reviewed and analyzed. Eight studies included karate programs, four studies included judo programs, two studies contained aikido programs, and two studies contained taekwondo programs. The total number of participants was 1615 (experimental group = 914, control group = 701). Based on the main findings, karate, judo, taekwondo, and aikido programs showed positive effects on the physical fitness of the experimental group of children. According to the results, the effects of these programs showed statistically significant improvements between the initial and final measurements of most of the examined experimental groups. Conclusion: We concluded that martial arts programs were helpful for improving the physical fitness of preschool and school children, especially for parameters such as cardiorespiratory fitness, speed, agility, strength, flexibility, coordination, and balance.

Keywords: combat sports; physical fitness; motor skills; cardiorespiratory fitness

Citation: Stamenković, A.; Manić, M.; Roklicer, R.; Trivić, T.; Malović, P.; Drid, P. Effects of Participating in Martial Arts in Children: A Systematic Review. *Children* **2022**, 9, 1203. https://doi.org/10.3390/children9081203

Academic Editor: Georgian Badicu

Received: 12 July 2022
Accepted: 1 August 2022
Published: 11 August 2022

Publisher's Note: MDPI stays neutral with regard to jurisdictional claims in published maps and institutional affiliations.

Copyright: © 2022 by the authors. Licensee MDPI, Basel, Switzerland. This article is an open access article distributed under the terms and conditions of the Creative Commons Attribution (CC BY) license (https://creativecommons.org/licenses/by/4.0/).

1. Introduction

Martial arts have been practiced for hundreds of years, and today, with modification, they are often used in the form of sports, self-defense, and recreation [1]. Applying well-organized martial arts programs to children can lead to an increase in physical fitness, although the test results are not unequivocal [2]. Many studies have explored the application of various martial arts programs in children in order to improve their motor skills and physical fitness [1–5]. Positive effects could sometimes be visible, especially in children who have been practicing martial arts since childhood. Regarding motor abilities, the development of explosive power, speed of movement, agility, strength, balance, and precision could be noticed [6,7], while on the other hand, in addition to specific physical fitness, aerobic and anaerobic endurance were developed [8–10]. In addition, the link between positive socio-psychological responses and involvement in martial arts has also been reported in children. Specifically, increased social skills and self-confidence, along with less aggressiveness, were evident among young martial artists [11]. Furthermore, martial arts can have a significant influence on mental health and on the formation of personal character in young people. Moreover, practicing martial arts helps young people to learn the elements of self-defense and improve their fitness level [12].

The progress of science and technology has led to a decrease in motor activity in favor of a sedentary lifestyle, which has negatively affected the physical development of children [7]. The percentage of children who meet the World Health Organization (WHO) criteria for the recommended level of physical activity [13] is very small and ranges from 2% to 14% and from 9% to 34% in European girls and boys, respectively [5]. Practicing martial arts greatly affects the improvement of the conative characteristics of a person, such as determination, consistency, and motivation [14], as a result of well-organized training, which can directly affect the improvement of physical fitness [15]. This correlates with the results of a study performed by Pinto-Escalona et al. [5]. Pinto-Escalona et al. conducted a one-year investigation in which the experimental group underwent karate intervention at school that led to improved cardiorespiratory fitness and balance in the children. This study was in line with the findings of Pavlova et al. [16] who conducted research with children aged between 5 and 6 years of age. The experimental group (EG) of children underwent a one-year karate program, compared to the classic program of preschool education that the control group (CG) underwent. The results showed that the EG had twice the overall growth rate of physical fitness compared to the CG, with the highest growth rate observed in children in the EG in the "Flamingo" pose (balance). In a study by Boguszewski and Socha [7], which included three groups of children from preschools (children that practiced karate, gymnasts, and inactive children), the best results in the upper extremity strength test were achieved by girls that practiced karate. In addition, based on the obtained results, Boguszewski et al. [17] reported significant improvements in deep squats, shoulder mobility, and torso stability (push-ups) in the experimental karate group. These two karate studies were correlated with the research of Kirpenko et al. [18], which showed that there was a significant improvement in the maximum aerobic capacity, strength, speed–power quality, endurance, and flexibility in the Kyokushin karate experimental group in which children between 10 and 12 years of age were examined. The existing literature reveals the positive impact of various martial arts and combat sports on children's well-being. The available evidence demonstrates a positive impact on the physiological and physical development of children who practice judo and meet the recommended standards of health-related physical activity [19]. Moreover, Lee and Kim [20] used 16-week taekwondo training in their work and found that children in the the experimental group showed increases in maximum strength and balance. Wasik et al. [21] reported positive effects on body posture; the number of occurrences of body asymmetry was reduced in children and adolescents following a taekwondo training program. In research by Falk and Mor [6], the experimental group of children, which contained children between 6 and 7 years of age, that underwent the 12-week resistance training and martial arts program, showed significant ($p < 0.05$) improvements compared with the control group in the sit-up and long jump tests.

Based on the authors' findings, there are a small number of review studies that have investigated the effects and impacts of martial arts programs on children's physical fitness [22]. Considering that the modern age has led to obesity and the deterioration of motor skills in children, this research could contribute to solving this global problem through the organization of martial arts programs, which could be performed as additional activities alongside regular physical education classes. Hence, this review aimed to determine the effects and influence that martial arts program intervention has on children's physical fitness, which includes motor skills and the aerobic and anaerobic abilities of children.

2. Materials and Methods

2.1. Inclusion Criteria

The following criteria were used to select the articles to be included in the review: (1) original scientific articles; (2) all studies that dealt with martial arts; (3) articles written in English; (4) articles where the participants in the sample were children of preschool and school age (between 4 and 18 years of age).

2.2. Exclusion Criteria

The following criteria were used to select the articles to be excluded from the research: (1) review studies; (2) studies in which the subjects were children younger than 4 years of age and older than 18 years of age; (3) articles with research in which the experimental groups of children were not subjected to martial arts training programs; (4) articles in which the results or investigated parameters were not adequately presented for further analysis.

2.3. Search Strategy and Study Selection

Electronic database searches were performed using Google Scholar, Pub Med, and Web of Science. The search terms covered the areas of the influence of martial arts training on physical fitness and martial arts programs for preschool and school children using a combination of the following keywords: "martial arts", "physical fitness", "children", "motor skills", "cardiorespiratory fitness". The results of a search of articles written in English and published between January 2006 and April 2021 were analyzed. Articles from the database list that were clearly not relevant were removed before assessing all other titles and abstracts using our predetermined inclusion and exclusion criteria. Inter-reviewer disagreements were resolved by consensus opinion or arbitration by a third reviewer. Reference lists of the selected manuscripts were also examined for any other potentially eligible articles.

2.4. Assessment of Bias

Two independent reviewers assessed the risk of bias. The agreement between the two reviewers was evaluated using k full-text screening statistics and an assessment of relevance and the risk of bias. When there was disagreement over the risk of bias, the third reviewer checked the data and formed the final decision. The agreement rate for k among reviewers was k = 0.95. Inter-observational reliability and agreement were calculated using the interclass correlation coefficient and Cohen's kappa test with values interpreted as follows: 0—no agreement; 0.01–0.20—slight agreement; 0.21–0.40—good agreement; 0.41–0.60—moderate agreement; 0.61–0.80—significant agreement; and 0.81–1.0—almost perfect agreement.

3. Results

Selection and Inclusion of Studies

We collected 159 studies through a database search in Google Scholar, PubMed, and Web of Science. Three more studies were found from other sources; thus, the total number of collected studies was 162. Next, we excluded 63 duplicate studies (duplicate search or publication of the thesis in a journal). After reviewing the research titles and abstracts of the 99 remaining selected studies, an additional 26 studies were excluded that included children with diseases and children older than 18 years of age. Another 10 studies were excluded because they included martial arts that were not classified as budo martial arts. In addition, seven other studies were excluded because full texts were not available. Therefore, we selected 67 studies after excluding 32 studies. We reviewed the full text of the 67 selected studies and classified them according to the PICOS criteria. After that, 24 studies with no control groups, seven studies with unclear statistics, and 18 studies with uncertain training durations were excluded. Therefore, we finally included 16 studies (8 studies included only male participants, 1 study included only female participants, and 7 studies included both sexes) (Table 1). The data selection method is presented within the following flow diagram (Figure 1).

Table 1. Results of the included studies.

Author (Year)	Subjects Program							Intervention	
	Age and Gender (y)	E (*n*)	C (*n*)	Duration (wk/m)	Frequency (Sessions/ (wk/)	Time (min/ Session)	E Training Content	Testing Content	Results after Intervention
Sekulic et al. (2006) [23]	7, M	Judo (*n* = 41)	Sports games (*n* = 57)	9 m	3	45	Judo training	COR (10 m polygon test); AGL (4 × 1.98 m shuttle run test); FLEX (maximal circumduction, sit and reach); SP (20 m sprint); LLES (standing long jump); CREN (3 min run); MEN (flexed-arm hang, 1 min of sit-ups).	All variables (E, C) ↑*; AGL, E↔*; FLEX, E↔*; MEN, E↔*; COR, (E, C) ↔.
Krstulović et al. (2010) [24]	7, F	Judo (*n* = 30)	Sports games (*n* = 49)	9 m	3	45	Judo training	COR (10 m polygon test); AGL (4 × 1.98 m shuttle run test); FLEX (maximal circumduction, sit and reach); SP (20 m sprint); LLES (standing long jump); CREN (3 min run); MEN (flexed-arm hang, 1 min of sit-ups).	All variables (E, C) ↑*; AGL, E ↔*; FLEX, E ↔*; MEN, E ↔*; MS, E ↔*; COR, (E, C) ↔.
Boguszevski and Socha (2011) [7]	4.5–6.5, M and F	Karate (*n* = 30)	Corrective gymnastics (*n* = 30); inactive (*n* = 28)	6 m	3	60	Karate training	RS (20 s of sit-ups); ULES (throwing 1 kg medicine ball); LLES (long jump from the start line); AGL (4 × 5 m shuttle run); FLEX (test "fingers-floor").	ULES, E (G) ↑*, E(B) ↑; LLES, E (G) ↔*; RS, ↔; AGL, E (G) ↑; FLEX, E (G) ↔*.
Demiral (2011) [25]	7–12, M and F	Judo, (*n* = 38),	Judo (*n* = 31)	12 m	ND	ND	Judo training	BAL ("Flamingo" balance test); COR (coordination test, rapidity test); LLES (standing long lump); ULES (medical ball throws); SP (running speed test); S (grip and back power test).	LLES, E (B, G) ↑*; (B, G) ↔*; ULES, E (B) ↑*; BAL, E (B, G) ↑*; COR, E (B, G) ↑*, G ↔*; SP, E (B, G) ↑*, B ↔*; S, E (B, G) ↑*.
Alesi et al. (2014) [4]	9–10, M and F	Karate (*n* = 19)	Inactive (*n* = 20)	8 wk	3	60	Karate training	AGL (coordination skills with hurdles); LLES (standing broad jump); SP (20 m sprint).	AGL, (E, C) ↔*; LLES, (E, C) ↔*; SP, (E, C) ↔*.
Padulo et al. (2014) [26]	8–12, M	High- intensity karate (*n* = 53)	Low- intensity karate (*n* = 20)	1 wk	14	60	Karate training	ULES (sitting medicine ball throw); LLES (standing long jump); AGL (frontal/lateral jumps); FLEX ("active" joint flexibility).	ULES, (E, C) ↔*, C ↑; LLES, (E, C) ↔*, E ↑*, C ↑; AGL, (E, C) ↔*, E ↑*, C ↑; FLEX, (E, C) ↔*, E ↑*, C ↑;

Table 1. *Cont.*

Author (Year)	Age and Gender (y)	Subjects Program					Intervention		
		E (n)	C (n)	Duration (wk/m)	Frequency (Sessions/ (wk/)	Time (min/ Session)	E Training Content	Testing Content	Results after Intervention
Mroczkowski (2015) [27]	7–10, M	Aikido and PE classes (n = 66)	PE classes (n = 41)	9 m	3	60	Aikido training	FLEX of hip ("samurai walk").	FLEX c1-TAOR (Lh), (E, C) ↔*, E ↑*; c2- AOER (Lh), E ↑; c3- AOIR (Lh), (E, C) ↔*, (E, C) ↑*; c4- TAOR (Rh), (E, C) ↔ *, E ↑*; c5- AOER (Rh), E ↑*; c6- AOIR (Rh), (E, C) ↑.
Pop et al. (2017) [28]	9–10, M	Aikido and PE classes (n = 5)	PE classes (n = 5)	8 m	3	60–90	Aikido training	BAL ("Flamingo" balance test); CREN (beep test, 20 m); SP (TRANSFER running 10 × 5 m at speed).	BAL, (E, C) ↔; CREN, (E, C) ↔; SP, (E, C) ↔*.
Ma and Qu, (2017) [29]	8–12, M	Karate (n = 51)	/	8 wk	7	5	Karate training	RS (30 s of sit-ups); LLES (standing long jump); ULES (throwing a medicine ball); AGL (4 × 5 m shuttle run); FLEX (sit and reach).	LLES, ↑*; RS, E ↑; AGL, ↑*; FLEX, ↔.
Pavlova et al. (2018) [16]	5–6, M	Karate (n = 33)	PE classes (n = 38)	12 m	3	45–60	Karate training	LLES (standing long jump); ULES (medical ball throws); EN (jumping with a rope to fatigue); RS (max. number of squats to fatigue); BAL ("flamingo" balance test).	LLES, (E, C) ↑*; ULES, E ↑*; RS, (E, C) ↑*; BAL, (E, C) ↑*; EN, (E, C) ↑*.
Pathare et al. (2018) [30]	6–12, M and F	Taekwondo HW (n = 11), OW (n = 6)	/	10 wk	2	50	Taekwondo training	LLES (just jump system, vertical jump); BAL (force plate (Watertown, Mass.) bipedal/normal stance eyes open and closed, single stance eyes open (30 s)); FLEX (sit and reach test); PH activity (Yamax SW-701 pedometer).	LLES, (HW, OW) ↔*; PH activity, (HW, OW) ↔; LLES, (HW, OW) ↔; BAL, (OW, HW) ↔, OW ↑; FLEX, (HW, OW) ↓*, ↔.
Top et al. (2018) [31]	7–10, M and F	Taekwondo B (n = 13), G (n = 9)	Inactive B (n = 12) G (n = 8)	12 wk	3	60	Taekwondo training	S (knee push-ups, sit-ups, wall sit); COR (touching nose with a finger); LLES (standing long jump); BAL (single leg balance test with closed and open eyes); AGL (shuttle run, stepping sideways, one-legged stationary hop, one-legged side hop, and two-legged side hop).	COR, (E, C) ↔, E ↑*; S, LLES, (E, C) ↔*, E ↑*; AGL, (E, C) ↔*.

Table 1. *Cont.*

Author (Year)	Age and Gender (y)	Subjects Program						Intervention	
		E (n)	C (n)	Duration (wk/m)	Frequency (Sessions/ (wk/)	Time (min/ Session)	E Training Content	Testing Content	Results after Intervention
Rutkowski et al. (2019) [2]	7.6 ± 0.4 M and F	Karate B/N (n = 15), G/N (n = 15), B/O (n = 15), G/O (n = 14).	/	10 wk	2	60	Karate training	EUROFIT FITNESS TEST: BAL; Arm SP; Abdominal MS; LLES; FLEX (sit and reach test).	BAL, B/N ↑*; Arm SP, B/N ↑*; (G/N, G/O) ↔*, (B/N, B/O) ↔*; (G/O, B/N) ↔*; Torso MS; (G/O, B/O) ↔*, G/O ↓; LLES, G/N ↔*; FLEX, B/N ↓*, G/N ↔*.
Brasil et al. (2020) [32]	8–13, M	Judo (n = 35), 20 Ob, 15 NOb.	/	12 wk	2	60	Judo training	CPET treadmill: 1. VO2peak; 2. HR at VO2peak.	VO2peak, E (Ob) ↓; HR at VO2peak, E ↔*.
Kyrpenko et al. (2020) [18]	10–12, M	Karate (n = 27)	Inactive (n = 29)	24 m	ND	ND	Karate training	CREN (1000 m running); Arm S (pull-up on the crossbar); LLES (long jump from place); AGL (4 × 9 m shuttle run); FLEX (body leaning forward); VO2max	CREN, E ↑*; LLES, E ↑*; AGL, (E, C) ↑*; FLEX, ↔; VO2max, E ↑*.
Pinto-Escalona et al. (2021) [5]	7–8, M and F	Karate (n = 388)	PE classes (n = 333)	12 m	2	60	Karate training	CREF (20 m shuttle run test); FLEX (frontal split test); BAL (Y balance test).	CREF, E; FLEX, E; BAL, E.

Legend: M—male; F—female; E—experimental group; C—control group; wk—week; m—month; ND—not defined; PE—physical education; B—boys; G—girls; B/N—boys of normal weight; G/N—girls of normal weight; B/O—overweight boys; G/O—overweight girls; Ob—obese; NOb—non-obese; PH—physical; HW—healthy weight; OW—overweight; S—strength; MS—muscle strength; ES—explosive strength; ULES—upper limb explosive strength; LLES—lower limb explosive strength; RS—repetitive strength; COR—coordination; AGL—agility; FLEX—flexibility; SP—speed; BAL—balance; EN—endurance; MEN—muscular endurance; CREN—cardiorespiratory endurance; CREF—cardiorespiratory fitness; FMS—functional movement screen; CPET—cardiopulmonary exercise testing; TAOR—total angle of rotation; AOER—angle of external rotation; AOIR—angle of internal rotation; Lh—left hip; Rh—right hip; ↑—increased (↑*– significantly increased); ↓—decreased (↓*– significantly decreased); ↔—no significant difference; ↔*—significant difference; (E, C)—between groups E and C.

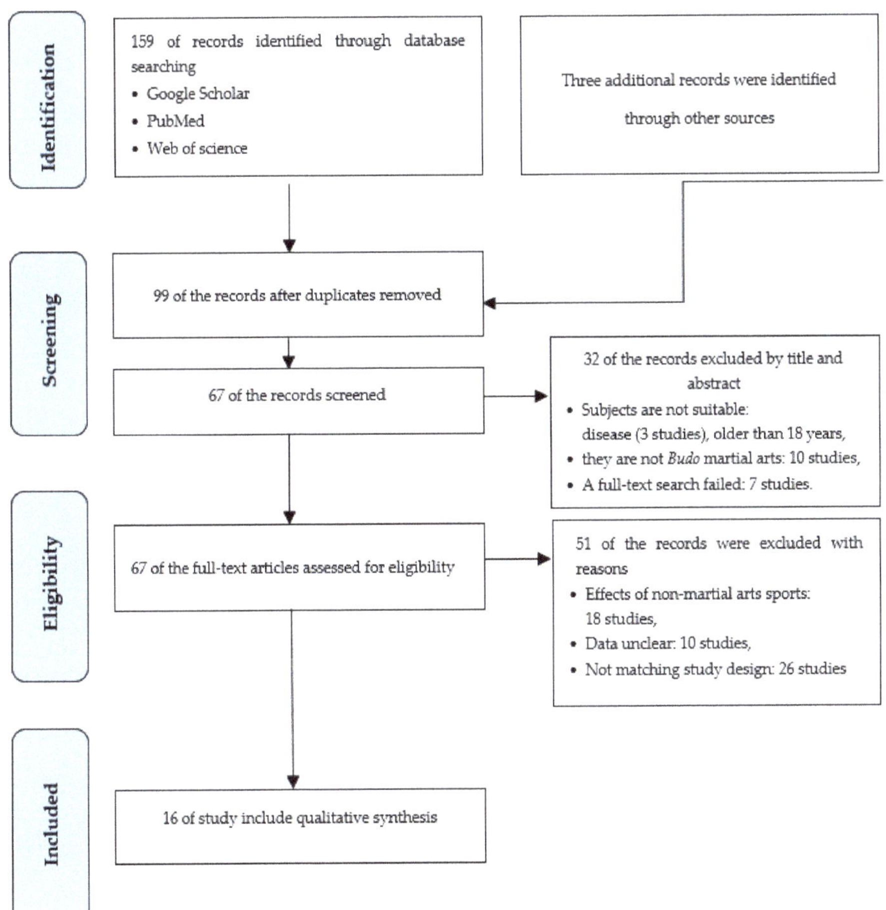

Figure 1. PRISMA flow diagram.

The study included 16 original research articles that tested the effects of martial arts programs in children aged between 4 and 13 years of age. The articles are presented in tabular form by year of publication from 2006 to 2021, respectively. There were no control groups in four articles [2,23,30,32]. The youngest participant was 4.5 years old [7], and the oldest participant was 13 years [33]. The total number of participants was 1615 (EG = 914, CG = 701). The sample sizes of the studies were analyzed and ranged from 10 [25] to 721 children [5]. In seven studies, both genders were included [2,4,5,7,25,30,31]. In eight studies, the participants were only male children [16,18,23,26–29,32], while one study included only girls [24]. The longest experimental treatment lasted two years [18] and the shortest only lasted a week [26]. Frequently, the duration of the sessions in all martial arts programs was 60 min [2,4,5,7,16,26–28,32]. Eight studies included experimental treatments of karate [2,4,5,7,16,18,26,29], four studies included judo treatments [23–25,32], in two articles taekwondo treatments were employed [26,27], and two studies included experimental aikido treatments [27,28].

4. Discussion

This study examined the effects of various martial arts programs on children's physical fitness. Based on the main findings, karate, judo, taekwondo, and aikido programs showed positive effects on physical fitness components. According to the results, the effects of these programs showed significant differences between the initial and final measurements of most of the examined experimental programs, but also when compared to the control groups. Cardiorespiratory fitness, speed, agility, strength, flexibility, coordination, and balance were used to assess physical fitness, while other parameters, such as body composition, mental conative, and cognitive capacities, were excluded.

4.1. Cardiorespiratory Fitness

Cardiorespiratory fitness as a parameter of physical fitness was represented in only three studies [18,28,32]. After a 24-month karate program, Kyrpenko et al. [18] found a large statistically significant improvement ($p < 0.01$) in the cardiorespiratory endurance parameter in the EG of karatekas, while in inactive children in the CG, this was not the case. The 1000 m test is a middle-distance running test that gives participants the opportunity to increase their cardiorespiratory fitness at the expense of muscle strength and to improve their running technique, unlike running for 20 m, 30 m, and 100 m, where the type of muscle fibers, despite the training, plays a major role. Since the treatment in the EG lasted 24 months, it was no wonder that there was a significant improvement in the final measurement, given that the subjects had enough time to increase their strength and improve their running technique. In addition, Brasil et al. [32] assessed the cardiorespiratory ability of obese and non-obese children after participating in a 12-week judo program. The authors concluded that there was a decrease in the VO2peak parameter in the obese children and statistically significant differences in HR at the VO2peak between non-obese and obese children. Given that obese children have an excess of inefficient adipose tissue that leads to accelerated fatigue, it was no wonder that HRs at the VO2peaks were higher in the obese children. Pop et al. [28] examined cardiorespiratory endurance, where the EG, in addition to regular classes of physical education, was subjected to an eight-month aikido program, while the control group attended only physical education classes. Testing was performed using the 20 m beep test, and, on that occasion, it was concluded that there were no statistically significant differences between the EG and the CG in the final measurements.

4.2. Speed and Agility

The shuttle run agility test has been the main tool to test this motor ability in many studies in which agility as a parameter of physical fitness was presented [7,18,23,24,29,31]. Specifically, using the nine-month judo program, Sekulic et al. [23] found statistically significant differences ($p < 0.05$) in the final measurements of the 4×1.98 m shuttle run agility test in an experimental group of subjects composed of boys. This was in correlation with the findings of Krstulović et al. [24], who also found statistically significant differences using the same test as the judo program of the same duration in female participants. In addition, using a six-month karate program, Boguszevski and Socha [7] used the 4×5 shuttle run test to assess agility, concluding that progress had been made in the final measurements, but they remained statistically insignificant. This was not correlated with Ma and Qu [29], who used the same 4×5 shuttle run test to identify the effects of a two-month karate program. They found a statistically significant difference ($p < 0.01$) in the final measurement of the EG that correlated with the results obtained by Top et al. [31] and Kyrpenko et al. [18], who tested the agility of the 4×9 shuttle run test using a two-year karate program and found a statistically significant improvement ($p < 0.01$) in the EG compared to the CG.

Padulo et al. [26] used frontal/side jumps in a quadratic agility assessment test to compare the effects of a one-week high- and low-intensity karate program. They found a statistically significant improvement in the EG. They also found a statistically significant difference between the EG and the CG in favor of the EG on the final measurement. Speed was one of the monitored parameters of physical fitness in five articles [4,23–25,28]. Various running speed tests were used to measure speed as a parameter of physical fitness. Sekulic et al. [23] concluded that the change in speed in boys did not occur in the initial and final EG measurements. This correlated with the findings of Krstulović et al. [24], who applied almost the same program for 9 months in female participants; however, this was not correlated with the findings of Demiral [25], whose research showed statistically significant differences between the EG and the CG in the 20 m sprint test. Perhaps the reason was that Demiral's study lasted 3 months longer than the previously mentioned studies [23,24], where the participants were subjected to experimental treatment for a longer period of time. A study conducted by Pop et al. [28] used 20 m sprint tests, while Alesi et al. [4] applied a transfer-running test to check the speed skill of participants and also found statistically significant differences between the EG and the CG in favor of the EG.

4.3. Strength

Strength, as a parameter of physical fitness, was assessed in 12 research articles [2,4, 7,16,18,23–26,29–31]. In most of the studies, the explosive strength of the legs and arms was measured using standing long jumps and medical ball throws tests. Specifically, in all studies in which explosive power was tested by these test protocols, statistically significant improvements in the EGs were present [2,4,7,16,18,22–26,29,31] and also in some CGs [16,22,24]. Since the ULES parameter (upper limb explosive strength) was presented in five articles [7,16,25,26,29], and the fact that in each of these articles medical ball throws were used, it was the main test for estimating this parameter. In four of the five studies, the karate program was applied, while the judo program was implemented in only one study [25]. Considering that karate is a striking martial art and sport, which requires highly developed explosiveness and quickness that is constantly emphasized during the training, it is absolutely justified and logical that this parameter improved. Repetitive strength was presented in three studies [7,16,29]. Based on the results of their research, Boguszevski and Socha [7] found no statistically significant differences between the EG and the CG on the initial and final measurements of the sit-up test in their study in which a six-month karate program was applied. This correlated with the results obtained by Ma and Qu [30], who tested the repetitive strength of karatekas conducting the same test and concluded that there were no statistically significant improvements in their EG at the final measurements. Interestingly, the authors of both articles came to the same conclusions, even though the karate program proposed by Boguszevski and Socha [7] lasted 24 weeks and the program of Ma and Qu [29] lasted only 8 weeks. It is probable that the main reason for this is the fact that there were no exercises in the experimental karate programs that affected the repetitive strength of the abdominal muscles. Pavlova et al. [16] found that after a 12-month karate program, there was a statistically significant improvement ($p < 0.01$) in the EG in repetitive leg strength. A significant improvement ($p < 0.01$) was also registered in the CG, where the participants attended only physical education classes. Since the children were aged between 5 and 6 years old at the beginning of the treatment, it is likely that one of the reasons for the great improvement was maturation.

4.4. Flexibility

Flexibility was one of the parameters for assessing physical fitness in 10 articles [2,5, 7,18,23,24,26,27,29,30]. The most common flexibility assessment test used in as many as five articles was the "sit and reach" test [2,23,24,29,30]. The results of this test showed statistically significant differences in flexibility between the EG and the CG in the final measurements in two articles by Krstulović et al. [24] and Sekulic et al. [23], in which 9-month judo programs were conducted. This was partially in line with the findings of Rutkowski et al. [2], who, by applying a 10-week karate program, found a statistically significant decrease in flexibility in normal-weight boys, while in normal-weight girls, there was a statistically significant difference. In addition, this did not correlate with the results of Pathare et al. [30] and Ma and Qu [29]; in these studies no statistically significant improvement in flexibility was observed. In addition, the author Mrockowski [27] used a nine-month aikido program, where the hip flexibility test "samurai walk" was used to examine flexibility, and the presence of statistically significant differences between the EG and the CG were evident. Boguszevski and Socha [7] used the "finger floor" test to assess flexibility and found that the participants in the EG group (girls) significantly improved their performances compared to the control group. This correlated with Padulo et al. [26], who obtained similar results, checking the flexibility of the joints after a one-week program of high-intensity karate (EG) and low-intensity karate (CG); however, this did not correlate with the findings of Kirpenko et al. [18] who concluded that 12 months of karate did not cause an improvement in flexibility.

4.5. Coordination

Coordination as a parameter of physical fitness was represented in four articles [23–25,31]. The results of the 10 m polygon test conducted by Krstulović et al. [24] and Sekulić et al. [23] did not show statistically significant differences in coordination between the EG and the CG in the final measurements of both studies. In his research, Demiral [25] applied the coordination test (balance skill) and rapidity test in the EG and the CG of judokas and found statistically significant improvements ($p < 0.01$) in both sexes of the EG. In addition, statistically significant differences ($p < 0.01$) in coordination were found in the EG of girls compared to the CG. However, these results were partially in line with the results of Top et al. [31], which indicated a statistically significant improvement ($p < 0.05$) in the EG, but no statistically significant differences between the EG and the CG in the final measurements were observed.

4.6. Balance

Balance was tested in the majority of studies using the "flamingo" balance test [16,25,28]. In addition, other articles used the following tests: the Y balance test [5], the single-leg balance test with closed and open eyes [2,31], and a test that involved balancing on a force plate device [30]. Demiral [25] found a statistically significant improvement in balance in the EG of boys and girls following a 12-month judo program. These findings were not correlated with the results of Pop et al. [28], who found no significant differences between the EG and the CG after an eight-month aikido program in addition to regular physical education classes. In addition, according to Demiral's findings, a partial correlation was found with the study of Pavlova et al. [16]; the EG achieved a statistically significant improvement in balance, which was the same as the CG, whose participants only attended regular physical education classes. Speaking of the Y balance test, Pinto-Escalona et al. [5] found a slight improvement in the EG group that was not statistically significant. In the research performed by Rutkowski et al. [2], in which all the subjects underwent experimental treatment, only boys of normal weight achieved a statistically significant improvement in balance. This was not consistent with the findings of Pathare et al. [30], who discovered that overweight children had greater improvements in balance than healthy-weight children.

4.7. Future Research

Important considerations were identified from this review to support the development of future research in this research field, such as the following:

- One needs to be cautious when adopting an intervention program model that has already been performed and has relevant methodological limitations.
- There is a lack of studies dealing with the impact of martial arts on children's cardiorespiratory and motor-skill parameters. In future studies, the emphasis should be on non-budo martial arts such as capoeira, muay Thai, Brazilian jiu-jitsu, wrestling, and other combat sports [33].
- The following studies might have investigate preschool children because, in most countries, children start practicing martial arts from an early age.
- Various martial arts programs for school and preschool children need to be analyzed and a larger number of physical fitness parameters with an emphasis on body composition, as well as the heterochronism of their development, need to be monitored.

4.8. Strengths and Limitations

This is one of the first review papers that combined several different martial arts and analyzed their effect on children, including a comprehensive search of studies and the assessment of their methodological qualities. The main limitation of this study is the fact that, in addition to judo, taekwondo, and aikido, the majority of the research papers were based on karate. Therefore, the karate program was the most effective in terms of positive impact on physical fitness. Additionally, a significantly larger number of male children were included in the monitored studies. Given that there are large differences in motor skills and cardiorespiratory fitness between boys and girls, the gender should be defined separately.

5. Conclusions

This review confirmed that martial arts programs lead to improved physical fitness in preschool and school children. Based on the results of the analysis, it could be concluded that cardiorespiratory fitness, speed, agility, strength, flexibility, coordination, and balance were the most important parameters of physical fitness, which demonstrated considerable improvement in the final measurements of EG participants. The obtained information could be very useful in promoting and advertising the positive aspects of these budo martial arts, which can directly affect children's and parents' choices when it comes to choosing the sport that they will practice.

Author Contributions: Conceptualization, A.S. and M.M.; methodology, R.R. and T.T.; software, M.M.; validation, P.M. and P.D.; formal analysis, M.M.; investigation, A.S.; resources, P.D.; data curation, M.M.; writing—original draft preparation, A.S. and M.M.; writing—review and editing, R.R., T.T., P.M. and P.D.; visualization, A.S.; supervision, P.D.; project administration, A.S.; funding acquisition, P.D. All authors have read and agreed to the published version of the manuscript.

Funding: This work was supported by the Serbian Ministry of Education, Science, and Technological Development and the Provincial Secretariat for Higher Education and Scientific Research (142-451-2594).

Institutional Review Board Statement: Not applicable.

Informed Consent Statement: Not applicable.

Data Availability Statement: Data are contained within the article.

Conflicts of Interest: The authors declare no conflict of interest.

References

1. Woodward, T.W. A review of the effects of martial arts practice on health. *Wis. Med. J.* **2009**, *108*, 40–43.
2. Rutkowski, T.; Sobiech, K.A.; Chwałczyńska, A. The effect of karate training on changes in physical fitness in school-age children with normal and abnormal body weight. *Physiother. Q.* **2019**, *27*, 28–33. [CrossRef]
3. Lakes, K.D.; Hoyt, W.T. Promoting self-regulation through school-based martial arts training. *J. Appl. Dev. Psychol.* **2004**, *25*, 283–302. [CrossRef]
4. Alesi, M.; Bianco, A.; Padulo, J.; Vella, F.P.; Petrucci, M.; Paoli, A.; Pepi, A. Motor and cognitive development: The role of karate. *Muscle Ligaments Tendons J.* **2014**, *4*, 114–120. [CrossRef]
5. Pinto-Escalona, T.; Gobbi, E.; Valenzuela, P.L.; Bennett, S.J.; Aschieri, P.; Martin-Loeches, M.; Martinez-de-Quel, O. Effects of a school-based karate intervention on academic achievement, psychosocial functioning, and physical fitness: A multi-country cluster randomized controlled trial. *J. Sport Health Sci.* 2021, in press. [CrossRef]
6. Falk, B.; Mor, G. The Effects of Resistance and Martial Arts Training in 6- to 8-Year-Old Boys. *Pediatr. Exerc. Sci.* **1996**, *8*, 48–56. [CrossRef]
7. Boguszewski, D.; Socha, M. Influence of karate exercises on motor development in preschool children. *J. Combat Sports Martial Arts* **2011**, *2*, 103–107. [CrossRef]
8. Mascarenhas, L.P.G.; Grzelczak, M.T.; de Souza, W.C.; Neto, A.S.; Chacón-Araya, Y.; de Campos, W. Aerobic power in prepubescent children with different levels of physical activity. *Retos. Nuevas Tend. Educ. Física Deporte Recreación* **2015**, *27*, 203–205.
9. Pocecco, E.; Faulhaber, M.; Franchini, E.; Burtscher, M. Aerobic power in child, cadet and senior judo athletes. *Biol. Sport* **2012**, *29*, 217–222. [CrossRef]

10. Alp, M.; Gorur, B. Comparison of Explosive Strength and Anaerobic Power Performance of Taekwondo and Karate Athletes. *Edu. Learn* **2020**, *9*, 149–155. [CrossRef]
11. Vertonghen, J.; Theeboom, M. Analysis of experiences of young martial artists. In *Sport, Culture & Society*; Doupona Topic, M., Licen, S., Eds.; University of Ljubljana: Ljubljana, Slovenia, 2008; pp. 149–154.
12. Warchol, K.; Korobeynikov, G.; Osiel, C.; Cynarski, W.J. Martial arts as a form of physical activity for children and young people in the opinion of adult inhabitants of Podkarpackie Voivodeship. *Ido Mov. Cult. J. Martial Arts Anthropol.* **2021**, *21*, 28–37.
13. Bull, F.C.; Al-Ansari, S.S.; Biddle, S.; Borodulin, K.; Buman, M.P.; Cardon, G.; Willumsen, J.F. World Health Organization 2020 guidelines on physical activity and sedentary behavior. *Br. J. Sports Med.* **2020**, *54*, 1451–1462. [CrossRef] [PubMed]
14. Levesque, R.J. *Encyclopedia of Adolescence*; Springer Science & Business Media: New York, NY, USA, 2011; pp. 846–847.
15. Fuller, J.R. Martial arts and psychological health. *Br. J. Med. Psychol.* **1988**, *6*, 317–328. [CrossRef]
16. Pavlova, I.; Bodnar, I.; Vitos, J. The role of karate in preparing boys for school education. *Phys. Act. Rev.* **2018**, *6*, 54–63. [CrossRef]
17. Boguszewski, D.; Jakubowska, K.; Adamczyk, J.G.; Białoszewski, D. The assessment of movement patterns of children practicing karate using the Functional Movement Screen test. *J. Combat. Sports Martial Arts* **2015**, *6*, 21–26. [CrossRef]
18. Kyrpenko, Y.V.; Budur, M.I.; Palevych, S.V.; Poddubny, O.G. The influence of Kyokushinkai Karate classes on the adaptive capabilities of adolescents. *HSR* **2020**, *5*, 48–56. [CrossRef]
19. Gutierrez-Garcia, C.; Astrain, I.; Izquierdo, E.; Gomez-Alonso, M.T.; Yague, J.M. Effects of judo participation in children: A systematic review. *Ido Mov. Cult. J. Martial Arts Anthrop.* **2018**, *18*, 63–73.
20. Lee, B.; Kim, K. Effect of Taekwondo training on physical fitness and growth index according to IGF-1 gene polymorphism in children. *Korean J. Physiol. Pharmacol.* **2015**, *19*, 341–347. [CrossRef] [PubMed]
21. Wąsik, J.; Motow-Czyz, M.; Shan, G.; Kluszczynski, M. Comparative analysis of body posture in child and adolescent taekwon-do practitioners and non-practitioners. *Ido Mov. Cult. J. Martial Arts Anthrop.* **2015**, *15*, 35–40.
22. Nam, S.S.; Lim, K. Effects of Taekwondo training on physical fitness factors in Korean elementary students: A systematic review and meta-analysis. *J. Nutr. Biochem.* **2019**, *23*, 36–47. [CrossRef] [PubMed]
23. Sekulic, D.; Krstulovic, S.; Katic, R.; Ostojic, L. Judo training is more effective for fitness development than recreational sports for 7-year-old boys. *Pediatr. Exerc. Sci.* **2006**, *18*, 329–338. [CrossRef]
24. Krstulović, S.; Kvesić, M.; Nurkić, M. Judo training is more effective in fitness development than recreational sports in 7 years old girls. *Facta Univ. Ser. Phys. Educ. Sport* **2010**, *8*, 71–79.
25. Demiral, S. The study of the effects of educational judo practices on motor abilities of 7–12 years aged judo performing children. *Asian Soc. Sci.* **2011**, *7*, 212–219. [CrossRef]
26. Padulo, J.; Chamari, K.; Chaabène, H.; Ruscello, B.; Maurino, L.; Sylos Labini, P.; Migliaccio, G.M. The effects of one-week training camp on motor skills in Karate kids. *J. Sports Med. Phys. Fit.* **2014**, *54*, 715–724.
27. Mroczkowski, A. Influence of aikido exercises on mobility of hip joints in children. In Proceedings of the 1st World Congress on Health and Martial Arts in Interdisciplinary Approach, Warsaw, Poland, 17–19 September 2015.
28. Pop, A.C.; Monea, G.; Barboş, I.P. Implementing aikido in physical activity programs with 9–10 years old children. *Stud. Univ. Babeş-Bolyai Educ. Artis Gymnast.* **2017**, *62*, 57–65. [CrossRef]
29. Ma, A.W.W.; Qu, L.H. Effects of Karate Training on Basic Motor Abilities of Primary School Children. *Adv. Phys. Educ.* **2017**, *7*, 130–139. [CrossRef]
30. Pathare, N.; Kimball, R.; Donk, E.; Kennedy, K.; Perry, M. Physical performance measures following ten weeks of taekwondo training in children: A pilot study. *J. Phys. Educ. Sport* **2018**, *7*, 1–11. [CrossRef]
31. Top, E.; Akil, M.; Aydin, N. The Effects of the Taekwondo Training on Children's Strength-Agility and Body Coordination Levels. *JTRM Kinesiol.* **2018**, *1*, 1–10.
32. Brasil, I.; Monteiro, W.; Lima, T.; Seabra, A.; Farinatti, P. Effects of judo training upon body composition, autonomic function, and cardiorespiratory fitness in overweight or obese children aged 8 to 13 years. *J. Sports Sci.* **2020**, *38*, 2508–2516. [CrossRef]
33. Barley, O.R.; Chapman, D.W.; Guppy, S.N.; Abbiss, C.R. Considerations when assessing endurance in combat sport athletes. *Front. Physiol.* **2019**, *10*, 205. [CrossRef]

Article

Physical Activity Design for Balance Rehabilitation in Children with Autism Spectrum Disorder

Andreea Maria Roșca [1,†], Ligia Rusu [1,*,†], Mihnea Ion Marin [2], Virgil Ene Voiculescu [1] and Carmen Ene Voiculescu [3,†]

1   Faculty of Physical Education and Sport, University of Craiova, 200585 Craiova, Romania; rosca.andreea.d5s@student.ucv.ro (A.M.R.); virgil.ene@anmb.ro (V.E.V.)
2   Faculty of Mechanics, University of Craiova, 200585 Craiova, Romania; mihnea_marin@yahoo.com
3   Faculty of Physical Education and Sport, Ovidius University of Constant, 900470 Constanța, Romania; carmenenevoiculescu@gmail.com
*   Correspondence: ligiarusu@hotmail.com
†   These authors contributed equally to this work.

Abstract: One of the characteristics of autism spectrum disorder (ASD) subjects is postural control deficit, which is significant when somatosensory perception is affected. This study analyzed postural stability evolution after physical therapy exercises based on balance training. The study included 28 children with ASD (average age 8 years, average weight 32.18 kg). The rehabilitation program involved performing balance exercises twice a week for three months. Subject assessment was carried out using the RSScan platform. The parameters were the surface of the confidence ellipse (A) and the length of the curve (L) described by the pressure center, which were evaluated before and after the rehabilitation program. Following data processing, we observed a significant decrease in the surface of the confidence ellipse by 92% from EV1 to EV2. Additionally, a decrease of 42% in the curve length was observed from EV1 to EV2. A t test applied to the ellipse surface showed a $p = 0.021$ and a Cohen's coefficient of 0.8 (very large effect size). A t test applied to the length L showed $p = 0.029$ and Cohen's coefficient of 1.27 mm. Thus, the results show a significant improvement in the two parameters. The application of the program based on physical exercise led to an improvement in the balance of children with autism under complex evaluation conditions.

Keywords: ASD; balance; postural control; rehabilitation

Citation: Roșca, A.M.; Rusu, L.; Marin, M.I.; Ene Voiculescu, V.; Ene Voiculescu, C. Physical Activity Design for Balance Rehabilitation in Children with Autism Spectrum Disorder. *Children* **2022**, *9*, 1152. https://doi.org/10.3390/children9081152

Academic Editors: Georgian Badicu, Ana Filipa Silva and Hugo Miguel Borges Sarmento

Received: 23 June 2022
Accepted: 27 July 2022
Published: 30 July 2022

Publisher's Note: MDPI stays neutral with regard to jurisdictional claims in published maps and institutional affiliations.

Copyright: © 2022 by the authors. Licensee MDPI, Basel, Switzerland. This article is an open access article distributed under the terms and conditions of the Creative Commons Attribution (CC BY) license (https://creativecommons.org/licenses/by/4.0/).

1. Introduction

World Health Organization (WHO) data published in April 2017 show that an estimated 1 in 160 children worldwide are suffering from an autism spectrum disorder (ASD). This represents more than 7.6 million years of life adjustments depending on the disability and 0.3% of the global burden of diseases. This estimation represents an average number, and the reported prevalence substantially varies across studies. However, some well-controlled studies have reported numbers that are substantially higher. In many countries with low and medium budgets, ASD prevalence levels are still unknown. Based on epidemiological studies conducted over the last 50 years, ASD prevalence seems to be increasing globally. There are many possible explanations for this apparent increase, including greater awareness, expanded diagnostic criteria, better diagnosis instruments, and better reporting.

Recent studies suggest that autism affects about one percent of European people, accounting for more than five million people in the European Union (EU). Over the last 30 years, the number of reported autism cases has increased rapidly in all countries where prevalence studies have been performed. This increase is partly due to increased awareness of this disorder among health professionals, parents, and the general population; changes in the diagnosis criteria for autism; and children being diagnosed from an early age.

Autism is also known as an early onset disorder of the central nervous system (CNS). Even though the underlying mechanisms are still unknown, autism is usually described as a disorder of the brain since many changes in the brain have been found and analyzed [1]. In fact, ASD symptoms have been associated with ubiquitous CNS atypicality.

Autism spectrum disorders (ASDs) are a group of complex disorders involving brain development [2]. This umbrella term includes autistic disorder, Asperger disorder, and atypical autism. These disorders are characterized by difficulties with interactions and social communication and a restrained range of interests and activities to those with a repetitive character [3]. ASDs begin in childhood and tend to persist into adolescence and adulthood. In most cases, ASD is identified in the first 5 years of life.

People with ASD often present other conditions, such as epilepsy, depression, anxiety, and attention deficit hyperactivity disorder (ADHD). The functional intelligence level of ASD persons is extremely variable, ranging from severe mental intellectual disability to very high IQ levels.

According to some nuclear magnetic resonance (NMR) studies [4], at the ages of 2–4 years, 90% of autistic patients have greater than average cerebral volumes.

Compared to the general population, autistic persons are at a higher risk of presenting with a range of concurrent medical disorders and premature mortality [5,6]. It has been observed that common genetic vulnerability and/or underlying biological mechanisms that involve more systems may contribute to a higher prevalence of somatic complications in autism [7,8].

Sensorial and motor deficits as well as fine and gross motor function disabilities are consistently reported in children with ASD and are correlated with the severity of the social communication deficiency [9,10]. Sensory abnormalities are often the earliest identifiable clinical features of developmental disorders [11,12]. Similar to social communication deficits, motor deficits may represent the basic features of autism when a larger spectrum of symptoms is taken into consideration [13].

Access to information from multiple sources provides many benefits, such as improving reaction times, the accuracy of identification, and processing efficiency. The ability to integrate visual and auditory stimuli can provide a base for social development, language, and communication.

In recent years, a high prevalence of sensory processing has been observed in autistic children that accounts for 30% to 100% of cases. For the first time, processing difficulties have been recognized as a diagnostic criterion for ASD. These difficulties have a negative impact on the development and learning capacities of the subjects as well as on their behavioral, cognitive, physical, and psychological functioning. This is the main reason behind the need to improve sensorial stimulation.

Many studies have shown a disturbance in the postural control efficiency of people with ASD with an increase in the disturbance of the center of pressure (CoP) parameters under two conditions: eyes open (EO) [14–19] and eyes closed (EC) [15–17].

Postural control during orthostatism also seems to be modified in children with ASD during difficult situations [15]. Postural control is based on the automatic system [15,20], involves paying attention during different simple or complex tasks, and provides postural control regulation [21].

Dual task (DT) is a paradigm that has been widely used to investigate the degree of automatic and attentional processing in postural control in children, as well as in older adults [14,21]. DT is capable of improving the capacity of a person to stand in a vertical position while it integrates afferents from all systems to reflect the degree of automatism in postural control [21]. DT contributes to observing the posture in a multimodal integration context, especially when combined with an associated cognitive task.

Skateboarding has a positive impact on the development of new motor abilities [22]. Scientists have shown the positive results of dancing on repetitive behavior, cognitive function, executive function [23], behavioral problems, physical fitness, and motor abilities [24,25] in children with ASD. The effect of gymnastic exercises on self-control was established in [15,26], and its effect on speech development and physical fitness indicators was shown in [26]. Exercise programs that involve cardio and fitness significantly increase fitness levels in children with ASD, improving aerobic resistance and muscle strength [27]. Exergaming use has been shown to reduce the number of stereotyped actions and improve cognitive and executive functions in children with ASD [23]. Outside games and training programs that mainly involve sport game elements increase PA [28] and positively affect motor abilities in children with ASD, including hand and body coordination, strength, and dexterity [29,30], as well as executive functions [31].

It has been demonstrated that many interventions could increase the balance capacities of children with ASD [29,30]. While many are non-specific (e.g., swimming or taekwondo [31]), we found only two studies that explored the possibility of using balance training in particular for the treatment of ASD [30].

Cheldavi H et al. [32] reported postural control improvement in two visual conditions (eye open and eye closed (EO, EC)) and on two surfaces (foam, hard) in 10-year-old children with ASD. In this study, 20 children with ASD undertook a training program based on a six-week-long balance training intervention with eyes open and eyes closed. The mean velocity and anteroposterior and mediolateral axis displacements were measured and they found that balance training improves the postural sway in different sensory conditions in children with ASD.

Travers [33] investigated the effect of a 6-week balance training program based on visual biofeedback in 29 children with ASD in EO, EC, and visual feedback conditions (the latter meaning they could see someone's center of pressure on a screen). The results of the study show that specific balance training programs are capable of improving postural control and suggest that this should be integrated into rehabilitation programs for children with ASD [30].

At the same time, the use of physical activity for balance rehabilitation could help to improve posture and ASD phenotype, not only by releasing attentional resources but also by addressing one of the many possible causes of ASD symptoms.

Additionally, Jabouille et al. [34] found that 4 weeks of balance rehabilitation using specific sensory stimulation such as balance foam increased cognitive load conditions. The dual task conditions consisted of presenting images representing a neutral condition, sadness, anger, and fear. The evaluation included the assessment of postural control by measuring the surface, velocity, mediolateral, and anteroposterior sway amplitudes of centers of pressure using a posturographic platform. The rehabilitation program resulted in a 30–96% improvement in postural control parameters in both participants.

Research by Caldani et al. [35] on postural control in children with ASD during a specific rehabilitation program was conducted using Balance Quest by Framiral on an unstable platform under three different viewing conditions. The participants were split into two groups. Group G1 spent 1 min on postural training and group G2 spent 6 min on postural training. They concluded that G2 showed a significant improvement in postural control as assessed by the Framiral platform and that new rehabilitation strategies need to consider incorporating postural rehabilitation in children with ASD.

Sensory information processes from visual, vestibular, and proprioceptive receptors contribute to postural stability in order to accomplish neuromuscular control, balance maintenance, and appropriate motor responses. Any disorders involving these processes or integration disorders affect an individual's balance, necessitating intervention. Ghafar et al. [36] propose an investigation into sensory integration and balance using the Biodex balance system (BBS) in children with autism spectrum disorder (ASD) during a static posture. They studied 74 children with ASD and evaluated their postural sway in four different situations: eyes open/firm surface, eyes closed/firm surface, eyes open/foam surface, and eyes closed/foam surface. Children with ASD showed a significant increase in postural sway and the results provide evidence that postural stability decreased in children with ASD. Additionally, they found that children diagnosed with ASD have postural control deficiencies, especially in cases in which visual and somatosensory input are disrupted.

The authors conclude that there is a need to conduct more research to find the best balance training program using different rehabilitation exercises and to provide a scientific basis for a training program. Balance rehabilitation in particular seems to have the potential to improve postural control in children with ASD. For this reason, and due to the lack of information that is currently available, we proposed this study.

The aim of our study was to analyze the role of a multisensorial stimulation intervention in children with autism spectrum disorder (ASD) during postural control and balance exercises.

2. Materials and Methods

2.1. Subjects

We studied 28 participants aged between 8 and 14 years old. Their average age was 8.6 years and their average weight was 32.18 kg with a standard deviation (SD) of 8.22. All participants had autism or autism spectrum disorder (ASD). They did not present aggressive behavior, intellectual disability, or ADHD. All subjects had similar levels of development, meaning that there were no significant differences between those who were 8 years old and those who were 14 years old. In Table 1, we present the demographic data. Figure 1 presents the gender distribution of the patients.

Table 1. Demographic data for subjects included in the study.

Subject No	Weight (kg)	Height (cm)	Gender	Level of Disability According to the Social Assistance Criteria
S1	37	134	F	Mild
S2	27	124	M	Mild
S3	36	134	M	Mild
S4	18	126	F	Mild
S5	22	125	M	Mild
S6	44	147	F	Mild
S7	44	138	M	Mild
S8	38	145	F	Mild
S9	25	128	M	Mild
S10	35	136	M	Mild
S11	20	137	M	Mild
S12	23	128	M	Mild
S13	45	150	F	Mild
S14	42	136	M	Mild
S15	40	140	F	Mild
S16	26	129	M	Mild
S17	37	136	M	Mild
S18	21	128	M	Mild
S19	24	124	M	Mild
S20	47	149	M	Mild
S21	41	151	M	Mild
S22	30	147	F	Mild
S23	26	132	M	Mild
S24	34	150	M	Mild
S25	17	135	M	Mild
S26	20	152	M	Mild
S27	43	142	F	Mild
S28	39	129	M	Mild

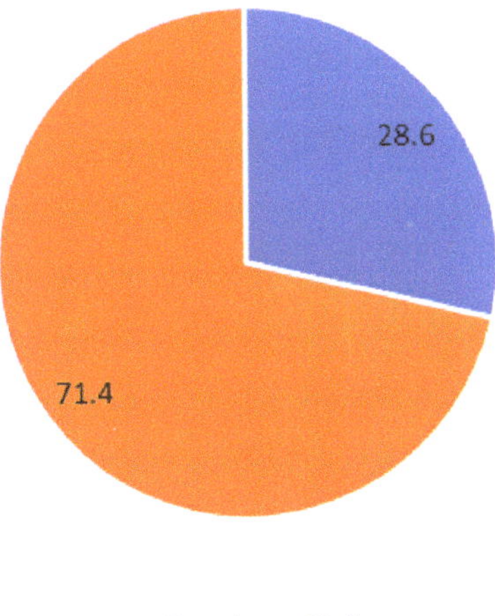

Figure 1. Gender distribution of patients. Participants are included in the special education system.

ASD is a genetically heterogenous group of neurobehavioral disorders characterized by impairment in three behavioral domains including communication, social interaction, and stereotypic repetitive behaviors. For this reason, it is possible to have different responses to tasks and training. We tried to select a homogenous group in order to ensure that specific aspects such as communication, social interaction, and stereotypic repetitive behaviors did not influence their participation in the training. However, all participants recruited in the present study had level one autism (mild) and were able to complete a series of motor competence assessments (i.e., Bruininks–Oseretsky Test of Motor Proficiency-2 and Movement Assessment Battery-2). The results of these assessments are not presented because this was not the aim of our research. We enrolled patients according to their medical and social assistance requirements. Children with level one ASD exhibit deficits in social communication without supports in place and inflexibility of behavior. Inclusion criteria: (1) diagnosed with level one ASD by a physician, (2) had the ability to understand and communicate with the physician and physiotherapist, and (3) had the ability to perform motor tasks. The exclusion criteria: had chronic medical disorders, visual impairments, physical impairments that could affect postural stability, attention deficit hyperactivity disorder (ADHD), intellectual disability, and had not participated in any physiotherapy programs to improve balance.

The participants are included in a special education system. The children attend a special education center and are diagnosed with ASD from authorized clinical personnel before they come to us to work on the rehabilitation of their balance and coordination. We receive their medical documents, which contain only diagnoses based on medical decisions. The grade of severity is assessed based on medical diagnosis, medical evaluation, and special criteria such as: biopsychosocial assessments, participation in social activities, physical activity, participation in the community, learning skills, selfcare, communication, and language mobility.

According to specific scales, children with ASD are evaluated using the Vineland scale (average 8.6/36 points); Harvey scale (average 29.75/93 points); and psychometric development scale (average value is 96.5%).

We excluded children with ADHD and intellectual disabilities because they do not show sufficient cooperation when participating in physiotherapy programs. The aim of our study was to increase participants' balance and coordination and we worked with children that had the capacity to understand and respond to the tasks in the training program.

2.2. Evaluation

The evaluation was conducted using the RSScan pressure and force platform to register the parameters, including the surface of the confidence ellipse (A) and the length of the curve (Lc) described by the pressure center and coefficient Lc/A. These parameters were registered before and after the physical exercise program in two evaluations: one before (EV1) and one after (EV2) the physical exercise (P.E.) program, which lasted 6 months.

The assessments were conducted in a quiet room with constant light and temperature levels. The children had to be barefoot during the assessment. The children had to stay calm on the platform while looking at an empty white wall. In this way, we avoided any type of distraction, and we used simple instructions such as "stay!" and "do not move!".

Posturographic data were obtained using the RSScan platform. The high-speed system performs accurate plantar pressure measurements with 4096 sensors at a scan rate of up to 500 Hz or 500 measurements per second. Raw data were filtered offline before counting the CoP parameters: the surface (the ellipse that contains 90% of the CoP coordinates), medium speed (medium speed of the CoP during the acquisition period, which means 30 s), and the CoP in the anteroposterior interval (AP) and mediolateral directions (ML) [16,17]. The CoP is the center of pressure in terms of body weight distribution, as recorded by force and pressure platforms.

The posturographic assessment is reliable and valid in children with typical development [17].

2.3. Statistical Analysis

The statistical analysis included descriptive analysis and a JB test (Jarque–Bera), which gives us information about the normal distribution of parameters. A Student's t test was applied to reveal any differences between parameter values from evaluation point 1 (EV1) to evaluation point 2 (EV2). The test indicates whether it is a significant difference.

We applied the Student's *t*-test for equal means. To analyze the effect size of the parameter evolution, we used the Cohen D coefficient. For correlations between parameters at the two points of evaluation—EV1 and EV2—we use Pearson's correlation and Spearman's correlation.

2.4. Training Program—Physical Activity (P.A.)

For therapy recommendations, based on the results of the evaluation, we proposed an initial P.A. therapy program based on a series of principles. We worked with each participant individually 3 times per week for 6 months using the program listed below. Each session lasted 30 min and followed the list of 6 exercises that we recommended in the exact order they are listed, from 1 to 7, starting with warm-up exercises.

We made our recommendations based on the evaluations performed with child interventions involving physical exercises for reducing the imbalances specific to autism, such as behavioral problems, stereotypical movements, lack of attention (same as for improving academic performance), social involvement, relations between peers, and motor perceptive skills.

Breathing exercises and stabilization exercises were used to restore muscle posture balance and the vertical axis of the body (in the same way as gymnastics), and the entire program was designed around sensory–motor coordination development, which is the most important step in forming the "cognitive–emotional brain" and intellectual abilities.

The objectives of the kinetic program were as follows:

- Positive influence on the body scheme representation;
- Recovery of laterality disturbances;
- Recovery of orientation, organizing, and spatial structure problems;
- Recovery of orientation and temporal structuring disorders;
- Recovery of balance and coordination disorders.

The modalities that are helpful towards achieving these objectives are the physical exercises that are proposed below.

Exercises that are useful for increasing both-leg and one leg balance as well as for improving coordination and stabilization were included. We recommend performing them 3 times per week.

Exercises and games were organized using simple equipment such as gymnastic benches and balance boards. For a complex and detailed rehabilitation program, we recommend the following:

Warm-up exercises for the neck and head (flexion–extensions), upper limbs (flexion–extensions, abductions, adductions, rotations), trunk (lateral left–right inclinations, trunk rotations), and lower limbs (flexion–extensions, circumductions, ball-like jumping and tip–toes–heels lifting) were recommended for the beginning of every training session.

We recommend patients walk on the wide side of a gymnastic bench (see Figure 2) for 4 repetitions 3 times per week. (This exercise is useful for increasing the balance and coordination of the lower limbs.)

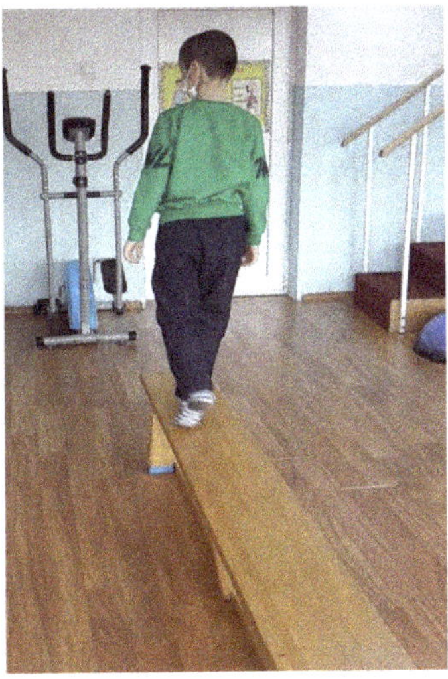

Figure 2. Walking on the wide side of a gymnastic bench.

1. Walking on a gymnastic bench, on the narrow side, as seen in Figure 3. (This exercise is useful for increasing the balance and coordination of the lower limbs; using the upper limbs to maintain balance means that we train those too.) We recommend that patients perform 4 repetitions of this exercise, 3 times per week.
2. Standing on one leg, like a stork, for 10 s per leg (this exercise improves balance and coordination). We recommend that patients perform 6 repetitions of this exercise, 3 times per week.
3. Standing on one leg, with the other leg in front, lateral, and behind without touching the ground (Figure 4). (This exercise is good for increasing dynamic balance.) We recommend that patients perform 6 repetitions of this exercise, 3 times per week.
4. Jumping from one leg to another while using circles on the ground to mark the place where the child must jump. (This exercise improves dynamic balance and coordination.) We recommend that patients perform 4 repetitions of this exercise, 3 times per week.
5. Balance board exercises. First, we need to help the child and keep him or her balanced by placing the board near a parallel bar, where they can maintain their own balance. When the child is confident enough, he or she can release the support and stay on the board by themselves (as shown in Figure 5). This should be performed 3 times per week. (This exercise is useful for increasing balance.)

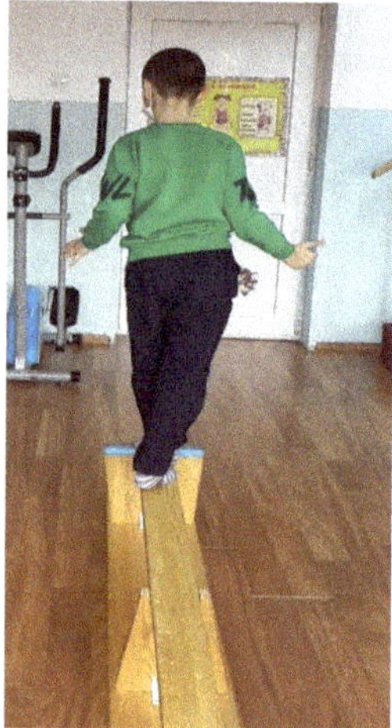

Figure 3. Walking on the narrow side of a gymnastic bench.

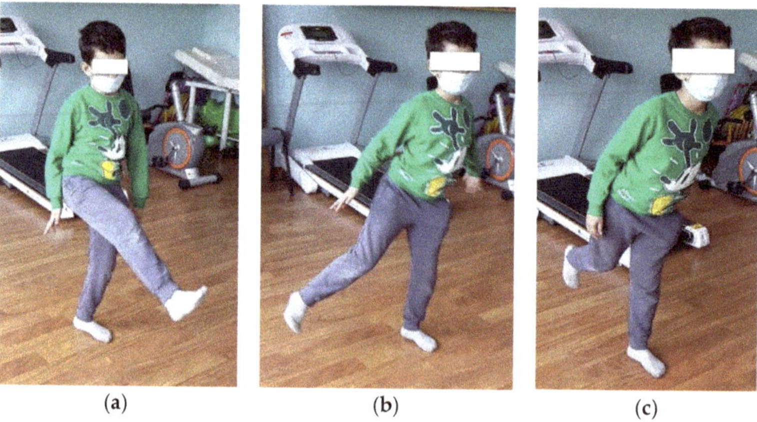

(a)   (b)   (c)

Figure 4. Standing on one leg while the other is in front (**a**), lateral (**b**), and behind (**c**).

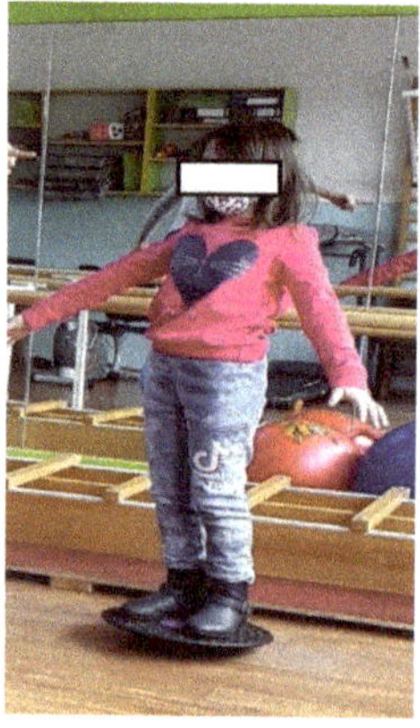

Figure 5. An image representing the exercise on the balance board.

3. Results

The initial evaluation results are presented in Table 2 and the descriptive statistics of the evaluation results are shown in Table 3.

Table 2. Data collected from subjects at the first assessment (EV1).

Subjects	Weight (Kg)	The Surface of the Confidence Ellipse A (mm^2)	The Length of the Curve (Lc) Described by COP (mm)	Coefficient Lc/A
S1	37	326	787	2.41
S2	27	357	525	1.47
S3	36	1489	407	0.27
S4	18	1437	900	0.63
S5	22	2935	1144	0.12
S6	44	30	480	16.16
S7	44	30	297	10.05
S8	38	328	788	2.40
S9	25	356	527	1.48
S10	35	1490	410	0.28
S11	20	1440	910	0.63
S12	23	2740	1150	0.12
S13	45	30	481	16.03
S14	42	29	300	10.34
S15	40	322	780	2.42
S16	26	353	520	1.47
S17	37	1480	417	0.28
S18	21	1430	912	0.64
S19	24	2816	1104	0.11
S20	47	30	470	15.77
S21	41	27	279	10.51
S22	30	336	786	2.34
S23	26	367	523	1.43
S24	34	1469	415	0.28
S25	17	1487	900	0.61
S26	20	2773	1094	0.11
S27	43	32	460	14.51
S28	39	28	279	10.13

COP =center of pressure.

Table 3. Min, max, and mean value and SD at the first evaluation.

	Weight (Kg)	The Surface of the Confidence Ellipse A (mm^2)	The Length of the Curve (Lc) Described by COP (mm)	Coefficient Lc/A
Minimum	17	26.55	279	0.11
Maximum	47	2935	1150	16.16
Mean	32.18	927.32	644.44	4.39
Standard deviation	9.50	977.42	283.43	5.75

A final evaluation (EV2) was conducted after 6 months, and the raw data are presented in Table 4. The descriptive statistics of the evaluation results are presented in Table 5.

Table 4. Data collected from subjects at the final assessment (EV2).

Subjects	Weight (Kg)	The Surface of the Confidence Ellipse A (mm^2)	The Length of the Curve (Lc) Described by COP (mm)	Coefficient Lc/A
S1	38	18	398	22.16
S2	28	3	268	89.77
S3	38	125	385	3.06
S4	21	37	274	7.33
S5	24	246	620	2.52
S6	43	18	351	19.25
S7	46	21	344	16.18
S8	33	20	391	19.58
S9	30	5	277	56.65
S10	40	127	384	3.01
S11	23	38	269	7.15
S12	28	240	617	2.57
S13	46	19	341	17.67
S14	42	19	334	17.34
S15	32	18	388	21.60
S16	25	4	277	75.22
S17	35	135	389	2.87
S18	26	35	249	7.04
S19	27	236	618	2.62
S20	43	18	349	19.25
S21	45	27	350	12.83
S22	39	17	388	22.87
S23	29	3	269	99.67
S24	34	129	371	2.88
S25	18	36	280	7.70
S26	22	264	613	2.32
S27	36	19	353	18.38
S28	37	22	349	16.04

Table 5. Min, max, and mean value and SD at the final evaluation.

	Weight (Kg)	The Surface of the Confidence Ellipse A (mm^2)	The Length of the Curve (L) Described by COP (mm)	Coefficient Lc/A
Minimum	18	2.70	249	2.32
Maximum	46	264	619.50	99.67
Mean	33.14	67.91	374.67	21.27
Standard deviation	8.22	84.23	110.66	26.32

The Jarque–Bera (JB) test provides us information about the normal distributions of parameters. We can see that we have a normal distribution for all parameters. For this reason, the Student's t test could be applied.

We applied the Student's t test for equal means. The Student's t test reveals whether there is any difference between the parameter values between EV1 and EV2 and whether these differences are significant. To analyze the effect size of parameter evolution, we used Cohen's D coefficient. The results of the Student's t test and Cohen's D test (coefficient) are presented in Table 6.

In Table 6, we can see that the Student's t tests showed us p values that did not meet the significance value of 0.05. Thus, the alternative hypothesis is accepted, meaning that the values of the surface of the confidence ellipse, the length of the curve described by the COP, and the coefficient Lc/A shows a significant improvement at EV2 compared with EV1.

Table 6. Student *t* test and Cohen's D coefficient values for parameters collected at EV1 and EV2.

	The Surface of the Confidence Ellipse	The Length of the Curve Described by the COP	Coefficient Lc/A
p values (results of Student's *t* test)	0.004	0.000	0.0016
Cohen's D test	0.8	1.25	−0.9

Significance level $p = 0.05$.

For correlations between parameters at EV1 and EV2, we used Pearson's correlation and Spearman's correlation and the results are presented in Tables 7 and 8.

Table 7. The coefficients of Pearson correlation for the values of parameters collected at EV1 and EV2.

Variables	EV2 Weight	EV2 The Surface of the Confidence Ellipse	EV2 The Length of the Curve Described by the COP	EV2 Coefficient Lc/A
EV1 Weight	0.918	−0.374	−0.128	−0.079
EV1 The surface of the confidence ellipse	−0.620	0.911	0.693	−0.456
EV1 The length of the curve described by the COP	−0.721	0.548	0.549	−0.258

Significance level $p = 0.05$. EV1 = evaluation moment 1; EV2 = evaluation moment 2.

Table 8. The coefficients of Spearman correlation for the values of parameters collected at EV1 and EV2.

Variables	EV2 Weight	EV2 The Surface of the Confidence Ellipse	EV2 The Length of the Curve Described by the COP	EV2 Coefficient Lc/A
EV1 Weight	0.903	−0.374	0.094	0.352
EV1 The surface of the confidence ellipse	−0.673	0.637	0.321	−0.616
EV1 The length of the curve described by the COP	−0.746	0.246	0.199	−0.215

Significance level $p = 0.05$.

Our results show the following:

The *p* values are less than the significance value and this means that the values for all 28 participants show significant differences between EV1 and EV2.

The average values of the surface of the confidence ellipse shows a decrease of 92%, from an average value of 927.32 mm^2 to 67.91 mm^2, and a decrease of 42% can be seen in terms of the length of the curve described by the COP (decreased from 644.44 mm^2 to 374.67 mm^2). The standard deviation decreased between EV1 and EV2, and this means that the values are more grouped around the mean value. We also observed a good correlation between the evolution of the surface of the confidence ellipse at EV1 and EV2 (coef Person = 0.911; coef Spearman = 0.637), but low correlations between the other parameters.

4. Discussion

The results of our research respond to the challenge regarding the improvement of balance and stability in people who suffer from ASD using multisensory stimulation. Positive effects were observed under the influence of exercises that required complex coordination and balance rehabilitation regarding improvements in postural control. According to [37] and Moseley and Pulvermüller [38], structural changes in ASD patients' brains may lead to a number of subtle deficits in motor control, including postural instability, which may eventually interfere with social and cognitive development by reducing opportunities to explore and interact with people and the environment. Brain plasticity is the main property that is involved [38,39]. Imbalance rehabilitation may focus therapeutically on early signs of central nervous system (CNS) anomalies as they appear in motor dysfunction, consequently attenuating social and cognitive deficits. For this reason, our proposal program influences the body scheme representation and involves restoring and improving brain activity in the field of motor control.

Morris and collaborators, in their 2015 study [18], speak about multi-sensorial integration, which is poor in children with ASD. Additionally, it has been observed that postural control in children with ASD is more affected by DT emotions and interaction than in typical development children [30,32]. Despite this, researchers report a higher CoP speed and a higher CoP surface during social image observations (meaning observing faces) than in images that contain neutral objects in children with ASD compared to a control group. Another study reported similar results by comparing balance performance in 30 children with ASD (12.1 ± 2.9 years old diagnosed by ADI-R, ADOS and DSM-5 criteria) with healthy children with average IQs when they were shown different emotional images of the face [30]. Impaired postural stability in children with ASD is well established in the literature [40], and the use of force plates by other authors demonstrate that it is also important to analyze the CoP. Children with ASD present greater CoP sway displacements [41], sway areas [42], standard deviations of COP coordinates [42], sway velocities [43], and root mean square of COP coordinates [44]. Li Y [45] investigated age's effect on postural stability in children with ASD in terms of comparing the amplitude and complexity of CoP sway during quiet standing in children with ASD among three different age groups: 6–8 years (under 8: U8), 9–11 years (U11), and 12–14 years (U14). The author found that the U14 group exhibited improved mediolateral postural stability compared to the U8 group, whereas no differences were found between the U8 and U11 or between the U11 and U14 groups. This study demonstrates that children with ASD have the possibility to develop postural stability slowly, but they observed significant changes over a long period of time. This is in accordance with our observations about the importance of introducing multisensorial training early to develop postural stability and increase balance and coordination.

We recommended early intervention programs specifically focused on improving the complexity of postural control as potentially beneficial for children with ASD. Future research is warranted to investigate postural control complexity in young autistic children and to evaluate the efficacy of early interventions at enhancing postural stability.

Many authors suggest that, due to increased cognitive challenges, children with ASD that present postural control problems are less capable of integrating emotional social cues while they are standing than unaffected children. The conclusions of this study demonstrate the importance of administering balance training combined with different sensorial stimuli to children with ASD.

Additionally, the training program significantly improved inferior limb strength, in contrast to superior limb strength, which presented no obvious improvements. This result may be explained by the training program that was used. In all training sessions, the exercises and the proposed content involved different types of static and dynamic balance, which made greater demands of inferior limbs. We should note that the upper limbs were used to maintain balance and the correct body posture, and this is in accordance with Laurenco's results [46].

Our results are also in accordance with those presented by Cheldavi et al. [32], which found that balance training (45 min/session; 3 sessions per week) improved postural control, which was assessed using a Bertec force plate. In terms of the training itself, our program is in accordance with the results of Najafabadi et al. [47], which proposes

Sports, Play, and Active Recreation for Kids (SPARK). Their proposal involves 36 sessions (3 sessions per week; 40 min per session), and the results are improved balance (static and dynamic), bilateral coordination, and social interaction.

In terms of innovation and significance, we consider research about postural sway to be complex in children with ASD, which necessitates the monitoring of the evolution of balance and postural control because the development of postural control is a dynamic process through which children learn to control multiple degrees of freedom of body segments to maintain balance. It is an innovation because we try to present an analysis of postural control in response to a specific training because, in recent years, many studies have investigated CoP complexity during quiet standing tasks in different populations, but not in children with ASD. Additionally, postural sway complexity can reflect the adaptive capacity of the postural control system, which is the result after this specific training regimen.

The limitations of this study include having a small number of children involved in the research, sometimes inconsistent participation in the training program, and poor communication with the physiotherapist. Because we had a small sample size, we had no control group, partly because only a small number of children and their parents consented to participate in the study, and also due to their inconsistent participation.

5. Conclusions

Future research is needed to test specific types of physical training in children with ASD due to the complexity of their responses.

Specific motor intervention that comprises different types of physical exercises and materials and that uses ludic exercises may potentially be a more effective strategy.

To improve patients' balance, postural control, and motor skills, we need a program whose aim is to work on fundamental motor skills based on multisensorial stimulation.

Specific types of physical exercise in this population improve their physical condition, cognitive and sensorial capacities, motor performance, and motor coordination (gait, balance, arm functions, and movement planning).

Physical exercise based on multisensory stimulation is a useful tool for the development of children with ASD and is becoming increasingly used, but we need to conduct evidence-based scientific research that supports this practice, giving it greater scientific robustness.

The motor intervention programs we propose for children with ASD provide benefits in many different domains.

Author Contributions: A.M.R.: data collection, study design, and methodology; L.R.: conceptualization, methodology, writing—original draft preparation, measurements, and investigation; M.I.M.: methodology, software, and visualization; V.E.V.: data curation, investigation, and software; C.E.V.: resources, writing—original draft preparation, and writing—review and editing. All authors have read and agreed to the published version of the manuscript.

Funding: This research received no external funding.

Institutional Review Board Statement: The study was conducted in accordance with the Declaration of Helsinki and approved by the Ethics Committee of the University of Craiova Research Centre for Human body motricity, study nr 235/18/10/2021.

Informed Consent Statement: Informed consent was obtained from all subjects involved in the study.

Data Availability Statement: Not applicable.

Conflicts of Interest: The authors declare no conflict of interest.

References

1. Hazlett, H.C.; Poe, M.D.; Gerig, G.; Styner, M.; Chappell, C.; Smith, R.G.; Vachet, C.; Piven, J. Early brain overgrowth in autism associated with an increase in cortical surface area before age 2 years. *Arch. Gen. Psychiatry* **2011**, *68*, 467. [CrossRef] [PubMed]
2. Walsh, C.A.; Morrow, E.M.; Rubenstein, J.L. Autism and brain development. *Cell* **2008**, *135*, 396–400. [CrossRef]
3. Noura, A.; Koushik, M. Classification of Autism Spectrum Disorder From EEG-Based Functional Brain Connectivity Analysis. *Neural Comput.* **2021**, *33*, 1914–1941. [CrossRef]
4. Courchesne, E.; Campbell, K.; Solso, S. Brain growth across the life span in autism: Age-specific changes in anatomical pathology. *Brain Res.* **2011**, *1380*, 138–145. [CrossRef] [PubMed]
5. Hirvikoski, T.; Mittendorfer-Rutz, E.; Boman, M.; Larsson, H.; Lichtenstein, P.; Bölte, S. Premature mortality in autism spectrum disorder. *Br. J. Psychiatry J. Ment. Sci.* **2016**, *208*, 232–238. [CrossRef] [PubMed]
6. Kohane, I.S.; McMurry, A.; Weber, G.; MacFadden, D.; Rappaport, L.; Kunkel, L.; Bickel, J.; Wattanasin, N.; Spence, S.; Murphy, S.; et al. The co-morbidity burden of children and young adults with autism spectrum disorders. *PLoS ONE* **2012**, *7*, e33224. [CrossRef]
7. Tye, B.W.; Commins, N.; Ryazanova, L.V.; Wühr, M.; Springer, M.; Pincus, D.; Churchman, L.S. Proteotoxicity from aberrant ribosome biogenesis compromises cell fitness. *eLife* **2019**, *8*, e43002. [CrossRef]
8. Vorstman, J.; Parr, J.R.; Moreno-De-Luca, D.; Anney, R.; Nurnberger, J.I., Jr.; Hallmayer, J.F. Autism genetics: Opportunities and challenges for clinical translation. *Nat. Rev. Genet.* **2017**, *18*, 362–376. [CrossRef]
9. Becker, E.B.; Stoodley, C.J. Autism spectrum disorder and the cerebellum. *Int. Rev. Neurobiol.* **2013**, *113*, 1–34. [CrossRef]
10. Bruchhage, M.; Bucci, M.P.; Becker, E. Cerebellar involvement in autism and ADHD. *Handb. Clin. Neurol.* **2018**, *155*, 61–72. [CrossRef] [PubMed]
11. Mosconi, M.W.; Sweeney, J.A. Sensorimotor dysfunctions as primary features of autism spectrum disorders. *Sci. China Life Sci.* **2015**, *58*, 1016–1023. [CrossRef] [PubMed]
12. Sacrey, L.A.; Zwaigenbaum, L.; Bryson, S.; Brian, J.; Smith, I.M.; Roberts, W.; Szatmari, P.; Roncadin, C.; Garon, N.; Novak, C.; et al. Can parents' concerns predict autism spectrum disorder? A prospective study of high-risk siblings from 6 to 36 months of age. *J. Am. Acad. Child Adolesc. Psychiatry* **2015**, *54*, 470–478. [CrossRef] [PubMed]
13. Lai, C.K.; Marini, M.; Lehr, S.A.; Cerruti, C.; Shin, J.E.; Joy-Gaba, J.A.; Ho, A.K.; Teachman, B.A.; Wojcik, S.P.; Koleva, S.P.; et al. Reducing implicit racial preferences: I. A comparative investigation of 17 interventions. *J. Exp. Psychol. Gen.* **2014**, *143*, 1765–1785. [CrossRef] [PubMed]
14. Bucci, M.P.; Doyen, C.; Contenjean, Y.; Kaye, K. The effect of performing a dual task on postural control in children with autism. *ISRN Neurosci.* **2013**, *2013*, 796174. [CrossRef]
15. Chen, M.H.; Lan, W.H.; Hsu, J.W.; Huang, K.L.; Su, T.P.; Li, C.T.; Lin, W.C.; Tsai, C.F.; Tsai, S.J.; Lee, Y.C.; et al. Risk of Developing Type 2 Diabetes in Adolescents and Young Adults with Autism Spectrum Disorder: A Nationwide Longitudinal Study. *Diabetes Care* **2016**, *39*, 788–793. [CrossRef] [PubMed]
16. Fournier, K.A.; Hass, C.J.; Naik, S.K.; Lodha, N.; Cauraugh, J.H. Motor coordination in autism spectrum disorders: A synthesis and meta-analysis. *J. Autism Dev. Disord.* **2010**, *40*, 1227–1240. [CrossRef] [PubMed]
17. Memari, A.H.; Ghaheri, B.; Ziaee, V.; Kordi, R.; Hafizi, S.; Moshayedi, P. Physical activity in children and adolescents with autism assessed by triaxial accelerometry. *Pediatric Obes.* **2013**, *8*, 150–158. [CrossRef]
18. Morris, M.C.; Tangney, C.C.; Wang, Y.; Sacks, F.M.; Bennett, D.A.; Aggarwal, N.T. MIND diet associated with reduced incidence of Alzheimer's disease. *Alzheimers Dement.* **2015**, *11*, 1007–1014. [CrossRef] [PubMed]
19. Radonovich, K.J.; Fournier, K.A.; Hass, C.J. Relationship between postural control and restricted, repetitive behaviors in autism spectrum disorders. *Front. Integr. Nneurosci.* **2013**, *7*, 28. [CrossRef]
20. Lacquaniti, F. Automatic control of limb movement and posture. *Curr. Opin. Neurobiol.* **1992**, *2*, 807–814. [CrossRef]
21. Massion, J. Postural control systems in developmental perspective. *Neurosci. Biobehav. Rev.* **1998**, *22*, 465–472. [CrossRef]
22. Woollacott, M.; Shumway-Cook, A. Attention and the control of posture and gait: A review of an emerging area of research. *Gait Posture* **2002**, *16*, 1–14. [CrossRef]
23. Thomas, D.T.; Erdman, K.A.; Burke, L.M. American College of Sports Medicine Joint Position Statement. Nutrition and Athletic Performance. *Med. Sci. Sports Exerc.* **2016**, *48*, 543–568. [CrossRef] [PubMed]
24. Anderson-Hanley, C.; Tureck, K.; Schneiderman, R.L. Autism and exergaming: Effects on repetitive behaviors and cognition. *Psychol. Res. Behav. Manag.* **2011**, *4*, 129–137. [CrossRef]
25. Zhao, M.; Chen, S. The Effects of Structured Physical Activity Program on Social Interaction and Communication for Children with Autism. *BioMed Res. Int.* **2018**, 1825046. [CrossRef]
26. Arzoglou, D.; Tsimaras, V.; Kotsikas, G.; Fotiadou, E.; Sidiropoulou, M.; Proios, M.; Bassa, E. The effect of a traditional dance training program on neuromuscular coordination of individuals with autism. *J. Phys. Educ. Sport* **2013**, *13*, 563–569. [CrossRef]
27. Zamani, M.; Ebrahimtabar, F.; Zamani, V.; Miller, W.H.; Alizadeh-Navaei, R.; Shokri-Shirvani, J.; Derakhshan, M.H. Systematic review with meta-analysis: The worldwide prevalence of Helicobacter pylori infection. *Aliment. Pharmacol. Ther.* **2018**, *47*, 868–876. [CrossRef]
28. Lochbaum, M.; Crews, D. Viability of Cardiorespiratory and Muscular Strength Programs for the Adolescent with Autism. *J. Evid. Based Integr. Med.* **2003**, *8*, 225–233. [CrossRef]

29. Pan, Y.; Chen, W.; Xu, Y.; Yi, X.; Han, Y.; Yang, Q.; Li, X.; Huang, L.; Johnston, S.C.; Zhao, X.; et al. Genetic Polymorphisms and Clopidogrel Efficacy for Acute Ischemic Stroke or Transient Ischemic Attack: A Systematic Review and Meta-Analysis. *Circulation* **2017**, *135*, 21–33. [CrossRef]

30. Gouleme, N.; Scheid, I.; Peyre, H.; Seassau, M.; Maruani, A.; Clarke, J.; Delorme, R.; Bucci, M.P. Postural Control and Emotion in Children with Autism Spectrum Disorders. *Transl. Neurosci.* **2017**, *8*, 158–166. [CrossRef]

31. Kim, Y.S.; Leventhal, B.L.; Koh, Y.J.; Fombonne, E.; Laska, E.; Lim, E.C.; Cheon, K.-A.; Kim, S.-J.; Kim, Y.-K.; Lee, H.; et al. Prevalence of autism spectrum disorders in a total population sample. *Am. J. Psychiatry* **2011**, *168*, 904–912. [CrossRef] [PubMed]

32. Cheldavi, H.; Shakerian, S.; Nahid, S.; Boshehri, S.; Zarghami, M. The effects of balance training intervention on postural control of children with autism spectrum disorder: Role of sensory information. *Res. Autism Spectr. Disord.* **2014**, *8*, 8–14. [CrossRef]

33. Travers, B.G.; Mason, A.H.; Mrotek, L.; Ellertson, A.; Dean, D.C.; Engel, C.; Gomez, A.; Dadalko, O.I.; McLaughlin, K. Biofeedback-based, videogame balance training in autism. *J. Autism Dev. Disord.* **2017**, *48*, 163–175. [CrossRef]

34. Jabouille, F.; Billot, M.; Hermand, E.; Lemonnier, E.; Perrochon, A. Balance rehabilitation for postural control in children with Autism Spectrum Disorder: A two-case report study. *Physiother. Theory Pract.* **2021**. [CrossRef] [PubMed]

35. Caldani, S.; Atzori, P.; Peyre, H.; Delorme, R.; Bucci, M.P. Short rehabilitation training program may improve postural control in children with autism spectrum disorders: Preliminary evidences. *Sci. Rep.* **2020**, *10*, 7917. [CrossRef] [PubMed]

36. Abdel Ghafar, M.A.; Abdelraouf, O.R.; Abdelgalil, A.A.; Seyam, M.K.; Radwan, R.E.; El-Bagalaty, A.E. Quantitative Assessment of Sensory Integration and Balance in Children with Autism Spectrum Disorders: Cross-Sectional Study. *Children* **2022**, *9*, 353. [CrossRef]

37. Gong, X.; Li, X.; Wang, Q.; Hoi, S.P.; Yin, T.; Zhao, L.; Meng, F.; Luo, X.; Liu, J. Comparing visual preferences between autism spectrum disorder (ASD) and normal children to explore the characteristics of visual preference of Children with ASD by improved visual preference paradigm: A case-control study. *Transl. Pediatr.* **2021**, *10*, 2006–2015. [CrossRef]

38. Moseley, R.L.; Pulvermüller, F. What can autism teach us about the role of sensorimotor systems in higher cognition? New clues from studies on language, action semantics, and abstract emotional concept processing. *Cortex* **2018**, *100*, 149–190. [CrossRef]

39. Brehmer, Y.; Li, S.C.; Müller, V.; von Oertzen, T.; Lindenberger, U. Memory plasticity across the life span: Uncovering children's latent potential. *Dev. Psychol.* **2007**, *43*, 465–478. [CrossRef]

40. Lim, Y.; Partridge, K.; Girdler, S.; Morris, S.L. Standing postural control in individuals with autism spectrum disorder: Systematic review and meta-analysis. *J. Autism Dev. Disord.* **2017**, *47*, 2238–2253. [CrossRef]

41. Smoot Reinert, S.; Jackson, K.; Bigelow, K. Using posturography to examine the immediate effects of vestibular therapy for children with autism spectrum disorders: A feasibility study. *Phys. Occup. Ther. Pediatr.* **2015**, *35*, 365–380. [CrossRef] [PubMed]

42. Doumas, M.; McKenna, R.; Murphy, B. Postural control deficits in autism spectrum disorder: The role of sensory integration. *J. Autism Dev. Disord.* **2016**, *46*, 853–861. [CrossRef] [PubMed]

43. Morris, S.L.; Foster, C.J.; Parsons, R.; Falkmer, M.; Falkmer, T.; Rosalie, S.M. Differences in the use of vision and proprioception for postural control in autism spectrum disorder. *Neuroscience* **2015**, *307* (Suppl. C), 273–280. Available online: http://www.sciencedirect.com/science/article/pii/S0306452215007721 (accessed on 23 June 2022). [CrossRef] [PubMed]

44. Memari, A.H.; Ghanouni, P.; Gharibzadeh, S.; Eghlidi, J.; Ziaee, V.; Moshayedi, P. Postural sway patterns in children with autism spectrum disorder compared with typically developing children. *Res. Autism Spectr. Disord.* **2013**, *7*, 325–332. Available online: http://www.sciencedirect.com/science/article/pii/S1750946712001183. (accessed on 23 June 2022). [CrossRef]

45. Li, Y.; Mache, M.A.; Todd, T.A. Complexity of center of pressure in postural control for children with autism spectrum disorders was partially compromised. *J. Appl. Biomech.* **2019**, *35*, 190–195. Available online: https://journals.humankinetics.com/view/journals/jab/35/3/article-p190.xml. (accessed on 23 June 2022). [CrossRef] [PubMed]

46. Laurenco, C.; Esteves, D.; Corredeira, R.; Seabra, A. The effect of a trampoline-based training program on the muscle strength of the inferior limbs and motor proficiency in children with autism spectrum disorders. *J. Phys. Educ. Sport.* **2015**, *15*, 592. [CrossRef]

47. Najafabadi, M.G.; Sheikh, M.; Hemayattalab, R.; Memari, A.H.; Aderyani, M.R.; Hafizi, S. The Effect of SPARK on Social and Motor Skills of Children with Autism. *Pediatr. Neonatol.* **2018**, *59*, 481–487. [CrossRef]

Article

Associations between Sedentary Time and Sedentary Patterns and Cardiorespiratory Fitness in Chinese Children and Adolescents

Ming Li [1,2], Xiaojian Yin [1,3,*], Yuqiang Li [1,2], Yi Sun [1,2], Ting Zhang [1,2], Feng Zhang [1,2], Yuan Liu [1,2], Yaru Guo [1,2] and Pengwei Sun [1,2]

[1] College of Physical Education and Health, East China Normal University, Shanghai 200241, China; liming23416@163.com (M.L.); liyuqiang-123@126.com (Y.L.); sunyi0084@163.com (Y.S.); noway1982@163.com (T.Z.); fzhang1988@126.com (F.Z.); yliu0809@163.com (Y.L.); yaruguo3299@163.com (Y.G.); 877523197@163.com (P.S.)

[2] Key Laboratory of Adolescent Health Assessment and Exercise Intervention of Ministry of Education, East China Normal University, Shanghai 200241, China

[3] College of Economics and Management, Shanghai Institute of Technology, Shanghai 201418, China

* Correspondence: xjyin1965@163.com; Tel.: +86-021-54342612

Abstract: The increase in sedentary behavior in children and adolescents has become a worldwide public health problem. This study aimed to explore the associations between sedentary time (ST) and sedentary patterns (SP) and the cardiorespiratory fitness (CRF) of Chinese children and adolescents. The CRF of 535 participants was determined using a 20-m shuttle run test. ST and SP were measured with accelerometers. Questionnaires were used to investigate the different types of ST. Multiple linear regression models were used to test the associations between ST and SP and CRF. In this study, only some ST and SP indicators were found to be significantly associated with CRF in girls. With each additional 10 min of screen time or passive traffic time, VO_{2max} decreases by 0.06 mL/kg/min ($B = -0.006$, 95% CI: $-0.010 \sim -0.001$) and 0.31 mL/kg/min ($B = -0.031$, 95% CI: $-0.061 \sim -0.002$), respectively, with MVPA control. With each additional 10 min of breaks in ST or duration of breaks in ST, VO_{2max} increases by 0.41 mL/kg/min ($B = 0.041$, 95% CI: $0.007 \sim 0.076$) and 0.21 mL/kg/min ($B = 0.021$, 95% CI: $0.007 \sim 0.035$), respectively, with control total ST. Breaks in ST ($B = 0.075$, 95% CI: $0.027 \sim 0.123$) and the duration of breaks in ST ($B = 0.021$, 95% CI: $0.012 \sim 0.146$) were positively correlated with CRF when controlling for LPA, but these associations were not significant when controlling for MVPA ($B = 0.003$, 95% CI: $-0.042 \sim 0.048$; $B = 0.001$, 95% CI: $-0.024 \sim 0.025$). The total ST of children and adolescents was found to not be correlated with CRF, but when ST was divided into different types, the screen time and passive traffic time of girls were negatively correlated with CRF. More breaks in ST and the duration of breaks in ST were positively associated with higher CRF in girls. MVPA performed during breaks in ST may be the key factor affecting CRF. Schools and public health departments should take all feasible means to actively intervene with CRF in children and adolescents.

Keywords: sedentary time; sedentary pattern; physical activity; cardiorespiratory fitness

Citation: Li, M.; Yin, X.; Li, Y.; Sun, Y.; Zhang, T.; Zhang, F.; Liu, Y.; Guo, Y.; Sun, P. Associations between Sedentary Time and Sedentary Patterns and Cardiorespiratory Fitness in Chinese Children and Adolescents. *Children* **2022**, *9*, 1140. https://doi.org/10.3390/children9081140

Academic Editors: Hugo Miguel Borges Sarmento, Georgian Badicu and Ana Filipa Silva

Received: 24 June 2022
Accepted: 27 July 2022
Published: 29 July 2022

Publisher's Note: MDPI stays neutral with regard to jurisdictional claims in published maps and institutional affiliations.

Copyright: © 2022 by the authors. Licensee MDPI, Basel, Switzerland. This article is an open access article distributed under the terms and conditions of the Creative Commons Attribution (CC BY) license (https://creativecommons.org/licenses/by/4.0/).

1. Introduction

Cardiorespiratory fitness (CRF) reflects the ability of the human body to absorb, transport, and utilize oxygen, which is the core element of physical health [1]. Many studies have confirmed that low CRF is highly associated with cardiovascular disease and all-cause mortality [2]. In recent decades, the levels of CRF of children and adolescents in the world have shown a continuous downward trend [3]. Therefore, urgent effective measures to improve CRF should be taken. Increasing physical activity, especially moderate-to-vigorous physical activity (MVPA), is an effective way to improve CRF in children and adolescents [4,5]. However, MVPA only accounts for a minor part of the waking time of children and adolescents every day; most of the day includes sedentary time (ST) [6]. Exploring the association between ST and CRF may lead to new breakthroughs for improving CRF in children and adolescents.

Sedentary behavior is any waking behavior characterized by an energy expenditure ≤ 1.5 METs while in a sitting or reclining position [7]. ST can be divided into screen time and non-screen time based on whether it is related to electronic screens [8]. For children and adolescents, screen time mainly includes watching TV/movies, using computers to play games, using mobile phones/tablets, etc. Non-screen time mainly includes attending class, writing extracurricular assignments, reading extracurricular books, attending extracurricular tutoring, passive traffic time, social sedentary time, etc. [8]. A previous study suggested that excessive ST may be an important risk factor for low CRF in children and adolescents [9]. However, the relationships between different types of ST and CRF in children and adolescents vary [7]. In order to reduce the health hazards caused by ST, an effective solution is to reduce the specific types of ST that cause health risks as much as possible [10]. However, most existing studies focus on total ST or a certain type of ST, which does not offer insights into whether a specific type of ST or total ST affects CRF. In addition, the relationship between the component characteristics of ST and CRF also deserves further consideration. The total ST measured by an accelerometer includes all components of ST (prolonged ST or non-prolonged ST). The relationship between different components of ST and CRF may not be consistent [11]. Therefore, further study is needed to comprehensively consider the impact of ST on CRF in children and adolescents from two aspects, type and component.

Along with ST, sedentary patterns (SP) may also be relevant for CRF [12]. SP reflects the manner of ST accumulation, which can be expressed by sedentary bouts, breaks in ST, and the duration of breaks in ST. The existing research results on the relationship between SP and CRF are not consistent. For example, Júdice et al. found that breaks in ST and non-prolonged sedentary bouts were positively associated with CRF [11]. However, Bailey et al. [13] and Denton et al. [14] did not find that breaks in ST were significantly associated with CRF. Furthermore, existing studies do not suggest which duration of breaks in ST is most beneficial for CRF.

In summary, this study selected representative nationwide samples from China, adopted a combination of subjective and objective methods to comprehensively analyze the associations between ST and SP and CRF in children and adolescents, and further identified the duration of breaks in ST that are beneficial to CRF, providing a reference for schools and public health departments to develop targeted interventions.

2. Materials and Methods

2.1. Participants

From March to July 2019, the research team recruited 840 children and adolescents aged 7–18 for the study. The children and adolescents came from seven Chinese cities: Shanghai, Taiyuan, Guangzhou, Changsha, Urumqi, Chengdu and Kunming. The research team randomly selected one elementary school, one middle school, and one high school in each city as study sites. Five boys and five girls were randomly selected from each age group. All students can complete a CRF test. 120 students were excluded for failing to wear accelerometers, and 185 were excluded for not completing the 20-m shuttle run test (SRT). A total of 535 students were finally included in this study.

2.2. CRF

The research team tested 20-m SRT in all participants. After warming up, participants stood on one of two horizontal lines placed 20 m apart. The participants, running back and forth between the lines, were required to increase their speed according to the music. The initial speed was 8.0 km/h, the second level was 9.0 km/h, and then the running speed was increased by 0.5 km/h for each level. When the participants could not maintain the speed set by the music, or could not follow the music's rhythm to reach the end line two consecutive times, the test was terminated. The total number was recorded as the final score [15]. Matsuzaka et al.'s [16] formula was used to estimate the maximum oxygen uptake: $VO_{2max} = 61.1 - 2.20 \times$ gender $- 0.462 \times$ age $- 0.862 \times$ BMI $+ 0.192 \times$ laps, where gender is expressed as 0 for boys and 1 for girls. VO_{2max} is usually used to reflect CRF, and 20-m SRT is highly correlated with VO_{2max} [17]. Therefore, we can use 20-m SRT to evaluate CRF in children and adolescents.

2.3. Sedentary Behavior

In this study, a subjective and objective approach was used to measure sedentary behavior (SB). Total ST, prolonged ST or non-prolonged ST, and SP were objectively measured using a GT3X+ (ActiGraph, Pensacola, FL, USA) accelerometer. During the measurement, the subjects were told to wear the accelerometer on the right hip for seven consecutive days—including five school days and two weekend days—and that it could not be removed except for bathing, swimming, and sleeping. The epoch duration was set at 5 s. The accelerometer began to record data from the early morning of the second day, and was retrieved by the researcher on the eighth day. The original data were downloaded through ActiLife version 6.10.2 software (ActiGraph, Pensacola, FL, USA). After the original data were downloaded, their validity was first screened: valid data needed to include at least three valid school days and one valid weekend day. A valid school day or weekend day meant at least 10 h of the device being effectively worn on the test day [18]. The cut-points of Evenson et al. [19] were adopted to classify SB (0–100 counts/min), light physical activity (LPA) (101–2295 counts/min), moderate physical activity (MPA) (2296–4011 counts/min), and vigorous physical activity (VPA) (4012 counts/min or more). These cut-points have high validity and reliability in the evaluation of the PA of children and adolescents [20]. The total ST was divided into prolonged ST (at least 20 min uninterrupted ST) and non-prolonged ST (less than 20 min ST). An experimental study has shown that children and adolescents who often sit for more than 20 min are at risk of metabolic diseases [21]. Based on this, 20 min was chosen as the cut-off point for distinguishing between prolonged ST and non-prolonged ST. The SP were expressed by prolonged sedentary bouts (the number of instances of at least 20 min uninterrupted ST), non-prolonged sedentary bouts (the number of instances of less than 20 min uninterrupted ST), breaks in ST (the number of interruptions in ST in which the accelerometer count raised above 100 counts/min, and which stayed for at least 1 min), and the duration of breaks in ST (the total time of breaks in ST). A questionnaire was used to measure the duration of the different types of ST. The questionnaire for children aged 7–9 was filled out with the help of their parents. Participants reported how much time they spent doing the following activities: watching TV/movies, using computers to play games, using mobile phones/tablets (screen time) from Monday to Friday and on Saturday and Sunday; writing extracurricular assignments, reading extracurricular books, attending extracurricular tutoring (these three behaviors are defined as learning behaviors that occur outside of class, and are collectively referred to as extracurricular learning time); sitting and chatting (social ST); and commuting to school (passive traffic ST). The time measured in the above questionnaire was the weighted average on school days (5/7) and the weekend (2/7). The test–retest reliability of each item in the questionnaire was between 0.79–0.91, which is acceptable.

2.4. Covariate

Urban or rural residence, socioeconomic status (SES), sleep time, and BMI were potential confounders. Questionnaires were conducted to collect information on urban or rural residence, SES, and sleep time. Parental education, parental occupation, and household income were used to assess the SES [22]. The sleep time duration, from the time of going to bed at night to getting up in the morning, was filled out by the individual. Body height was determined using a mechanical height gauge, and body weight was measured using an electronic scale. The values were accurate to 1 decimal place.

2.5. Statistics Analysis

The normality of all variables was tested using a histogram and Q–Q plot. An independent sample *t*-test (for normalvariables) and Mann–Whitney *U* test (for skewedvariables) were used to compare the gender differences of each variable. A chi-squared test was used to compare the gender differences of residence. Spearman's correlation was used to test the correlations between various variables. Multiple linear regression models were used to test the associations between different types of ST and SP and CRF in children and adolescents. In the analysis of the relationship between CRF and ST, CRF was included in the model as the dependent variable, and ST indexes as the independent variable. Model 1 adjusted for the age, urban and rural areas, SES, sleep time, and BMI; Model 2 further adjusted for the MVPA based on Model 1. In the analysis of the relationship between CRF and SP, CRF was included in the model as the dependent variable, and SP indexes as the independent variable. Model 1 adjusted for the age, urban and rural areas, SES, sleep time, and BMI; Model 2 further adjusted for the total ST based on Model 1. In the analysis of the intensity attributes of the duration of breaks in ST, CRF was included in the model as the dependent variable, and breaks in ST or the duration of breaks in ST as the independent variable. Model 1 adjusted for the age, urban and rural areas, SES, sleep time, BMI, total ST, and LPA; Model 2 adjusted for the age, urban and rural areas, SES, sleep time, BMI, total ST, and MVPA. Statistical significance was set at 0.05, and all analyses were conducted using IBM SPSS version 25.0 for Windows (IBM Corp., Armonk, NY, USA).

3. Results

3.1. Descriptive Characteristics of the Sample

Table 1 presents the descriptive characteristics of the sample, including demographic, CRF, and MVPA data. A total of 535 participants (47.8% boys, 52.2% girls) were included in this study. The table shows that the BMI, sleep time, VO_{2max}, and MVPA values of boys were higher than those of girls, and the difference was statistically significant ($p < 0.05$).

Table 1. Descriptive characteristics of the sample.

Parameters	Boys (*n* = 255)	Girls (*n* = 280)	*p*	Cohen's *d*
Age (y) [a]	12.4 ± 3.4	12.7 ± 3.3	0.20	0.09
Residence (%) [b]				
Urban	61.2	65.4		
			0.32	0.04
Rural	38.8	34.6		
Height (cm) [a]	153.2 ± 19.7	150.6 ± 15.2	0.08	0.15
Weight (kg) [a]	45.3 ± 17	41.7 ± 12.3	<0.001	0.24
BMI (kg/m^2) [a]	18.6 ± 3.2	18.0 ± 2.9	0.02	0.20
SES [a]	0.05 ± 1.1	−0.03 ± 0.9	0.42	0.08
Sleep time (h/d) [a]	8.4 ± 1.3	8.0 ± 1.6	<0.001	0.28
VO_{2max} (mL/kg/min) [a]	46.5 ± 4.1	42.9 ± 3.6	<0.001	0.93
MVPA (min/d) [a]	62.0 ± 19.3	49.5 ± 15.5	<0.001	0.71

Note: Values are presented as mean ± SD and percent; [a] independent sample *t*-test; [b] chi-squared test. BMI, body mass index; SES, socioeconomic status; MVPA, moderate-to-vigorous physical activity.

Figure 1 presents that the proportion of prolonged ST (42.1%) of boys was lower than that of girls (45.8%), the proportions of MVPA (7.0%) and LPA (19.2%) were higher than those of girls (5.5% and 17.1%), and the proportion of non-prolonged ST (31.7%) was close to that of girls (31.6%).

3.2. Descriptive Characteristics of ST and SP

Table 2 presents the descriptive characteristics of the different types of ST and SP. In the comparison of ST, the total ST (655.1 min/d) and prolonged ST (373.7 min/d) of boys were higher than those of girls (699.5 min/d and 414.0 min/d), and the difference was statistically significant ($p < 0.001$). There were no significant gender differences for other types of ST. In the comparison of SP, non-prolonged sedentary bouts (30.0 number/d) and breaks in ST (41.2 number/d) for boys were higher than those for girls (22.8 number/d and 33.7 number/d), and the difference was statistically significant ($p < 0.001$). There was no significant sex difference in the number of prolonged sedentary bouts.

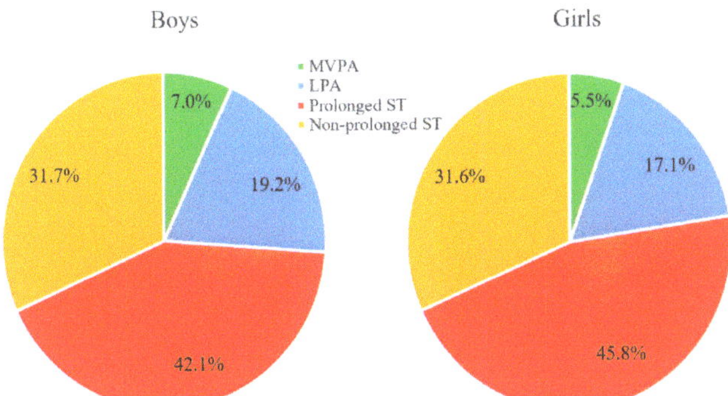

Figure 1. The proportions of prolonged ST, non-prolonged ST, LPA, and MVPA, stratified by gender.

Table 2. Descriptive characteristics of ST and SP.

Parameters	Boys (n = 255)	Girls (n = 280)	p	Cohen's d
Total ST (min/d) [a]	655.1 ± 119.9	699.5 ± 106.7	<0.001	0.30
Prolonged ST (min/d) [a]	373.7 ± 157.4	414.0 ± 145.4	<0.001	0.27
Non-prolonged ST (min/d) [a]	281.5 ± 92.3	285.5 ± 94.3	0.62	0.04
Screen time (min/d) [b]	31.4 (4.29, 88.9)	30.0 (2.86, 95.7)	0.81	0.02
Extracurricular learning time (min/d) [a]	350.7 (259.3, 490.7)	345.7 (250.7, 500.0)	0.99	<0.01
Passive traffic ST (min/d) [a]	10.7 (5.7, 22.9)	12.1 (7.1, 17.4)	0.97	<0.01
Social ST (min/d) [a]	22.9 (12.9, 51.3)	22.9 (15.2, 42.9)	0.64	0.05
Prolonged sedentary bouts (number/d) [a]	10.1 ± 3.2	9.9 ± 3.0	0.43	0.06
Non-prolonged sedentary bouts (number/d) [a]	30.0 ± 14.2	22.8 ± 12	<0.001	0.55
Breaks in ST (number/d) [a]	41.2 ± 13.8	33.7 ± 11.5	<0.001	0.59
Duration of breaks in ST (min/d) [a]	98.1 ± 31.1	78.3 ± 24.0	<0.001	0.71

Note: Values are presented as mean ± SD and median (25th, 75th percentiles). [a] independent sample *t*-test; [b] Mann–Whitney *U* test. ST, sedentary time.

3.3. Correlation Analysis of SB Variables and CRF

Table 3 shows that the VO_{2max} values were positively associated with non-prolonged sedentary bouts ($r = 0.17$, $p < 0.01$), breaks in ST ($r = 0.19$, $p < 0.01$), and the duration of breaks in ST ($r = 0.19$, $p < 0.01$).

3.4. Associations between ST and CRF in Children and Adolescents

Table 4 shows that there were negative associations between screen time ($B = -0.005$, 95% *CI*: $-0.010 \sim -0.001$, adjusted $R^2 = 0.59$) and passive traffic ST ($B = -0.030$, 95% *CI*: $-0.061 \sim -0.001$, adjusted $R^2 = 0.62$) and CRF in girls, and the associations were still significant after further controlling for MVPA ($B = -0.006$, 95% *CI*: $-0.010 \sim -0.001$, adjusted $R^2 = 0.61$; $B = -0.031$, 95% *CI*: $-0.061 \sim -0.002$, adjusted $R^2 = 0.64$). This means that for each additional 10 min of screen time or passive traffic time, VO_{2max} decreases by 0.06 mL/kg/min and 0.31 mL/kg/min, respectively, with MVPA control. There was no statistically significant association between ST and CRF in boys.

Table 3. Correlation analysis of each variable.

Parameters	1	2	3	4	5	6	7	8	9	10	11	12
1. VO_{2max}	-											
2. Total ST	−0.09	-										
3. Prolonged ST	0.05	0.07	-									
4. Non-prolonged ST	−0.06	−0.05	−0.83 **	-								
5. Screen time	−0.04	−0.06	−0.05	0.07	-							
6. Extracurricular learning ST	−0.06	0.19 **	−0.03	−0.01	−0.05	-						
7. Passive traffic ST	−0.13	−0.07	−0.14 *	0.11	0.07	0.03	-					
8. Social ST	−0.001	−0.08	0.03	−0.04	0.24 **	−0.01	−0.04	-				
9. Prolonged sedentary bouts	0.03	0.06	0.91 **	−0.73 **	−0.05	−0.04	−0.16 *	0.06	-			
10. Non-prolonged sedentary bouts	0.17 **	−0.62 **	−0.24 **	0.16 **	−0.02	−0.14 **	−0.05	−0.01	−0.25 **	-		
11. Breaks in ST	0.19 **	−0.62 **	−0.03	−0.01	−0.04	−0.15 **	−0.10	0.001	−0.02	0.97 **	-	
12. Duration of breaks in ST	0.19 **	−0.51 **	−0.05	0.02	0.07	−0.17 **	−0.09	0.05	−0.05	0.84 **	0.86 **	-

Note: ST, sedentary time. * $p < 0.05$, ** $p < 0.01$.

Table 4. Associations between ST and CRF in children and adolescents.

Independent Variables	Boys		Girls	
	Model 1	Model 2	Model 1	Model 2
Total ST	−0.001 (−0.005, 0.004)	0.001 (−0.004, 0.005)	0.003 (−0.001, 0.007)	0.005 (−0.001, 0.009)
Prolonged ST	−0.001 (−0.003, 0.002)	−0.001 (−0.002, 0.002)	0.001 (−0.001, 0.003)	0.001 (−0.001, 0.003)
Non-prolonged ST	−0.001 (−0.005, 0.004)	−0.001 (−0.005, 0.004)	−0.001 (−0.004, 0.002)	−0.001 (−0.004, 0.002)
Screen time	−0.002 (−0.007, 0.003)	−0.002 (−0.008, 0.003)	−0.005 (−0.010, −0.001) *	−0.006 (−0.010, −0.001) **
Extracurricular learning ST	0.001 (−0.002, 0.003)	0.001 (−0.001, 0.003)	−0.001 (−0.003, 0.001)	−0.001 (−0.002, 0.001)
Passive traffic ST	0.001 (−0.033, 0.035)	0.001 (−0.033, 0.035)	−0.030 (−0.061, −0.001) *	−0.031 (−0.061, −0.002) *
Social ST	0.004 (−0.006, 0.015)	0.004 (−0.007, 0.015)	0.003 (−0.004, 0.010)	0.003 (−0.003, 0.010)

Note: The values are presented as *B* (95% *CI*). Model 1 adjusted for age, urban and rural areas, SES, sleep time, and BMI; Model 2 adjusted for Model 1 + MVPA. CRF, cardiorespiratory fitness; BMI, body mass index; SES, socioeconomic status; ST, sedentary time. * $p < 0.05$, ** $p < 0.01$.

3.5. Associations between SP and CRF in Children and Adolescents

Table 5 shows that there were no statistically significant association between breaks in ST (*B* = 0.031, 95% *CI*: −0.003~0.064) and CRF in girls, but these were positively correlated after controlling for total ST (*B* = 0.041, 95% *CI*: 0.007~0.076, adjusted R^2 = 0.59). Further, the duration of sedentary breaks (*B* = 0.015, 95% *CI*:0.001~0.028, adjusted R^2 = 0.59) was positively associated with CRF, and the association was still significant after further controlling for the total ST (*B* = 0.021, 95% *CI*:0.007~0.035, adjusted R^2 = 0.60). This means that for each additional 10 min of breaks in ST or the duration of breaks in ST, VO_{2max} increases by 0.41 mL/kg/min and 0.21 mL/kg/min, respectively, with control total ST. There was no statistically significant association between SP and CRF in boys.

Table 5. Associations between sedentary patterns and CRF in children and adolescents.

Independent Variables	Boys		Girls	
	Model 1	Model 2	Model 1	Model 2
Prolonged sedentary bouts	0.001 (−0.129, 0.129)	0.001 (−0.128, 0.130)	0.066 (−0.035, 0.167)	0.056 (−0.046, 0.158)
Non-prolonged sedentary bouts	0.025 (−0.012, 0.061)	0.025 (−0.014, 0.064)	0.020 (−0.011, 0.051)	0.030 (−0.002, 0.063)
Breaks in ST	0.027 (−0.011, 0.064)	0.028 (−0.013, 0.068)	0.031 (−0.003, 0.064)	0.041 (0.007, 0.076) *
Duration of breaks in ST	0.010 (−0.004, 0.025)	0.011 (−0.004, 0.027)	0.015 (0.001, 0.028) *	0.021 (0.007, 0.035) **

Note: The values in the table are *B* (95% *CI*). Model 1 adjusted for age, urban and rural areas, SES, sleep time, and BMI; Model 2 adjusted for Model 1 + total ST. CRF, cardiorespiratory fitness; BMI, body mass index; SES, socioeconomic status; ST, sedentary time. * $p < 0.05$. ** $p < 0.01$.

3.6. Exploration of the Intensity Attribute of the Duration of Breaks in ST in Girls

Table 6 shows that breaks in ST (*B* = 0.075, 95% *CI*: 0.027~0.123) and the duration of breaks in ST (*B* = 0.021, 95% *CI*:0.012~0.146) were positively correlated with CRF when controlling for LPA in girls, but these associations were not significant when controlling for

MVPA ($B = 0.003$, 95% *CI*: -0.042~0.048; $B = 0.001$, 95% *CI*: -0.024~0.025). This means that MVPA in the duration of breaks in ST is a key component affecting CRF.

Table 6. Exploration of the intensity attribute of the duration of breaks in ST in girls.

Independent Variables	Model 1	Model 2
Breaks in ST	0.075 (0.027, 0.123) **	0.003 (-0.042, 0.048)
Duration of breaks in ST	0.021 (0.012, 0.046) **	0.001 (-0.024, 0.025)

Note: The values in the table are B (95% *CI*). Model 1 adjusted for age, urban and rural areas, SES, sleep time, BMI, total sedentary time, and LPA; Model 2 adjusted for age, urban and rural areas, SES, sleep time, BMI, total sedentary time, and MVPA. CRF, cardiorespiratory fitness; BMI, body mass index; SES, socioeconomic status; ST, sedentary time. ** $p < 0.01$.

4. Discussion

The purpose of this study was to explore the associations between different types of ST and SP and CRF in children and adolescents. The study found that there was no significant correlation between total ST and CRF in children and adolescents, but when ST was divided into different types, the screen time and passive traffic ST of girls were negatively correlated with CRF and independent of MVPA. Breaks in ST and the duration of breaks in ST were positively correlated with CRF in girls. MVPA performed during breaks in ST may be the key factor affecting CRF.

Most studies have confirmed that MVPA is positively correlated with CRF in children and adolescents [4,5]. However, in view of the extremely low proportion of MVPA performed in a day, exploring the relationship between ST and CRF may provide new breakthroughs for improving CRF in children and adolescents. For example, Santos et al. found that the effect of ST on CRF was independent of physical activity, and even an additional increase in MVPA could not offset the adverse effect of prolonged ST on CRF over a long period of time. Reducing ST is an important means to improve CRF in children and adolescents [23]. Other studies have also found a negative correlation between ST and CRF, but the difference in their findings was that increasing MVPA was found to be able to weaken [9,24] or offset [25] the adverse effect of ST on CRF. However, some studies did not find a significant correlation between ST and CRF in children and adolescents [14,26,27]; the results of the present study also support this conclusion. In addition, after further dividing the total ST into prolonged ST and non-prolonged ST, this study found that distinguishing between the components of ST (prolonged ST and non-prolonged ST) had no effect on the significance of the association. The inconsistency of research conclusions described here may be related to the inconsistency of the cut-points of sedentary behavior in different studies, resulting in the incomparability of ST [14]. Therefore, in follow-up studies, when investigating the relationship between ST and CRF in children and adolescents, the cut-off point of sedentary behavior should include the widely used and reliable boundary value, so as to facilitate the horizontal comparison of different studies. It is important to note that although this study and previous studies did not find a significant association between the duration of sedentary breaks and CRF in children and adolescents, there is evidence that higher levels of ST are negatively associated with health outcomes in adults [28]. Therefore, we cannot ignore the possible long-term adverse effects of ST.

Due to the popularity of portable devices with electronic screens, the screen time of children and adolescents has shown a rapid upward trend in recent decades [10]. Most studies support the theory that screen time is negatively correlated with CRF, and several studies have indicated that more than 2 h of screen time per day is related to a decline in CRF. Consistent with previous studies, this study found that although the association between total ST and CRF was not significant, when ST was divided into different types, the screen time of females was negatively correlated with CRF and independent of MVPA. It should be noted that the screen time investigated in this study was only recreational ST. In recent years, children and adolescents' learning ST has been increasing due to the widespread use of multimedia technology in daily teaching and the online teaching methods adopted during the COVID-19 pandemic [10]. The sedentary guidelines of existing countries clearly propose that online teaching time should be limited to avoid possible health threats caused by excessive screen time [8]. However, the current evidence on the relationship between learning ST and CRF in children and adolescents is still unclear. Future research should further distinguish the nature of ST, and accurately identify risk factors for CRF.

Previous studies have focused on the relationship between active transportation to school and CRF, and most studies support the theory that active transportation, especially cycling to school, is positively correlated with CRF in children and adolescents [29]. Other studies have confirmed that the CRF of children and adolescents who travel to school by passive transportation is lower than that of individuals who travel to school by active transportation [30]. The findings of this study further improve the chain of evidence that suggests that passive traffic-sitting in girls is negatively correlated with CRF and independent of MVPA. Based on the above evidence chain, we can speculate that children and adolescents who travel to school by active transportation have higher CRF, whereas those who travel to school by passive transportation have lower CRF. The reason for this may be that active transportation to school increases physical activity, which is positively correlated with CRF, whereas passive transportation increases ST, which is negatively correlated with CRF.

This study did not find a significant association between extracurricular learning ST and CRF. Although there is no clear evidence that too much extracurricular learning ST has a detrimental effect on CRF, given the negative impact of too much extracurricular learning ST on other movement behaviors [10] and the positive correlation between movement behavior and CRF [31], we still cannot ignore the potential long-term adverse effects of a heavy schoolwork load on CRF in children and adolescents. Moreover, this study found that the extracurricular learning ST of Chinese children and adolescents (about 360 min/d) is much higher than the recommended amount of international school-based sedentary behavior recommendations (10~60 min/d) [8]. Reducing the schoolwork load of Chinese children and adolescents should be targeted.

This study found that breaks in ST and the duration of breaks in ST were positively associated with CRF in girls. Combined with the high correlation between breaks in ST and the duration of breaks in ST (r = 0.86), we can speculate that more breaks in ST can lead to longer durations of breaks in ST, which may be a key factor in improving CRF in children and adolescents. This result also confirms previous research findings. Judice et al.'s study showed that the benefit of sedentary bout interruptions did not come from the interruptions themselves, but from the physical activity performed during the sedentary bout interruptions [11]. The results of this study suggest that breaks in ST can have a beneficial effect on CRF. Breaks in ST are essentially a change from sedentary behavior to physical activity. Intensity is an important aspect to consider. This study found that breaks in ST and the duration of breaks in ST were positively associated with CRF after further controlling for LPA. However, the association was not significant after controlling for MVPA. We can speculate that in this study, MVPA was the determinant factor affecting CRF in the accumulated duration of breaks in the ST of girls. The results of this study suggest that although directly increasing MVPA is an important means to promote CRF, considering the generally low MVPA levels of children and adolescents [32], increasing MVPA by increasing breaks in ST and the duration of breaks in ST may also be a feasible method to improve CRF. The latest release of the International School-based Sedentary Behavior Recommendations also suggests that prolonged ST should be interrupted as much as possible to reduce the harm of sedentary behavior to children and adolescents [8].

We further evaluated the effects of screen time, passive traffic time, breaks in ST, and the duration of breaks in ST on the goodness-of-fit of the models using a hierarchical regression method. The results showed that the adjusted R^2 was increased by 0.012, 0.016, 0.009, and 0.014, respectively (Supplementary Materials). Although the above significant variables had little influence on the goodness-of-fit of the model, we still need to pay attention to them as an important means of CRF intervention in children and adolescents. Previous studies pointed out that although MVPA or ST showed low-to-medium correlations with CRF in children and adolescents, it did not mean that they were not important [5,7]. Therefore, schools and public health departments should take all feasible means to actively intervene with CRF in children and adolescents.

5. Strengths and Limitations

A major strength of this study was that it selected a representative sample from six administrative regions in China. Furthermore, this study adopted a combination of objective and subjective measures to evaluate sedentary behavior, which is conducive to a more comprehensive understanding of the relationship between the ST and SP and CRF in children and adolescents. There are still several limitations which should be considered. First, as this was a cross-sectional study, it was impossible to determine the continuous impact of ST and SP on CRF. Second, the ST objectively measured using an accelerometer in this study may have been overestimated, as lower-intensity behaviors, such as standing or lying down, may have been identified as sedentary behavior.

6. Conclusions

There was no significant correlation between total ST and CRF in children and adolescents, but when ST was divided into different types, the screen time and passive traffic ST of girls were negatively correlated with CRF and independent of MVPA. Breaks in ST and the duration of breaks in ST were positively correlated with CRF in girls. MVPA performed during breaks in ST may be the key factor affecting CRF. Schools and public health departments should take all feasible means to actively intervene with CRF in children and adolescents.

Supplementary Materials: The following supporting information can be downloaded at: https://www.mdpi.com/article/10.3390/children9081140/s1, Table S1: Association between screen time and CRF in girls. Table S2: Association between passive traffic ST and CRF in girls. Table S3: Association between breaks in ST and CRF in girls. Table S4: Association between duration of breaks in ST and CRF in girls.

Author Contributions: Conceptualization, M.L. and X.Y.; methodology, Y.L. (Yuqiang Li); software, Y.S. and P.S.; formal analysis, Y.S.; investigation, Y.G. and F.Z.; data curation, T.Z. and Y.L. (Yuan Liu); writing—original draft preparation, M.L.; writing—review and editing, X.Y. All authors have read and agreed to the published version of the manuscript.

Funding: This work was supported by the National Philosophy Social Science Fund of China (Grant No.: 21BTY121).

Institutional Review Board Statement: This study was approved by the Human Experimental Ethics Committee of East China Normal University (No. HR 006-2019).

Informed Consent Statement: Before the investigation, students and parents were told that the purpose of the study was to understand the physical activity of the students. Written informed consent was obtained from all students and parents. This investigation was conducted as an anonymous survey.

Data Availability Statement: The data that support the findings of this study are available from Xiaojian Yin but restrictions apply to the availability of these data, which were used under license for the current study and so are not publicly available.

Acknowledgments: We would like to acknowledge the students and parents who participated in the test. We are grateful for their contributions.

Conflicts of Interest: The authors declare no conflict of interest.

References

1. Sui, X.; LaMonte, M.J.; Blair, S.N. Cardiorespiratory fitness and risk of nonfatal cardiovascular disease in women and men with hypertension. *Am. J. Hypertens.* **2007**, *20*, 608–615. [CrossRef] [PubMed]
2. Blair, S.N.; Kohl, H.R.; Paffenbarger, R.J.; Clark, D.G.; Cooper, K.H.; Gibbons, L.W. Physical fitness and all-cause mortality. A prospective study of healthy men and women. *JAMA* **1989**, *262*, 2395–2401. [CrossRef] [PubMed]

3. Tomkinson, G.R.; Leger, L.A.; Olds, T.S.; Cazorla, G. Secular trends in the performance of children and adolescents (1980-2000): An analysis of 55 studies of the 20m shuttle run test in 11 countries. *Sports Med.* **2003**, *33*, 285–300. [CrossRef]

4. Ried-Larsen, M.; Grontved, A.; Moller, N.C.; Larsen, K.T.; Froberg, K.; Andersen, L.B. Associations between objectively measured physical activity intensity in childhood and measures of subclinical cardiovascular disease in adolescence: Prospective observations from the European Youth Heart Study. *Br. J. Sports Med.* **2014**, *48*, 1502–1507. [CrossRef] [PubMed]

5. Kristensen, P.L.; Moeller, N.C.; Korsholm, L.; Kolle, E.; Wedderkopp, N.; Froberg, K.; Andersen, L.B. The association between aerobic fitness and physical activity in children and adolescents: The European youth heart study. *Eur. J. Appl. Physiol.* **2010**, *110*, 267–275. [CrossRef]

6. Ruiz, J.R.; Ortega, F.B.; Martinez-Gomez, D.; Labayen, I.; Moreno, L.A.; De Bourdeaudhuij, I.; Manios, Y.; Gonzalez-Gross, M.; Mauro, B.; Molnar, D.; et al. Objectively measured physical activity and sedentary time in European adolescents: The HELENA study. *Am. J. Epidemiol.* **2011**, *174*, 173–184. [CrossRef] [PubMed]

7. Tremblay, M.S.; Leblanc, A.G.; Kho, M.E.; Saunders, T.J.; Larouche, R.; Colley, R.C.; Goldfield, G.; Gorber, S.C. Systematic review of sedentary behaviour and health indicators in school-aged children and youth. *Int. J. Behav. Nutr. Phy.* **2011**, *8*, 98–119. [CrossRef] [PubMed]

8. Saunders, T.J.; Rollo, S.; Kuzik, N.; Demchenko, I.; Belanger, S.; Brisson-Boivin, K.; Carson, V.; Da, C.B.; Davis, M.; Hornby, S.; et al. International school-related sedentary behaviour recommendations for children and youth. *Int. J. Behav. Nutr. Phys. Act.* **2022**, *19*, 39–52. [CrossRef] [PubMed]

9. Marques, A.; Santos, R.; Ekelund, U.; Sardinha, L.B. Association between Physical Activity, Sedentary Time, and Healthy Fitness in Youth. *Med. Sci. Sport. Exer.* **2014**, *47*, 575–580. [CrossRef]

10. Kuzik, N.; Costa, B.G.G.D.; Hwang, Y.; Verswijveren, S.J.J.M.; Rollo, S.; Tremblay, M.S.; Bélanger, S.; Carson, V.; Davis, M.; Hornby, S.; et al. School-Related Sedentary Behaviours and Indicators of Health and Well-Being Among Children and Youth_ A Systematic Review. *Int. J. Behav. Nutr. Phy.* **2022**, *19*, 40–71. [CrossRef] [PubMed]

11. Judice, P.B.; Silva, A.M.; Berria, J.; Petroski, E.L.; Ekelund, U.; Sardinha, L.B. Sedentary patterns, physical activity and health-related physical fitness in youth: A cross-sectional study. *Int. J. Behav. Nutr. Phys. Act.* **2017**, *14*, 25–34. [CrossRef] [PubMed]

12. Jones, M.A.; Skidmore, P.M.; Stoner, L.; Harrex, H.; Saeedi, P.; Black, K.; Barone, G.B. Associations of accelerometer-measured sedentary time, sedentary bouts, and physical activity with adiposity and fitness in children. *J. Sports Sci.* **2020**, *38*, 114–120. [CrossRef] [PubMed]

13. Bailey, D.P.; Charman, S.J.; Ploetz, T.; Savory, L.A.; Kerr, C.J. Associations between prolonged sedentary time and breaks in sedentary time with cardiometabolic risk in 10-14-year-old children: The HAPPY study. *J. Sports Sci.* **2017**, *35*, 2164–2171. [CrossRef] [PubMed]

14. Denton, S.J.; Trenell, M.I.; Plotz, T.; Savory, L.A.; Bailey, D.P.; Kerr, C.J. Cardiorespiratory fitness is associated with hard and light intensity physical activity but not time spent sedentary in 10-14 year old schoolchildren: The HAPPY study. *PLoS ONE* **2013**, *8*, e61073. [CrossRef]

15. Leger, L.A.; Mercier, D.; Gadoury, C.; Lambert, J. The multistage 20 metre shuttle run test for aerobic fitness. *J. Sports Sci.* **1988**, *6*, 93–101. [CrossRef]

16. Matsuzaka, A.; Takahashi, Y.; Yamazoe, M.; Kumakura, N.; Ikeda, A.; Wilk, B.; Bar-Or, O. Validity of the Multistage 20-M Shuttle-Run Test for Japanese Children, Adolescents, and Adults. *Pediatri. Exerc. Sci.* **2014**, *16*, 113–125. [CrossRef]

17. Tomkinson, G.R.; Lang, J.J.; Tremblay, M.S.; Dale, M.; LeBlanc, A.G.; Belanger, K.; Ortega, F.B.; Leger, L. International normative 20 m shuttle run values from 1 142 026 children and youth representing 50 countries. *Br. J. Sports Med.* **2017**, *51*, 1545–1554. [CrossRef]

18. Ekelund, U.; Sardinha, L.B.; Anderssen, S.A.; Harro, M.; Franks, P.W.; Brage, S.; Cooper, A.R.; Riddoch, C.; Froberg, K. Associations between objectively assessed physical activity and indicators of body fatness in 9- to 10-y-old European children: A population-based study from 4 distinct regions in Europe (the European Youth Heart Study). *Am. J. Clin. Nutr.* **2004**, *80*, 584–590. [CrossRef] [PubMed]

19. Evenson, K.R.; Catellier, D.J.; Gill, K.; Ondrak, K.S.; McMurray, R.G. Calibration of two objective measures of physical activity for children. *J. Sports Sci.* **2008**, *26*, 1557–1565. [CrossRef] [PubMed]

20. Trost, S.G.; Loprinzi, P.D.; Moore, R.; Pfeiffer, K.A. Comparison of accelerometer cut points for predicting activity intensity in youth. *Med. Sci. Sports Exerc.* **2011**, *43*, 1360–1368. [CrossRef]

21. Belcher, B.R.; Berrigan, D.; Papachristopoulou, A.; Brady, S.M.; Bernstein, S.B.; Brychta, R.J.; Hattenbach, J.D.; Tigner, I.J.; Courville, A.B.; Drinkard, B.E.; et al. Effects of Interrupting Children's Sedentary Behaviors With Activity on Metabolic Function: A Randomized Trial. *J. Clin. Endocrinol. Metab.* **2015**, *100*, 3735–3743. [CrossRef] [PubMed]

22. Guo, Y.; Yin, X.; Sun, Y.; Zhang, T.; Li, M.; Zhang, F.; Liu, Y.; Xu, J.; Pei, D.; Huang, T. Research on Environmental Influencing Factors of Overweight and Obesity in Children and Adolescents in China. *Nutrients* **2021**, *14*, 35. [CrossRef] [PubMed]

23. Santos, R.; Mota, J.; Okely, A.D.; Pratt, M.; Moreira, C.; Coelho-E-Silva, M.J.; Vale, S.; Sardinha, L.B. The independent associations of sedentary behaviour and physical activity on cardiorespiratory fitness. *Br. J. Sports Med.* **2014**, *48*, 1508–1512. [CrossRef] [PubMed]

24. Martinez-Gomez, D.; Ortega, F.B.; Ruiz, J.R.; Vicente-Rodriguez, G.; Veiga, O.L.; Widhalm, K.; Manios, Y.; Béghin, L.; Valtueña, J.; Kafatos, A.; et al. Excessive sedentary time and low cardiorespiratory fitness in European adolescents: The HELENA study. *Arch. Dis. Child.* **2011**, *96*, 240–246. [CrossRef]

25. Cabanas-Sánchez, V.; Martínez-Gómez, D.; Esteban-Cornejo, I.; Pérez-Bey, A.; Castro Piñero, J.; Veiga, O.L. Associations of total sedentary time, screen time and non-screen sedentary time with adiposity and physical fitness in youth: The mediating effect of physical activity. *J. Sports Sci.* **2019**, *37*, 839–849. [CrossRef] [PubMed]

26. Joensuu, L.; Syväoja, H.; Kallio, J.; Kulmala, J.; Kujala, U.M.; Tammelin, T.H. Objectively measured physical activity, body composition and physical fitness: Cross-sectional associations in 9- to 15-year-old children. *Eur. J. Sport Sci.* **2018**, *18*, 882–892. [CrossRef]

27. Porter, A.K.; Matthews, K.J.; Salvo, D.; Kohl, H.W. Associations of Physical Activity, Sedentary Time, and Screen Time With Cardiovascular Fitness in United States Adolescents: Results From the NHANES National Youth Fitness Survey. *J. Phys. Act. Health* **2017**, *14*, 506–512. [CrossRef] [PubMed]

28. Biswas, A.; Oh, P.I.; Faulkner, G.E.; Bajaj, R.R.; Silver, M.A.; Mitchell, M.S.; Alter, D.A. Sedentary Time and Its Association With Risk for Disease Incidence, Mortality, and Hospitalization in Adults: A Systematic Review and Meta-analysis. *Ann. Intern. Med.* **2015**, *162*, 123–132. [CrossRef] [PubMed]

29. Larouche, R.; Saunders, T.J.; Faulkner, G.E.; Colley, R.; Tremblay, M. Associations between active school transport and physical activity, body composition, and cardiovascular fitness: A systematic review of 68 studies. *J. Phys. Act. Health.* **2014**, *11*, 206–226. [CrossRef] [PubMed]

30. Ostergaard, L.; Kolle, E.; Steene-Johannessen, J.; Anderssen, S.A.; Andersen, L.B. Cross sectional analysis of the association between mode of school transportation and physical fitness in children and adolescents. *Int. J. Behav. Nutr. Phy.* **2013**, *10*, 91–97. [CrossRef] [PubMed]

31. Scott, R.; Olga, A.; Tremblay, M.S. The whole day matters: Understanding 24-hour movement guideline adherence and relationships with health indicators across the lifespan. *J. Sport Health Sci.* **2020**, *9*, 493–510. [CrossRef]

32. Guthold, R.; Stevens, G.A.; Riley, L.M.; Bull, F.C. Global trends in insufficient physical activity among adolescents: A pooled analysis of 298 population-based surveys with 1.6 million participants. *Lancet Child. Adolesc. Health* **2020**, *4*, 23–35. [CrossRef]

Article

Static and Dynamic Balance Indices among Kindergarten Children: A Short-Term Intervention Program during COVID-19 Lockdowns

Einat Yanovich * and Salit Bar-Shalom

The Academic College at Wingate, Netanya 4290200, Israel; salinka27@gmail.com
* Correspondence: einaty@wincol.ac.il

Abstract: The COVID-19 pandemic outbreak had a negative impact on kindergarten activities. These young children, who had been compelled to stay home during lockdowns, suffered a lack of movement and loss of mobility, resulting in deteriorated physical motor skills. Lack of sufficient motor experience in early childhood can impair children's motor and cognitive development. Balance skills are fundamental to all other motor abilities, from the most basic movements to the most complex motor skills. The purpose of this study was to implement a short-term physical activity program, which may have a direct effect on children's fundamental balance ability. Ninety-six kindergarten children (45 boys and 51 girls), aged 4–6 years, participated in the study. Data were analyzed using three-way ANOVA and interaction analyses. The results suggest that short, focused, and dedicated balance training programs have a beneficial influence on the static balance of preschoolers and can mitigate some of the negative physical outcomes of lockdowns. In conclusion, this study indicates that a short-term physical training program had a positive effect on the motor abilities of preschoolers after COVID-19-related lockdowns. More research is needed in order to fully understand the complete impact of the worldwide health crisis and the best ways in which to address it.

Keywords: COVID-19; lockdown; motor learning; motor skills; balance

Citation: Yanovich, E.; Bar-Shalom, S. Static and Dynamic Balance Indices among Kindergarten Children: A Short-Term Intervention Program during COVID-19 Lockdowns. *Children* **2022**, *9*, 939. https://doi.org/10.3390/children9070939

Academic Editor: Georgian Badicu

Received: 7 April 2022
Accepted: 20 June 2022
Published: 22 June 2022

Publisher's Note: MDPI stays neutral with regard to jurisdictional claims in published maps and institutional affiliations.

Copyright: © 2022 by the authors. Licensee MDPI, Basel, Switzerland. This article is an open access article distributed under the terms and conditions of the Creative Commons Attribution (CC BY) license (https://creativecommons.org/licenses/by/4.0/).

1. Introduction

The term "balance" refers to a type of motor movement coordination in which the visual and kinesthetic components of the body muscles work together with the balance sensors that are located in the middle ear in order to maintain the body's stability without unnecessary movements or falling [1]. The ability of body balance depends on internal systems, such as the vestibular system (equilibrium sensors), the proprioceptive system (responsible for motion sensors), and the visual system. Balance also depends on certain external factors, such as the support base, the center of gravity, and the body's structure and weight [2].

In the field of movement, the ability of two-legged creatures to stabilize themselves is especially important and requires being perpendicular from the center of gravity and its supporting area, i.e., standing on their legs and the area between their legs. Moving from standing on one leg to standing on the other requires advancement along with a narrow basis while maintaining stability and activating the balance system as part of the posture [3]. In addition, basic movements such as running, changing direction, stopping, and advancing on elevated surfaces require the maintaining of the person's stability and necessitate training for developing the balance system [4].

Balance skills serve as the basis of all other motor abilities, from basic movements to the most complex motor skills [5]. The muscular-nervous system even constitutes a connection point for certain cognitive processes, such as attention, concentration, and mental imagery [6]. A correlation was found between body balance and mental states, such as the sense of self-efficacy, depression, and anxiety [7].

The ability to maintain balance is subject to change during a person's lifespan, such as changes in body structure, experience in the given activity, and complex challenges [8].

1.1. Types of Balance

There are two types of balance: static and dynamic. Static balance takes place when the center of gravity is maintained vertically above the base, without changing the base lengthwise. Static balance is maintained as long as the pressure on the organs that are carrying the body weight is consistently located at the center of the body mass [9]. Humans can choose the base of support, its size, and how the body is structured on it [10]. Maintaining balance is more challenging when the base of support is smaller, has a greater slope, and is less stable, and if the individual has a higher center of gravity [4]. Examples of static balance include standing on one foot and standing in an elevated position on one's feet.

The term dynamic balance relates to the ability to maintain the center of gravity above the base during movement, with the body exiting the center of gravity. In dynamic balance, the main process is the coordination between holding the torso above the center of gravity and various advancement movements, which allows for stability and reflectivity in reaction to changes that occur during the movement. To successfully maintain dynamic balance, one must be prepared with responses to expected alterations [11]. Dynamic balance is part of any skill of progress and is manifested in the moment of transition from base to base when there is a detachment of a part of the body moving from the ground. Examples of detachment are walking, running, jumping, and landing.

1.2. The Importance of Balance in Preschool Children

Significant maturity of the nervous system cannot be achieved in preschool children, even when they are going through normal and substantial motor development processes if routine experiences of certain key movements are lacking (as in the case of lockdown). The muscular nervous system, which enables movement coordination, is acquired and developed from the embryonic stage onwards [12]. Early childhood presents a window of opportunity for motor development, yet insufficient motor experience can impair motor development among children and even later in life. The system's requirements for creating efficient movements are complex, yet the prevailing assumption is that such requirements serve as the main factors in the development of the balance system during childhood and in the importance of abundant movement experience [5].

Balance ability is the basic condition for the performance of motor skills, and it depends on internal and external systems. Motor experience in ages 3–7 is a necessary condition and a key factor for developing synchronization between various systems. Nevertheless, it is difficult to indicate the exact age at which the ability to balance ability reaches maturity or peak functionality since the developmental processes of balance are not consistent. Balance sensors, for example, are the last to evolve due to the complexity of this system and its dependence on other sensory systems. As such, toddlers and young children mainly use motion and vision sensors to maintain balance, only adding balance sensors later on in their development [13].

Throughout this developmental process, there are certain milestones that should be pointed out. During the first weeks of life, babies practice using their balance sensors by trying to lift the upper part of their torso, neck, and head while lying on their stomachs. From the second half of infancy (i.e., from six months and onwards) and towards the end of this period, children are able to perform activities such as crawling, standing, and slow walking. However, these activities cannot be performed without initial coordination between vision, motion, and balance sensors. From the moment infants begin walking, this system receives a significant developmental boost since this skill requires constant adaptation of the body to movement alterations in space and to different situations, resulting in the development of balance sensors [14].

In independent walking, children demonstrate control of their torso volatility for the first time, as previous skills required them to concentrate on controlling their body balance above the base [15]. The important development of the child's balance ability at the kindergarten age advances at a rapid pace following ample practice. Educational frameworks for ages 3–7, therefore, should monitor the development of this ability among the children in order to ensure continuous improvement [16].

1.3. The Effect of the COVID-19 Pandemic

Following the outbreak of the COVID-19 pandemic in Israel in March 2020, educational institutions across Israel closed their doors—including kindergartens, rendering more than half a million children aged 3–6 homebound and in social isolation. As kindergartens provide a supportive physical and cognitive environment for the acquisition and development of cognitive, emotional, social, and physical skills among young children, short-term effects of the lockdown on educational, personal, and social aspects rapidly emerged [17–19]. The long-term effects, however, are yet to be fully observed and explored.

Quarantine phases have a widespread effect on the acquisition and establishment of knowledge and skills. Significant developmental milestones and processes that take place in kindergarten could have been damaged by the long lockdown, with the possible forming of gaps and inequality processes. There is already evidence as to longer screen time and less physical activity among children [19], as well as poorer balance [20]. Gaps in social, sensory-motor, language, life, and motor skills have also been documented [17]. Long periods of quarantine may also alter the way children play and affect their social behavior [21]. There is also concern that if not adequately addressed, these gaps in development may affect children's acquisition of developmental indicators in the future as well.

Static balance tends to decrease between the end of kindergarten and the end of the first year at school (ages 5–7) [22]. Deterioration may worsen in times of emergency lockdown. Without noteworthy interventions, such gaps may continue to expand, in addition to primary deficits that are already showing indications. The education system in Israel faced three remote-learning periods during the crisis, with 4–8 weeks of grace between each of these periods. The first period lasted 8 weeks, the second 6 weeks, and the third remote-learning period lasted 5 weeks. During these periods, all parks, playgrounds, and recreation areas were closed. As such, educators found themselves attempting to achieve as much as possible during the infrequent periods when the children physically attended kindergartens and schools.

In light of this background, the aim of this study was to implement a short, targeted program that could have a direct effect on children's fundamental balance ability. The study hypothesizes that this short-term program will make a beneficial contribution to children's static and dynamic balance.

2. Materials and Methods

2.1. Participants

The study included 96 kindergarten children (45 boys and 51 girls), aged 4–6 years, from four different kindergartens in a central district in Israel. The four kindergartens were all situated on one site and belonged to the same cluster, which was chosen through convenience sampling. This kindergarten cluster ("Eshkol Ganim" in Hebrew) is characterized by a medium-high socioeconomic status and serves the majority of children in this geographical area. All four kindergartens are structured similarly. Each building housing a kindergarten consists of two large rooms; one has a table and chairs for educational activities such as painting and art, and the other is used for gatherings and for free play. The outside yard of each kindergarten is equipped with a sandbox, a ladder and slide, and a swing.

The four kindergartens were cluster-randomized, whereby two kindergartens were randomly assigned to the control group (n = 46; age: 4.8 ± 1.63), and the remaining two were assigned to the study group (n = 50; age: 4.63 ± 1.12). This allocation was mandatory since the study was conducted by the physical education teacher as part of the regular curriculum in which each kindergarten receives only one hour per week of physical education, and since the COVID-19 health regulations disallowed the mixing of classes. Anthropometric data (height and weight) were not collected since the study was conducted as part of the kindergarten's routines and had to follow the formal restrictions. The Israeli Ministry of Education prohibits taking any body measurements in all the institutions under its inspection. The inclusion criterion was participating in all study sessions. Children who missed one or more sessions were not included in the sample, although they continued to participate in the classes. The kindergarten teachers had no previous knowledge of the study's goals.

2.2. Procedure

After receiving the Institutional review board agreement, parents' consent forms, and permission from the district's kindergartens supervisor, the study began. Data collection in all kindergartens was conducted by the same experimenter during the day's activities, while the daily routine was kept fixed.

In the pre-intervention phase (T0—one session), the experimenter met with the kindergarten teachers and children and evaluated the static and dynamic balance of each child via a battery of motor tests described below. The tests were conducted in an isolated room in the kindergarten, which had been specifically pre-assigned for this process. The resting time between each motor test was two minutes (passive break) to allow the children to relax. Two children were tested intermittently; one performed the motor tests, while the other rested. The order of the motor tests was identical for all children, and the kindergarten staff was not involved in the data collection process. The encounters lasted 45 min each and were conducted throughout the regular kindergarten day, from 8 am to 1:30 p.m.

The intervention phase (T1—three sessions) entailed balance training, as presented in Appendix A. Both the experimental group and the control group underwent three 45-minute sessions, with seven days between each session. This study design was selected as a weekly session of 45-minute physical education is the standard curriculum. During these interventions, both dynamic and static balance training was taught to the experimental group, while general physical activities (e.g., walking at a changing pace, dancing, moving, and stopping) with no accentuation of balance were carried out by the control group. The balance training consisted of 10 minutes of warming up, 30 min of balance routines (3–5 repetitions of each routine), and 5 minutes of relaxing and summarizing the session. A physical education teacher for preschoolers conducted the interventions in both study groups and maintained the same methodology.

Post-intervention assessments (T2—one session) were carried out during the fifth and final session. These assessments were identical to those carried out during the first session in the pre-intervention phase. Following the completion of the study, and after all measurements were taken, the control group then received the balance training, and the experimental group received the regular physical activity sessions as part of their physical education classes.

2.3. Research Tools

To examine the impact of the intervention, the pre- and post-intervention tests included two motor tests: the one-leg stance (stork balance stand test) [23] and the one-leg hop test [24]. The one-leg stance evaluates the individual's ability to stand on one leg for as long as possible, as measured in seconds. The test's reliability depends on how strictly the test is conducted combined with the individual's level of motivation to perform. Test validity, calculated based on published tables regarding fitness level, indicated high correlations [25]. In this study, the children were instructed to stand on one leg for as long as possible, with parallel thighs and with the opposite knee elevated to $90°$. The best measurement out of three was documented for each leg. The one-leg hop test evaluates dynamic balance and has been shown to have a strong intrarater reliability, with an intraclass correlations coefficient of 0.85 for both legs. In this study, the children were asked to hop on one leg as many times as they could without stopping between hops and then to repeat the exercise on the other leg. The best measurement out of three was documented for each leg.

2.4. Statistical Analysis

Data analysis was performed using SPSS® version 26.0 for Windows (IBM Corp, Armonk, NY, USA). Comparisons between the groups at the different time points were conducted using a three-way analysis of variance (3-way ANOVA; groups, time, and leg; $\alpha < 0.05$) followed by a two-way interaction analysis ($\alpha < 0.05$). Analysis of the static and dynamic balance produced descriptive statistics, including the mean, standard deviation (SD), and standard error (SE).

3. Results

One hundred and two children started the program. Ninety-six of them attended all intervention sessions and were included in the study (94.1%).

The motor test results are presented in Table 1.

Table 1. Motor test results pre- and post-intervention (mean ±SD).

Motor Test	Intervention	Experimental Group (N = 50)	Control Group (N = 46)
One-leg stance, right leg (seconds)	Pre	16.2 ± 8.6	16.3 ± 7.2
	* Post	*** 27.7 ± 9.9	19.8 ± 7.9
One-leg stance, left leg (seconds)	Pre	9.4 ± 5.1	9.1 ± 3.9
	* Post	*** 19.5 ± 5.6	13.9 ± 5.2
One-leg hop, right leg (repetitions)	Pre	3.4 ± 1.3	3.7 ± 1.1
	** Post	**** 5.3 ± 1.8	4.8 ± 1.5
One-leg hop, left leg (repetitions)	Pre	2.6 ± 1.5	2.0 ± 1.0
	** Post	3.9 ± 1.5	2.9 ± 1.3

* Significant time–group interaction. ** Significant time–leg interaction. *** Significant within group. **** Significant between groups.

From the three-way ANOVA on the one-leg stance tests, no three-way interaction between groups, time (T0 vs. T2), or leg (right vs. left) was found ($F(1,94) = 3.669$, $p = 0.058$, $\eta^2 = 0.038$). Moreover, no two-way interaction was found between leg and time ($F(1,94) = 0.004$, $p > 0.05$, $\eta^2 = 0.000$) or between leg and group ($F(1,94) = 0.4$, $p > 0.05$, $\eta^2 = 0.004$).

An important finding is the time–group interaction ($F(1,94) = 64.466$, $p < 0.001$, $\eta^2 = 0.407$), whereby the improvement in the one-leg stance in the experimental group following their dynamic and static balance training was greater than the improvement seen among the control group, which did not receive such training, as presented in Figure 1.

Additionally, participants in the experimental group significantly improved their one-leg stance on the right and left leg by 71% and 108%, respectively ($p < 0.05$), while participants in the control group tended to improve their one-leg stance on the right and left leg by only 22% and 53%, respectively.

Three-way ANOVA on the one-leg hop tests showed a tendency of an interaction between groups, time, and leg ($F(1,94) = 1.291$, $p > 0.05$, $\eta^2 = 0.014$). However, a two-way interaction was seen between leg and time ($F(1,94) = 4.113$, $p < 0.05$, $\eta^2 = 0.042$). Moreover, a significant time and group interaction ($F(1,94) = 9.378$, $p < 0.05$, $\eta^2 = 0.091$) was evident, whereby the improvement in the one-leg hop test in the experimental group was greater than that of the control group (Figure 1). Additionally, participants in the experimental group improved their one-leg hop results on the right and left leg by 56% and 50%, respectively, while the participants in the control group improved their one-leg hop results on the right and left leg by 30% and 45%, respectively.

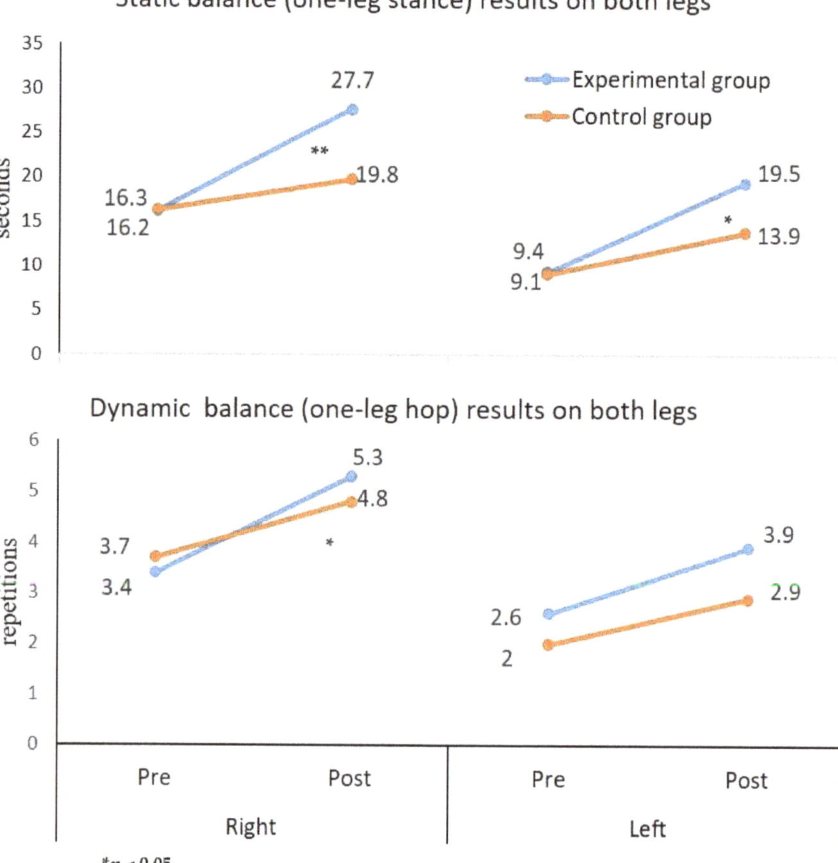

Figure 1. Results of time–group interaction of the static balance (one-leg stance in seconds) and dynamic balance (one-leg hop in repetitions) of both legs, pre- and post-intervention in both study groups (experimental vs. control).

4. Discussion

A great amount of research has been conducted in the past few years in order to explore the impact of the COVID-19 pandemic on different aspects of life [17,19]. This study aimed to examine a means for improving the important skill of balance among young children who returned to kindergarten after a number of long lockdowns and periods of social isolation due to this emergency situation. The study hypothesized that specific balance training could contribute to improved static and dynamic balance performance among those children. The results of this study suggest that a short, designated treatment focused on balance has a beneficial influence on the static balance of preschoolers.

The COVID-19 pandemic crisis created a novel situation in which educational institutions were forced to close their doors, and children were required to stay indoors for long periods of time. As a result, deterioration was observed in various physical and cognitive aspects, with the sedentary pattern—which had already been documented prior to the crisis—increasing alarmingly [17,19]. Furthermore, children began to show significantly shorter single-leg standing time and a significantly greater number of falls per month following the onset of the emergency situation [20].

The short-term intervention program presented in this study was able to reverse some of the negative physical outcomes created by the long lockdowns. For example, the results of the one-leg stance showed an interaction between group and treatment, whereby the static stability of the experimental group improved significantly following the intervention training. This change cannot be explained by normal development alone, as the intervention only lasted five weeks, and such a change was not seen in the control group. As there is evidence that static balance tends to decrease between the end of kindergarten and the end of the first year at school (up to age 7) [22], these findings offer an easy and cost-effective solution. As such, after a long sedentary period, a short, specifically designed program may contribute to significant balance improvement.

Both static and dynamic balance were tested in this study. While the static balance of the experimental group improved significantly compared to that of the control group, an improvement was seen in the dynamic balance tests in both groups. It has been suggested that dynamic balance may improve through core stability training [26]. The control group in this study performed general physical activities that may have had a positive influence on some core muscles and, in turn, on the participants' dynamic balance. More research may be needed to further address this point.

Balance skills are fundamental to all motor abilities, from the most basic movements to the most complex motor skills [5], and even constitute a connection point for certain cognitive processes, such as attention, concentration, and mental imagery [6]. The need to balance the body in stationary and mobile situations occurs frequently throughout the day. As such, it is important to train this skill from an early age. Intervention programs where balance plays a key role (e.g., yoga and "Minds in Motion") were found to also contribute to the academic skills of elementary school children [27]. As such, when choosing a motor skill on which to focus, balance skills should be given preference.

The practical outcome of this study may be the possibility of designing short, cost-effective programs consisting of only a few sessions that can be used to restore deteriorated physical abilities. These programs may be used in times of lockdowns or other emergency situations. Although this study was focused on balance, further studies may target other motor abilities, such as coordination, kinesthesis, or even speed. Furthermore, future studies may be focused on restoring or improving balance or other abilities among children diagnosed with delay in their motor development, such as developmental coordination disorder (DCD).

Despite the importance of this research study and its outcomes, some limitations should be addressed. Although four kindergartens were included in the study, providing an adequate sample size, they were all part of the same educational institution. Moreover, this framework catered to families of strong financial and social standing. As such, future studies could benefit from comparing children from different institutions and from different backgrounds and statuses. In a time of uncertainty, such as during the COVID-19 pandemic, children tend to miss many days of school, especially when they or their family members test positive for the virus. As such, future studies should monitor the attendance of the children in the intervention program and offer an intervention program to those who are forced to stay remote.

5. Conclusions

In conclusion, this study showed that even a short-term targeted program can positively improve the motor abilities of preschoolers, especially after long periods of lockdown and social isolation. Moreover, such interventions can be applied with ease, even by kindergarten teachers who are not trained in physical education, without incurring additional resources such as time and money. Further research, however, is needed to fully understand the complete impact of the crisis on children's balance and motor skills and how to best address this issue.

Author Contributions: Conceptualization, E.Y.; Methodology, E.Y.; Investigation, E.Y.; Data Curation E.Y.; Data Interpretation, E.Y. and S.B.-S.; Writing—Original Draft Preparation, E.Y. and S.B.-S.; Writing—Review and Editing, E.Y. and S.B.-S. All authors have read and agreed to the published version of the manuscript.

Funding: This research received no external funding.

Institutional Review Board Statement: Ethical review and approval were waived for this study due to the fact that the activity was part of the daily routine of the kindergarten preschoolers and was conducted by the kindergarten teachers.

Informed Consent Statement: Not applicable.

Data Availability Statement: The data presented in this study are available on request from the corresponding author.

Acknowledgments: The authors would like to acknowledge the teachers of the kindergartens who took part in the study.

Conflicts of Interest: The authors declare no conflict of interest.

Appendix A

Appendix A contains detailed information on the interventions and motor skills tests that were conducted in the study.

Table A1. Outline of the balance and control procedures.

S#	Group Focusing on Non-Balance Topics (control)	Group Focusing on Balance
1	**Meeting goals:** Learn concepts of space and spatial boundaries. **Sample tasks:** Playing the song "In our yard". Each time the word "peace" is heard in the song, the children will enter a hoop—the teacher explains that this is their personal space and asks what can be done in a personal space. The children stand "in their homes" (the hoops). When the music plays the children go out to skip in the general space (between the hoops), a space that belongs to everyone, and they learn behavioral rules in the common space.	**Meeting goals:** Improving static and dynamic balance using various bases and heights. **Sample tasks:** 1. Move in space according to the music, stop in any way you like, and at any stop on other body parts. 2. In an "all-four position" move on the hands and feet, change the directions of progress backwards, sideways. Raise another limb. Find other ways to balance your body on four support bases. Try to move on them. 3. Try, in a small leg-spread, to detach the legs and turn a quarter turn, a half turn, and land without losing balance. 4. Balance the body on one of the body parts (not on the legs!)– Buttocks, abdomen, back.
2	**Meeting goals:** The children will experience movement at different rhythms and at different height levels. **Sample tasks:** 1. Progress in space at a slow pace at a low altitude level. 2. Move forward at a rapid pace on the toes. Upon clapping move in a stooped position. Move forward in a creative way at a medium-height level	**Meeting goals:** Static balance with eyes closed and on an elevated platform. **Sample tasks:** 1. Move between the stools. One knock—balance yourself on the stool. Two knocks—balance yourself on the floor. 2. Stand on one leg on the stool and together we clap our hands up to the sky. 3. Stand with one leg next to the stool, try to lift the stool and place it without lowering the other leg.

Table A1. *Cont.*

S#	Group Focusing on Non-Balance Topics (control)	Group Focusing on Balance
3	**Meeting goals:** The children will move their bodies in different ways and practice how to stop the movement. The children will move in space while considering their kindergarten counterparts. **Sample tasks:** 1. Walking around the kindergarten. When the music stops, everyone stands like a statue, motionless. 2. Walking backwards and standing next to a friend without contact when the music stops. 3. Imitation dance—mimicking the teachers' movements. Each time the song stops the children must stand still.	**Meeting goals:** Balancing the body on different fulcrums in different situations than usual. **Sample tasks:** 1. Find a partner, try different swings. 2. Sit back-to-back, combine elbows with your partner, bring your knees close to your chest, try to get up together. 3. Lie on the floor and try to roll together like a "sausage" (straight body). 4. Standing alone, try with a large legs-spread to balance on both hands and one knee. 5. Stand and create a statue on the narrowest base while others (children or the teacher) try to unbalance you by pushing slightly.

Note. S#—session number.

References

1. Goddard Blyth, S. *Attention, Balance and Coordination: The A.B.C. of Learning Success*, 2nd ed.; Wiley-Blackwell: Hoboken, NJ, USA, 2017. [CrossRef]
2. Davlin, C.D. Dynamic balance in high level athletes. *Percept Motor Skill* **2004**, *98*, 1171–1176. [CrossRef] [PubMed]
3. Hof, A.L. The "extrapolated center of mass" concept suggests a simple control of balance in walking. *Hum. Movement Sci.* **2008**, *27*, 112–125. [CrossRef] [PubMed]
4. Haddad, J.M.; Rietdyk, S.; Claxton, L.J.; Huber, J. Task-dependent postural control throughout the lifespan. *Exerc. Sport Sci. Rev.* **2013**, *41*, 123–132. [CrossRef] [PubMed]
5. Frick, A.; Möhring, W. A matter of balance: Motor control is related to children's spatial and proportional reasoning skills. *Front. Psychol.* **2016**, *6*, 2049. [CrossRef] [PubMed]
6. Taube, W.; Mouthon, M.; Leukel, C.; Hoogewoud, H.M.; Annoni, J.M.; Keller, M. Brain activity during observation and motor imagery of different balance tasks: An fMRI study. *Cortex* **2015**, *64*, 102–114. [CrossRef]
7. Mirelman, A.; Herman, T.; Nicolai, S.; Zijlstra, A.; Zijlstra, W.; Becker, C.; Hausdorff, J.M. Audio-biofeedback training for posture and balance in patients with Parkinson's disease. *J. Neuro Eng. Rehabil.* **2011**, *8*, 35. [CrossRef]
8. Gubbels, J.S.; Kremers, S.P.; Stafleu, A.; Goldbohm, R.A.; de Vries, N.K.; Thijs, C. Clustering of energy balance-related behaviors in 5-year-old children: Lifestyle patterns and their longitudinal association with weight status development in early childhood. *Int. J. Behav. Nutr. Phys.* **2012**, *9*, 77. [CrossRef]
9. Rogers, M.W.; Wardman, D.L.; Lord, S.R.; Fitzpatrick, R.C. Passive tactile sensory input improves stability during standing. *Exp. Brain Res.* **2001**, *136*, 514–522. [CrossRef]
10. Paillard, T.H.; Noé, F. Effect of expertise and visual contribution on postural control in soccer. *Scand. J. Med. Sci. Sport* **2006**, *16*, 345–348. [CrossRef]
11. Hatzitaki, V.; Zlsi, V.; Kollias, I.; Kioumourtzoglou, E. Perceptual-motor contributions to static and dynamic balance control in children. *J. Motor Behav.* **2002**, *34*, 161–170. [CrossRef]
12. Adolph, K.E.; Berger, S.E. Physical and motor development. In *Developmental Science: An Advanced Textbook*, 5th ed.; Bornstein, M.H., Lamb, M.E., Eds.; Psychology Press: Livingston, NJ, USA, 2010; pp. 223–281.
13. Ferber-Viart, C.; Ionescu, E.; Morlet, T.; Froehlich, P.; Dubreuil, C. Balance in healthy individuals assessed with Equitest: Maturation and normative data for children and young adults. *Int J Pediatr Otorhi* **2007**, *71*, 1041–1046. [CrossRef] [PubMed]
14. Singh, D.K.A.; Rahman, N.N.A.A.; Rajikan, R.; Zainudin, A.; Nordin, N.A.M.; Karim, Z.A.; Yee, Y.H. Balance and motor skills among preschool children aged 3 to 4 years old. *Malays. J. Med. Health Sci.* **2015**, *11*, 63–68.
15. Claxton, L.J.; Melzer, D.K.; Ryu, J.H.; Haddad, J.M. The control of posture in newly standing infants is task dependent. *J. Exp. Child Psychol.* **2012**, *113*, 159–165. [CrossRef] [PubMed]
16. Williams, H.G. *Perceptual and Motor Development*; Prentice Hall: Englewood Cliffs, NJ, USA, 1983.
17. Alonso-Martínez, A.M.; Ramírez-Vélez, R.; García-Alonso, Y.; Izquierdo, M.; García-Hermoso, A. Physical activity, sedentary behavior, sleep and self-regulation in Spanish preschoolers during the COVID-19 lockdown. *Int. J. Environ. Res. Public Health* **2021**, *18*, 693. [CrossRef]
18. Brooks, S.K.; Webster, R.K.; Smith, L.E.; Woodland, L.; Wessely, S.; Greenberg, N.; Rubin, G.J. The psychological impact of quarantine and how to reduce it: Rapid review of the evidence. *Lancet* **2020**, *395*, 912–920. Available online: https://www.thelancet.com/pdfs/journals/lancet/PIIS0140-6736(20)30460-8.pdf (accessed on 6 March 2022). [CrossRef]

19. Chambonniere, C.; Lambert, C.; Fearnbach, N.; Tardieu, M.; Fillon, A.; Genin, P.; Duclos, M. Effect of the COVID-19 lockdown on physical activity and sedentary behaviors in French children and adolescents: New results from the ONAPS national survey. *Eur. J. Integr. Med.* **2021**, *43*, 101308. [CrossRef] [PubMed]
20. Ito, T.; Sugiura, H.; Ito, Y.; Noritake, K.; Ochi, N. Effect of the COVID-19 emergency on physical function among school-aged Children. *Int. J. Environ. Res. Public Health* **2021**, *18*, 9620. [CrossRef] [PubMed]
21. Graber, K.M.; Byrne, E.M.; Goodacre, E.J.; Kirby, N.; Kulkarni, K.; O'Farrelly, C.; Ramchandani, P.G. A rapid review of the impact of quarantine and restricted environments on children's play and the role of play in children's health. *Child Care Health Dev.* **2021**, *47*, 143–153. [CrossRef]
22. Reisberg, K.; Riso, E.M.; Jürimäe, J. Physical activity, fitness, and cognitive performance of Estonian first-grade schoolchildren according their MVPA level in kindergarten: A longitudinal study. *Int. J. Environ. Res. Public Health* **2021**, *18*, 7576. [CrossRef]
23. Cadenas-Sanchez, C.; Martinez-Tellez, B.; Sanchez-Delgado, G.; Mora-Gonzalez, J.; Castro-Piñero, J.; Löf, M.; Ortega, F.B. Assessing physical fitness in preschool children: Feasibility, reliability and practical recommendations for the PREFIT battery. *J. Sci. Med. Sport* **2016**, *19*, 910–915. [CrossRef]
24. Sawle, L.; Freeman, J.; Marsden, J. Intra-rater reliability of the multiple single-leg hop-stabilization test and relationships with age, leg dominance and training. *Int. J. Sport Phys. Ther.* **2017**, *12*, 190–198.
25. Johnson, B.L.; Nelson, J.K. *Practical Measurements for Evaluation in Physical Education*, 4th ed.; McMillan: New York, NY, USA, 1986.
26. Sadeghi, H.; Shariat, A.; Asadmanesh, E.; Mosavat, M. The effects of core stability exercise on the dynamic balance of volleyball players. *Int. J. Appl. Exerc. Phys.* **2013**, *2*, 1–10.
27. de Paleville, D.G.T.; Immekus, J.C. A randomized study on the effects of Minds in Motion and Yoga on motor proficiency and academic skills among elementary school children. *J. Phys. Act Health* **2020**, *17*, 907–914. [CrossRef] [PubMed]

Article

The Effects of Verbal Encouragement during a Soccer Dribbling Circuit on Physical and Psychophysiological Responses: An Exploratory Study in a Physical Education Setting

Bilel Aydi [1,†], Okba Selmi [1,*,†], Mohamed A. Souissi [2,3], Hajer Sahli [1], Ghazi Rekik [4,5], Zachary J. Crowley-McHattan [6], Jeffrey Cayaban Pagaduan [7], Antonella Muscella [8], Makram Zghibi [1,‡] and Yung-Sheng Chen [5,9,10,*,‡]

1   Research Unit, Sportive Performance and Physical Rehabilitation, High Institute of Sports and Physical Education of Kef, University of Jendouba, EI Kef 7100, Tunisia; aydibilel18@yahoo.fr (B.A.); sahlihajer2005@yahoo.fr (H.S.); makwiss@yahoo.fr (M.Z.)
2   Physical Activity, Sport and Health, Research Unit (UR18JS01), National Observatory of Sport, Tunis 1003, Tunisia; gaddoursouissi@yahoo.com
3   High Institute of Education and Continuing Training of Tunis, Virtual University of Tunisia, Montplaisir 1073, Tunisia
4   Research Laboratory: Education, Motricity, Sport and Health (LR19JS01), High Institute of Sport and Physical Education, Sfax University, Sfax 3000, Tunisia; ghazi.rek@gmail.com
5   Tanyu Research Laboratory, Taipei 112, Taiwan
6   Discipline of Sport and Exercise Science, Faculty of Health, Southern Cross University, Lismore 2480, Australia; zac.crowley@scu.edu.au
7   Faculty of Physical Culture, Palacký University, 77900 Olomouc, Czech Republic; jcpagaduan@gmail.com
8   Department of Biological and Environmental Science and Technologies (DiSTeBA), University of Salento, 73100 Lecce, Italy; antonella.muscella@unisalento.it
9   Department of Exercise and Health Sciences, University of Taipei, Taipei 111, Taiwan
10 Exercise and Health Promotion Association, New Taipei City 241, Taiwan
*   Correspondence: okbaselmii@yahoo.fr (O.S.); yschen@utaipei.edu.tw (Y.-S.C.); Tel.: +886-2-2871-8288 (Y.-S.C.)
†   These authors contributed equally to this work.
‡   These authors contributed equally to this work.

Citation: Aydi, B.; Selmi, O.; Souissi, M.A.; Sahli, H.; Rekik, G.; Crowley-McHattan, Z.J.; Pagaduan, J.C.; Muscella, A.; Zghibi, M.; Chen, Y.-S. The Effects of Verbal Encouragement during a Soccer Dribbling Circuit on Physical and Psychophysiological Responses: An Exploratory Study in a Physical Education Setting. *Children* **2022**, *9*, 907. https://doi.org/10.3390/children9060907

Academic Editor: Lise Hestbæk

Received: 6 May 2022
Accepted: 14 June 2022
Published: 17 June 2022

Publisher's Note: MDPI stays neutral with regard to jurisdictional claims in published maps and institutional affiliations.

Copyright: © 2022 by the authors. Licensee MDPI, Basel, Switzerland. This article is an open access article distributed under the terms and conditions of the Creative Commons Attribution (CC BY) license (https://creativecommons.org/licenses/by/4.0/).

Abstract: Verbal encouragement (VE) can be used by physical education (PE) practitioners for boosting motivation during exercise engagement. The purpose of this study was to investigate the effects of VE on psychophysiological aspects and physical performance in a PE context. Twenty secondary school male students (age: 17.68 ± 0.51 yrs; height: 175.7 ± 6.2 cm; body mass: 67.3 ± 5.1 kg, %fat: 11.9 ± 3.1%; PE experience: 10.9 ± 1.0 yrs) completed, in a randomized order, two test sessions that comprised a soccer dribbling circuit exercise (the Hoff circuit) either with VE (CVE) or without VE (CNVE), with one-week apart between the tests. Heart rate (HR) responses were recorded throughout the circuit exercise sessions. Additionally, the profile of mood-state (POMS) was assessed pre and post the circuit exercises. Furthermore, rating of perceived exertion (RPE), traveled distance, and physical activity enjoyment (PACES) were assessed after the testing sessions. Furthermore, the CVE trial resulted in higher covered distance, %HRmax, RPE, PACES score, (Cohen's coefficient d = 1.08, d = 1.86, d = 1.37, respectively; all, $p < 0.01$). The CNVE trial also showed lower vigor and higher total mood disturbance (TMD) (d = 0.67, d = 0.87, respectively, $p < 0.05$) and was associated with higher tension and fatigue, compared to the CVE trial (d = 0.77, d = 1.23, respectively, $p < 0.01$). The findings suggest that PE teachers may use verbal cues during soccer dribbling circuits for improving physical and psychophysiological responses within secondary school students.

Keywords: verbal encouragement; physical education; Hoff circuit; enjoyment; mood; performance

1. Introduction

When designing teaching situations in team ball sports such as soccer, most physical education (PE) teachers rely heavily on diverse dribbling circuits [1]. Due to the integrated work (training with ball), these training circuits solicit various physical and cognitive performances in practitioners (students/players) [2,3]. Previous works in the literature have reported that soccer-specific training circuits are a commonly used training method that can simultaneously improve the physical, physiological, and technical aspects of players [4–6]. The set-up of training circuits depends on several factors that determine its success, namely distance, duration, the space provided, and the encouragement of the PE teacher [4].

Continuous stimuli provided by PE teachers is crucial to improve student's physical and cognitive performances as well as their motivational beliefs [7]. It is believed that verbal encouragement (VE) from the practitioners can enhance intrinsic motivation [7–11], which clearly affects physical commitment and positive interaction with physical exertion and desire to do exercises [7,12]. This, in turn, leads to better technical, physical, and emotional responses [8,13,14].

Several studies about integrated training have reported the importance of the encouragement provided by the physical trainers on training intensity, expressed as perceived exertion and physiological responses [15]. In fact, the motivation resulting from integrated training sessions may be related to good mood state, positive physical pleasure, and vigorous exercise intensity [7]. For example, Selmi et al. [12] reported that encouragement cues result in optimal motivation and physical enjoyment, subsequently leading to enhancement of psychophysiological responses in participants during integrated exercises. Additionally, Selmi et al. [16] suggested that encouragement cues from practitioners ensures mood balance during integrated training in soccer players.

During PE sessions, VE provided by a course teacher is recognized as a form of external motivation that advances physical engagement [7]. However, limited information addresses soccer exercise training regarding the importance of VE on participant's performance. To our knowledge, no study has tackled the effects of VE on psycho-physiological and emotional responses by utilizing a soccer specific intervention during technical circuits in school-based PE sessions. Given the positive benefits of encouragement on athletes' motivation, as well as the potential influences of exercise intensity and positive affective responses, research to fill this gap in the literature is warranted.

Therefore, the purpose of this study was to examine the impact of the PE teacher's VE on physical performance and students' perception of exertion, physical enjoyment, and mood while completing a training circuit with a ball (Hoff circuit). Using soccer-specific training drills in physical education sessions is important and beneficial for students as the use of these types of drills is educationally more effective and motivating than conventional drills. This specific intervention provides physical and psychological benefits due to positive feelings and high exercise intensity [17].

2. Materials and Methods

2.1. Participants

The study was approved by the research ethics committee of the High Institute of Sports and Physical Education of Kef, Tunisia (approval no. 011/2020). The experiments were conducted in accordance with the Declaration of Helsinki. Twenty male students enrolled in one secondary school in Tunisia were involved in this study (age: 17.68 ± 0.51 years; PE experience: 10.9 ± 1. years; height: 175.7 ± 6.2 cm; body mass: 67.3 ± 5.1 kg, %fat: 11.9 ± 3.1%). The participants were from the same study class and practiced two physical education sessions per week (Tuesday and Friday). All participants had no reported injuries or illnesses before or during the study. A researcher of our study group informed all risks and benefits associated with participation to participants, and their parents voluntarily agreed to participate in the research and gave written informed written consent after a detailed explanation about the objectives, procedures, and risks involved in the study.

A priori power analysis was used to estimate the sample size (G*Power Version 3.1.9.4., Düsseldorf, Germany), based on the t test family (Means: difference between two dependent means). The analysis output showed that a sample size of 19 subjects would be sufficient to identify significant differences (effect size = 0.887, power $(1 - \beta) = 0.95$ with an actual power of 95.46 in this study.

2.2. Data Collection and Analysis

2.2.1. Anthropometric Measurements

The participants' height and body mass were measured using a standard protocol that the variability of measurement was within 0.2 kg and 5 mm. Height and body mass were measured to the nearest 0.1 cm and 0.1 kg with a digital scale (OHAUS, Florhman Park, NJ, USA), respectively. Four-sides skinfold thickness was used to determine the percentage of body fat (biceps, triceps, subscapular and suprascapular), using a calibrated Harpenden skinfold caliper (Holtain Instruments, Crosswell, Pembrokeshire, UK). A well-trained sports scientist conducted the anthropometric measurements in this study. The percentage of body fat (%body fat) was calculated via a validated method: %body fat = (495/body density) − 450; D = 1.1533 − (0.0643 × L) with D = body density (g/mL), and L = log of the total of the 4 skinfolds (mm) [18,19].

2.2.2. Vameval Test

The Vameval test was conducted on a 200 m running track. The testing field was set with ten cones placed every 20 m at specific sites on the pitch following a preprogrammed auditory signal (i.e., a beep). The initial running speed was determined at 8 km.h^{-1} and subsequently increased by 0.5 km.h^{-1} every minute until exhaustion. The students controlled their running pattern between the cones. The test was terminated when a participant failed to maintain the running speed guided by the beep consecutive shuttles or felt they could not continue the run [12]. The herat rate (HR) was recorded via a Polar Team Sport System (Polar-Electro OY, Kempele, Finland). The highest HR average value over 5 s during the Vameval test was recorded as Vameval-HRmax. The reliability of the Vameval protocol to detect the maximal HR has been shown previously (a Cronbach's α value of 0.83) [20].

2.2.3. Heart Rate

During the Hoff circuit exercises, the HR was recorded via portable HR sensors (Polar Team Sport System, Polar-Electro OY, Kempele, Finland). The HR detection was recorded every 5 s. The participants frequently checked the position of the HR sensor throughout the exercise. Subsequently, the HR data were expressed as a percentage of Vameval-HRmax (%HRmax) and mean HR (HRmean). The %HRmax was calculated by the formula: %HRmax = (HRmean/Vameval-HRmax) × 100 [21].

2.2.4. Rating of Perceived Exertion (RPE)

The internal load was measured directly after the Hoff circuit exercises. The Borg CR-10 RPE was used to assess the level perceived effort during the Hoff circuit exercise in this study [22]. The validity and reliability of this method were reported ed in previous studies [23]. The participants were familiarized with the RPE-scale in the first week of the study to maximize the accuracy of the answers.

2.2.5. Profile of Mood States (POMS)

The Profile of Mood State (POMS) was used to evaluate mood state before and after the Hoff circuit exercises [24]. After a period of familiarization, the survey was administered to all participants 15 min before and 5 min after each training circuit to assess mood state [16]. The questionnaire consists of 37 items, rated on a 5-point Likert scale ranging (from 0, not at all to 4, extremely). The POMS assess six states: tension-anxiety, depression-dejection, anger-hostility, vigor-activity, fatigue-inertia, and confusion-bewilderment. The POMS subscales can be combined into a total mood score (TMD) score by adding the scores for the five negative states and subtracting the positive mood (vigor) score. Adding a constant of 100 to avoid negative numbers, [TMD = (Anger + Confusion + Depression + Fatigue + Tension) − Vigor + 100]. The Cronbach's α of the POMS test ranged from 0.84 to 0.92 in the present study. The participants completed the POMS on papers outside where they completed the Hoff circuit exercise.

2.2.6. Physical Enjoyment

The 8-item Physical Activity Enjoyment Scale (PACES) was completed for the assessment positive feelings associated with exercise training in this study [25]. The participants were asked to rate "how you feel at the moment regarding the exercise training you experienced" via a 7-points scale (ranging from 1, it's very enjoyable to 7, it's not fun at all). The final score was calculated by a sum of the 8 items scores. The score ranges from 8 to 56 points. The large number of PACES scores indicates high level of physical enjoyment [25]. The Cronbach's α value of the PACES test was 0.89 in the present study.

2.3. Procedure

Within the first week, the anthropometric characteristics were measured, and all participants performed the Vmaeval test [12] to obtain individual Hrmax (Figure 1). In the experimental days, two Hoff circuit sessions were performed separated by a one-week interval during regular PE sessions. The order of sessions was randomized and counterbalanced such that half of the participants completed the Hoff circuit exercise with VE (CVE) and the other half completed the Hoff circuit exercise without VE (CNVE) first and vise versa. Each subject performed one CVE trial and one CNVE trial.

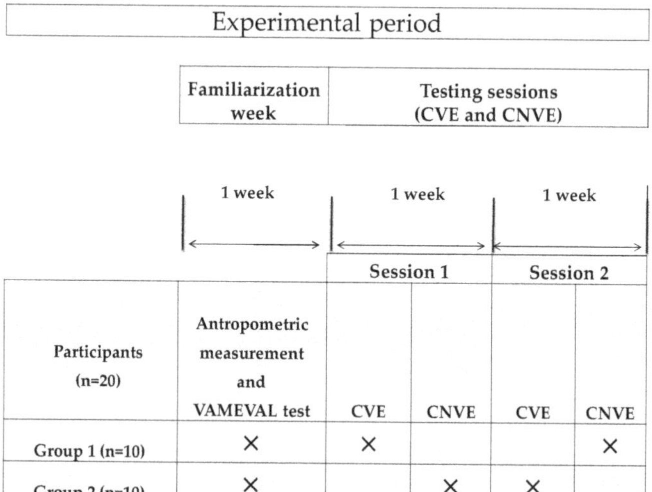

Experimental period				
Familiarization week	**Testing sessions (CVE and CNVE)**			
1 week	1 week		1 week	
	Session 1		Session 2	
Participants (n=20)	Antropometric measurement and VAMEVAL test	CVE	CNVE	CVE	CNVE
Group 1 (n=10)	✗	✗			✗
Group 2 (n=10)	✗		✗	✗	

Figure 1. Representative diagram of the experimental protocol. n: number of participants; CVE: circuit exercise with verbal encouragement; CNVE: circuit exercise without verbal encouragement.

The experimental sessions (Figure 2) were carried out on an artificial grass pitch in the morning physical education sessions (between 9:00 and 10:30 a.m.). The duration of each circuit exercise was 10 min. The HR was continuously monitored during each trial (CVE and CNVE). The distance traveled of the Hoff circuit exercise, RPE, and PACES were recorded immediately after the tests. Furthermore, the POMS was measured before and after each session. The participants reported the scales independently to avoid hearing the responses of their colleagues.

A standardized 12-min warm-up exercises was given to all participants (consisting of jogging, muscle coordination exercises, dynamic stretching, passing drills with ball, and ended with three 10 m sprints). No static passive stretching exercise was performed during the warm-up activities [26]. The participants only performed dynamic stretches before the experimental sessions [16]. Three minutes passive recovery was given after prior to the Hoff circuit exercise. All participants refrained from strenuous physical effort at least 48 h prior to experiments. All participants familiarized with the RPE, the PACES, the POMS, and the Hoff circuit during the familiarization week. A group of researchers and a PE teacher who collected all data.

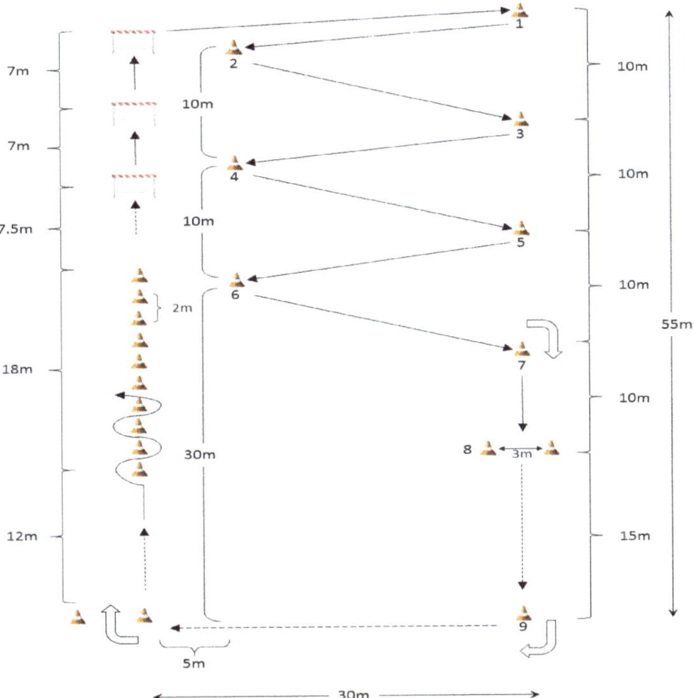

Figure 2. Illustration of the Hoff circuit exercise. The participants dribble the ball through the circuit. The circuit length is set of 51.5 m on left-hand side and 55 m on the right-hand side and the width is set of 35 m. The participants were required to perform backward dribbling between the cone 7 to the gate 8. Three hurdles (30–35 cm height) and 22 cones (two for the starting line and two cones for the backward run gate) were set on the field. Total distance of per lap is 290 m. The distance between hurdle 3 and cone 1 is 30.5 m. The distance between cones (1–2, 2–3, 3–4, 4–5, 5–6 and 6–7) is 25.5 m each [27].

2.4. Hoff Circuit Exercise

The Hoff circuit exercise was performed outdoor. The size of the pitch and the duration of the circuit training have been strictly implemented in other studies [19]. The track distance is 290 m. Participants move a soccer ball across the circuit by dribbling. The goal of the exercise was to do the maximum distance possible for 10 min. Each participant was informed the spent time at 5 min and at 9 min [19]. Five participants were grouped for the tests at the same time. Every minute, the testing signal is given to one participant. Thus, the researchers accounting the test had 4 min for starting the five participants and then switched to the halfway test signal that occurred in the successive minute for the first running participant. When the researchers indicated the halfway test signal for the fifth running participant (minute 9), at the same time began the last minute signal for the first running participant. The participants wore colored bibs that were always assigned in the same order to the running participants numbered 1 to 5 [27].

The participants were received VE from the PE teacher in the CVE trial, while they performed the test without VE in the CNVE trial. The PE teacher moved around the testing field while encouraging the participants (instruction such as Go Go Go, Again Again, Move, Again, Faster, More active, A bit of willpower, More effort, Courage and so on) [7,15]. The encouragement was spontaneous based on the behavior of the participants, according to his effort and movements. During the CNVE trial, the PE teacher stood next to the circuit and controlled the participants but did not provide VE cues.

Statistical Analyses

All data are presented as mean $\pm$ standard deviation (SD). The normality of the data was confirmed using the Kolmogorov-Smirnov test. Paired *T*-test was used to compare the Hrmax, Hrmean, physical variables (distance traveled), physical enjoyment (PACES score), RPE, and Hrmax. The Cohen's coefficient (*d*) was used to determine the magnitude of differences between CVE and CNVE trials [28]. The sales of magnitude were considered trivial: $d \leq 0.20$; small: $0.20 < d \leq 0.50$; medium: $0.50 < d \leq 0.80$; and large: $d > 0.80$ [29].

Regarding the POMS scores, a two-way Analysis of Variance (ANOVA) was used to study the effect of the "exercise method" (CVE and CNVE), "effort" (pre- and post-exercise) and their interaction (exercise method $\times$ effort) on the scores of the six subscales (depression, anger, confusion, fatigue, anxiety, and vigor) and the TMD. Partial Eta-Squared (η^2) was used from two-way ANOVA outputs. When a significant interaction effect was found, the analysis was completed with a post-hoc Bonferroni test. The level of statistical significance was set at $p < 0.05$.

3. Results

3.1. Physical and Physiological Performance

Results presented in Table 1 indicate significant changes in the distance traveled (D), Hrmax, and RPE variables between the trials (all, $p < 0.001$).

Table 1. Comparison of the distance traveled, Hrmax, and RPE variables between circuit exercise with and without verbal encouragement.

Variables	CVE	CNVE	*d*	Rating
Distance traveled (m)	1688.1 $\pm$ 206.7	1483.7 $\pm$ 180.7 ***	1.08	Large
Hrmax (%)	88.31 $\pm$ 2.45	84.06 $\pm$ 2.27 ***	1.86	Large
RPE (AU)	7.5 $\pm$ 0.82	6.3 $\pm$ 0.97 ***	1.37	Large

Hrmax: maximal heart rate; RPE: rating of perceived exertion; CI 95%, d: Cohen's coefficient, CVE: circuit exercise with verbal encouragement, CNVE: circuit exercise without verbal encouragement, *** $p < 0.001$.

3.2. Physical Enjoyment

The PACES score is significantly higher ($p < 0.001$, $d = 1.36$, Large) in the CVE trial (39.95 $\pm$ 3.17), compared to that of CNVE trial (46.65 $\pm$ 6.41).

3.3. Mood State

No significance of main effect (condition and time) and interaction on depression, anger and confusion scores were observed. These mood parameters were not significantly affected by either "effort" or "training method". However, there was a significant main effect of training method on TMD, fatigue, and vigor and a significant main effect of effort on fatigue (Table 2). Additionally, a significant interaction effect was found for tension, fatigue and TMD (Table 1). In Figure 3, Bonferroni's post-hoc comparisons shows that only CVE trial the TMD scores decreased significantly (from 102.65 $\pm$ 12.43 to 97.45 $\pm$ 11.16) and those for vigor increased significantly (from 16.95 ± 4.03 to 13.45 ± 5.92) while anxiety (from 1.65 ± 2.47 to 1.45 ± 2.33) and fatigue scores increased significantly for the CNVE trial (from 4.9 ± 2.51 to 8.6 ± 4.07).

Table 2. Analysis of Variance results with 2 × 2 repeated measures (exercise method: circuit with verbal encouragement and circuit without verbal encouragement) × effort: pre- and post-exercise).

Variables	Effort (Main Effect)		Exercise Method (Main Effect)		Interaction	
	$F_{(1,19)}$	η^2	$F_{(1,19)}$	η^2	$F_{(1,19)}$	η^2
Depression	0.012	0.001	0.40	0.02	1.22	0.06
Anger	0.18	0.009	2.20	0.11	1.23	0.06
Anxiety	0.08	0.005	2.66	0.12	8.19 **	0.30
Confusion	2.48	0.11	0.99	0.05	3.11	0.14
Fatigue	9.87 ***	0.34	8.94 ***	0.32	9.94 ***	0.34
Vigor	0.66	0.03	8.55 ***	0.31	0.21	0.01
TMD	1.57	0.07	6.02 *	0.24	4.27 *	0.18

TMD: total disturbance of mood. * $p < 0.05$, ** $p < 0.01$, *** $p < 0.001$.

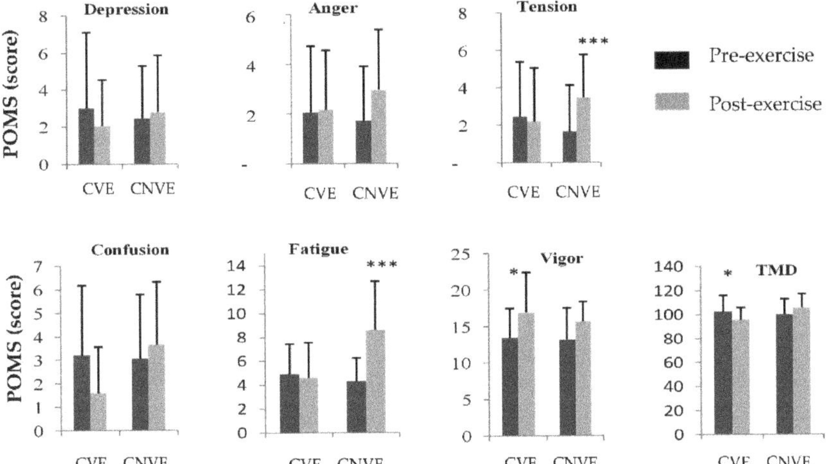

Figure 3. The profile of mood states (POMS) scores during the Hoff circuit exercise with and without verbal encouragement during pre-exercise and post-exercise measures. CVE, circuit exercise with verbal encouragement; CNVE, circuit exercise without verbal encouragement; TMD, total mood disturbance. Error bars indicate within-subject standard deviation. * a significant difference between pre-exercise and post-exercise values. * $p < 0.05$, *** $p < 0.001$.

4. Discussion

This study investigated the effect of the PE teacher's VE on the HR, RPE, physical performance, physical enjoyment, and mood state of school male students during a soccer dribbling circuit exercise. The results of the present study were as follows: (1) the CVE condition increased distance traveled, %Hrmax and internal intensity to a greater extent than that of CNVE condition; (2) the physical enjoyment was greater after the CVE; and (3) The CVE condition resulted in positive mood state, compared to that of CNVE condition.

In term of internal factors, the result of this study indicated that the RPE was significantly greater during the CVE. This indicates that they performed the Hoff exercise with high intensity which results in a high solicitation of the cardiorespiratory demands [5,23]. During the CVE trial, the RPE is very hard (>7), compared to that of the CNVE trial. Therefore, VE from the PE teacher motivates students to exert great effort during the Hoff circuit exercise. This result suggests that the high level of perceived effort is associated with the effort produced by students during physical activity [15,16].

Our results have been confirmed by Rampinini et al. [15] showed that VE caused higher values in terms of RPE compared to the lack of encouragement during specific soccer training (4 vs. 4) by modifying the dimensions of the field (small, medium, and large). The results gave the following RPE values: small filed: 7.6 ± 0.5, 6.3 ± 0.5; middle filed: 7.9 ± 0.5, 6.6 ± 0.6; large field: 8.1 ± 0.5, 6.8 ± 0.5, respectively. For example, Brandes and Elvers [30] reported the effectiveness of VE in promoting physical effort, training exercise adherence and internal intensity. This result is in agreement with Sahli et al. [7], studied the effects of VE on psychophysiological aspects during small-sided games (SSG). They reported that HR and RPE values were significantly higher in SSG with VE. Additionally, Selmi et al., [12] compared the effects of VE during specific training soccer on physiological variables and RPE in youth soccer players. The authors reported that RPE score and %Hrmax were higher in exercise training with VE than in training exercise without VE.

Regarding physical performance, this study indicated that VE from the PE teacher has a positive impact on the distance traveled on the Hoff circuit exercise. The incentive variable leads to an improvement in running distance. These findings suggest that the PE teacher's VE can motivate the male school students to put in a great level of commitment, physical engagement, and run a further distance. Overall, the results of the present study suggested that the CVE induced higher physical contributions to energy demands and exercise intensity, thus confirming the significance of encouragement in improving the physical condition of participants.

Affective responses to training activity are an essential characteristic of performance [5,31,32]. In addition to focus and engagement, mood and enjoyment can predict training adherence [7]. In the present study, we observed that the physical enjoyment score (PACES) measuring the CVE trial from the teacher was significantly higher than that measured after the CNVE trial, confirming with a recent study of Selmi et al. [16] who reported that players synchronized with VE had higher PACES scores during the SSG than that of the high intensity interval training (HIIT). Selmi et al. [12] claimed that reduced game exercises with VE led to greater physical enjoyment than games without verbal encouragement. Additionally, Sahli et al. [7] reported that PACES scores measured after PE sessions with the teacher's encouragement were higher than that without encouragement. This result may explain that motivating factors may clarify a high level of PACES [7,33,34]. In this context, other studies have indicated that the students who are most motivated in PE sessions are those who have a higher level of enjoyment [7,12,35]. These findings also suggest that the physical enjoyment induced by the training methods may vary depending on the modality of exercise, the behavior of the PE teacher, the results, and the desire of the students [36]. Interestingly, our result was relatively supported by previous studies [5,12,15,34], indicating that the physical activity sessions with VE is also more enjoyable for students. Furthermore, these indications suggested that the encouragement factor was easily motivating and enjoyable for the students and could therefore be a more effective method for improving the psychological responses during exercise engagement.

Regarding the comparison of the mood's response (POMS) between CVE and CNVE, the present study recorded that VE resulted in an improvement in the mood state. Our hypothesis was supported by the result showing VE from the PE teacher contributed to positive improvement in the mood state of the students.

The POMS is a common psychological measure for athletes during athletic training [7,12,18,32,37]. During motivational training exercises, participants generally reported positive changes in the mood state [7,16]. As shown in our study, the score of the POMS variables was positively altered with the presence of VE from the PE teacher. The anxiety score (11.36%) and fatigue (6.52%) were significantly reduced, while vigor was increased (26.02%). The significant increase in vigor and decrease in negative variables (anxiety and fatigue) of POMS in students leaded to a reduction in the TMD score. These findings are consistent with several studies that have investigated the effect of teacher/coach's VE on psychological aspects during integrated training. Indeed, we have shown that VE ensures the stability of the mood state during intense exercises and reduced games (3 vs. 3) [16]. Thus, Sahli et al. [7] reported that training of SSG (4 vs. 4) with VE from the PE teacher resulted in a significant improvement in vigor, and a significant decrease in anxiety and TMD scores. The authors indicated that VE is a key determinant of positive mood in students. Along the same lines, Selmi et al. [16] claimed that lack of motivation in physical exercise leads to mood disturbance with higher fatigue, tension, and decreased energy. Additionally, Andradeet al. [38] stated that mood disorders decrease with the presence of motivation of the PE teacher.

Based on our findings, the absence of VE during a PE session causes an inability to concentrate and a lack of physical engagement which results in changes in the mood, the behavior, and the anxiety [7]. Lane and Terry [39] showed that vigor played a crucial role in positive improvement of the mood, while in our study we recorded a more remarkable increase in this condition. This outcome suggests that an increase in the vigor subscale score is likely to lead to a positive feeling during PE sessions. Other research has shown that the vigor that follows motivational training exercises causes improvements in mood, increased energy, and physical engagement [40,41]. It has been suggested to explain this that reduced vigor and increased TMD during physical activity sessions are associated with lack of motivation and physical pleasure which are associated with unpleasant psychological responses [7,16,31]. These results suggest that lack of motivation during physical activity causes negative feelings in students unlike motivational exercises [42]. They could be related to recent studies regarding the affirmation of the central role of VE in the intrinsic motivation of students [8,38,43].

During vigorous exercise such as the Hoff circuit exercise, motivational factors may explain the positive change in some of the different POMS scores. However, studies have reported that engaging in motivational sport training improves participants' mood [43–45]. The present study indicated that the behavior of a PE teacher can influence student behavior, intensity of exercise, affecting psychophysiological and emotional responses, indicating that the performance is associated with a positive teaching style.

Several limitations should be considered when interpreting the current results. First, the study sample was small due to the difficulty of recruiting large numbers of homogeneous participants. Furthermore, the tests used only one circuit format and used only one average age of students. Finally, it would be interesting to relate these aspects with technical aspects such as time motion variables (zones of running intensity, etc.), since these aspects are important variables of performance.

This investigation has practical implications. To our knowledge, this investigation is the first to examine physical performance, perceived exertion, physical enjoyment, and mood state during a soccer-specific training circuit among students. The VE from a PE teacher can be considered as a necessary variable during PE sessions since it induces important physical aspects, and positive sensations. This is the reason why PE teachers should verbally encourage their students during specific physical activity sessions to increase the intensity of play, create positive emotional states, and improve the student's performance.

5. Conclusions

The present study indicates that the RPE, the physical enjoyment, the distance traveled, Hrmax and the positive mood are higher during the Hoff circuit with the PE teacher's VE in comparison with Hoff's circuit without VE. The VE is suggested as an effective intervention to improve exercise intensity, physical performance, mood state, and physical enjoyment when a soccer-specific exercise is used during PE sessions. The findings demonstrated that PE teachers should utilize VE to improve the motivation, physical engagement, and commitment in school students. Future investigations examining VE should be conducted during other exercise training or sports game (individual or collective) of the PE session.

Other aspects (i.e., physical, physiological, technical, and emotional aspects) and students with different fitness levels, gender, and age should be further explored to extend the applicability of the results.

Author Contributions: Conceptualization, B.A., O.S. and M.A.S.; methodology, B.A., O.S., M.A.S., H.S. and M.Z.; validation, B.A. and O.S.; formal analysis, B.A., O.S., M.A.S., H.S., A.M. and M.Z.; investigation, B.A., H.S. and M.Z.; data curation, H.S., A.M. and M.Z.; writing—original draft preparation, B.A., O.S., M.A.S., H.S., G.R., Z.J.C.-M., J.C.P., A.M., M.Z. and Y.-S.C.; writing—review and editing, B.A., O.S., M.A.S., H.S., G.R., Z.J.C.-M., J.C.P., A.M., M.Z. and Y.-S.C.; visualization, B.A. and O.S.; supervision, O.S.; project administration, B.A., O.S., M.A.S., H.S., A.M. and M.Z. All authors have read and agreed to the published version of the manuscript.

Funding: This research received no external funding.

Institutional Review Board Statement: The study was conducted in accordance with the Declaration of Helsinki, and the protocol was approved by the research ethics committee of the High Institute of Sports and Physical Education of Kef, Tunisia (approval no. 011/2020 and date of approval 15 March 2020).

Informed Consent Statement: Informed consent was obtained from all participants involved in the study.

Data Availability Statement: The data presented in this study are available on request from the corresponding author.

Acknowledgments: The authors thank all students and physical education teachers who participated in this study.

Conflicts of Interest: The authors declare no conflict of interest.

References

1. Setiawan, R.R. A circuit-based football heading exercise model in football school ages 14–17 years. *ACTIVE J. Phys. Educ. Sport Health Rec.* **2019**, *8*, 129–133.
2. Hill-Haas, S.V.; Dawson, B.; Impellizzeri, F.M.; Coutts, A.J. Physiology of small-sided games training in football. *Sports Med.* **2011**, *41*, 199–220. [CrossRef] [PubMed]
3. Zein, M.I.; Saryono, S.; Laily, I.; Garcia-Jimenez, J.V. The effect of short period high-intensity circuit training-modified FIFA 11+ program on physical fitness among young football players. *J. Sports Med. Phys. Fit.* **2020**, *60*, 11–16. [CrossRef]
4. Zagatto, A.M.; Papoti, M.; Da Silva, A.S.R.; Barbieri, R.A.; Campos, E.Z.; Ferreira, E.C.; Chamari, K. The Hoff circuit test is more specific than an incremental treadmill test to assess endurance with the ball in youth soccer players. *Biol. Sport* **2016**, *3*, 263–268. [CrossRef]
5. Selmi, O.; Ouergui, I.; Levitt, D.E.; Nikolaidis, P.T.; Knechtle, B.; Bouassida, A. Small-sided games are more enjoyable than high-intensity interval training of similar exercise intensity in soccer. *J. Sports Med.* **2020**, *11*, 77. [CrossRef]
6. Nugroho, S.; Nasrulloh, A.; Karyono, T.H.; Dwihandaka, R.; Pratama, K.W. Effect of intensity and interval levels of trapping circuit training on the physical condition of badminton players. *J. Phys. Educ. Sport* **2021**, *21*, 1981–1987.
7. Sahli, H.; Selmi, O.; Zghibi, M.; Hill, L.; Rosemann, T.; Knechtle, B.; Clemente, F.M. Effect of the verbal encouragement on psychophysiological and affective responses during small-sided games. *Inter. J. Environ. Res. Public. Health* **2020**, *17*, 8884. [CrossRef]
8. Pacholek, M.; Zemková, E. Effects of verbal encouragement and performance feedback on physical fitness in young adults. *Sustainability* **2022**, *14*, 1753. [CrossRef]
9. Andreacci, J.L.; Lemura, L.M.; Cohen, S.L.; Urbansky, E.A.; Chelland, S.A.; Duvillard, S.P.V. The effects of frequency of encouragement on performance during maximal exercise testing. *J. Sports Sci.* **2002**, *20*, 345–352. [CrossRef]
10. Engel FSahli, A.; Faude, O.; Kölling, S.; Kellmann, M.; Donath, L. Verbal encouragement and between-day reliability during high-intensity functional strength and endurance performance testing. *Front. Physiol.* **2019**, *10*, 460. [CrossRef]
11. Kilit, B.; Arslan, E. Effects of high-intensity interval training vs. on-court tennis training in young tennis players. *J. Strength Cond. Res.* **2019**, *33*, 188–196. [CrossRef]
12. Selmi, O.; Khalifa, W.B.; Ouerghi, N.; Amara, F.; Zouaoui, M. Effect of verbal coach encouragement on small sided games intensity and perceived enjoyment in youth soccer players. *J. Athl. Enhanc.* **2017**, *3*, 16–17. [CrossRef]
13. Neto, J.M.D.; Silva, F.B.; De Oliveira, A.L.B.; Couto, N.L.; Dantas, E.H.M.; de Luca Nascimento, M.A. Effects of verbal encouragement on performance of the multistage 20 m shuttle run. *Acta Sci. Health Sci.* **2015**, *37*, 25–30. [CrossRef]
14. Sahli, H.; Haddad, M.; Jebabli, N.; Sahli, F.; Ouergui, I.; Ouerghi, N.; Zghibi, M. The Effects of verbal encouragement and compliments on performance and psychophysiological responses during the repeated change of direction sprint test. *Front. Psychol.* **2022**, *12*, 698673. [CrossRef]

15. Rampinini, E.; Impellizzeri, F.M.; Castagna, C.; Abt, G.; Chamari, K.; Sassi, A.; Marcora, S.M. Factors influencing physiological responses to small-sided soccer games. *J. Sports Sci.* **2007**, *25*, 659–666. [CrossRef]

16. Selmi, O.; Haddad, M.; Majed, L.; Khalifa, B.; Hamza, M.; Chamari, K. Soccer training: High-intensity interval training is mood disturbing while small, sided games ensure mood balance. *J. Sports Med. Phys. Fit.* **2018**, *58*, 1163–1170. [CrossRef]

17. Hoff, J.; Wisløff, U.; Engen, L.C.; Kemi, O.J.; Helgerud, J. Soccer specific aerobic endurance training. *Br. J. Sports Med.* **2002**, *36*, 218–221. [CrossRef]

18. Durnin, J.V.; Womersley, J.V.G.A. Body fat assessed from total body density and its estimation from skinfold thickness: Measurements on 481 men and women aged from 16 to 72 years. *Br. J. Nut.* **1974**, *32*, 77–97. [CrossRef]

19. Siri, W.E. Body composition from fluid spaces and density: Analysis of methods. *Nutrition* **1993**, *9*, 480–492.

20. Buchheit, M.; Mendez-Villanueva, A. Reliability and stability of anthropometric and performance measures in highly-trained young soccer players: Effect of age and maturation. *J. Sports Sci.* **2013**, *31*, 1332–1343. [CrossRef] [PubMed]

21. Dellal, A.; Chamari, K.; Pintus, A.; Girard, O.; Cotte, T.; Keller, D. Heart rate responses during small-sided games and short intermittent running training in elite soccer players: A comparative study. *J. Strength Cond. Res.* **2008**, *22*, 1449–1457. [CrossRef] [PubMed]

22. Foster, C.; Florhaug, J.A.; Franklin, J.; Gottschall, L.; Hrovatin, L.A.; Parker, S.; Dodge, C. A new approach to monitoring exercise training. *J. Strength Cond. Res.* **2001**, *15*, 109–115. [PubMed]

23. Coyne, J.O.; Gregory Haff, G.; Coutts, A.J.; Newton, R.U.; Nimphius, S. The current state of subjective training load monitoring—A practical perspective and call to action. *Sports Med. Open* **2018**, *4*, 1–10. [CrossRef] [PubMed]

24. Baker, F.; Denniston, M.; Zabora, J.; Polland, A.; Dudley, W.N. A POMS short form for cancer patients: Psychometric and structural evaluation. *Psychooncology* **2002**, *11*, 273–281. [CrossRef]

25. Mullen, S.P.; Olson, E.A.; Phillips, S.M.; Szabo, A.N.; Wójcicki, T.R.; Mailey, E.L.; McAuley, E. Measuring enjoyment of physical activity in older adults: Invariance of the physical activity enjoyment scale (paces) across groups and time. *Int. J. Behav. Nutr. Phys. Act.* **2011**, *8*, 1–9. [CrossRef]

26. Haddad, M.; Dridi, A.; Chtara, M.; Chaouachi, A.; Wong, D.P.; Behm, D.; Chamari, K. Static stretching can impair explosive performance for at least 24 hours. *J. Strength Cond. Res.* **2014**, *28*, 140–146. [CrossRef]

27. Chamari, K.; Hachana, Y.; Kaouech, F.; Jeddi, R.; Moussa-Chamari, I.; Wisløff, U. Endurance training and testing with the ball in young elite soccer players. *Br. J. Sports Med.* **2005**, *39*, 24–28. [CrossRef]

28. Cohen, J. Statistical power analysis. *Curr. Dir. Psychol. Sci.* **1992**, *1*, 98–101. [CrossRef]

29. Hopkins, W.; Marshall, S.; Batterham, A.; Hanin, J. Progressive statistics for studies in sports medicine and exercise science. *Med. Sci. Sports Exerc.* **2009**, *41*, 3. [CrossRef]

30. Brandes, M.; Elvers, S. Elite youth soccer players' physiological responses, time-motion characteristics, and game performance in 4 vs. 4 small-sided games: The influence of coach feedback. *J. Strength Cond. Res.* **2017**, *31*, 2652–2658. [CrossRef]

31. Los Arcos, A.; Vázquez, J.S.; Martín, J.; Lerga, J.; Sánchez, F.; Villagra, F.; Zulueta, J.J. Effects of small-sided games vs. interval training in aerobic fitness and physical enjoyment in young elite soccer players. *PLoS ONE* **2015**, *10*, e0137224. [CrossRef]

32. Berger, B.G.; Motl, R.W. Exercise and mood: A selective review and synthesis of research employing the profile of mood states. *J. Appl. Sport Psychol.* **2000**, *12*, 69–92. [CrossRef]

33. Garcia-Mas, A.; Palou, P.; Gili, M.; Ponseti, X.; Borras, P.A.; Vidal, J.; Sousa, C. Commitment, enjoyment and motivation in young soccer competitive players. *Span. J. Psychol.* **2010**, *13*, 609–616. [CrossRef] [PubMed]

34. Toh, S.H.; Guelfi, K.J.; Wong, P.; Fournier, P.A. Energy expenditure and enjoyment of small-sided soccer games in overweight boys. *Hum. Mov. Sci.* **2011**, *30*, 636–647. [CrossRef]

35. Alvarez, M.S.; Balaguer, I.; Castillo, I.; Duda, J.L. Coach autonomy support and quality of sport engagement in young soccer players. *Span. J. Psychol.* **2009**, *12*, 138–148. [CrossRef]

36. Bartlett, J.D.; Close, G.L.; MacLaren, D.P.; Gregson, W.; Drust, B.; Morton, J.P. High-intensity interval running is perceived to be more enjoyable than moderate-intensity continuous exercise: Implications for exercise adherence. *J. Sports Sci.* **2011**, *29*, 547–553. [CrossRef]

37. Miranda, R.E.E.P.C.; Antunes, H.K.M.; Pauli, J.R.; Puggina, E.F.; Da Silva, A.S.R. Effects of 10-week soccer training program on anthropometric, psychological, technical skills and specific performance parameters in youth soccer players. *Sci. Sports* **2013**, *28*, 81–87. [CrossRef]

38. Andrade, A.; Cruz, W.M.D.; Correia, C.K.; Santos, A.L.G.; Bevilacqua, G.G. Effect of practice exergames on the mood states and self-esteem of elementary school boys and girls during physical education classes: A cluster-randomized controlled natural experiment. *PLoS ONE* **2020**, *15*, e0232392. [CrossRef]

39. Lane, A.M.; Terry, P.C. The nature of mood: Development of a conceptual model with a focus on depression. *J. Appl. Sport Psychol.* **2000**, *12*, 16–33. [CrossRef]

40. Duncan, G.J.; Engel, M.; Claessens, A.; Dowsett, C.J. Replication and robustness in developmental research. *Devel. Psychol.* **2014**, *50*, 2417. [CrossRef]

41. Kageta, T.; Tsuchiya, Y.; Morishima, T.; Hasegawa, Y.; Sasaki, H.; Goto, K. Influences of increased training volume on exercise performance, physiological and psychological parameters. *J. Sports Med. Phys. Fit.* **2015**, *56*, 913–921.

42. Lewis, A.D.; Huebner, E.S.; Malone, P.S.; Valois, R.F. Life satisfaction and student engagement in adolescents. *J. Youth Adolesc.* **2011**, *40*, 249–262. [CrossRef]

43. Andrade, A.; Correia, C.K.; Cruz, W.M.D.; Bevilacqua, G.G. Acute effect of exergames on children's mood states during physical education classes. *Games Health J.* **2019**, *8*, 250–256. [CrossRef] [PubMed]
44. Cecchini, J.; González, C.; Carmona, Á.; Arruza, J.; Escartí, A.; Balagué, G. The influence of the physical education teacher on intrinsic motivation, self-confidence, anxiety, and pre-and post-competition mood states. *Eur. J. Sport Sci.* **2001**, *1*, 1–11. [CrossRef]
45. Legey, S.; Aquino, F.; Lamego, M.K.; Paes, F.; Nardi, A.E.; Neto, G.M.; Machado, S. Relationship among physical activity level, mood and anxiety states and quality of life in physical education students. *Clin. Pract. Epidemiol. Ment. Health* **2017**, *13*, 82. [CrossRef]

Article

Mediterranean Diet Adherence, Body Mass Index and Emotional Intelligence in Primary Education Students—An Explanatory Model as a Function of Weekly Physical Activity

Eduardo Melguizo-Ibáñez [1], Gabriel González-Valero [2], Georgian Badicu [3,*], Ana Filipa-Silva [4,5,6], Filipe Manuel Clemente [4,5,7], Hugo Sarmento [8], Félix Zurita-Ortega [1] and José Luis Ubago-Jiménez [1]

[1] Faculty of Education Sciences, Department of Didactics of Musical, Plastic and Corporal Expression, University of Granada, 18071 Granada, Spain; edumeliba@correo.ugr.es (E.M.-I.); felixzo@ugr.es (F.Z.-O.); jlubago@ugr.es (J.L.U.-J.)
[2] Faculty of Education and Sport Sciences (Melilla), Department of Didactics of Musical, Plastic and Corporal Expression, University of Granada, 52071 Melilla, Spain; ggvalero@ugr.es
[3] Department of Physical Education and Special Motricity, Faculty of Physical Education and Mountain Sports, Transilvania University of Braşov, 500068 Braşov, Romania
[4] Escola Superior Desporto e Lazer, Instituto Politécnico de Viana do Castelo, Rua Escola Industrial e Comercial de Nun'Álvares, 4900-347 Viana do Castelo, Portugal; anafilsilva@gmail.com (A.F.-S.); filipe.clemente5@gmail.com (F.M.C.)
[5] Research Center in Sports Performance, Recreation, Innovation and Technology (SPRINT), 4960-320 Melgaço, Portugal
[6] The Research Centre in Sports Sciences, Health Sciences and Human Development (CIDESD), 5001-801 Vila Real, Portugal
[7] Instituto de Telecomunicacoes, Delegacao da Covilha, 1049-001 Lisboa, Portugal
[8] University of Coimbra, Research Unit for Sport and Physical Activity (CIDAF), Faculty of Sport Sciences and Physical Education, 3040-248 Coimbra, Portugal; hg.sarmento@gmail.com
* Correspondence: georgian.badicu@unitbv.ro

Citation: Melguizo-Ibáñez, E.; González-Valero, G.; Badicu, G.; Filipa-Silva, A.; Clemente, F.M.; Sarmento, H.; Zurita-Ortega, F.; Ubago-Jiménez, J.L. Mediterranean Diet Adherence, Body Mass Index and Emotional Intelligence in Primary Education Students - An Explanatory Model as a Function of Weekly Physical Activity. *Children* **2022**, *9*, 872. https://doi.org/10.3390/children9060872

Academic Editor: Odysseas Androutsos

Received: 26 April 2022
Accepted: 10 June 2022
Published: 11 June 2022

Publisher's Note: MDPI stays neutral with regard to jurisdictional claims in published maps and institutional affiliations.

Copyright: © 2022 by the authors. Licensee MDPI, Basel, Switzerland. This article is an open access article distributed under the terms and conditions of the Creative Commons Attribution (CC BY) license (https://creativecommons.org/licenses/by/4.0/).

Abstract: Adolescence is a key developmental period from a health, physical and psychological perspective. In view of this, the present research aimed to establish the relationship between emotional intelligence, Mediterranean diet adherence, BMI and age. In order to address this aim, (a) an explanatory model is developed of emotional intelligence and its relationship with Mediterranean diet adherence, BMI and age, and (b) the proposed structural model is examined via multi-group analysis as a function of whether students engage in more than three hours of physical activity a week. To this end, a quantitative, non-experimental (ex post facto), comparative and cross-sectional study was carried out with a sample of 567 students (11.10 ± 1.24). The instruments used were an ad hoc questionnaire, the KIDMED questionnaire and the TMMS-24. Outcomes reveal that participants who engage in more than three hours of physical activity a week score more highly for emotional intelligence than those who do not meet this criterion. Furthermore, it was also observed that, whilst the majority of the sample was physically active, improvement was required with regards to Mediterranean diet adherence.

Keywords: Mediterranean diet; emotional intelligence; physical activity; body mass index

1. Introduction

In a physical and psychological sense, adolescence marks a sensitive growth phase [1], which starts during puberty and finishes with bio-psychosocial maturity [2]. It typically takes place between 10 and 19 years of age [3]. With regards to psychological aspects, adolescents are yet to fully configure their personality, leading them to be easily influenced [3]. Thus, it is important to encourage them to adopt positive habits with regards to physical activity and health [4]. Whilst the physical and psychological benefits of an active and healthy lifestyle are clear, during adolescence, a reduction is seen in the time spent engaged in physical activity and the quality of dietary patterns [5].

The Mediterranean diet is considered to be a healthy dietary pattern, not only in terms of the quantity of food intake but, also, in terms of quality, food preparation and nutrient suitability [6]. Currently, a change in dietary patterns is taking place in Western societies, with an increase in pre-cooked dishes with a high caloric level, increasing the incidence of cardiovascular diseases. [7]. Food availability is one of factors leading to a worsening of dietary habits [8]. The Mediterranean diet is characterized by the intake of specific food varieties, especially wholegrains, olive oil, dairy products, bread, fruits and vegetables [9]. Caprara [10] also revealed that Mediterranean diet adherence combined with high levels of physical activity helped to increase life expectancy [11]. Positive adherence to a dietary pattern also brings numerous benefits in physical and mental areas [7]. In the physical area, a reduction in BMI, a reduction in waist circumference and the prevention of cardiovascular diseases are observed [11], whereas in the mental area, notable improvements in self-concept and emotional intelligence are observed [7].

Physical activity has been defined as any exercise performed by the skeletal muscles involving energetic expenditure [12]. Villasana et al. [13] showed that weekly physical activity engagement falls during adolescence. Likewise, Hernández-Álvarez et al. [14] state that this drop in physical exercise is due to a greater preference for sedentary habits over active activities. WHO [15] states that, in order to follow an active lifestyle, children and adolescents aged 5 to 17 years old should engage in moderate and vigorous physical activity for at least 60 min a day on three days a week. Following these guidelines brings physical health benefits such as a reduction of waist circumference, improvement of blood pressure and reduced BMI, as well as numerous improvements to mental health, such as better emotional control [16,17].

Emotional control plays a key role in leading a healthy lifestyle [18]. It can be defined as a multidimensional experience with three response systems: cognitive/subjective, behavioural/expressive and physiological/adaptive [19]. Further, educational settings must provide emotional education as a key element to the integral development of adolescents [20]. Thus, within the classroom, emotional intelligence is conceived as a key element to emotional control [21]. Among the different definitions proposed for the concept of emotional intelligence, the most widely accepted is that proposed by Salovey et al. [19]. These authors state that emotional intelligence describes the ability to accurately perceive, value and express emotions in a personal way, whilst also understanding the emotions experienced by others. Moreover, this model outlines three stages that make up emotional intelligence. The first of these is emotional perception, which pertains to the ability to perceive one's own emotions and others' emotions [22]. The second, emotional understanding, corresponds to the ability to understand and process information garnered from emotions [19] and emotional regulation and the ability to promote understanding and personal growth [19]. In this case during adolescence, emotional intelligence plays a fundamental role in overcoming disruptive states [20]; however, when adulthood is reached, the construct that makes up emotional intelligence (attention, clarity and repair) helps people to become emotionally competent [18,22]. Finally, another factor that can affect the development of emotional intelligence is the mental image perceived by the subjects of themselves [7]. In this case, it has been observed that people who are overweight show worse levels of emotional intelligence than those who reflect a normal weight and are satisfied with their physical appearance [17].

Based on the above, the present research outlines the following hypotheses:

- Participants who report more than three hours of physical activity a week will reveal better associations between BMI, Mediterranean diet adherence and emotional intelligence.
- Participants who report engaging in less than three hours of physical activity a week will reflect worse associations between BMI, Mediterranean diet adherence and emotional intelligence.

- Participants who engage in more than three hours of physical activity a week will have better scores for emotional intelligence, Mediterranean diet adherence and BMI than those who do not claim to meet these criteria.

In conclusion, the present research aims to establish the relationship between emotional intelligence, Mediterranean diet adherence, BMI and age. In order to address this aim, (a) an explanatory model of emotional intelligence and its relationship with Mediterranean diet adherence, BMI and age will be developed, and (b) the structural model will be tested via multi-group analysis as a function of whether students engage in more than three hours of physical activity a week.

2. Materials and Methods

2.1. Participants and Design

A non-experimental (ex post facto), descriptive and cross-sectional study was carried out with students from different public (state-funded) schools in the city of Granada. Schools were selected to participate randomly. The sample was composed of a total of 567 students aged between 9 and 13 years old (11.10 $\pm$ 1.24). With regards to sex, 53% were male (n = 303) and 47% were female (n = 264). Data collection was carried out after obtaining consent from legal guardians. A cover letter was sent to parents informing them about the aims and nature of the study and, subsequently, written informed consent was obtained. In terms of sample size, the overall sample of primary school students in the city of Granada is 83,940. Setting a 5.0% sampling error gave a confidence level of 97% for the recruited sample size.

2.2. Instruments

Sociodemographic questionnaire: Designed to collect information on age and sex. This tool was also used to collect data on sport or physical activity engagement. Participants were asked whether or not they engaged in more than 3 h of physical activity a week (do you engage in more than 3 h of physical activity outside of school hours?). In this case, this question is used to check whether the participants are physically active or not [23]. Participating Physical Education teachers measured and weighed participants for subsequent calculation of BMI. In this case, the mathematical formula of dividing the participant's weight by the squared height was used to calculate the BMI. The cut-off points used in the present investigation are the specific values for children according to their age that correspond to the following values for adults: Obesity (scores above 30 kg/m^2), overweight (scores between 25 kg/m^2 and 29.99 kg/m^2), normal weight (values between 24.99 kg/m^2 and 18 kg/m^2), underweight (scores below 18.5 kg/m^2) and severely underweight (less than 16.5 kg/m^2) [24,25].

KIDMED questionnaire: The Spanish version developed by Serrá-Majem et al. [26] was used for the present study. This questionnaire includes a total of 16 items that can be answered negatively or positively. Items 5, 11, 13 and 15 are negatively framed and, therefore, positive responses are scored as -1. All remaining questions are negatively framed, with affirmative responses being scored as +1. Potential final scores range from 4 to 12. Scores are then classified according to three groups describing Mediterranean diet adherence. The following groups are defined: optimal diet (8 and above), needs improvement (2–7) and poor diet (1). Reliability analysis of the data obtained for this questionnaire yielded acceptable results (α = 0.771).

Trait Meta-Mood Scale 24: Developed by Salovey et al. [19]. The Spanish version developed by Fernández-Berrocal et al. [27] was used in the present study. In this case, the version used for data collection is adapted for the study population [27]. This questionnaire comprises a total of 24 items that are rated on a five-point Likert scale (1 = "Disagree" to 5 = "Strongly agree"). On this scale, emotional intelligence is evaluated as a construct made up of three factors. These factors are emotional attention (items 1, 2, 3, 4, 5, 6, 7 and 8), understanding of emotional states (items 9, 10, 11, 12, 13, 14, 15 and 16) and emotional regulation (17, 18, 19, 20, 21, 22, 23 and 24). Items are summed to produce an overall score.

In the present study, reliability indices for emotional attention (α = 0.902), emotional clarity (α = 0.845) and emotional repair (α = 0.891) were all acceptable.

2.3. Procedure

During data collection, an information pack was sent out by the Department of the Didactics of Musical, Artistic and Corporal Expression of the University of Granada. This letter informed school management about the aims and scope of the study. Following this, an e-mail was sent by the schools to students' legal guardians to inform them of the study aims and request their consent for participation of their children in the research project. Once this permission was obtained, schools were provided with a link to a Google Form which laid out the previously described aims. Due to the COVID-19 pandemic, the project investigators were unable to personally attend the school, as during data collection (March 2021), entry to schools was restricted. Teachers received instructions from researchers around how questionnaires should be completed. Further, two questions were intentionally repeated on the questionnaire in order to check for random responding and ensure that questionnaires were completed correctly. As a result, a total of 19 questionnaires were discarded due to incorrect completion. Finally, the present study complied with the ethical standards for research involving human subjects set out in the Helsinki Declaration of 1975 and was supervised by the Research Ethics Committee of the University of Granada (1230/CEIH/2020).

2.4. Data Analysis

The IBM SPSS Statics 25.0 program (IBM Corp, Armonk, NY, USA) was used for data analysis. A descriptive analysis was carried out according to frequencies. Comparative analysis was performed using contingency tables and Student t-tests for independent samples. Statistically significant differences were determined using Pearson's Chi-square statistics, with a significance level of 95%. In order to calculate the statistical power achieved for t-statistic calculation, Cohen's standardized d-index was used [28]. In this sense, effects are interpreted as null ($\leq$0.19), small (0.20–0.49), medium (0.50–0.79) and large ($\geq$0.80). Phi and Cramer's V was used for interpretation of contingency table outcomes, with effects being interpreted as weak (<0.2), moderate (0.2–0.59) and large ($\geq$0.6) [29]. The Kolmogorov-Smirnov test was used to examine data distribution, with a normal distribution being confirmed.

The IBM SPSS Amos 26.0 program (IBM Corp, Armonk, NY, USA) was used to develop structural equation models. This model allows existing relationships to be established between participants who engage in more than 3 h of weekly physical activity and those who do not meet this physical-sport criterion. In this case, each model is composed of three endogenous variables (MDA; AGE; BMI) and three exogenous variables (EA; EC; ER). For the endogenous variables, causal relationships are examined in consideration of observed associations between the indicators and the degree of measurement reliability. This enables the error originated by the measurement of observed variables to be included in the model. Figure 1 was developed in which unidirectional arrows are produced from regression weights and represent lines of influence. In addition, a significance level of 0.05 was established. In this case, relationships between the three-factor construct of emotional intelligence and age, body mass index and Mediterranean diet adherence were examined.

Finally, model fit was evaluated following visualisation of its different parameters. According to Bentler [30] and McDonald [31], goodness of fit should be evaluated according to chi-square, with non-significant *p*-values indicating good fit. With regards to the comparative fit index (CFI), values above 0.95 reflect good fit. With regards to the goodness of fit index (GFI), scores above 0.90 show good fit. With regards to the incremental reliability index (IFI), values above 0.90 reflect good fit. Finally, in the case of root mean square approximation (RMSEA), scores below 0.1 evidence adequate model fit.

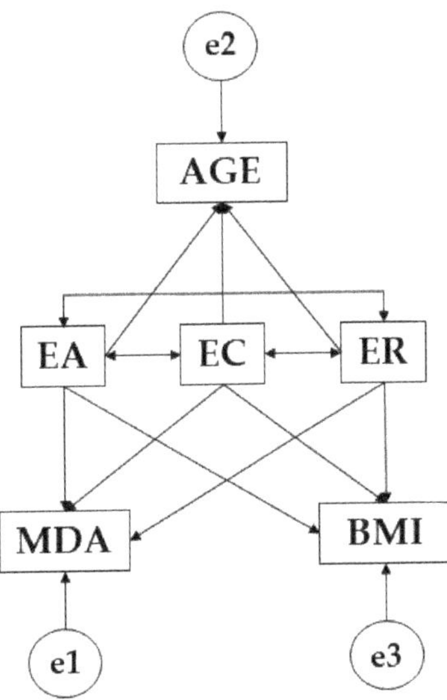

Figure 1. Structural equation model. Note: Emotional attention (EA); emotional clarity (EC); emotional repair (ER); body mass index (BMI); Mediterranean diet adherence (MDA); age (AGE).

3. Results

With regards to emotional attention, the outcomes presented in Table 1 show that higher scores were recorded for individuals who met the established the sport criteria (M = 3.496) relative to participants who did not meet this criterion (M = 3.423). Turning attention to emotional clarity, higher scores were reported by participants who engaged in more than 3 h of physical activity a week (M = 3.612) relative to those who spent less time engaged in such activity (M = 3.429). Finally, in terms of emotional repair, higher scores were observed for participants who met the sport criterion (M = 3.792) relative to those who did not (M = 3.423).

Table 1. Comparative analysis between physical activity (Yes/No) and emotional intelligence.

			Levene Test				**t-Test**			**ES (d)**	**95% CI**
		N	**M**	**SD**	**F**	**Sig**	**T**	**df**	**P**		
EA	No	69	3.423	0.825	0.352	0.553	−0.684	86.915	>0.05	0.525	[0.161; 0.342]
	Yes	498	3.496	0.805							
EC	No	69	3.429	0.877	0.009	0.926	−1.628	0.107	>0.05	0.214	[0.038; 0.466]
	Yes	498	3.612	0.851							
ER	No	69	3.423	0.815	0.763	0.383	−3.520	87.380	≤0.05	0.458	[0.205; 0.711]
	Yes	498	3.792	0.805							

Note: 1 Emotional attention (EA); emotional clarity (EC); emotional repair (ER). Note 2: Sample (N); Average (M); Standard Deviation (SD); F-Snedecor for equality of variances (F); Significance Level (Sig); T-value (T); Degrees of Freedom (df); Bilateral Significance (P); Cohen's standardized d-index (ES (d)); 95% Confidence Interval of the Difference (95% CI).

Table 2 shows the descriptive analysis for the BMI variable. In this case, it is observed that 29.6% (*n* = 168) show as underweight, whereas more than two thirds of the total show as normal weight (69.1%; *n* = 392). Only 1.3% were overweight (*n* = 7).

Table 2. BMI descriptive analysis.

		Frequency	Percentage
	Underweight	168	29.6%
Body Mass Index	Normal weight	392	69.1%
	Overweight	7	1.3%
Total		567	100%

Results presented in Table 3 ($p \leq 0.05$) reveal that participants who did engage in physical activity were more likely to be underweight ($n = 156$; 31.3%) than overweight ($n = 12$; 2.4%). In the case of normal weight, higher scores are observed for those who did not engage in more than 3 h of physical activity a week ($n = 57$; 82.6%).

Table 3. Comparative analysis of body mass index and physical activity engagement.

ES	Phi = 0.119		Body Mass Index			Total
	Cramer's V = 0.119		Underweight	Normal Weight	Overweight	
PA	No	Count	12	57	0	69
		% PA	17.4%	82.6%	0.0%	100%
	Yes	Count	156	330	12	498
		% PA	31.3%	66.3%	2.4%	100%

Note 1: Physical Activity (PA).

Table 4 pertains to the relationship between physical activity and Mediterranean diet adherence, with statistically significant differences emerging at the level $p \leq 0.05$. In this case, it is observed that participants who met the physical activity/sport criterion ($n = 159$; 89.8%) were more likely to follow an optimal diet than those who did not ($n = 18$; 10.2%).

Table 4. Comparative analysis of Mediterranean diet adherence and physical activity engagement.

ES	Phi = 0.083		PA		Total
	Cramer's V = 0.083		No	Yes	
MDA	Poor quality	Count	3	18	21
		% MDA	14.3%	85.7%	100%
	Needs improvement	Count	48	321	369
		% MDA	13.0%	87.0%	100%
	Optimal diet	Count	18	159	177
		% MDA	10.2%	89.8%	100%

Note 1: Physical Activity (PA); Mediterranean Diet Adherence (MDA).

The structural equation models (Figure 2) demonstrated good fit for each of the different examined indices. Chi-square analysis produced a significant p-value ($X^2 = 1.611$; df = 3; pl = 0.657); however, these data cannot be interpreted in isolation due to the sensitivity of this statistic to sample size [32]. For the general model, other standardised fit indices were used which are less sensitive to sample size. Comparative fit (CFI), normalised fit (NFI) and incremental fit (IFI) indices were 0.998, 0.994 and 0.998, respectively, whereas the Tucker-Lewis index (TLI) was 0.946. All of these indices demonstrate excellent fit. In addition, the root mean of square error of approximation (RMSEA) value obtained was 0.010.

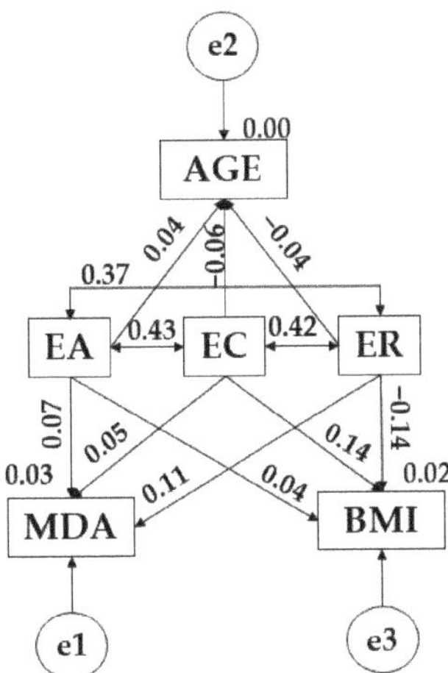

Figure 2. SEM pertaining to the entire sample. Note: Emotional attention (EA); emotional clarity (EC); emotional repair (ER); body mass index (BMI); Mediterranean diet adherence (MDA); age (AGE).

In this case, Figure 2 and Table 5 show the relationships between the variables for the whole sample. A positive relationship is observed between age and emotional attention (r = 0.441), whereas negative relationships are observed for emotional clarity (r = −0.059) and emotional repair (r = −0.004). Continuing with the relationship between the emotional domain and adherence to the Mediterranean diet, a positive relationship was observed with all the variables that make up emotional intelligence (r = 0.066; r = 0.046), showing a level of significance with emotional repair ($p \leq 0.05$; r = 0.111). Finally, body mass index shows a positive relationship with emotional clarity ($p \leq 0.05$; r = 0.143) and emotional attention (r = 0.004) and a negative relationship with emotional repair ($p \leq 0.05$; r = −0.140).

Table 5. SEM belonging to the entire sample.

Association between Variables	RW				SRW
	Estimation	SE	CR	*p*	Estimation
AGE ← EA	0.075	0.097	0.771	0.441	0.037
AGE ← EC	−0.115	0.094	−1.218	0.223	−0.059
AGE ← ER	−0.008	0.094	−0.087	0.931	−0.004
MDA ← EA	0.059	0.042	1.406	0.160	0.066
MDA ← EC	0.038	0.041	0.947	0.343	0.046
BMI ← EC	0.109	0.037	2.954	**	0.143
BMI ← ER	−0.110	0.037	−2.975	**	−0.140
MDA ← ER	0.096	0.041	2.366	**	0.111
BMI ← EA	0.003	0.038	0.084	0.933	0.004
EC ← → EA	0.169	0.018	9.332	***	0.426
EC ← → ER	0.170	0.019	9.198	***	0.419
EA ← → ER	0.143	0.017	8.296	***	0.372

Note 1: Regression weights (RW); standardized regression weights (SRW); standard error (SE); critical ratio (CR). Note 2: Emotional attention (EA); emotional clarity (EC); emotional Repair (ER); body mass index (BMI); Mediterranean diet adherence (MDA); age (AGE). Note 3: ** $p \leq 0.05$; *** $p < 0.001$.

Turning attention to the structural equation models, the model developed for participants who meet the physical-sporting criteria produced a significant *p*-value ($X^2 = 6.072$;

df = 3; pl = 0.108). As with before, the CFI produced was 0.992, whereas NFI and IFI values were 0.986 and 0.993, respectively. A TLI of 0.946 was also obtained, with all of these aforementioned values pointing to excellent fit. Finally, the RMSEA value obtained was 0.045.

Both Figure 3 and Table 6 present the regression weights associated with the theoretical model, with statistically significant relationships being indicated at $p < 0.05$ and $p < 0.001$. In this case, emotional attention was positively related with Mediterranean diet adherence ($p < 0.001$; r = 0.186), age (r = 0.067), emotional clarity ($p < 0.001$; r = 0.482) and emotional repair ($p < 0.001$; r = 0.435), with a negative relationship being found with body mass index (r = −0.067). Moving on to consider emotional clarity, a positive relationship is observed with Mediterranean diet adherence (r = 0.045), body mass index ($p \leq 0.05$; r = 0.165) and emotional repair; however, a negative relationship is shown with age (r = −0.112). In the case of emotional repair, positive relationships are observed with Mediterranean diet adherence (r = 0.071) and age (r = 0.021), whereas a negative relationship is seen with body mass index ($p < 0.001$; r = −0.209).

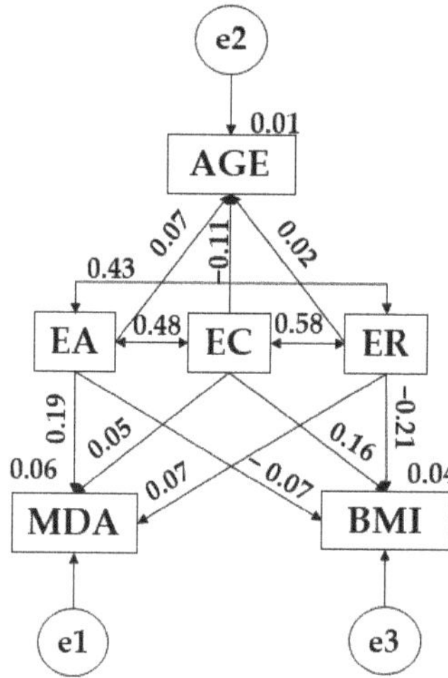

Figure 3. SEM pertaining to participants who engage in more than 3 h of PA a week. Note: Emotional attention (EA); emotional clarity (EC); emotional repair (ER); body mass index (BMI); Mediterranean diet adherence (MDA); age (AGE).

Turning attention to the structural equation models, the model developed to participants who do not engage in more than 3 h showed good fit in terms of all of the different examined indices. Chi-square analysis produced a significant *p*-value (X^2 = 8.330; df = 3; pl = 0.040). As with before, the CFI produced was 0.955, whereas NFI and IFI values were 0.912 and 0.928, respectively. A TLI of 0.934 was also obtained, with all of these aforementioned values pointing to excellent fit. Finally, the RMSEA value obtained was 0.056.

Figure 4 and Table 7 present the regression weights associated with the theoretical model, with statistically significant relationships being indicated at the level of $p < 0.05$ and $p < 0.001$. In this case, emotional attention was positively related with Mediterranean diet adherence (r = 0.130), body mass index ($p < 0.05$; r = 0.383), emotional clarity ($p < 0.001$; r = 0.464) and emotional repair ($p < 0.001$; r = 0.510), whereas it was negatively related with age ($p < 0.05$; r = −0.289). Emotional clarity produced positive associations with age ($p < 0.05$; r = 0.302) and emotional repair ($p < 0.001$; r = 0.519); however, negative relationships were produced Mediterranean diet adherence (r = −0.045) and body mass index (r = −0.056). Finally, with regards to emotional repair, positive relationships emerged Mediterranean diet adherence (r = 0.129), age (r = 0.041) and body mass index (r = 0.07).

Table 6. SEM pertaining to participants who engage in more than 3 h of PA per week.

Association between Variables	RW				SRW
	Estimation	SE	CR	*p*	Estimation
MDA ← EA	0.032	0.009	3.666	***	0.186
AGE ← EA	0.106	0.082	1.293	0.196	0.067
BMI ← EA	−0.042	0.032	−1.307	0.191	−0.067
MDA ← EC	0.007	0.009	0.808	0.419	0.045
AGE ← EC	−0.166	0.086	−1.937	0.053	−0.112
BMI ← EC	0.097	0.034	2.891	**	0.165
MDA ← ER	0.012	0.010	1.297	0.195	0.071
AGE ← ER	0.033	0.088	0.380	0.704	0.021
BMI ← ER	−0.131	0.035	−3.775	***	−0.209
EC ← → EA	0.330	0.034	9.680	***	0.482
EC ← → ER	0.398	0.036	11.199	***	0.581
EA ← → ER	0.282	0.032	8.888	***	0.435

Note 1: Regression weights (RW); standardized regression weights (SRW); standard error (SE); critical ratio (CR). Note 2: Emotional attention (EA); emotional clarity (EC); emotional Repair (ER); body mass index (BMI); Mediterranean diet adherence (MDA); age (AGE). Note 3: ** $p \leq 0.05$; *** $p < 0.001$.

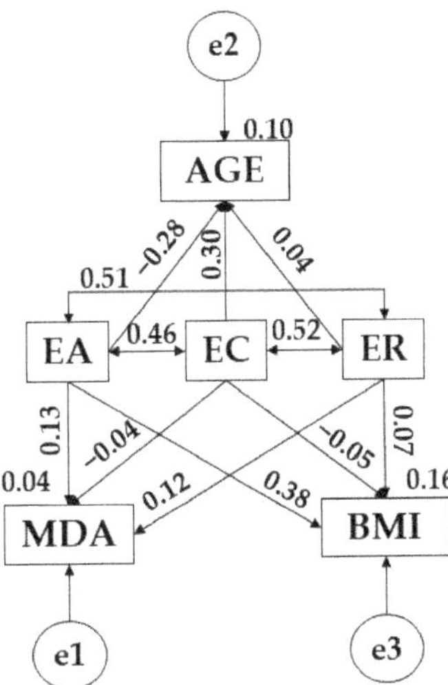

Figure 4. SEM pertaining to participants who do not engage in more than 3 h of PA a week. *Note:* Emotional attention (EA); emotional clarity (EC); emotional repair (ER); body mass index (BMI); Mediterranean diet adherence (MDA); age (AGE).

Table 7. SEM pertaining to participants who do not engage in more than 3 h of PA a week.

Association between Variables	RW				SRW
	Estimation	SE	CR	*p*	Estimation
MDA ← EA	0.027	0.030	0.906	0.365	0.130
AGE ← EA	−0.403	0.194	−2.077	**	−0.289
BMI ← EA	0.177	0.062	2.853	**	0.383
MDA ← EC	−0.009	0.028	−0.312	0.755	−0.045
AGE ← EC	0.396	0.184	2.156	**	0.302
BMI ← EC	−0.024	0.059	−0.413	0.680	−0.056
MDA ← ER	0.027	0.031	0.866	0.386	0.129
AGE ← ER	0.058	0.204	0.285	0.776	0.041
BMI ← ER	0.035	0.065	0.531	0.595	0.074
EC ← → EA	0.332	0.095	3.473	***	0.464
EC ← → ER	0.366	0.096	3.800	***	0.519
EA ← → ER	0.339	0.090	3.748	***	0.510

Note 1: Regression weights (RW); standardized regression weights (SRW); standard error (SE); critical ratio (CR). Note 2: Emotional attention (EA); emotional clarity (EC); emotional Repair (ER); body mass index (BMI); Mediterranean diet adherence (MDA); age (AGE). Note 3: ** $p \leq 0.05$; *** $p < 0.001$.

4. Discussion

The present study describes existing relationships between emotional intelligence, adherence to the Mediterranean diet, age and body mass index as a function of the time spent engaged in weekly physical activity. Findings met the proposed objectives and, therefore, the following discussion serves to compare obtained outcomes with those reported in previous research.

Present findings show that individuals who report engaging in more than three hours of weekly physical activity also report higher scores for attention, clarity and emotional repair than those who engage in such activity for less than three hours. In view of these results, Vaquero-Solís et al. [33] argue that physical activity engagement is beneficial for emotional control. Likewise, Angarita-Ortiz et al. [34] uphold that physical activity engagement contributes to better emotional intelligence, since physical exercise is used as a means to achieving emotional well-being.

With regards to the relationship between BMI and physical activity engagement, it was observed that students who claimed to engage in more than three hours of physical activity a week were more likely to be overweight or underweight than those who did not meet the aforementioned criteria. Hugely contrasting findings were reported by Chavesa et al. [35] and Power et al. [36]. Explanations for this can be found in the study conducted by Salcin et al. [37] who identified during adolescence a lack of control over diet exists, with an increase in fat intake and an increase in the consumption of pre-cooked dishes with a high calorie content. Further, Melguizo-Ibáñez et al. [38] and Padial-Ruz et al. [39] argue that family functioning plays a key role in diet control and encouraging physical activity.

Turning attention to Mediterranean diet adherence and physical activity, it was observed that more than half of participants who engaged in more than three hours of physical activity a week were in need of improving their diet. Similar findings were reported by Melguizo-Ibáñez et al. [40] and Morales-Camacho et al. [41] who found that dietary patterns worsen during adolescence. Moreover, Muros et al. [42] argue that adherence tends to reduce due to the fact that adolescents tend to consume more saturated fats in place of foods rich in proteins and carbohydrates.

With regards to the present associative analysis, students who engaged in more than three hours of physical activity a week were observed to demonstrate a stronger relationship between Mediterranean diet adherence and attention and emotional clarity. Similar outcomes were obtained by Marfil-Carmona et al. [43] who stated that healthy lifestyles in general have a positive impact on the overall development of adolescents, including that of the emotional sphere [44]. On the other hand, with regards to the relationship between Mediterranean diet adherence and emotional repair, participants who engaged in less than three hours of physical activity a week were observed to obtain better outcomes. Contrasting results were obtained by Castro-Sánchez et al. [45] who concluded that engagement in physical exercise helps to channel disruptive states and, consequently, improve emotional outcomes.

With regards to BMI and emotional intelligence, participants who engaged in more than three hours of physical activity a week demonstrated a negative relationship between BMI, attention and emotional clarity. In view of these findings, Cuesta-Zamora et al. [46] conclude that obsession with physical fitness can lead to a continuous state of body dissatisfaction, leading to a state of non-acceptance at the detriment to the emotional state of adolescents. A positive relationship between BMI and emotional clarity was also observed in participants who engaged in more than three hours of physical activity a week. These findings are supported by Andrei et al. [47] who argue that emotional clarity helps to improve mental health and, as a result, improve physical performance and physical condition.

Outcomes pertaining to emotional intelligence reveal that participants who engage in more than three hours of physical activity a week demonstrate better outcomes. In this sense, Vaquero-Solís et al. [33] conclude that emotional clarity, together with the attention given to the negative feelings experienced during physical activity, helps to channel disruptive emotions due to the division of neurotransmitters [48]. In contrast, better relationships between emotional attention and repair were shown in participants who do not engage in more than three hours of physical activity a week. Contrasting findings have been reported by Wang et al. [49] who stated that physical activity engagement provides benefits for the perception of feelings and coping. Further, research by Trigueros et al. [50] concluded that an active lifestyle helps to improve emotional and physical well-being.

Finally, the present research reports that leading an active lifestyle brings benefits in the emotional and physical area [51]. In this case, the present study highlights the need to create a motivation and positive attitude towards the practice of physical activity at an early age [50] that will last into adulthood and help to improve people's health [52] and the decrease of cardiovascular diseases in different societies [53].

Limitations and Future Perspectives

The present study has a number of limitations. The main limitation is that the present study is cross-sectional and based on a single measurement in a specific population at a determined timepoint. This only allows for the identification of relationships between variables at the examined timepoint and prevents causal relationships from being established. Another limitation is that the sample consisted of primary school pupils from a very specific geographical area. This prevents findings from being generalised to wider geographical areas at a national or regional level. In addition, the presence of COVID-19 and the resultant mobility restrictions imposed had an impact on data collection, as researchers were unable to access schools, leading to a smaller sample being recruited. It should also be noted that, although reliable instruments were used to collect weight and height data, these instruments carry an implicit measurement error. Despite recruiting a sizeable final sample which enabled a confidence level of 97%, it was not possible to collect data from other schools. As a result, external variables such as socioeconomic level, education or school ownership could not be considered.

In terms of future perspectives, it would be interesting to carry out a comprehensive experimental study targeting nutritional and emotional education, via physical activity engagement.

5. Conclusions

Descriptive analysis showed that participants who engaged in more than three hours of physical activity a week have better emotional intelligence than those who do not meet this criterion. Furthermore, it was also observed that the majority of the sample was physically active; however, improvements were required with regards to Mediterranean diet adherence.

Turning attention to the structural equation models, better relationships between Mediterranean diet adherence and attention and emotional clarity were observed in participants who reported more than 3 h of physical activity a week; however, better relationships were seen between Mediterranean diet adherence and emotional repair in those who did not achieve this level of physical activity. A better relationship between BMI and emotional repair and attention was also observed in participants who engaged in less than 3 h of physical activity a week.

Finally, a stronger relationship was observed between the variables comprising emotional intelligence in participants who engaged in more than 3 h of weekly physical activity, whereas participants who do not meet this criterion demonstrated a stronger relationship between emotional attention and emotional repair.

Author Contributions: Conceptualization, E.M.-I., G.G.-V. and F.Z.-O.; methodology, J.L.U.-J.; formal analysis, E.M.-I.; data curation, G.G.-V.; writing—original draft preparation, F.Z.-O., G.B., J.L.U.-J., E.M.-I., G.G.-V. and F.M.C.; writing—review and editing, A.F.-S., G.B., F.M.C., F.Z.-O., H.S., G.G.-V. and J.L.U.-J. All authors have read and agreed to the published version of the manuscript.

Funding: This research received no external funding.

Institutional Review Board Statement: The study was conducted in accordance with the Declaration of Helsinki, and approved by the Research Ethics Committee of the University of Granada (1230/CEIH/2020), approval date is 2020/4/23.

Informed Consent Statement: Informed consent was obtained from all subjects involved in the study.

Data Availability Statement: The data used to support the findings of current study are available from the corresponding author upon request.

Conflicts of Interest: The authors declare no conflict of interest.

References

1. Allen, M.S.; Robson, D.A.; Vella, S.A.; Laborde, S. Extraversion development in childhood, adolescence and adulthood: Testing the role of sport participation in three nationally-representative samples. *J. Sports Sci.* **2021**, *39*, 2258–2265. [CrossRef] [PubMed]
2. Brylka, A.; Wolke, D.; Ludyga, S.; Bilgin, A.; Spiegler, J.; Trower, H.; Gkiouleka, A.; Gerber, M.; Brand, S.; Grob, A.; et al. Physical activity, mental health, and well-being in very pre-term and term born adolescents: An individual participant data meta-analysis of two accelerometry studies. *Int. J. Environ. Res. Public Health* **2021**, *18*, 1735. [CrossRef] [PubMed]
3. Branje, S.; de Moor, E.L.; Spitzer, J.; Becht, A.I. Dynamics of identity development in adolescence: A decade in review. *J. Res. Adolesc.* **2021**, *31*, 908–927. [CrossRef] [PubMed]
4. Mollborn, S.; Lawrence, E.M.; Saint-Onge, J.M. Contributions and challenges in health lifestyles research. *J. Health Soc. Behav.* **2021**, *62*, 388–403. [CrossRef] [PubMed]
5. Morelli, C.; Avolio, E.; Galluccio, A.; Caparello, G.; Manes, E.; Ferraro, S.; Caruso, A.; De Rose, D.; Barone, I.; Adornetto, C.; et al. Nutrition education program and physical activity improve the adherence to the Mediterranean diet: Impact on inflammatory biomarker levels in healthy adolescents from the DIMENU longitudinal study. *Front. Nutr.* **2021**, *8*, 685247. [CrossRef] [PubMed]
6. Medina, F.X.; Solé-Sedeno, J.M.; Bach-Faig, A.; Aguilar-Martínez, A. Obesity, Mediterranean diet, and public health: A vision of obesity in the Mediterranean context from a sociocultural perspective. *Int. J. Environ. Res. Public Health* **2021**, *18*, 3715. [CrossRef] [PubMed]
7. Sirkka, O.; Fleischmann, M.; Abrahamse-Berkeveld, M.; Halberstadt, J.; Olthof, M.; Seidell, J.; Corpeleijn, E. Dietary patterns in early childhood and the risk of childhood overweight: The GECKO Drenthe birth cohort. *Nutrients* **2021**, *13*, 2046. [CrossRef] [PubMed]
8. Min, K.; Wang, J.; Liao, W.; Astell-Burt, T.; Feng, X.; Cai, S.; Liu, Y.; Zhang, P.; Su, F.; Yang, K.; et al. Dietary patterns and their associations with overweight/obesity among preschool children in Dongcheng District of Beijing: A cross-sectional study. *BMC Public Health* **2021**, *21*, 223. [CrossRef] [PubMed]

9. Sotos-Prieto, M.; Del Rio, D.; Drescher, G.; Estruch, R.; Hanson, C.; Harlan, T.; Hu, F.B.; Loi, M.; McClung, J.P.; Mojica, A.; et al. Mediterranean diet—Promotion and dissemination of healthy eating: Proceedings of an exploratory seminar at the Radcliffe institute for advanced study. *Int. J. Food Sci. Nutr.* **2022**, *73*, 158–171. [CrossRef]
10. Caprara, G. Mediterranean-type dietary pattern and physical activity: The winning combination to counteract the rising burden of non-communicable diseases (NCDs). *Nutrients* **2021**, *13*, 429. [CrossRef]
11. Menotti, A.; Puddu, P.E.; Maiani, G.; Catasta, G. Cardiovascular and other causes of death as a function of lifestyle habits in a quasi-extinct middle-aged male population. A 50-year follow-up study. *Int. J. Cardiol.* **2016**, *210*, 173–178. [CrossRef] [PubMed]
12. Guthold, R.; Stevens, G.A.; Riley, L.M.; Bull, F.C. Global trends in insufficient physical activity among adolescents: A pooled analysis of 298 population-based surveys with 1·6 million participants. *Lancet Child Adolesc. Health* **2020**, *4*, 23–35. [CrossRef]
13. Villasana, M.V.; Pires, I.M.; Sá, J.; Garcia, N.M.; Zdravevski, E.; Chorbev, I.; Lameski, P.; Flórez-Revuelta, F. Promotion of healthy nutrition and physical activity lifestyles for teenagers: A systematic literature review of the current methodologies. *J. Pers. Med.* **2020**, *10*, 12. [CrossRef] [PubMed]
14. Hernández-Alvarez, J.L.; Velázquez-Buendía, R.; Martínez-Gorroño, M.E.; Garoz-Puerta, I. Lifestyle and physical activity in Spanish children and teenagers: The impact of psychosocial and biological factors: Physical activity and psychosocial and biological factors. *J. Appl. Biobehav. Res.* **2009**, *14*, 55–69. [CrossRef]
15. World Health Organization. *WHO Guidelines on Physical Activity and Sedentary Behaviour: At a Glance*; World Health Organization: Geneva, Switzerland, 2020.
16. Alviz, M.M.; del Barco, B.L.; Lázaro, S.M.; Gallego, D.I. Rol predictivo de la inteligencia emocional y la actividad física sobre el autoconcepto físico en escolares. *Sport. Sci. J. Sch. Sport Phys. Educ. Psychomot.* **2020**, *6*, 308–326. [CrossRef]
17. Carton, A.; Barbry, A.; Coquart, J.; Ovigneur, H.; Amoura, C.; Orosz, G. Sport-related affective benefits for teenagers are getting greater as they approach adulthood: A large-scale French investigation. *Front. Psychol.* **2021**, *12*, 738343. [CrossRef]
18. Dai, W.; Zhou, J.; Li, G.; Zhang, B.; Ma, N. Maintaining normal sleep patterns, lifestyles and emotion during the COVID-19 pandemic: The stabilizing effect of daytime napping. *J. Sleep Res.* **2021**, *30*, e13259. [CrossRef]
19. Salovey, P.; Mayer, J.D.; Goldman, S.L.; Turvey, C.; Palfai, T.P. Emotional attention, clarity, and repair: Exploring emotional intelligence using the Trait Meta-Mood Scale. In *Emotion, Disclosure and Health*; Pennebaker, J.W., Ed.; American Psychological Association: Washington, DC, USA, 1995; pp. 125–154.
20. Ogren, M.; Johnson, S.P. Factors facilitating early emotion understanding development: Contributions to individual differences. *Hum. Dev.* **2021**, *64*, 108–118. [CrossRef]
21. Denervaud, S.; Mumenthaler, C.; Gentaz, E.; Sander, D. Emotion recognition development: Preliminary evidence for an effect of school pedagogical practices. *Learn. Instr.* **2020**, *69*, 101353. [CrossRef]
22. Fernández-Berrocal, P.; Cabello, R.; Castillo, R.; Extremera, N. Gender differences in emotional intelligence: The mediating effect of age. *Behav. Psychol.* **2012**, *20*, 77–89.
23. Arufe-Giráldez, V.; Zurita-Ortega, F.; Padial-Ruz, R.; Castro-Sánchez, M. Association between Level of Empathy, Attitude towards Physical Education and Victimization in Adolescents: A Multi-Group Structural Equation Analysis. *Int. J. Environ. Res. Public Health* **2019**, *16*, 2360. [CrossRef] [PubMed]
24. Hales, C.M.; Fryar, C.D.; Carroll, M.D.; Freedman, D.S.; Ogden, C.L. Trends in obesity and severe obesity prevalence in US youth and adults by sex and age, 2007–2008 to 2015–2016. *JAMA* **2018**, *319*, 1723. [CrossRef] [PubMed]
25. Cole, T.J.; Lobstein, T. Extended international (IOTF) body mass index cut-offs for thinness, overweight and obesity: Extended international BMI cut-offs. *Pediatr. Obes.* **2012**, *7*, 284–294. [CrossRef]
26. Serra-Majem, L.; Ribas, L.; Ngo, J.; Ortega, R.M.; García, A.; Pérez-Rodrigo, C.; Aranceta, J. Food, youth and the Mediterranean diet in Spain. Development of KIDMED, Mediterranean diet quality index in children and adolescents. *Public Health Nutr.* **2004**, *7*, 931–935. [CrossRef]
27. Fernandez-Berrocal, P.; Extremera, N.; Ramos, N. Validity and reliability of the Spanish modified version of the Trait Meta-Mood Scale. *Psychol. Rep.* **2004**, *94*, 751–755. [CrossRef]
28. Cohen, J. A power primer. *Psychol. Bull.* **1992**, *112*, 155–159. [CrossRef]
29. López-Roldán, P.; Fachelli, S. *Metodología de la Investigación Social Cuantitativa*, 1st ed.; Campus de la UAB: Barcelona, Spain, 2015.
30. Bentler, P.M. Comparative fit indexes in structural models. *Psychol. Bull.* **1990**, *107*, 238–246. [CrossRef] [PubMed]
31. McDonald, R.P.; Marsh, H.W. Choosing a multivariate model: Noncentrality and goodness of fit. *Psychol. Bull.* **1990**, *107*, 247–255. [CrossRef]
32. Tenenbaum, G.; Eklund, R.C. *Handbook of Sport Psychology*; John Wiley & Sons: Hoboken, NJ, USA, 2007.
33. Vaquero-Solís, M.; Amado-Alonso, D.; Sánchez-Oliva, D.; Sánchez-Miguel, P.A.; Iglesias-Gallego, D. Inteligencia emocional en la adolescencia: Motivación y actividad física. *RIMCAFD* **2020**, *20*, 119. [CrossRef]
34. Angarita-Ortiz, M.F.; Calderón-Suescún, D.P.; Carrillo-Sierra, S.M.; Rivera-Porras, D.; Cáceres-Delgado, M.; Rodríguez-González, D. Factores de protección de la salud mental en universitarios: Actividad física e inteligencia emocional. *AVFT* **2020**, 753–759. [CrossRef]
35. Chavesa, A.M.; Cardoso, M.O.; de Abreua, P.R.; Maneschy, M.S.; Passos, R.P.; Lima, B.N.; Junior, G.B.V.; Novo, A.F.M.P.; Almeida, K.S. Analysis of the levels of physical activity and body mass index of college students: A systematic review. *Soc. Med.* **2021**, *14*, 140–152.

36. Power, T.G.; Fisher, J.O.; O'Connor, T.M.; Micheli, N.; Papaioannou, M.A.; Hughes, S.O. Self-regulatory processes in early childhood as predictors of Hispanic children's BMI z-scores during the elementary school years: Differences by acculturation and gender. *Appetite* **2022**, *168*, 105778. [CrossRef] [PubMed]
37. Salcin, L.O.; Karin, Z.; Miljanovic Damjanovic, V.; Ostojic, M.; Vrdoljak, A.; Gilic, B.; Sekulic, D.; Lang-Morovic, M.; Markic, J.; Sajber, D. Physical activity, body mass, and adherence to the Mediterranean diet in preschool children: A cross-sectional analysis in the Split-Dalmatia County (Croatia). *Int. J. Environ. Res. Public Health* **2019**, *16*, 3237. [CrossRef] [PubMed]
38. Melguizo-Ibáñez, E.; Viciana-Garófano, V.; Zurita-Ortega, F.; Ubago-Jiménez, J.L.; González-Valero, G. Physical Activity Level, Mediterranean Diet Adherence, and Emotional Intelligence as a Function of Family Functioning in Elementary School Students. *Children* **2021**, *8*, 6. [CrossRef]
39. Padial-Ruz, R.; Viciana-Garófano, V.; Palomares-Cuadros, J. Adherencia a la dieta mediterránea, la actividad física y su relación con el IMC, en estudiantes universitarios del grado de primaria, mención de educación física, de Granada. *ESPHA* **2018**, *2*, 30–49. [CrossRef]
40. Melguizo-Ibáñez, E.; Zurita-Ortega, F.; Ubago-Jiménez, J.L.; González-Valero, G. Niveles de adherencia a la dieta mediterránea e inteligencia emocional en estudiantes del tercer ciclo de educación primaria de la provincia de Granada. *Retos* **2020**, *40*, 264–271. [CrossRef]
41. Camacho, W.J.M.; Zambrano, S.E.O.; Camacho, M.A.M.; Contreras, A.C.H.; Acevedo, A.R.; Valencia, E.J.D.; Cárdenas, A.C.; Alarcón, L.X.N.; Munar, L.C.A.; Sánchez, A.M.N.; et al. Nutritional status and high adherence to the Mediterranean diet in Colombian school children and teenagers during the COVID-19 pandemic according to sex. *J. Nutr. Sci.* **2021**, *10*, 1–18. [CrossRef]
42. Muros, J.J.; Cofre-Bolados, C.; Arriscado, D.; Zurita, F.; Knox, E. Mediterranean diet adherence is associated with lifestyle, physical fitness, and mental wellness among 10-y-olds in Chile. *Nutrition* **2017**, *35*, 87–92. [CrossRef]
43. Marfil-Carmona, R.; Ortega-Caballero, M.; Zurita-Ortega, F.; Ubago-Jiménez, J.L.; González-Valero, G.; Puertas-Molero, P. Impact of the mass media on adherence to the Mediterranean diet, psychological well-being and physical activity. Structural equation analysis. *Int. J. Environ. Res. Public Health* **2021**, *18*, 3746. [CrossRef]
44. Zach, S.; Eilat-Adar, S.; Ophir, M.; Dotan, A. Differences in the association between physical activity and people's resilience and emotions during two consecutive Covid-19 lockdowns in Israel. *Int. J. Environ. Res. Public Health* **2021**, *18*, 13217. [CrossRef]
45. Castro-Sánchez, M.; Lara-Sánchez, A.J.; Zurita-Ortega, F.; Chacón-Cuberos, R. Motivation, anxiety, and emotional intelligence are associated with the practice of contact and non-contact sports: An explanatory model. *Sustainability* **2019**, *11*, 4256. [CrossRef]
46. Cuesta-Zamora, C.; González-Martí, I.; García-López, L.M. The role of trait emotional intelligence in body dissatisfaction and eating disorder symptoms in preadolescents and adolescents. *Pers. Individ. Differ.* **2018**, *126*, 1–6. [CrossRef]
47. Andrei, F.; Nuccitelli, C.; Mancini, G.; Reggiani, G.M.; Trombini, E. Emotional intelligence, emotion regulation and affectivity in adults seeking treatment for obesity. *Psychiatry Res.* **2018**, *269*, 191–198. [CrossRef] [PubMed]
48. Ubago-Jiménez, J.L.; Zurita-Ortega, F.; San Román-Mata, S.; Puertas-Molero, P.; González-Valero, G. Impact of physical activity practice and adherence to the Mediterranean Diet in relation to Multiple Intelligences among university students. *Nutrients* **2020**, *12*, 2630. [CrossRef] [PubMed]
49. Wang, K.; Yang, Y.; Zhang, T.; Ouyang, Y.; Liu, B.; Luo, J. The relationship between physical activity and emotional intelligence in college students: The mediating role of self-efficacy. *Front. Psychol.* **2020**, *11*, 967. [CrossRef]
50. Trigueros, R.; Aguilar-Parra, J.M.; Cangas, A.J.; Bermejo, R.; Ferrándiz, C.; López-Liria, R. Influence of emotional intelligence, motivation and resilience on academic performance and the adoption of healthy lifestyle habits among adolescents. *Int. J. Environ. Res. Public Health* **2019**, *16*, 2810. [CrossRef]
51. Ramirez-Granizo, I.A.; Sánchez-Zafra, M.; Zurita-Ortega, F.; Puertas-Molero, P.; González-Valero, G.; Ubago-Jiménez, J.L. Multidimensional Self-Concept Depending on Levels of Resilience and the Motivational Climate Directed towards Sport in Schoolchildren. *Int. J. Environ. Res. Public Health* **2020**, *17*, 534. [CrossRef]
52. Castro-Sánchez, M.; Zurita-Ortega, F.; García-Marmol, E.; Chacón-Cuberos, R. Motivational Climate in Sport Is Associated with Life Stress Levels, Academic Performance and Physical Activity Engagement of Adolescents. *Int. J. Environ. Res. Public Health* **2019**, *16*, 1198. [CrossRef]
53. Conde-Pipó, J.; Melguizo-Ibáñez, E.; Mariscal-Arcas, M.; Zurita-Ortega, F.; Ubago-Jiménez, J.L.; Ramírez-Granizo, I.; González-Valero, G. Physical Self-Concept Changes in Adults and Older Adults: Influence of Emotional Intelligence, Intrinsic Motivation and Sports Habits. *Int. J. Environ. Res. Public Health* **2021**, *18*, 1711. [CrossRef]

Article

Adolescents with Higher Cognitive and Affective Domains of Physical Literacy Possess Better Physical Fitness: The Importance of Developing the Concept of Physical Literacy in High Schools

Barbara Gilic [1,2], Pavle Malovic [3], Mirela Sunda [2], Nevenka Maras [4] and Natasa Zenic [1,*]

[1]   Faculty of Kinesiology, University of Split, 21000 Split, Croatia; bargil@kifst.hr
[2]   Faculty of Kinesiology, University of Zagreb, 10000 Zagreb, Croatia; mirela.sunda@skole.hr
[3]   Faculty for Sport and Physical Education, University of Montenegro, 81400 Niksic, Montenegro; pavle.malovic93@live.ac.me
[4]   Faculty of Teacher Education, University of Zagreb, 10000 Zagreb, Croatia; nevenka.maras@ufzg.hr
*   Correspondence: natasazenic@gmail.com

Abstract: Physical literacy (PL) is thought to facilitate engagement in physical activity, which could lead to better physical fitness (PF). The aim of this study was to examine the reliability of the Croatian version of two frequently applied PL questionnaires that evaluate knowledge and understanding, perceived competence, environment, and value for literacy, numeracy, and PL and validity regarding correlation with objectively evaluated PF in adolescents. Five hundred forty-four high school students (403 females, 141 males) from Croatia were tested on PF (standing long jump, sit-ups for 30 s, sit-and-reach test, multilevel endurance test) and two PL questionnaires. The reliability of the Croatian version of the Canadian Assessment of Physical Literacy knowledge and understanding (CAPL-2-KU) and PLAYself was good ($\alpha = 0.71$–0.81 for PLAYself subscales, $\kappa = 0.39$–0.69 for CAPL-2-KU). Genders differed in the self-description dimension of PLAYself, with higher results in boys ($Z = 3.72$, $p < 0.001$). CAPL-2-KU and PLAYself total score were associated with PF in boys and girls, with PLAYself having stronger associations with PF. This research supports the idea of PL as an essential determinant for the development of PF, highlighting the necessity of the development of cognitive and affective domains of PL in physical education throughout a specifically tailored pedagogical process.

Keywords: adolescent; exercise; sport pedagogy; physical education

Citation: Gilic, B.; Malovic, P.; Sunda, M.; Maras, N.; Zenic, N. Adolescents with Higher Cognitive and Affective Domains of Physical Literacy Possess Better Physical Fitness: The Importance of Developing the Concept of Physical Literacy in High Schools. *Children* **2022**, *9*, 796. https://doi.org/10.3390/children9060796

Academic Editors: Georgian Badicu, Ana Filipa Silva and Hugo Miguel Borges Sarmento

Received: 13 May 2022
Accepted: 26 May 2022
Published: 28 May 2022

Publisher's Note: MDPI stays neutral with regard to jurisdictional claims in published maps and institutional affiliations.

Copyright: © 2022 by the authors. Licensee MDPI, Basel, Switzerland. This article is an open access article distributed under the terms and conditions of the Creative Commons Attribution (CC BY) license (https://creativecommons.org/licenses/by/4.0/).

1. Introduction

Lack of physical activity (PA) is considered a major public health problem in the 21st century [1,2]. Moreover, most of the young population is not physically active; that is, 81% of children and adolescents do not have sufficient levels of PA [3]. Accordingly, the physical fitness (PF) of adolescents is also very low, as it has direct connections with PA [4]. Due to the high incidence of insufficient PA worldwide, researchers hypothesized that some determinant or link was affecting this PA deficiency [5,6]. Movement competence, defined as skill development that assures efficient performance in various movements and activities [7], might affect PA, but only a weak relationship was found between the two variables [8]. Consequently, it led to the conclusion that a more complex construct than movement competence is related to PA engagement [9]. One of the theoretically important elements for developing and achieving lifelong participation in PA is physical literacy (PL).

PL is broadly defined as a "disposition to capitalize on our human-embodied capability wherein the individual has the motivation, confidence, physical competence, knowledge, and understanding to value and take responsibility for maintaining purposeful physical pursuits and activities throughout the life course" [10]. PL consists of four domains: the physical domain (physical competence), cognitive domain (knowledge and understanding), affective domain (motivation and confidence), and behavioral domain (engagement in PA) [10]. The main principle of PL is an individual's "ability to capitalize on the interaction between physical competence and affective characteristics" [11]. What is important is that it is theorized that the knowledge and understanding (K&U) facet can actually positively influence other domains of PL, as it supports the awareness and valuing of developing physical competence and can increase motivation and confidence for participating in PA [12]. Since PL supports the progress of movement competence through psychological components such as motivation and confidence [13], it was logical to consider PL as an important determinant of PA as well. Supportively, it was evidenced that children with higher PL scores had increased odds of meeting PA guidelines [14].

Although the importance and idea of PL are widely accepted, there is no global consensus about the most proper form of the PL evaluation. As a result, authorities around the globe have developed unique PL concepts and, consequently, specific PL assessment tools [15–18]. Among others, the Canadian Assessment of Physical Literacy (CAPL) and the Physical Literacy Assessment of Youth (PLAY) assessment tools are the most popular and commonly used in research [15]. Although the PL concept is multidimensional, and scientists believe that every component that determines PL should be examined, it is often challenging to have the conditions and time to assess the overall PL. Therefore, questionnaires that at least approximately determine the PL and individual PL domains are also used [15,19].

The most commonly used questionnaires are parts of (i) CAPL-2 and (ii) PLAY tools; (i) CAPL-2 knowledge and understanding questionnaire (CAPL-2-KU) and (ii) PLAYself questionnaire [20,21]. Specifically, CAPL-2-KU assesses knowledge and understanding by questions evaluating: (i) understanding of physical activity and sedentary behavior recommendations, (ii) knowledge of movement and fitness parameters and procedures for improving them, and (iii) perceptions of health [18,19]. Meanwhile, PLAYself includes items assessing: (i) confidence and motivation, (ii) knowledge and understanding, and (iii) environmental engagement ability [22]. Thus, from the brief overview of each questionnaire, PLAYself and CAPL-2-KU most likely do not assess the same domain of PL and in the same way.

It is evidenced that PL may facilitate an increase in PA and, therefore, directly impact health [23,24]. For example, it has been recorded that children with higher PL had higher physical fitness (PF) [25], which is an important indicator of health status [26]. Additionally, PF is associated with lower abdominal obesity, decreased risk of cardiovascular disease, and improved bone and mental health [27], and there is evidence that PF has strong relations to metabolic risks in younger children [28]. Therefore, it could be theorized that PL can influence PF and improve health in general in high school students who are still in the developmental phase of their lives. However, although previous studies evidenced positive correlations between PL and certain indices of overall health status, guidelines for PL promotion in the context of promoting health are missing [23,29,30]. Thus, this connection remained mostly theoretical, which means that future studies investigating, for instance, the associations between PF (as a highly important indicator of health status in youth) and PL domains are warranted.

Moreover, boys and girls tend to differ in PA and also in PF and PL. Specifically, boys are generally more active, are more involved in sports activities, and, therefore, possess better fitness status compared with girls [31]. Moreover, boys have higher scores in PL but mostly in the physical competence and behavioral domains of PL [32,33]. Indeed, several studies did not record the difference between boys and girls in the knowledge and understanding, motivation, and confidence parts of PL [20,34]. Moreover, a study on adolescents showed that girls had higher scores in the knowledge and understanding part of PL than boys ($t = -2.29$, $p < 0.05$; 6.6 ± 1.2 vs. 6.3 ± 1.3) [35]. However, most of the previous studies examining gender differences in PL have been conducted on younger children (8–12 years old), while only a few studies investigated PL in older children and adolescents—high school students [35,36].

School authorities around the globe embraced the idea of developing PL as a primary goal in physical education (PE), as students gain the knowledge and competence needed to have an active and healthy lifestyle [11,37]. However, in the territory of southeastern Europe, including Croatia, the PL concept has not been implemented thus far. At the same time, indicators of PA and obesity in children and adolescents from Croatia and neighboring countries are showing devastating figures, with youth from the territory being regularly in the highest 10 percentiles when it comes to physical inactivity and obesity/overweight in Europe [38]. Therefore, it is crucial to change the perspective and focus on alternative concepts in the PE curriculum, which may include orienting toward PL as an important determinant of overall health in children and adolescents. As a first step, it is necessary to evaluate the applicability of the PL concept in a specific sociodemographic environment in the territory while highlighting currently used standards of achievement within the PE curriculum (i.e., PF standards) [39].

Therefore, the aims of the study were (i) to evaluate the reliability and applicability of the Croatian version of two common PL measurement tools (e.g., PLAYself and CAPL-2-KU) and (ii) to establish the validity of the applied questionnaires while establishing (ii-a) gender differences in applied tools and (ii-b) the associations between the cognitive and affective domains of PL and objectively measured health-related PF in high school adolescents. Initially, we hypothesized that adolescents with better cognitive and affective domains of PL would have better PF. Knowing the importance of the PL concept and the sociocultural background in this region (including the similarity of the spoken languages), we believed that the findings of the study would be broadly applicable in educational systems of southeastern European countries.

2. Materials and Methods

2.1. Participants and Study Design

This research included 544 adolescents (403 females, 141 males) aged 14–18 years attending two high schools in Osijek-Baranja County, Croatia. All students were in good health and did not have any injury or illness during the investigation, which was determined by regular medical examination at the beginning of the school year. Students that had any medical condition or injury were excluded from the study. Students regularly participated in physical education classes twice a week. Students signed informed consent before the study began and were introduced to the study's aims and procedures. For students under 18, parents or legal guardians signed informed consent. The study was approved by the Ethical Board Faculty of the Kinesiology University of Zagreb, Croatia (Ref. no.25/2021, date of approval 16 July 2021).

2.2. Variables and Measurements

The study included anthropometric variables, PF tests, and an assessment of the cognitive and affective domains of PL. Anthropometrics and PF tests were performed in a closed facility (school gymnasium) from 8:00 to 14:00 in the morning. All tests were assessed by experienced evaluators: PE teachers with the highest levels of specialization and experience in testing PF and cognitive and affective domains of PL.

Anthropometric variables included measurement of body mass (in 0.1 kg), body height (in cm), and calculated body mass index (BMI = mass (kg)/height^2 (m)) [40].

PF tests used in this study were part of the standard Croatian tests assessed in high school and included the standing long jump (as a measure of power), sit-ups for 30 s (as a measure of strength), sit-and-reach test (as a measure of flexibility), and a multilevel endurance fitness test (as a measure of cardiorespiratory endurance). The reliability of the tests in similar samples had been previously proven [41].

The standing long jump test was conducted on a standardized jumping mat with an accuracy of one centimeter (Ghia Sport, Pazin, Croatia). Students had to perform a maximal forward jump from a standing position by bending their knees and using an arm swing.

The test was performed for three trials with 20–30 s of rest in between, and the best result (longest jump) was taken in the analysis [42,43].

For the sit-up test, students had to perform a maximal number of sit-ups in 30 s. The test started with students lying on their backs, with bent knees and palms on their thighs and partners sitting on their feet. They had to lift their torso to pass the level of their kneecaps with their hands. The result was the number of correct sit-ups in the 30 s [44,45].

The sit-and-reach test was conducted on a standardized wooden box. Participants were sitting on the floor with extended legs, with the soles of their feet placed flat against the box. They were instructed to bend forward and reach as far as possible on the measuring tape placed on the box and hold a maximal position for 3 s. The test was performed for three trials, and the best score (in cm) was taken as a result [46].

A multilevel endurance fitness test (beep test) was conducted using an alternative 15 m protocol. This test is usually performed on 20 m lines, but its utility has been proven even at 15 m distances in children and adolescents [47]. Students had to run between two points (cones) 15 m apart in time to recorded beeps. They started at the first interval at a speed of 8.5 km/h, which increased by 0.5 km/h with each level. When participants failed to reach the cone in time, the test was ended for them. The highest level reached was taken as the result of this test [48].

The CAPL-2 knowledge and understanding questionnaire (CAPL-2-KU) and PLAYself questionnaire were used to assess the cognitive and affective domains of PL. Questionnaires were completed on the online platform Survey Monkey (SurveyMonkey Inc., San Mateo, CA, USA). Original versions of CAPL-2-KU and PLAYself were translated into the Croatian language by two experienced researchers. The third researcher back-translated the Croatian version into the English language, and the English-speaking researcher evaluated the back-translated version. Items that were not clear to two experienced PE teachers were corrected, and the final Croatian versions of PL questionnaires were made.

The CAPL-2-KU questionnaire consisted of 12 questions, including guidelines for daily physical activity and daily sedentary time, the definition of cardiorespiratory fitness and muscular strength, understanding of fitness and impact on physical activity, methods of skill, and fitness improvement. Each question had four provided answers; a correct answer was scored as 1, and an incorrect answer was scored as 0. The maximum possible result was 12 points. This questionnaire was proven to be feasible, reliable, and valid in Canadian children [19]. However, since this was the first study where CAPL-2-KU was used in the Croatian language (please see before where we explained translation-back translation), the questionnaire was applied twice in a time frame of 7 days in order to evaluate the test–retest reliability of the measurement tool.

The PLAYself questionnaire is part of the PLAY tools and is used to establish the perceived level of PL of children and adolescents. PLAYself has four subsections: (i) environment, assessing the degree of movement confidence in different environments (e.g., activities in the gym, in and on the water, on the snow); (ii) PL self-description measure of affective and cognitive aspects related to PL that determines an individual's self-efficacy and its relation to their participation in PA, including questions such as "It does not take me long to learn new skills, sports or activities", "I think that being active is important for my health and well-being", and "I understand the words that coaches and PE teachers use"; (iii) relative ranking of literacy, numeracy, and physical literacy in different settings including school, at home, and with friends, which examines how much an individual values each literacy; and (iv) fitness, which is determined by the question "My fitness is good enough to let me do all the activities I choose" and is not included in the final score. The final score consists of subtotals from the first three subsections divided by the number of questions. The maximum PLAYself score is 100, representing high self-perceived PL [21]. PLAYself demonstrated good psychometric properties in adolescents [22], but in this study, the Croatian version of the questionnaire was applied twice (test–retest) in a period of 7 days in order to evaluate the reliability of the applied Croatian version of the questionnaire.

2.3. Statistical Analyses

The normality of the variables was tested with the Kolmogorov–Smirnov test. The internal consistency of subsets of the PLAYself questionnaire (i.e., environment, self-description, relative ranking of literacies) was estimated by Cronbach's alpha coefficients (α) for both test and retest. In general, α-values indicate the correlation between items; thus, high α-values justify summarizing the items into one subscale. Accordingly, α-values were considered as: unacceptable ≤ 0.5; poor ≥ 0.5–0.60; questionable ≥ 0.60–0.7; acceptable ≥ 0.70–0.8; good ≥ 0.80–0.9; excellent ≥ 0.90 [49]. The test–retest reliability of PLAYself total scores and subsets was checked by intraclass correlation coefficients (ICCs) followed by 95% confidence intervals (CIs) and a two-way mixed-effect model with absolute agreement. ICCs were interpreted as: poor ≤ 0.5, moderate $= 0.5$–0.75, good $= 0.75$–0.9, excellent ≥ 0.90 [50].

For estimating the test–retest reliability of the CAPL-2-KU questionnaire, due to response data being dichotomous, weighted Cohen's kappa coefficients (κ) with 95% CI and percent of the overall agreement (p0) were calculated for each question. κ-values were interpreted as slight $= 0.00$–0.20, fair $= 0.21$–0.40, moderate $= 0.41$–0.60, substantial $= 0.61$–0.80, and almost perfect $= 0.81$–1.00, and p0 $\geq 80\%$ was considered as acceptable [51].

After checking the reliability of the questionnaires, further statistical analyses were applied in order to evaluate the validity of the applied PL questionnaires. In the first phase, the construct validity was determined by factor analysis utilizing the principal component analysis extraction—Guttman–Kaiser criterion of extraction with a Varimax raw rotation. The Mann–Whitney U test was used to determine the differences between boys and girls in all variables. The discriminant validity of the PL questionnaires was assessed by determining whether boys and girls deferred in the PL scores. Then, Pearson's correlation coefficients were calculated to determine the association between CAPL-2-KU and PLAYself (total score) questionnaires (convergent validity). Further, Pearson's correlation coefficients were calculated to determine the association of both questionnaires with PF indices, which was done for the total sample and gender-stratified. All correlation analyses were controlled for age as a covariate, knowing the possible influence of students' age on PF and PL. Pearson's R was interpreted as very weak $= 0.00$–0.19, weak $= 0.20$–0.39, moderate $= 0.40$–0.69, strong $= 0.70$–0.89, very strong correlation $= 0.90$–1.00 [52].

Statistical package Statistica ver.13 (Palo Alto, CA, USA) was used for all analyses, and a *p*-level of 0.05 was applied.

3. Results

3.1. Reliability and Validity of the PLAYself and CAPL-2-KU Questionnaire

The reliability of the PLAYself is shown in Table 1. The PLAYself had acceptable-to-good internal consistency. Precisely, α-values for the environment subsection consisting of six items were acceptable at test and retest. Self-description subset (consisting of 12 items) had acceptable and good α-values at test and retest, respectively. The subsection relative ranking of literacy, numeracy, and physical literacy had good α-values at test and retest.

Table 1. Internal consistency and test–retest reliability of the PLAYself questionnaire.

Variable	Test	Retest	Test–Retest Reliability	
	α	α	ICC	95% CI
PLAY environment	0.71	0.77	0.82	0.76 to 0.86
PLAY self-description	0.76	0.81	0.87	0.82 to 0.90
PLAY literacy	0.76	0.74	0.59	0.47 to 0.69
PLAY numeracy	0.79	0.75	0.66	0.56 to 0.74
PLAY physical literacy	0.77	0.71	0.53	0.41 to 0.64
PLAYself total score			0.85	0.79 to 0.89

Legend: α = Cronbach's alpha coefficient; ICC = intraclass correlation; 95% CI = 95% confidence interval.

The intraclass correlation coefficient (ICC) was good for the total PLAYself score (ICC = 0.85). Environment subsection had good test–retest reliability (ICC = 0.82), self-description subset had good reliability (ICC = 0.87), while the subsection relative ranking of literacy, numeracy, and physical literacy had moderate reliability (ICC = 0.59, 0.66, 0.53, respectively).

The reliability of the CAPL-2-KU is shown in Table 2. Following dichotomization in the CAPL-2-KU, of 12 items, two had substantial (κ = 0.67 to 0.69), four had moderate (κ = 0.44 to 0.49), four had fair (κ = 0.30 to 0.39), and two had slight (κ = 0.14 to 0.20) test–retest reliability. According to the percent of absolute agreement, six items had acceptable reliability (p0 = 84.87 to 93.42%), while the other six items had somewhat lower agreement (p0 = 70.39 to 77.63%). It is important to note that several items with slight κ-values had high p0 (e.g., κ = 0.14 with p0 = 93.42).

Table 2. Test–retest reliability of the CAPL-2-KU questionnaire.

Item	κ (95% CI)	Test–Retest % Agreement (p0)
Q1	0.69 (0.57 to 0.81)	85.53
Q2	0.67 (0.55 to 0.80)	84.87
Q3	0.30 (0.08 to 0.44)	77.63
Q4	0.20 (−0.03 to 0.43)	86.18
Q5	0.44 (0.30 to 0.59)	73.03
Q6	0.46 (0.31 to 0.62)	77.63
Q7	0.39 (0.24 to 0.55)	73.68
Q8	0.32 (0.16 to 0.47)	70.39
Q9	0.36 (0.19 to 0.53)	76.97
Q10	0.14 (−0.14 to 0.43)	93.42
Q11	0.49 (0.30 to 0.68)	86.84
Q12	0.45 (0.24 to 0.65)	86.84

Legend: κ = Weighted kappa coefficient; 95% CI = 95% confidence interval; Q1 = Physical activity guidelines; Q2 = Sedentary behavior and screen time guidelines; Q3 = Cardiovascular fitness definition; Q4 = Musculoskeletal fitness definition; Q5 = Importance of having fun during physical activity; Q6 = Importance of physical activity in general; Q7 = Knowledge of muscle endurance; Q8 = Knowledge of muscle strength exercises; Q9 = Knowledge of when to perform stretching exercises; Q10 = Knowledge of the meaning of pulse, heartbeat; Q11 = Knowledge of how to improve sports skills; Q12 = Knowledge of how to get in better shape.

The construct validity of the PLAYself questionnaire, i.e., its five subscales, was confirmed. Precisely, factor analysis extracted two significant factors. The first factor explained 37.53%, and the second factor explained 31.46% of the variance (Table 3).

Table 3. Factor analysis of the PLAYself questionnaire.

	Factor 1	Factor 2
PLAYself environment	0.84	0.00
PLAYself self-description	0.91	0.09
PLAYself literacy	0.06	0.83
PLAYself numeracy	0.07	0.83
PLAYself physical literacy	0.58	0.44
FV	1.88	1.57
PT	0.38	0.31

FV—Factor variance; PT—Proportion of the explained variance.

3.2. Gender Differences in Fitness and Physical Literacy Results

Gender differences in anthropometry, fitness, and physical literacy results are shown in Table 4. Boys were taller and had greater body mass and body mass index than girls. Boys achieved better results in all fitness tests except for flexibility, where girls reached better scores (Table 4).

Table 4. Descriptive statistics and differences between boys and girls in physical literacy, anthropometry, and fitness variables.

	Boys (195)		Girls (403)		MW–U Test	
	Mean	SD	Mean	SD	Z-Value	p
Body height (cm)	180.65	7.2	167.58	8.05	16.32	0.001
Body mass (kg)	73.3	15.18	60.5	12.48	11.01	0.001
Body mass index (kg/m^2)	22.42	4.27	21.52	3.95	3.21	0.01
Standing long jump (cm)	213.2	33.46	168.87	25.95	14.76	0.001
Sit-and-reach (cm)	7.22	8.04	12.71	11.18	−7.54	0.001
Sit-ups (n)	68.68	38.83	51.79	12.22	11.26	0.001
Beep test (level)	11.6	3.24	7.89	2.47	12.53	0.001
CAPL-2-KU	8.63	2.19	8.52	2.22	0.52	0.6
PLAYself total score	69.29	13.05	67.66	12.96	1.94	0.05
PLAYself environment	52.07	17.57	49.87	16.54	1.62	0.11
PLAYself self-description	74.26	17.22	68.77	17.12	3.72	0.001
PLAYself literacy	74.5	20.45	82.97	18.06	−4.72	0.001
PLAYself numeracy	63.06	24.61	64.19	23.89	−0.36	0.72
PLAYself physical literacy	84.88	17.62	87.02	17.3	−1.49	0.17

MW–U = Mann–Whitney U test; SD = Standard deviation.

Boys and girls achieved similar scores in CAPL-2-KU (scores of 8.63 and 8.52, $p > 0.6$) and PLAYself total score (scores of 69.26 and 67.66, $p > 0.05$) (see Table 4). Significant gender differences were found in PLAYself subdomains: boys had greater results in the subdomain of self-description (scores of 74.26 and 68.77, respectively; $p = 0.001$), while girls had greater results in the subset of ranking of literacy (scores of 82.92 and 74.5, respectively; $p = 0.001$), which confirmed the discriminant validity of the questionnaire. PLAYself and CAPL-2-KU were not intercorrelated in the total sample (R = 0.03), in boys (R = 0.1), and in girls (R = 0.01).

For the total sample, CAPL-2-KU was associated with sit-and-reach (R = 0.10) and sit-ups (R = 0.11). PLAYself had stronger associations with fitness variables (R = 0.27 to 0.48). However, due to previously mentioned significant differences between boys and girls in fitness, more specific insight on the associations between fitness and PL is evidenced in gender-stratified correlational analyses (Table 5).

Correlations between PL questionnaires and fitness status in boys is presented in Table 5. Specifically, CAPL-2-KU was significantly associated only with the sit-and-reach test (4% of the common variance) in boys. Meanwhile, fitness tests (standing long jump, sit-ups, and beep test) were significantly associated with PLAYself total score (3% to 17% of the common variance), subsection of environment (4% to 17% of the common variance), self-description (5% to 25% of the common variance), and ranking of physical literacy (8% to 8% of the common variance).

For girls, CAPL-2-KU had only low correlation (R = 0.11) with the beep test. Further, all physical fitness tests were associated with PLAYself total score (3–18% of the common variance), subsection of environment (3% to 12.5%), self-description (4–26% of the common variance), and ranking of physical literacy (1% to 5% of the common variance) in girls (Table 5).

Table 5. Pearson's correlation coefficients between physical literacy questionnaires.

Variable	Body Height	Body Mass	BMI	Standing Long Jump	Sit-and-Reach	Sit-Ups	Beep Test
				Total sample ($n = 544$)			
CAPL-2-KU	0.00	0.02	0.02	0.05	0.10 *	0.11 **	0.11
PLAYself total score	0.02	−0.06	−0.10 *	0.29 ***	0.11 **	0.37 ***	0.42
PLAYself environment	0.04	−0.04	−0.07	0.27 ***	0.10 *	0.34 ***	0.36
PLAYself self-description	0.10 *	−0.05	−0.13 ***	0.38 ***	0.08	0.46 ***	0.51
PLAYself literacy	−0.20 ***	−0.08	0.01	−0.15 ***	0.09 *	−0.09 *	−0.10
PLAYself numeracy	−0.04	0.00	0.01	0.03	0.05	−0.03	0.01
PLAYself physical literacy	−0.08	−0.05	−0.02	0.06	0.06	0.10 **	0.18
				Boys ($n = 141$)			
CAPL-2-KU	0.05	0.09	0.08	0.03	0.19 **	0.13	0.13
PLAYself total score	−0.14	−0.04	0.01	0.16 *	0.06	0.40 ***	0.45 ***
PLAYself environment	−0.06	−0.01	0.02	0.19 **	0.05	0.37 ***	0.44 ***
PLAYself self-description	−0.11	−0.10	−0.07	0.23 ***	0.05	0.47 ***	0.52 ***
PLAYself literacy	−0.09	0.10	0.14	−0.11	−0.01	−0.04	−0.07
PLAYself numeracy	−0.06	0.03	0.05	−0.09	0.01	−0.13	−0.11
PLAYself physical literacy	−0.16	0.00	0.06	0.07	0.10	0.29 ***	0.28 ***
				Girls ($n = 403$)			
CAPL-2-KU	−0.04	−0.03	−0.02	0.06	0.07	0.10	0.11 *
PLAYself total score	0.01	−0.15 **	−0.18 ***	0.40 ***	0.19 ***	0.38 ***	0.48 ***
PLAYself environment	0.02	−0.13 **	−0.15 **	0.35 ***	0.18 ***	0.35 ***	0.35 ***
PLAYself self-description	0.05	−0.16 **	−0.20 ***	0.44 ***	0.19 ***	0.44 ***	0.53 ***
PLAYself literacy	−0.11 **	−0.06	−0.02	0.00	0.05	0.04	0.06
PLAYself numeracy	−0.03	−0.02	−0.02	0.11	0.08	0.04	0.11
PLAYself physical literacy	−0.03	−0.06	−0.06	0.12	0.03	0.07	0.23 ***

Legend: * $p < 0.05$; ** $p < 0.01$; *** $p < 0.001$; BMI = Body mass index.

4. Discussion

The most important findings in our research are as follows. First, Croatian versions of the CAPL-2-KU and PLAYself are appropriately reliable. Next, results obtained for CAPL-2-KU and PLAYself questionnaires are not intercorrelated. Moreover, there were no differences between boys and girls in the applied PL questionnaire, except in the dimension of self-description where boys had higher results than girls.

4.1. CAPL-2-KU and PLAYself—Reliability and Construct Validity of the Croatian Versions

Both CAPL-2-KU and PLAYself questionnaires had appropriate reliability. Specifically, CAPL-2-KU total score had moderate test–retest reliability, while individual questions had fair-to-substantial reliability and an acceptable overall percentage of agreement. Precisely, questions regarding physical activity (Q1) and screen time guidelines (Q2) had substantial reliability, while questions regarding musculoskeletal fitness (Q4) and knowledge of the meaning of pulse–heartbeat (Q10) had slight reliability. Most probably, tested children have appropriate knowledge of physical activity guidelines (Q1 and Q2), while their theoretical knowledge of musculoskeletal fitness (Q4) is somewhat lower and, thus, less consistent. Supportively, the original version of CAPL-2-KU showed moderate test–retest reliability in Canadian children aged 8–12 years [19]. They also reported low reliability of the question regarding how to get in better shape, and the authors explained it by possibly misinterpreting the question.

The Croatian version of the PLAYself questionnaire had acceptable-to-good internal consistency and moderate-to-good test–retest reliability. Such results are also in line with previous studies where the authors examined the reliability of the PLAYself [22,53]. Specifically, good internal consistency and moderate test–retest reliability were reported for PLAYself in children aged 8–14 years [22]. Meanwhile, the study by Caldwell, Di Cristofaro, Cairney, Bray, and Timmons [53] recorded questionable-to-good internal consistency of PLAYself environment, self-description, and ranking of literacies subsections in children aged 8.4–13.7 years.

The factor analysis confirmed the construct validity of the PLAYself questionnaire and its subscales. Two factors were extracted. Factor one was correlated with the environment and self-description subscales, indicating that these two subscales define similar constructs of PL. The second factor was associated with the ranking of two literacies—numeracy and literacy—meaning that they determine the ranking of literacies. Thus, it can be concluded that the Croatian version of the PLAYself questionnaire has good construct validity, which is in accordance with a study that also showed good psychometric properties, including the convergent validity, of the PLAYself questionnaire in Canadian children and youth [22].

4.2. CAPL-2 Knowledge and Understanding and PLAYself

As PL has become an important and widely investigated concept, numerous definitions of PL exist to date [15]. Accordingly, as PL assessment depends on how PL is defined, various PL assessment tools are also used. Although our finding on the lack of correlation between CAPL and PLAYself may seem surprising at first glance, it is actually in line with the main idea that PL is a generally complex concept and that various components/domains contribute to overall PL [10]. Having that in mind, in the explanation of the evident independence of CAPL-2-KU and PLAYself, a brief overview of definitions and assessment procedures of the two observed measurement tools is provided in the following text.

CAPL-2 was developed by the Healthy Active Living and Obesity Research Group, and the authors used Whitehead's definition of PL: "the motivation, confidence, physical competence, knowledge and understanding that individuals develop in order to maintain physical activity at an appropriate level throughout their life" [13]. Knowledge and understanding are considered the core elements of the cognitive domain of PL, and Ennis [54] argued that knowledge and understanding provide the basis for recognizing and knowing what, when, and how to perform physical activity and believed that they comprise the heart of the PL concept. The CAPL-2-KU includes questions on physical activity and sedentary behavior recommendations, knowledge of fitness, and related terms; that is, it is mainly based on the theoretical knowledge of PA and its importance [19].

On the other hand, PLAYself is a part of the Physical Literacy Assessment for Youth (PLAY) tools developed by the Canadian Sport for Life Society [21]. The authors of PLAY tools believe that "people who are physically literate have the competence, confidence, and motivation to enjoy a variety of sports and physical activities" [21]. PLAY consists of several tools, and one of them is a PLAYself questionnaire used for the self-description of PL in children and youth. Specifically, as explained in Methods, PLAYself consists of three subscales that make up the final score (environment, self-description, ranking of literacies) [22]. Thus, PLAYself assesses the cognitive and affective domains of PL, with a special emphasis on perceived competence, which is related to participating in a greater variety of activities and sports [34].

Therefore, although both are oriented toward PL in general, CAPL-2-KU and PLAYself seem to assess different domains of PL in different ways. Briefly, CAPL-2-KU mainly assesses theoretical knowledge of PA, sedentary behavior, and definitions regarding various aspects and forms of PA and is primarily oriented toward the cognitive domain of PL. On the other hand, PLAYself is more oriented toward the cognitive and affective domains of PL (i.e., self-description and perceived competence for PA). Supportively, it is already theorized that PLAYself does not properly assess the understanding of PA [15]. Therefore, the lack of correlation between these two measurement tools is understandable. This is additionally discussed later when correlations between fitness status and PL are contextualized.

4.3. Physical Literacy and Gender

Considering the significant gender differences in fitness status (e.g., boys achieved better results in all tests but flexibility), a result showing no evident gender differences in PL could be somewhat surprising. Briefly, because of the evident superiority of boys in fitness status, it would be expected that girls would (objectively) perceive their PL as (also) lower (in Croatia, boys and girls participate together in PE classes). However, irrespective of mixed PE classes, fitness norms (standards) in the Croatian educational system are standardized for each gender, which allows children to objectively evaluate their achievement in comparison to their gender. Therefore, it is possible that both boys and girls self-evaluated and reported even their PL while comparing themselves within their own gender.

Indeed, it has been reported that children and adolescents are good at judging themselves against others from their age and gender groups and are probably forming more precise appraisals of their own ability [55]. A study of Caldwell et al. (2021) did not record differences between boys and girls in confidence, motivation, and knowledge about PL, as assessed by the PLAYself questionnaire [53]. Additionally, there were no gender differences in the motivation and confidence and knowledge and understanding domains assessed with CAPL-2 in Canadian children [20], which altogether support our findings and discussion.

Despite the non-significant difference in overall PL, when the self-description subset of PLAYself was specifically observed, boys had higher scores than girls. In the study of Kozera [56], boys had significantly higher PL self-description scores than females (mean difference 2.54), mainly due to lower scores for questions related to competence and enjoyment in females. Moreover, the study by Jefferies, Bremer, Kozera, Cairney, and Kriellaars [22] reported that PLAYself self-description was associated with general sport competence in adolescents aged 8–14 years [22]. However, this is understandable because the self-description subset consists of questions covering self-perceived competence for playing sports or engaging in physical activities, while boys are consistently more involved in sports and physical activities than girls and logically feel more competent in sporting activities [57]. One could argue that the previously explained mechanism of "within-gender comparison" could appear here also, but this is not likely because of the mixed PE classes in Croatia, where boys and girls often play sports together, which leaves no doubt about better competence in boys.

4.4. Physical Literacy and Its Association with Physical Fitness

Our results show that PL is positively associated with PF in high school children, which is generally in accordance with previous studies conducted with somewhat younger adolescents and children [25,34]. Specifically, a study by Caldwell et al. (2020) reported an association between PL assessed by PLAY tools and aerobic fitness in children 9–12 years old [34]. Additionally, Lang et al. (2018) showed a significant association between cardiorespiratory fitness and all four domains of PL assessed by CAPL-2 in Canadian children 8–12 years old [25]. The association between PL composite scores and PF is logical and supports the importance of PL in improving PF and health in general. However, in our study, PLAYself displayed higher associations with PF indices than CAPL-2-KU. The possible explanation is discussed in further text.

Previous studies displayed an association between the PLAYself subset of self-description and PF. In brief, cardiovascular fitness, jumping capacity, and abdominal strength were significantly associated with PLAYself, meaning that adolescents with better fitness status have higher self-perception of their physical capabilities [22]. Moreover, PLAYself was related to health-related fitness assessments, including speed and 20 m shuttle run tests in children and adolescents [56]. This supports the notion of an interrelationship between actual competence (i.e., objectively measured PF) and self-perceived competence assessed by the PLAYself. In the meantime, CAPL-2-KU is oriented more toward the cognitive domain of PL, which also should be related to PF as well (i.e., better knowledge on PF would logically be related to better objectively measured PF). However, in our study, the association between CAPL-2 KU and PF is evidently low (although statistically significant, but this was due to a large number of subjects and a large number of degrees of freedom). Similarly, in the study conducted on Canadian children aged 8–12 years, the physical competence domain had the strongest, while the K&U domain had the weakest associations with fitness status [25]. The reason for the relatively low association should be found in the fact that the Croatian PE curriculum still ignores the necessity of improving knowledge and understanding as one of the crucial domains of PL itself. The authors of the study as PE teachers and academicians can witness that the PE curriculum in Croatia is mainly focused on the development of motor competencies (i.e., motor skills and fitness status), while the improvement of overall "theoretical" knowledge related to PA, its overall importance in everyday life, and its associations with health status is lacking. Therefore, it is not surprising that the association between the K&U domain of PL and PF is lower than the association between PLAYself and PF in Croatian adolescents.

4.5. Limitations and Strengths

The main limitation of this study is that we conducted the PL assessment only by questionnaires that evaluated only the cognitive and affective domains of PL. The reason for this is that in Croatia, thus far, the concept of PL practically does not exist, and this is one of the first studies investigating this issue in the country and region. Thus, at first, we included only questionnaires for the preliminary assessment of PL in Croatian adolescents. Indeed, this study is a preliminary investigation of PL in Croatia, which will provide a basis and act as a cornerstone for further studies investigating PL in more detail and including other (all) domains of PL.

This is one of the first studies in southeastern Europe and probably the first one in Croatia that assessed PL in adolescents using two different PL questionnaires. Furthermore, one of the strengths of this research is that we validated applied questionnaires by correlating them to an extensive battery of objectively measured fitness tests. Additionally, the included fitness tests could be considered as an assessment of the physical competence domain of PL in future studies (at least as a part of the physical competence domain of PL assessments). As authors are directly involved in the PE curriculum in the country, we believe that those standardized fitness tests that are already implemented in the Croatian PE classes could be considered valid for assessing the physical domain of PL, as they are similar to several PL physical tests already implemented in PL assessment tools (CAPL-2 and PLAY tools).

5. Conclusions

Croatian versions of PLAYself and CAPL-2-KU assessment tools showed appropriate reliability. Therefore, applied PL questionnaires can be used in the evaluation of different PL domains in children from Croatia but also in the whole region of southeastern Europe where similar languages are spoken.

The validity of the applied questionnaires is confirmed by throughout analysis of associations between PL assessment tools and objectively measured PF. In brief, since PF is associated with PL in Croatian adolescents, we may support the assumption/theory that PL is an essential determinant for the development of PF.

42. Thomas, E.; Petrigna, L.; Tabacchi, G.; Teixeira, E.; Pajaujiene, S.; Sturm, D.J.; Sahin, F.N.; Gómez-López, M.; Pausic, J.; Paoli, A. Percentile values of the standing broad jump in children and adolescents aged 6–18 years old. *Eur. J. Transl. Myol.* **2020**, *30*, 9050. [CrossRef]

43. Ab Rahman, Z.; Kamal, A.A.; Noor, M.A.M.; Geok, S.K. Reliability, Validity, and Norm References of Standing Broad Jump. *Rev. Geinetecgestao Inov. E Tecnol.* **2021**, *11*, 1340–1354. [CrossRef]

44. Chandana, A.; Xubo, W. The Test Battery: Evaluate Muscular Strength and Endurance of the Abdominals and Hip-Flexor Muscles. In Proceedings of the General Sir John Kotelawala Defence University—14th International Research Conference, Colombo, Sri Lanka, 9–10 September 2021; pp. 23–28.

45. Ojeda, Á.H.; Maliqueo, S.G.; Barahona-Fuentes, G. Validity and reliability of the Muscular Fitness Test to evaluate body strength-resistance. *Apunt. Sports Med.* **2020**, *55*, 128–136. [CrossRef]

46. Miyamoto, N.; Hirata, K.; Kimura, N.; Miyamoto-Mikami, E. Contributions of hamstring stiffness to straight-leg-raise and sit-and-reach test scores. *Int. J. Sports Med.* **2018**, *39*, 110–114. [CrossRef]

47. McClain, J.J.; Welk, G.J.; Ihmels, M.; Schaben, J. Comparison of Two Versions of the PACER Aerobic Fitness Test. *J. Phys. Act. Health* **2006**, *3*, S47–S57. [CrossRef]

48. Bagchi, A.; Nimkar, N.; Yeravdekar, R. Development of Norms for Cardiovascular Endurance Test for Youth Aged 18–25 Years. *Mortality* **2019**, *1*, 3–4. [CrossRef]

49. Connelly, L.M. Cronbach's alpha. *Medsurg. Nurs.* **2011**, *20*, 45–47. [PubMed]

50. Koo, T.K.; Li, M.Y. A Guideline of Selecting and Reporting Intraclass Correlation Coefficients for Reliability Research. *J. Chiropr. Med.* **2016**, *15*, 155–163. [CrossRef]

51. McHugh, M.L. Interrater reliability: The kappa statistic. *Biochem. Med.* **2012**, *22*, 276–282. [CrossRef]

52. Akoglu, H. User's guide to correlation coefficients. *Turk. J. Emerg. Med.* **2018**, *18*, 91–93. [CrossRef]

53. Caldwell, H.; Di Cristofaro, N.A.; Cairney, J.; Bray, S.R.; Timmons, B.W. Measurement properties of the Physical Literacy Assessment for Youth (PLAY) Tools. *Appl. Physiol. Nutr. Metab.* **2021**, *46*, 571–578. [CrossRef]

54. Ennis, C.D. Knowledge, transfer, and innovation in physical literacy curricula. *J. Sport Health Sci.* **2015**, *4*, 119–124. [CrossRef] [PubMed]

55. Stodden, D.F.; Goodway, J.D.; Langendorfer, S.J.; Roberton, M.A.; Rudisill, M.E.; Garcia, C.; Garcia, L.E. A Developmental Perspective on the Role of Motor Skill Competence in Physical Activity: An Emergent Relationship. *Quest* **2008**, *60*, 290–306. [CrossRef]

56. Kozera, T.R. Physical Literacy in Children and Youth. Ph.D. Thesis, University of Manitoba, Winnipeg, MB, Canada, 2017.

57. Telford, R.M.; Telford, R.D.; Cochrane, T.; Cunningham, R.B.; Olive, L.S.; Davey, R. The influence of sport club participation on physical activity, fitness and body fat during childhood and adolescence: The LOOK Longitudinal Study. *J. Sci. Med. Sport* **2016**, *19*, 400–406. [CrossRef] [PubMed]

Article

The Correlation between Psychological Characteristics and Psychomotor Abilities of Junior Handball Players

Vlad-Alexandru Muntianu *, Beatrice-Aurelia Abalașei , Florin Nichifor and Iulian-Marius Dumitru

Faculty of Physical Education and Sports, University „Alexandru Ioan Cuza", 700506 Iași, Romania;
beatrice.abalasei@uaic.ro (B.-A.A.); florin.nichifor@uaic.ro (F.N.); iulian.dumitru@uaic.ro (I.-M.D.)
* Correspondence: vlad.muntianu@uaic.ro

Abstract: The general development of the sports world has guided researchers in sports science to study excellence in sports performance, namely, the study of the characteristics and requirements specific to each sport. However, in order to meet these requirements, each individual must have a set of specific characteristics similar to those of the group to which he/she belongs. The variables in the study are related to the psychomotor abilities and psychological aspects that could influence the overall performance of junior III handball players. The main work instruments are related to field testing and psychological characteristics measurement. For psychomotor abilities, we used means such as the TReactionCo software (eye–hand coordination), Just Jump platform (dynamic balance), Tractronix system (general dynamic coordination), and Illinois test (spatial-temporal orientation), and for the psychological characteristics, we used the Motivational Persistence Questionnaire. In addition, the result of the study is represented by new software that we created in order to better observe the level of development of these characteristics in junior handball players. From a statistical point of view, we calculated the correlations between psychomotor abilities and psychological characteristics using ANOVA in order to see field position differences and performed linear regression for the variables of this study.

Keywords: handball; motor capacity; psychological characteristics; junior players; software creation

Citation: Muntianu, V.-A.; Abalașei, B.-A.; Nichifor, F.; Dumitru, I.-M. The Correlation between Psychological Characteristics and Psychomotor Abilities of Junior Handball Players. *Children* **2022**, *9*, 767. https://doi.org/10.3390/children9060767

Academic Editors: Hugo Miguel Borges Sarmento, Georgian Badicu and Ana Filipa Silva

Received: 6 April 2022
Accepted: 20 May 2022
Published: 24 May 2022

Publisher's Note: MDPI stays neutral with regard to jurisdictional claims in published maps and institutional affiliations.

Copyright: © 2022 by the authors. Licensee MDPI, Basel, Switzerland. This article is an open access article distributed under the terms and conditions of the Creative Commons Attribution (CC BY) license (https://creativecommons.org/licenses/by/4.0/).

1. Introduction

The handball game, due to its complexity, requires and is equally part of the improvement of the motor manifestation mode during matches from the perspective of two components.

Players must be mentally prepared to cope with psychological discomfort resulting from the stressful effects of competitions, training periods in isolated areas, the monotony of training, affected interpersonal relationships, or even conflict situations.

Sport psychology, as a branch of psychology applied in our field, has the objective of studying the adaptations of people's mental processes, needs for competitive activity, and training periods.

All stages of sports training, especially in the current period, contain an increased level of difficulty in the context of interdisciplinary and psychological contributions [1,2].

An individual or collective investigation from a psychological point of view can establish the level at which the athlete is able to fulfill or demonstrate as effectively as possible the skills, knowledge, and degree of training, all of which are related to the level of competition.

The psychology of sports activities, as a precisely directed scientific discipline, addresses and studies the educational and intellectual processes of athletes engaged in sports activities, with behavior and motor reaction being precisely pursued goals.

In order to obtain the best possible performance, motor activity presents physical and mental demands, and the maximum effort alternates between these two spectra of the performance of players.

The main purpose of this work is to highlight correlations between the psychological characteristics and psychomotor abilities of junior handball players. The main mental aspects that we chose were represented by motivational persistence, self-discipline, and planning capacity.

Firstly, we chose to assess motivational persistence because it can be an important factor that can help players to continue their activity at a higher level of performance, even if we are talking about junior players. Motivation mainly concerns behaviors focused on fulfilling personal goals such as joy, curiosity, satisfaction, and interest. All of these are integrated with the inner self and correspond to the value system of the athlete [3].

Secondly, self-discipline can be an important aspect in team sports and is characterized by the capacity of players to assess the situation that they face daily according to the necessities of the game.

This psychological characteristic has a much greater influence than the close follow-up of training, level of training, and skill development because a handball player must have mental self-discipline in order to make the best decisions in unpredictable environments where competitiveness is extremely high [4].

Thirdly, planning capacity has great importance in playing team handball because of the influence that it can have in daily situations that a team and the players experience. According to sports psychologists such as [5], planning should be conducted in accordance with physical training. First of all, together with the athlete and coaches, sports psychologists study the main objectives of each training cycle in order to be able to fulfill a series of mental objectives that are established.

The game of handball, as a performance sport, requires intense effort from players due to the extremely difficult demands of training and competitions that require maximum concentration, in addition to all of the physical, volitional, and mental abilities.

Sports performance, like human performance in general, is determined by logical, emotional, and creative activities of the spirit of athletes and plays a key role in obtaining the best indices of motor manifestation [6].

The main goal of high performance in sports activities is considered the borderline activity of human possibilities and multilateral mental and physical processes that evolve simultaneously with the athlete.

In recent years, the study of performance in sports based on anthropometry has shown [7]:

1. How morphological prototypes are important for success within and among the sports phenomenon;
2. An increased morphological variability in some sports compared to other disciplines or sports;
3. Athletes who own or who by specific means obtain an optimal anthropometric profile for a specific sporting event are more likely to achieve success;
4. Morphological optimization is useful for assessing training status and talent selection at both female and male levels.

The intensity of this sport varies between walking, moderate running, short-distance sprints, lateral movements, and forward/backward movements, so an increased level of specific endurance is critically needed to maintain an optimal degree of play throughout the match [8,9].

We can say that, along with body mass, other anthropometric characteristics have been shown to be crucial for sports activities and thus achieving performance in the game of handball [10], as well as in other team sports such as volleyball, football, and rugby [8]. For example, throwing can be seen as the most important technical action that players perform [11–13] and is dependent on the arm's ability to achieve a sufficient degree of acceleration so that the ball leaves the hand at the maximum possible speed.

Based on the anthropometric measurements of handball players presenting different levels of practicing this sport, it was concluded that, among the players analyzed, those who showed a higher level of performance were the tallest with a large wingspan and an optimal body mass with low proportions of fat [14].

The analysis of the entire kinematic chain at the time of throwing is seen as an impractical means of evaluation, with joint mobility being a common method of quantification using traditional goniometric measurements of the area of movement. Few studies have deepened these types of measurements and their influence on mobility, and statistically insignificant results can be found in the reports of specialized studies [7,15].

Skills such as obtaining visual information about a moving object (ball and opponents) and a high degree of oculomotor coordination are advantages for players, as it allows them to react to external stimuli more effectively by adapting their whole body according to the situation in the field [14]. Other studies conducted showed that perceptual skills combined with a good ability to anticipate movements and distribute attention can lead to success in this team sport [16].

Competitions in sports games take place in environments that differ in space and time, aspects that condition the wave of information provided by opponents, and a good decision-making ability allows athletes to move on the playing surface in the best circumstances.

Authors have stated the importance of the reaction time arising from oculomotor coordination for obtaining victories [17–19]. The evaluation of the reaction time of athletes revealed that the teams that obtained positive results had a much higher index of reaction time, which led to winning matches.

New training methods are focused on good physical training, which is achieved by using methods and means similar to game situations and from the perspective of the psychological aspects required by the modern version of this sport [20].

Experts in this sport and coaches analyze the efficiency of the players that they train through standardized methods for each competition, so they can accurately determine the contribution of each player to the success of the team, as well as the team in general [21].

The explosive force on the lower limbs is responsible for performing various variations of jumping, speed running, and acyclic movements in the handball game. In this sport, a player must generate forces with different variations in intensity (throwing or fighting for positions on the field) [22]. This is why a differentiated load in each training session must be included for the development of maximum strength. Vertical jumps are predominant in performing most technical and tactical tasks in handball. Impulse jumps on both legs are much more common in the defensive phases (blocking the balls thrown by the opponent).

The application of different methods and means for improving the explosive force had a positive result in the case of juniors but also for those who practice this sport at a high level. The most common method for improving vertical jumping is plyometric training [1,23], followed by neuromuscular and proprioceptive training.

If we are to analyze the awareness of sports specialists, we can notice that it has increased with regard to the role of personality in developing sports performance. In competitive sports, such as handball, the players and trainers are on the lookout for better performance; therefore, great efforts are made in order to enhance the player's achievements. While trying to obtain better performance, the actual performance of the athletes is constantly evolving alongside their psychological characteristics [24].

Further research has focused on the diversity of psychological profiles and psychological characteristics such as motivation, mechanisms involved in coping with stress, and the adaptation skills of different athletes by age, gender, and sport type [6,25–28].

In our domain (sports science), a nomothetic approach is more frequently used. Domain specialists that have the same view focus on the socio-demographics and sports characteristics of athletes and carry out comparisons by age, gender and sport type [29–32].

Alongside the assessment of some basic somatic parameters [12,33,34], motor skills [35,36], and sports seniority [36,37], it seems imperative to realize evaluations of psychomotor abilities.

Aspects such as the prediction of the adversary's movements with or without the ball, attention, choosing the appropriate response, perception, and high levels of sensory and motor fitness are some elements that can help a team win the game [19,38]. With the aid of these psychomotor abilities and by gathering visual information through a high level of eye–hand coordination, players can react to external stimuli with more efficiency and adapt their movements according to situations that happen on the court [39].

As work objectives, we identified the most important ones, such as:

- Assessing the level of development of certain psychomotor characteristics;
- Identifying the most important psychological characteristics that could influence the activity of junior handball players;
- Creating a performance profile of junior III handball players;
- Creating a software program that could help the training process and the selection phase by observing the general development of the players.

Firstly, we assumed that the most important psychological characteristics for junior handball players are motivational persistence, planning capacity, self-discipline, implementation, and recurrence.

Secondly, we assumed, through the second hypothesis, that some psychomotor characteristics of the subjects have correlations with psychological aspects.

Lastly, the final hypothesis was to see if there is any correlation between the psychological aspects of these junior handball players, which was tested with multivariate tests.

2. Materials and Methods

2.1. Sample

For this study, we chose to apply our evaluation protocol on 181 junior III handball players from 10 teams in Romania. The research group was an average of 13.5 ± 0.5 years, and they came from different living areas (rural/urban). All of the findings were based on the results that we gathered from tests and measurements, and the analysis was conducted using statistical software.

2.2. Design

For the research design, we chose a set of field tests in order to assess specific psychomotor components of junior III handball players, further applying a questionnaire for some psychological characteristics such as motivational persistence, self-discipline, and planning.

The main psychomotor characteristics that we focused on were eye–hand coordination, dynamic balance, spatial orientation, and general dynamic coordination, and for each of these elements, we applied a field test with specific elements from the handball game.

All of the field tests had the main objective of testing the hypotheses of the present research study.

2.3. Instruments

As instruments, we used TReactionCo Sofware, laptops, special keyboards, tripods, Just Jump platform, infrared sensors, cones, and the Motivational Persistence Questionnaire for the psychological aspects; all of their actual usage in the research is presented in the Procedure subsection.

2.4. Data Analysis

The data presented are based on correlations, means and standard deviations, 95% confidence intervals, p values, Durbin Watson statistic, R-value, ANOVA analysis, and also Cronbach's alpha.

After obtaining data from the players, all of the information was input into the statistical analysis software SPSS v. 26.

2.5. Procedure

Before the beginning of field testing, all players were informed about the structure of the evaluation, and the tests were explained and demonstrated so that they would have an overall view of the process. The evaluation started with the questionnaire to measure their levels of motivational persistence, self-discipline, and planning capacity. The questionnaire used is a validated one, and we also computed Cronbach's alpha in SPSS to obtain a score reliability coefficient.

Eye–hand coordination: For the system setting, the upper limb option is selected. The athlete sits on a chair and positions the keyboard resting on his/her thighs. The laptop is placed in front of the subject at a distance allowing better observation of markings that appear on the screen. The evaluation begins when the athlete presses one of the three keys. A red dot appears on the left or right side of the screen, and the athlete has the task of pressing the key on the side where that mark appears as soon as possible after the appearance of the stimulus. Each subject has to analyze 20 successive images, and the average reaction time at the level of the upper limbs is recorded. It is considered a correct assessment if the athlete has more than 10 correct hits/touches. The average time should be as short as possible.

Dynamic balance: Depending on the type of jump that needs to be performed, the player positions him/herself on the jumping platform, performing the movement necessary to achieve the goal in order to achieve the best values (SJ, FJ, CMJ, and 4X). Depending on the distance to be covered, the distance is calculated with the help of a roulette wheel, positioning the sides on either side of the running lane to position the gates with infrared photocells. The athlete positions him/herself at the starting line and starts running at free speed and enters the deceleration process after passing the last area with the whole cell with photocells.

Spatial-temporal orientation—Illinois test description: With the help of roulette, a distance of 5 m in a straight line is measured. Heads are positioned at point 0, and 2.5 m and 5 m distances are measured. From the level of the head from a distance of 2.5 m, 3 successive distances of 3.3 m are measured with the help of the roulette wheel, at the level of which milestones are placed. At the start and finish points, two gates with infrared sensors are positioned, which start automatically after the athlete leaves, and his/her final time is recorded. The subject has to complete the route as shown below.

General dynamic coordination: Distances of 5/10/15/20 m are measured on a straight line, and the markings for the positioning of the tripods with photocells are positioned on either side of the corridor on which the athlete will run. At the beginning of the test, the athlete must run at a high tempo without major deviations in the direction of travel in order to avoid inconsistencies in the values obtained.

Agility—505 test description: The athlete must run to the 15 m marker in order to accumulate sufficient speed to move; after crossing the 5 m mark and crossing the imaginary drawn line, the athlete must run back the same distance of 5 m. The recorded time is the time in which the athlete travels a distance of 5 m (round trip). The twisting ability of each leg is tested, and the subjects are instructed that they should not exceed the 5 m line by much so as not to waste too much time.

Choosing the psychological variables in the present study was preceded by analyzing research papers, in which we tried to identify the elements that might have implications in sports practice. Generally, we identified some common aspects, such as motivation, and we considered that even at a young age, the capacity to maintain a high level of motivation can represent a true advantage in the future careers of these junior handball players. This action was followed by identifying other characteristics that we considered to be important (self-discipline and planning capacity); these two have the ability to improve the general level of the athletes. It is necessary, at any age, when practicing team sports, to have the capacity to self-organize for the tasks that are presented in training sessions, in official competitions, and during free time between these events. Continuous growth and the maintenance of a linear and constant evolution can be influenced by such factors. In addition, the capacity to plan, even for a junior player, is important, as it means that if the objectives and the purposes of the physical activities are clearly stated, the sports life of the individual can improve.

Regarding the psychological characteristics, they were evaluated by applying the Motivational Persistence Questionnaire, which contains 30 items divided into 5 categories (with 5 questions for each dimension), with a 5-step answer scale (1—to a very small extent; 5—to a very large extent) and can determine the levels of the following characteristics: motivational persistence, planning, self-discipline, determination, recurrence, and ambition [40].

- Model questions:
- For motivational persistence: "I maintain my motivation even in activities that lasts for months";
- For planning: "I plan in detail what I have to do for the next day";
- For self-discipline: "Even if it's not necessary, I put my things in order."

3. Results

In the first phase, we computed reliability statistics in order to obtain the Cronbach's alpha value for the Motivational Persistence Questionnaire that we applied in our research group. As is shown in Table 1, the 622 value puts the motivational persistence, self-discipline and planning capacity in the upper range of confidence for the specific items.

Table 1. Reliability statistics of the questionnaire applied.

Item 1	0.617
Item 2	0.606
Item 3	0.608
Item 4	0.594
Item 5	0.626
Item 6	0.590
Item 7	0.598
Item 8	0.629
Item 9	0.596
Item 10	0.609
Item 11	0.619
Item 12	0.625
Item 13	0.630
Item 14	0.624
Item 15	0.615
Cronbach's Alpha	0.622

In the table below (Table 2) are presented the means of the psychological aspects after the evaluation of players alongside the number of players and standard deviation. In addition, it presents the overall value of Cronbach's alpha for the questionnaire applied and the individual values for the items attributed to the motivational persistence, planning capacity, and self-discipline.

Table 2. Mean and standard deviation of the general scores of psychological characteristics.

	Motivational Persistence	Planning	Self-Discipline
Mean	6.31	6.39	6.44
N	181	181	181
SD	2.52	2.27	2.30

N—number of subjects; SD—standard deviation.

We also calculated the general evaluation results for the psychological aspects as well as the mean and standard deviation of the group. If we analyze the levels presented in the table above, we can see that the whole group can be qualified in the upper limit of development for the three psychological characteristics measured (Table 2).

Furthermore, we performed an ANOVA analysis (Table 3) in order to observe the main differences between the means of specific field positions for the psychological characteristics. As is shown in the table above, we obtained multiple results that are statistically significant, which led us to the conclusion that for different players that occupy a certain area in the field, the psychological characteristics are present but to different extents. This variation can be connected to the general psychological state of the junior players, and outside factors can influence their motor capacities at some point.

Table 3. Post hoc analysis. Statistically significant means of the psychological characteristics.

Dependent Variable			Sig.
Motivational persistence	Right wing	Right back	0.764
		Center back	0.003
		Left wing	0.049
		Goalkeeper	0.001
	Right back	Right wing	0.764
		Center back	0.008
		Goalkeeper	0.000
	Center back	Right wing	0.003
		Right back	0.008
		Left back	0.004
		Pivot	0.000
		Goalkeeper	0.000
	Left back	Right wing	0.931
		Center back	0.004
		Goalkeeper	0.001
	Left wing	Right wing	0.049
		Pivot	0.003
		Goalkeeper	0.000
		Left wing	0.000
		Pivot	0.031

Table 3. *Cont.*

Dependent Variable			Sig.
Planning	Right wing	Right back	0.777
		Center back	0.005
		Left wing	0.020
		Goalkeeper	0.003
	Right back	Right wing	0.777
		Center back	0.013
		Left wing	0.043
		Goalkeeper	0.001
	Center back	Right wing	0.005
		Right back	0.013
		Left back	0.007
		Pivot	0.015
		Goalkeeper	0.000
	Left back	Right wing	0.936
		Center back	0.007
		Left wing	0.027
		Goalkeeper	0.002
	Left wing	Right wing	0.020
		Right back	0.043
		Left back	0.027
		Pivot	0.046
		Goalkeeper	0.000
Self-discipline	Right wing	Right back	0.777
		Center back	0.005
		Left wing	0.000
		Goalkeeper	0.000
	Right back	Right wing	0.777
		Centre back	0.012
		Left wing	0.001
		Pivot	0.712
		Goalkeeper	0.000
	Center back	Right wing	0.005
		Right back	0.012
		Left back	0.007
		Pivot	0.007
		Goalkeeper	0.000
	Left back	Right wing	0.935

In the table above (Table 4), we show the Pearson correlations calculated for the field tests that we used in this study, and the significant values are highlighted. As we can observe, there is a high negative correlation between the countermovement jump and the 505 test (-0.229) and a high positive correlation between the 505 test and the 10 m run

(0.418). The 10 m run also has a strong negative correlation with the countermovement jump (-0.477).

Table 4. Main correlations between psychomotor testing results.

		Agility (S)	Dynamic Balance (INCH)	General Dynamic Coordination (S)
Agility (S)	Pearson Correlation	1	-0.229 **	0.418 **
	Sig. (2-tailed)		0.002	0.000
	N	181	181	181
Dynamic Balance (INCH)	Pearson Correlation	-0.229 **	1	-0.477 **
	Sig. (2-tailed)	0.002		0.000
	N	181	181	181
	Sig. (2-tailed)	0.157	0.581	0.783
	N	181	181	181
General Dynamic Coordination (S)	Pearson Correlation	0.418 **	-0.477 **	1
	Sig. (2-tailed)	0.000	0.000	
	N	181	181	181

Legend: N—number of players. ** Correlation is significant at the 0.01 level (2-tailed).

For the agility and general dynamic coordination, the r-value (0.418) signifies that there is a directly proportional influence and growth between them. This means that better indices or low results on the agility course can increase the results for the psychomotor ability mentioned. This led us to a primary conclusion that the training process of the junior handball players must be performed properly, and it must influence, in balanced proportions, all of the motor and psychomotor areas.

The dynamic balance presents a negative r-value (-0.477) with general dynamic coordination. This result means that if the dynamic balance results increase, the psychomotor ability decreases in a beneficial way by reducing the time when taking the test. The handball game, by its growth, is known as a very dynamic sport in which the players must perform multiple motor tasks, such as running and passing the ball, running–stopping–throwing, and avoiding the opponents. Thus, all of these elements can be positively influenced by the effective development of these psychomotor abilities.

Another statistical analysis that we conducted was related to the effects of the psychological aspects of the players, which were analyzed and tested through the questionnaire applied in our study (Table 5). As we can notice, there is a significance of 0.00 between motivational persistence, planning, and self-discipline, while planning capacity has a strong correlation of 0.00 with self-discipline. By gathering these results, we can state the fact that besides the motor capacities of these junior handball players, psychological aspects such as motivational persistence, self-discipline, and planning capacity are some factors that need to be taken under consideration in reference to these players and in general in the handball game.

Table 5. Multivariate tests for the psychological aspects.

	F	Sig.
Motivational persistence * planning	411.89	0.000
Motivational persistence *	603.56	0.000
self-discipline	2.85	0.001
Planning * self-discipline	4262.17	0.000

Lastly, we performed linear regression (Table 6) in order to see which characteristics tend to have a greater influence when taken into consideration with the other variables of the study. Our output shows a relationship between psychomotor abilities (general dynamic coordination, eye–hand coordination, spatial-temporal orientation, and dynamic balance) and psychological aspects (self-discipline and motivational persistence) of 83%, which can influence the general evolution of performance when practicing this team sport.

Table 6. Linear regression for the variables of the study.

Model Summary [b]				
Model	R	R Square	Adjusted R Square	Std. Error of the Estimate
1	0.917 [a]	0.840	0.833	0.595

[a]. Predictors: (constant), self-discipline, general dynamic coordination (s), eye–hand coordination, spatial-temporal orientation(s), dynamic balance, and motivational persistence. [b]. Dependent variable: planning.

The conclusion from these values is that in a dynamic sport that involves the continuous movement of players with and without the ball, a good level of these psychomotor abilities alongside a good psychological capacity can improve the overall output of the players; these aspects need to be taken into consideration in the training activity.

To show the influence of the variables and their weight in the overall linear regression, we highlight the standardized coefficients and the p values of the psychomotor and psychological aspects (Table 7). As is shown in the table, statistical significance is indicated for eye–hand coordination ($p = 0.02$), general dynamic coordination ($p = 0.04$), and self-discipline ($p = 0.03$). This led us to the conclusion that these three aspects could have a greater influence on the overall process of the evolution of junior handball players.

Table 7. Statistical significance of variables in the general regression.

	Standardized Coefficients	Sig.
	Beta	
Eye–hand coordination	−0.584	0.022
Dynamic balance	0.005	0.950
General dynamic coordination	0.899	0.045
Motivational persistence	0.091	0.653
Planning	0.021	0.904
Self-discipline	0.312	0.039

Another important aspect that can be highlighted relates to the software that we created in order to have a better perspective of the psychomotor and psychological characteristics of junior III handball players. The "Skills" software, by using mathematical algorithms, can create an overview of the above-mentioned aspects, and it can find utility both in the training process and in talent identification actions.

4. Discussion

Firstly, we assessed the level of development of psychomotor abilities of junior handball players by applying motor tests that mimic specific elements of the handball game, and the results were registered and further analyzed from a statistical point of view. We focused our attention upon eye–hand coordination, dynamic balance, general dynamic coordination, and spatial-temporal orientation, these being important abilities that can influence the handball game and can represent an important factor in the talent identification phase.

As we know, of great importance in this stage of their development, psychological aspects can influence their general evolution both in the sports area and in the social environment.

After applying the questionnaire, we can observe the level of development of these psychological characteristics, for which the players scored above average (on a scale from 1 to 10) in the case of motivational persistence (6.31), planning (6.39), and self-discipline (6.44), but for better performance, we believe that the average scores of the three components could be improved.

Furthermore, we tried to create a performance profile of junior handball players by selecting these psychomotor abilities alongside psychological aspects, these having the overall capacity of creating an overview for the players.

Some studies have neem related to the influence of supplements and their interaction with motor capacity [41], and others are related to the optimization of motor skills such as hand–eye coordination [42] and biomechanics [43], but we tried to first assess more psychomotor abilities that could influence the handball game in order to have a better overview of them.

In regard to talent identification, there are studies that have focused mainly on playing position and the month of birth [44], and the difference in our study is represented by the motor and psychological aspects that we tried to identify as important to talent identification. Other researchers have tried to identify factors for talent identification by interviewing coaches and studying psychological characteristics such as coordination and carefully planning the activity of the players [45]. This information can support our research by highlighting the importance of some psychological aspects regarding the identification of gifted players.

The level of performance of the players can be compared to a series of studies that have connected anthropometric characteristics with physical components such as agility and the ability to jump in the identification of talented players [46]. Other studies from a series of disciplines have also included some of the performance characteristics mentioned previously, these being considered important in the general assessment of junior players in order to identify talented athletes [47–51].

After computing linear regression results, we obtained a good relationship between psychomotor abilities and psychological aspects, and in terms of the statistical value, it was shown that the R-value (0.83/83%) puts these two aspects as general elements that can condition the level of the other variables presented in this study.

A practical application of the results of this study is the creation of the "Skills" software through which we tried to aid the training, selection, and talent identification processes. In a general overview (as can be seen in the figures below), the pages of the software are related firstly to the playing positions on the handball field, and after choosing players from the uploaded database, the evaluator or coach can choose any player that they want in order to assess their general level of psychomotor manifestation (Figure 1). There are more important aspects that can be seen from this besides psychomotor abilities; as is shown in Figure 2, there are also some somatic measurements that we consider to be important in the handball game. Lastly, the psychological aspects of the players are presented (Figure 3), all of which make it possible to observe and make the best decision for the team in terms of training, selection, and even talent identification. The software was created for that purpose using IT technologies (Python), and other studies that involve its usage are set to be written.

Some limitations of the study could be represented by the number of players analyzed; a much larger study could be conducted and might include a greater number of athletes. In addition, another limitation is related to the specialty of the literature in the sense that we could not find more research articles to compare our results with.

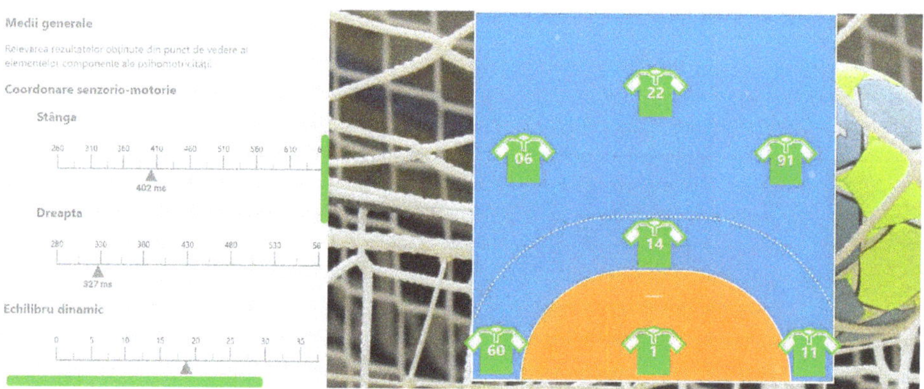

Figure 1. Representation of the created software's first page.

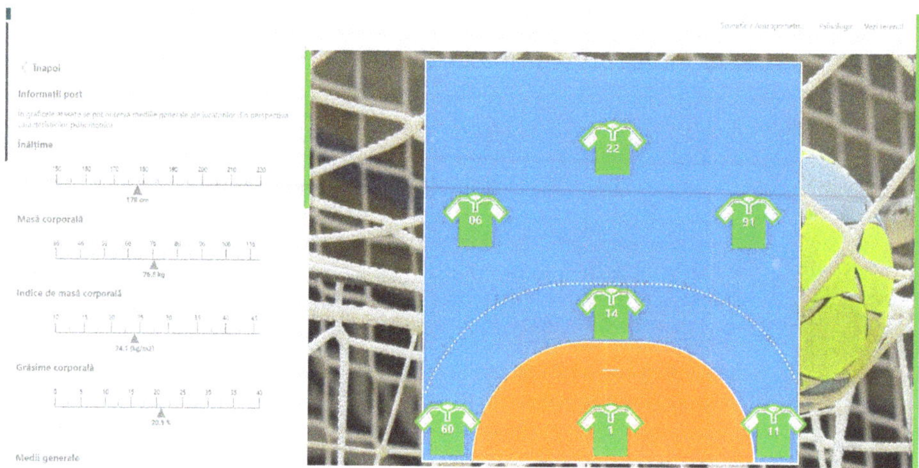

Figure 2. Representation of the software's second page.

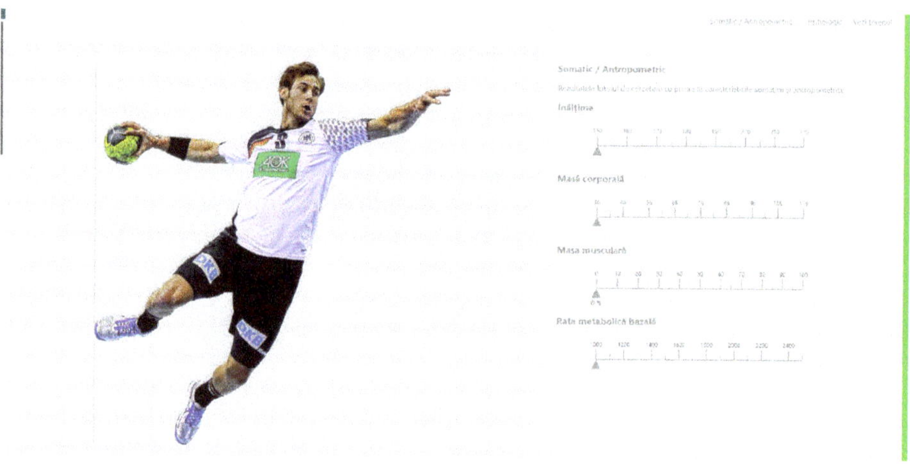

Figure 3. Representation of the software's third page.

5. Conclusions

From all of the data obtained from the statistical analysis, we can highlight the need to make this kind of comparison between psychomotor and psychological factors in junior handball players. The presence of significant correlations supports the veracity of the control tests used in outlining the main characteristics of these junior players.

On the other hand, the results obtained can be seen as a starting point for further leading the training process of these junior handball players by concentrating on and leading the activity towards improving the elements presented in the study.

Author Contributions: Conceptualization, V.-A.M. and B.-A.A.; investigation, V.-A.M.; writing—original draft preparation; V.-A.M. and F.N., writing—review and editing; B.-A.A., supervision; B.-A.A., methodology; F.N. and I.-M.D.; resources—I.-M.D. All authors have read and agreed to the published version of the manuscript.

Funding: This research received no external founding.

Institutional Review Board Statement: The study was conducted in accordance with the Declaration of Helsinki and approved by the Ethics Committee of University "Alexandru Ioan Cuza" from Iași, Faculty of Physical Education and Sports (protocol code 454/ 3 May 2021).

Informed Consent Statement: Each participant provided written, informed consent to participate in the study through their legal representatives.

Data Availability Statement: Not applicable.

Conflicts of Interest: The authors declare no conflict of interest.

References

1. Gabbett, T.J. Physiological characteristics of junior and senior rugby league players. *Br. J. Sports Med.* **2002**, *36*, 334–339. [CrossRef] [PubMed]
2. Ghuman, P.S.; Godara, H.L. The analysis of plyometric training program on university handball players. *J. Sports Phys. Educ.* **2013**, *1*, 37–41.
3. Alesi, M.; Gómez-López, M.; Chicau-Borrego, C.; Monteiro, D.; Granero-Gallegos, A. Effects of a Motivational Climate on Psychological Needs Satisfaction, Motivation and Commitment in Teen Handball Players. *Int. J. Environ. Res. Public Health* **2019**, *16*, 2702. [CrossRef] [PubMed]
4. Dosil, J. *The Sport Psychologist's Handbook a Guide for Sport-Specific Performance Enhancement*; John Wiley & Sons Ltd.: Chichester, UK, 2006.
5. Dosil, J. *Psicología de la Actividad Física y del Deporte*; McGraw-Hill: Madrid, Spain, 2004.
6. Hermassi, S.; Gabbett, T.J.; Spencer, M.; Khalifa, R.; Chelly, M.S.; Chamari, K. Relationship between explosive performance measurements of the lower limb and repeated shuttle-sprint ability in elite adolescent handball players. *Int. J. Sports Sci. Coach.* **2014**, *9*, 1191–1204. [CrossRef]
7. Massuça, L.; Fragoso, I. Morphological characteristics of adult male handball players considering five levels of performance and playing position. *Coll. Antropol.* **2015**, *39*, 109–118. [PubMed]
8. Michalsik, L.; Madsen, K.; Aagaard, P. Match Performance and Physiological Capacity of Female Elite Team Handball Players. *Int. J. Sports Med.* **2013**, *35*, 595–607. [CrossRef]
9. Povoas, S.C.; Seabra, A.F.; Ascensao, A.M.; Magalhaes, J.; Soares, J.M.; Rebelo, A.N. Physical and physiological demands of elite team handball. *J. Strength Cond. Res.* **2012**, *26*, 3365–3375. [CrossRef]
10. Manchado, C.; Tortosa-Martínez, J.; Vila, H.; Ferragut, C.; Platen, P. Performance factors in women's team handball: Physical an physiological aspects—A review. *J. Strength Cond. Res.* **2013**, *27*, 1708–1719. [CrossRef]
11. Campa, F.; Silva, A.M.; Talluri, J.; Matias, C.N.; Badicu, G.; Toselli, S. Somatotype and bioimpedance vector analysis: A new target zone for male athletes. *Sustainability* **2020**, *12*, 4365. [CrossRef]
12. Gorostiaga, E.M.; Granados, C.; Ibáñez, J.; Izquierdo, M. Differences in physical fitness and throwing velocity among elite and amateur male handball players. *Int. J. Sport. Med.* **2005**, *26*, 225–232. [CrossRef]
13. Granados, C.; Izquierdo, M.; Ibañez, J.; Bonnabau, H.; Gorostiaga, E. Differences in physical fitness and throwing velocity among elite and amateur female handball players. *Int. J. Sports Med.* **2007**, *28*, 860–867. [CrossRef] [PubMed]
14. Van den Tillaar, R. Comparison of range of motion tests with throwing kinematics in elite team handball players. *J. Sports Sci.* **2016**, *34*, 1976–1982. [CrossRef] [PubMed]
15. Schwesig, R.; Hermassi, S.; Wagner, H.; Fischer, D.; Fieseler, G.; Molitor, T.; Delank, K.S. Relationship Between the Range of Motion and Isometric Strength of Elbow and Shoulder Joints and Ball Velocity in Women Team Handball Players. *J. Strength Cond. Res.* **2016**, *30*, 3428–3435. [CrossRef] [PubMed]

16. Paul, M.; Biswas, S.K.; Sandhu, J.S. Role of sports vision and eye hand coordination training in performance of table tennis players. *Braz. J. Biomotricity* **2011**, *5*, 106–116.

17. Mankowska, M.; Poliszczuk, T.; Poliszczuk, D.; Johne, M. Visual perception and its effect on reaction time and time-movement anticipation in elite female basketball players. *Pol. J. Sport Tour.* **2015**, *22*, 3–8. [CrossRef]

18. Abbott, A.; Button, C.; Pepping, G.J.; Collins, D. Unnatural selection: Talent identification and development in sport. *Nonlinear Dyn. Psychol. Life Sci.* **2005**, *9*, 61–88.

19. Kida, N.; Oda, S.; Matsumura, M. Intensive baseball practice improves the Go/Nogo reaction time, but not the simple reaction time. *Cogn. Brain Res.* **2005**, *22*, 257–264. [CrossRef]

20. Yüksel, M.; Tunç, G. Examining the Reaction Times of International Level Badminton Players Under 15. *Sports* **2018**, *6*, 20. [CrossRef]

21. Bojić, I.; Kocić, M.; Veličković, M.; Nikolić, D. Correlation between morphological characteristics and situational—Motor abilities of young female handball players. In Proceedings of the 20 Scientific Conference "FIS Communications 2017" in Physical Education, Sport and Recreation, Nis, Serbia, 19–21 October 2017; pp. 317–322.

22. Ohnjec, K.; Vuleta, D.; Milanović, D.; Gruić, I. Performance indicators of teams at the 2003 World Handball Championship for women in Croatia. *Kinesiology* **2008**, *40*, 69–79.

23. Spieszny, M.; Zubik, M. Modification of strength training programs in handball players and its influence on power during the competitive period. *J. Hum. Kinet.* **2018**, *63*, 149–160. [CrossRef]

24. Hammami, M.; Gaamouri, N.; Aloui, G.; Shephard, R.J.; Chelly, M.S. Effects of combined plyometric and short sprint with change-of-direction training on athletic performance of male U15 handball players. *J. Strength Cond. Res.* **2018**, *33*, 662–675. [CrossRef] [PubMed]

25. Tóth, L. The emergence, development and importance of the self and self-image. *Magy. Sporttudományi Szle.* **2006**, *7*, 6–10.

26. Bebetsos, E.; Bebetsos, G. Greek youth team handball players and their satisfaction levels. *Eur. Handball Mag.* **2006**, 1–9.

27. Diehm, R.; Armatas, C. Surfing: An avenue for socially acceptable risk-taking, satisfying needs for sensation seeking and experience seeking. *Personal. Individ. Differ.* **2004**, *36*, 663–667. [CrossRef]

28. Kais, K.; Raudsepp, L. Intensity and direction of competitive state anxiety, self-confidence and athletic performance. *Kinesiology* **2005**, *37*, 13–20.

29. Révész, L. The Features of the Hungarian Swimming Sports in Terms of the Choice of the Branch of Sport, Selection and the Care of Talented Persons. Ph.D. Thesis, Semmelweis Egyetem, Budapest, Hungary, 2008.

30. Géczi, G. Success and Talent Development as Indicated by Motor Tests and Psychometric Variables of u18 Ice Hockey Players. Ph.D. Thesis, Semmelweis Egyetem, Budapest, Hungary, 2009.

31. Csáki, I.; Fózer-Selmeci, B.; Bognár, J.; Szájer, P.; Zalai, D.; Géczi, G.; Révész, L.; Tóth, L. New measurement method: Examination of psychological factors by the Vienna Test System among football players. *Testnev. Sport Tudomány* **2016**, *1*, 8–20. [CrossRef]

32. Gyömbér, N.; Lénart, A.; Kovács, K. Differences between Personality Characteristics and Sport Performance by Age and Gender. *Acta Fac. Educ. Phys. Univ. Comenianae. Publ.* **2013**, *53*, 5–17.

33. Tóth, L.; Géczi, G.; Bognár, J. Examination of competitive state anxiety, athletic coping strategies, and emotional attributes in Hungarian ice hockey players. *Kalokagathia* **2011**, *49*, 129–147.

34. Kőnig-Görögh, D.; Gyömbér, N.; Szerdahelyi, Z.; Laoues, N.; Balogh, Z.O.; Tóth-Hosnyánszki, A.; Csaba, Ö. Personality profiles of junior handball players: Differences as a function of age, gender, and playing positions. *Cogn. Brain Behav.* **2017**, *21*, 237–247. [CrossRef]

35. Massuça, L.; Fragoso, I. Study of Portuguese handball players of different playing status. A morphological and biosocial perspective. *Biol. Sport* **2011**, *28*, 37. [CrossRef]

36. Justin, I.; Vuleta, D.; Pori, P.; Kajtna, T.; Pori, M. Are taller handball goalkeepers better? Certain characteristics and abilities of Slovenian male athletes. *Kinesiol. Int. J. Fundam. Appl. Kinesiol.* **2013**, *45*, 252–261.

37. Hoff, J.; Almåsbakk, B. The effects of maximum strength training on throwing velocity and muscle strength in female team-handball players. *J. Strength Cond. Res.* **1995**, *9*, 255–258.

38. Nakamoto, H.; Mori, S. Sport-specific decision-making in a Go/NoGo reaction task: Difference among nonathletes and baseball and basketball players. *Percept. Mot. Ski.* **2008**, *106*, 163–170. [CrossRef] [PubMed]

39. Kioumourtzoglou, E.; Derri, V.; Tzetzls, G.; Theodorakis, Y. Cognitive, perceptual, and motor abilities in skilled basketball performance. *Percept. Mot. Ski.* **1998**, *86*, 771–786. [CrossRef] [PubMed]

40. Constantin, T.; Holman, A.; Hojbotă, A. Development and Validation of a Motivational Persistence Scale. *Psihologija* **2011**, *45*, 99–120. [CrossRef]

41. Kaczka, P.; Maciejczyk, M.; Batra, A.; Tabęcka-Łonczyńska, A.; Strzała, M. Acute Effect of Caffeine-Based Multi-Ingredient Supplement on Reactive Agility and Jump Height in Recreational Handball Players. *Nutrients* **2022**, *14*, 1569. [CrossRef]

42. Badau, D.; Badau, A.; Ene-Voiculescu, C.; Larion, A.; Ene-Voiculescu, V.; Mihaila, I.; Fleacu, J.L.; Tudor, V.; Tifrea, C.; Cotovanu, A.S.; et al. The Impact of Implementing an Exergame Program on the Level of Reaction Time Optimization in Handball, Volleyball, and Basketball Players. *Int. J. Environ. Res. Public Health* **2022**, *19*, 5598. [CrossRef]

43. Tuquet, J.; Cartón, A.; Marco-Contreras, L.A.; Mainer-Pardos, E.; Lozano, D. Analysis of the Steps Cycle in the Action of Throwing in Competition in Men's Elite Handball. *Sustainability* **2022**, *14*, 5291. [CrossRef]

44. Gomez-Lopez, M.; Granero-Gallegos, A.; Feu Molina, S.; Luis, J.; Rios, C. Relative age effect during the selection of young handball player. *J. Phys. Educ. Sport* **2017**, *17*, 418–423.
45. Bjørndal, C.T.; Ronglan, L.T.; Andersen, S.S. Talent development as an ecology of games: A case study of Norwegian handball. *Sport Educ. Soc.* **2015**, *22*, 864–877. [CrossRef]
46. Gómez-López, M.; Angosto, S.; Ruiz-Sánchez, V. Relative age effect in the selection process of handballplayers of the regional selection teams. *E-Balonmanocom J. Sport Sci.* **2017**, *13*, 3–14.
47. Mohamed, H.; Vaeyens, R.; Matthys, S.; Multael, M.; Lefevre, J.; Lenoir, M.; Philppaerts, R. Anthropometric and performance measures for the development of a talent detection and identification model in youth handball. *J. Sports Sci.* **2009**, *27*, 257–266. [CrossRef] [PubMed]
48. Lidor, R.; Falk, B.; Arnon, M.; Cohen, Y.; Segal, G.; Lander, Y. Measurement of talent in team handball: The questionable use of motor and physical tests. *J. Strength Cond. Res.* **2005**, *19*, 318–325. [CrossRef] [PubMed]
49. Cobley, S.; Baker, J.; Wattie, N.; McKenna, J. Annual age-grouping and athlete development: A meta-analytical review of relative age effects in sport. *Sports Med.* **2009**, *39*, 235–256. [CrossRef] [PubMed]
50. Matthys, S.P.; Fransen, J.; Vaeyens, R.; Lenoir, M.; Philippaerts, R. Differences in biological maturation, anthropometry and physical performance between playing positions in youth team handball. *J. Sports Sci.* **2013**, *31*, 1344–1352. [CrossRef]
51. Reilly, T.; Bangsbo, J.; Franks, A. Anthropometric and physiological predispositions for elite soccer. *J. Sports Sci.* **2000**, *18*, 669–683. [CrossRef]

children

Article

Reliability of International Fitness Scale (IFIS) in Chinese Children and Adolescents

Ran Bao [1,†] , Sitong Chen [2,†] , Kaja Kastelic [3,4] , Clemens Drenowatz [5] , Minghui Li [6] , Jialin Zhang [7] and Lei Wang [8,*]

1 Centre for Active Living and Learning, School of Education, The University of Newcastle, Newcastle 2308, Australia; Ran.Bao@uon.edu.au
2 Institute for Health and Sport, Victoria University, Melbourne 8001, Australia; sitong.chen@live.vu.edu.au
3 Andrej Marušič Institute, University of Primorska, Muzejski Trg 2, 6000 Koper, Slovenia; kaja.kastelic@s2p.si
4 InnoRenew CoE, Livade 6, 6310 Izola, Slovenia
5 Division of Sport, Physical Activity and Health, University of Education Upper Austria, 4020 Linz, Austria; clemens.drenowatz@ph-ooe.at
6 Department of Sports Science and Physical Education, The Chinese University of Hong Kong, Hong Kong SAR, China; venuslmh@link.cuhk.edu.hk
7 School of Physical Education and Sport Science, Fujian Normal University, Fuzhou 350007, China; zhangjialinx@126.com
8 School of Physical Education and Sport Training, Shanghai University of Sport, Shanghai 200438, China
* Correspondence: wl2119@163.com
† These authors contributed equally to this work.

Abstract: Background and Objectives: It has previously been shown that the International Fitness Scale (IFIS) is a reliable and valid instrument when used in numerous regions and subgroups, but it remains to be determined whether the IFIS is a reliable instrument for use with Chinese children and adolescents. If the reliability of the IFIS can be verified, populational surveillance and monitoring of physical fitness (PF) can easily be conducted. This study aimed to test the reliability of the IFIS when used with Chinese children and adolescents. Methods: The convenience sampling method was used to recruit study participants. In total, 974 school-aged children and adolescents between 11 and 17 years of age were recruited from three cities in Southeast China: Shanghai, Nanjing and Wuxi. The study participants self-reported demographic data, including age (in years) and sex (boy or girl). The participants completed the questionnaire twice within a two-week interval. Results: A response rate of 95.9% resulted in a sample of 934 participants (13.7 ± 1.5 years, 47.4% girls) with valid data. On average, the participants were 13.7 ± 1.5 years of age. The test–retest weighted kappa coefficients for overall fitness, cardiorespiratory fitness, muscle fitness, speed and agility and flexibility were 0.52 (Std. errs. = 0.02), 0.51 (Std. errs. = 0.02), 0.60 (Std. errs. = 0.02), 0.55 (Std. errs. = 0.02) and 0.55 (Std. errs. = 0.02), respectively. Conclusions: The International Fitness Scale was found to have moderate reliability in the assessment of (self-reported) physical fitness in Chinese children and adolescents. In the future, the validity of the IFIS should be urgently tested in Chinese subgroup populations.

Keywords: the International Fitness Scale (IFIS); reliability; children; adolescents; Chinese

Citation: Bao, R.; Chen, S.; Kastelic, K.; Drenowatz, C.; Li, M.; Zhang, J.; Wang, L. Reliability of International Fitness Scale (IFIS) in Chinese Children and Adolescents. *Children* **2022**, *9*, 531. https://doi.org/10.3390/children9040531

Academic Editors: Georgian Badicu, Ana Filipa Silva and Hugo Miguel Borges Sarmento

Received: 28 February 2022
Accepted: 7 April 2022
Published: 8 April 2022

Publisher's Note: MDPI stays neutral with regard to jurisdictional claims in published maps and institutional affiliations.

Copyright: © 2022 by the authors. Licensee MDPI, Basel, Switzerland. This article is an open access article distributed under the terms and conditions of the Creative Commons Attribution (CC BY) license (https://creativecommons.org/licenses/by/4.0/).

1. Introduction

Physical fitness (PF) is an important indicator of an individual's capability to perform physical activity and maintain good health [1]. PF can be defined as the ability of body systems to work together efficiently to allow a human to be healthy and perform daily living activities [2]. Accordingly, PF is a significant marker of health [3,4]. PF is a significant predictor of mortality and morbidity in all-cause [5] and cardiovascular diseases [1,6] and adiposity [7], and the negative impacts of these diseases during childhood and adolescence have negative effects on one's health in adulthood [1]. In addition, PF was also shown to be related to mental health [8,9], including cognitive functions [10] (e.g., academic performance [11]), depression, anxiety, psychological stress [12] and well-being [7,13]. In addition to health-related PF, which includes cardiorespiratory endurance (CRF), muscular strength and endurance (i.e., muscle fitness, MF), body composition and flexibility (FL), there is also skill-related PF, which includes balance, coordination, speed, power, reaction time and speed and agility [2].

Generally, laboratory and field measurements have been used to evaluate PF [4]. A recent review demonstrated that CRF and MF have been most frequently evaluated in children and adolescents [14]. It was, however, also concluded that standard PF assessments would be needed in the future [14]. Moreover, laboratory or field measurements require time, facilities and equipment, and thus may be less feasible in population-based studies [4,15]. Alternatively, self-reported PF or survey-based methods may be more suitable for the assessment of PF in epidemiological studies. There are several existing fitness scale instruments, such as the Physical Self-Perception Profile (PSPP) [16] or the Self-Reported Fitness (SRFit) scale [17], but the limitation of these scales includes having too many items and targeting specific sub-groups of the population [18].

Therefore, a simple self-administered instrument with no limitations in terms of populations might be suitable for use in population-based surveys. Ortega et al. developed a self-administrated scale to evaluate PF in the general population, which is known as the International Fitness Scale (IFIS) [19]. The IFIS uses a five-point Likert scale ("very good", "good", "average", "poor" and "very poor") to assess various components of PF [19]. It has been translated into nine languages and consists of five parts, assessing overall PF, cardiorespiratory fitness, muscle fitness, speed and agility as well as flexibility [19]. Previous research demonstrated acceptable construct reliability and validity in European and South American countries in children and adolescents [4,19–22] as well as adults [18,23–26]. In comparison with adolescents, the IFIS was found to be more reliable and higher levels of PF were reported in children (3–10 years) [4]. In addition, gender differences in self-reported PF were also observed in the previous study [21]. The differences in age and gender in the reliability and validity of the IFIS suggest that future studies should be directed toward this topic. Although the IFIS has been shown to be a reliable and valid instrument to use to assess self-rated PF in numerous regions and subgroups, it remains to be determined whether the IFIS is a reliable instrument with Chinese children and adolescents. National data from China, however, indicated that only a small percentage (approximately 31.75%) of school-aged children and adolescents rated their PF as "excellent" or "good" [27]. In addition, there has been a decline in PF in young adults over the last several decades [28–30]. The decline in PF negatively affects youth health, as discussed above; the monitoring of PF is significant in the design of a strategy to promote a level of PF in Chinese youth. Another justification for this study is that the reliability of the IFIS can be verified, and populational surveillance and the monitoring of PF can easily be conducted. Even though a PF testing system has been built in recent years, a population-based survey is still urgently needed [31]. Considering China's large population, a Chinese version of the IFIS would be beneficial in the monitoring and promotion of PF. The application of similar methods can also facilitate international comparison of PF.

Therefore, researchers need a simple and useful instrument to evaluate the levels of PF in various subgroups to monitor and promote the health of Chinese populations. Cultural adaptation, however, requires the reliability of the IFIS to be tested with Chinese children and adolescents. This study, therefore, aimed to determine the reliability of the International Fitness Scale, Chinese-version (IFIS-C), in children and adolescents.

2. Materials and Methods

2.1. Participants and Sampling

A pilot study was conducted with children and adolescents, which aimed to evaluate the test–retest reliability and construct the validity of the International Fitness Scale in Chinese children and adolescents. The convenience sampling method was used to select participants, and 1170 school-aged children and adolescents were recruited from Shanghai,

Nanjing and Wuxi for this study. An invitation letter was sent to the potential schools, and 7 schools were interested in and agreed to participate in this study. According to previous studies, the sample size in the present study met the standard of reliability [4,20]. In total, 974 school-aged children and adolescents between 11 and 17 years of age provided valid data and were included in the final analysis. This study was approved by the Institutional Review Board (IRB) of the Shanghai University of Sport (Code No.: 102772021RT071). Prior to the questionnaire survey, children and adolescents signed assent forms, and their parents or guardians signed informed consent.

2.2. Measures

2.2.1. Demographics

Children and adolescents were required to self-report demographic data, including age (year) and gender (boy or girl). Participants were separated by age into children and adolescents using a cut point of 13 years [32]; in this study, adolescents were 13 years old and above [32].

2.2.2. Self-Reported Fitness

The International Fitness Scale (IFIS) was used to evaluate the self-estimations of PF using a 5-point Likert scale ("very poor", "poor", "average", "good" and "very good"). Specifically, the IFIS evaluates overall fitness, cardiorespiratory fitness (CRF), muscular strength, speed and agility and flexibility [19]. The English version of the IFIS was separately translated into Chinese (IFIS-C) by two authors who are both proficient in Chinese and English and have research backgrounds in physical activity and fitness promotion. Disagreements on translations between the two authors were resolved by a third author. In addition, back-translation for the IFIS-C into English was conducted by two persons whose source language (English) was their mother tongue [33]. These two persons had no background in PF [33].

2.2.3. Data Collection

The reliability of the International Fitness Scale (IFIS-C) was evaluated through a test–retest design. In order to prevent recall bias, the children and adolescents who participated were required to complete the test–retest within a two-week interval [34]. The procedure used for the reliability test adhered to the COSMIN methodology for systematic reviews of PROMs—user manual (box 6 (reliability)) [35]. On both occasions, the participants completed the measurement under the guidance of the same physical education teacher. The measurement times, timing of the measurement (e.g., before physical education class), measurement place and instructions were similar during both assessment time points.

2.2.4. Statistical Analysis

The sample size for the test–retest reliability was identified according to previous studies [4,20]. The test–retest reliability of the IFIS-C was calculated for categorical variables using weighted kappa coefficients [36]. Kappa coefficients of less than 0 indicated "no agreement"; kappa = 0.0 to 0.20 indicated "slight agreement"; kappa = 0.21 to 0.40 indicated "fair agreement"; kappa = 0.41 to 0.60 indicated "moderate agreement"; kappa = 0.61 to 0.80 indicated "substantial agreement"; and kappa = 0.81 to 1.00 indicated "almost perfect agreement" [37]. The level of statistical significance was set at $p < 0.05$. SPSS software version 26.0 and Stata MP version 14.1 (Stata Corp LP, College Station, TX, USA) were used to calculate descriptive characteristics and the weighted kappa coefficients in this study. A Confirmatory Factor Analysis (CFA) for the IFIS-C was analyzed using IBM SPSS AMOS 26.0 Graphics. A Confirmatory Factor Analysis (CFA) was conducted to evaluate whether the perceived PF measured using the IFIS-C was consistent with the nature of the construct of the IFIS-C. Factor loadings below 0.3 were not interpreted, and a factor loading of 0.5 or higher was accepted [38].

3. Results

After deleting questionnaires with missing data, a response rate of 95.9% resulted in a sample of 934 children and adolescents (47.4% girls) with valid data that were included in the analyses. The sample consisted of 390 children (41.8% girls) and 544 adolescents (58.2% girls). The average age was 13.7 ± 1.5 years.

Figures 1 and 2 show the results of self-rated fitness. Overall, most children rated their fitness as "good", and most adolescents rated their fitness as "average". Meanwhile, no adolescents reported their SP–AG as "very poor".

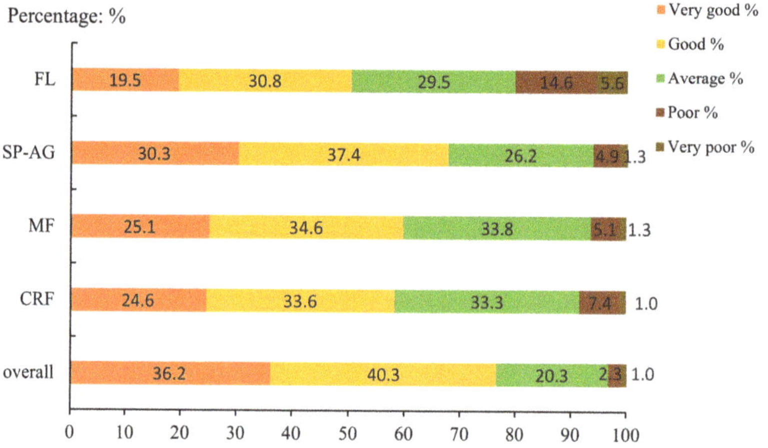

Figure 1. Distribution of responses by categories of self-reported physical fitness in children (CRF = cardiorespiratory fitness; MF = muscular fitness; SP–AG = speed and agility; FL = flexibility).

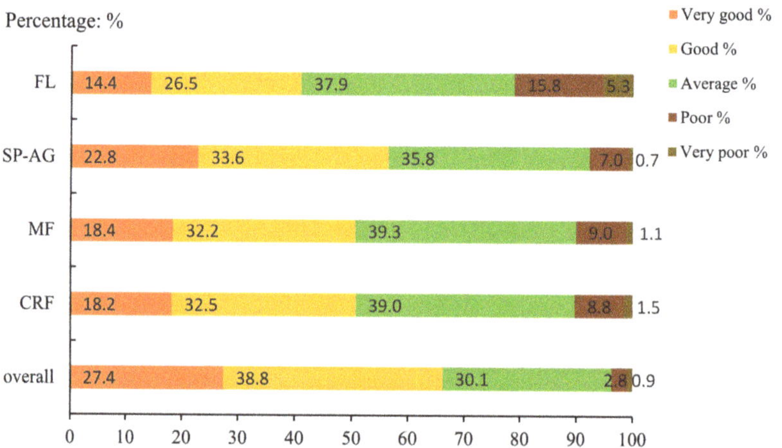

Figure 2. Distribution of responses by categories of self-reported PF in adolescents (CRF = cardiorespiratory fitness; MF = muscular fitness; SP–AG = speed and agility; FL = flexibility).

Table 1 shows the overall weighted kappa values of the test–retest reliability of the self-reported IFIS-C. Test–retest weighted kappa coefficients for overall, CRF, MF, SP–AG and FL were 0.52 (Std. errs. = 0.02), 0.51 (Std. errs. = 0.02), 0.60 (Std. errs. = 0.02), 0.55 (Std. errs. = 0.02) and 0.55 (Std. errs. = 0.02), respectively, which indicate the moderate reliability of the IFIS-C [36]. The highest weighted kappa value of IFIS-C was found for the MF score (weighted kappa = 0.60).

Table 1. Reliability of the International Fitness Scale.

	Test One, Mean (SD)	Test Two, Mean (SD)	Weighted Kappa	Std. Errs.
Overall	3.97 ± 0.88	3.95 ± 0.92	0.52	0.02
CRF	3.64 ± 0.94	3.71 ± 0.93	0.51	0.02
MF	3.66 ± 0.93	3.67 ± 0.95	0.60	0.02
SP–AG	3.79 ± 0.93	3.84 ± 0.91	0.55	0.02
FL	3.35 ± 1.10	3.40 ± 1.09	0.55	0.02

SD = standard deviation; CRF = cardiorespiratory fitness; MF = muscular fitness; SP–AG = speed and agility; FL = flexibility.

Table 2 shows the weighted kappa of the IFIS-C by gender. The overall weighted kappa of the IFIS-C showed moderate reliability, with a somewhat better weighted kappa value observed in girls (0.54) than in boys (0.48). Specifically, the highest weighted kappa value observed in boys was 0.58 (Std. errs. = 0.03) for MF, and the lowest weighted kappa value was 0.48 (Std. errs. = 0.03) for overall fitness. In girls, the highest weighted kappa value was 0.60 (Std. errs. = 0.03) for FL and MF, and the lowest weighted kappa value was 0.49 (Std. errs. = 0.03) for CRF.

Table 2. Weighted kappa of International Fitness Scale by gender.

		Test One, Mean (SD)	Test Two, Mean (SD)	Weighted Kappa	Std. Errs.
Boys	overall	4.11 ± 0.82	4.07 ± 0.91	0.48	0.03
	CRF	3.76 ± 0.96	3.84 ± 0.95	0.51	0.03
	MF	3.78 ± 0.96	3.81 ± 0.99	0.58	0.03
	SP–AG	3.95 ± 0.96	3.96 ± 0.95	0.55	0.03
	FL	3.23 ± 1.10	3.26 ± 1.12	0.51	0.03
Girls	overall	3.82 ± 0.90	3.81 ± 0.93	0.54	0.03
	CRF	3.51 ± 0.91	3.56 ± 0.89	0.49	0.03
	MF	3.52 ± 0.88	3.51 ± 0.87	0.60	0.03
	SP–AG	3.62 ± 0.86	3.70 ± 0.84	0.53	0.03
	FL	3.47 ± 1.08	3.57 ± 1.04	0.60	0.03

SD = standard deviation; CRF = cardiorespiratory fitness; MF = muscular fitness; SP–AG = speed and agility; FL = flexibility.

The weighted kappa of the IFIS-C in children and adolescents are shown in Table 3. Moderate reliability was indicated for all of the components of the IFIS-C in the group of children and adolescents. Nevertheless, higher weighted kappa values were observed in children compared to adolescents. In children, the highest coefficient of weighted kappa was 0.69 (Std. errs. = 0.04) for MF, and the lowest coefficient of weighted kappa was 0.61 for overall fitness and CRF. In adolescents, the highest coefficient of weighted kappa was 0.52 (Std. errs. = 0.03) for MF, and the lowest coefficient of weighted kappa was 0.42 (Std. errs. = 0.03) for overall fitness and CRF. In addition, the internal consistency was accepted, and the alpha coefficient was 0.719 (data not shown in the tables).

The goodness-of-fit for the IFIS-C is outlined in Table 4. The fit indices were 0.95 or higher, and the RMSEA and SRMR were below 0.08, which indicate a good model fit [39]. Indices of model fit indicated that the IFIS-C showed a good model fit in Chinese children and adolescents. However, the RMSEA was 0.114 in girls, which was not acceptable according to the cutoff of below 0.08. Table 5 shows the factor loadings, CR and AVE of the IFIS-C. Factor loadings of CRF, MF and SP–AG were all above 0.5, which indicated acceptable values. However, the factor loading of flexibility was below 0.5. Regarding CR and AVE, CR was accepted by gender and age subgroups. However, values of AVE were only accepted in boys and adolescents.

Table 3. Weighted kappa of International Fitness Scale in children and adolescents.

		Test One, Mean (SD)	Test Two, Mean (SD)	Weighted Kappa	Std. Errs.
	overall	4.08 ± 0.86	4.11 ± 0.90	0.61	0.04
	CRF	3.73 ± 0.95	3.84 ± 0.93	0.61	0.03
Children (n = 390)	MF	3.77 ± 0.93	3.81 ± 0.93	0.69	0.04
	SP–AG	3.91 ± 0.93	3.98 ± 0.90	0.64	0.04
	FL	3.44 ± 1.13	3.48 ± 1.12	0.66	0.03
	overall	3.89 ± 0.87	3.83 ± 0.92	0.44	0.03
	CRF	3.57 ± 0.94	3.62 ± 0.93	0.42	0.03
Adolescents (n = 544)	MF	3.58 ± 0.93	3.57 ± 0.95	0.52	0.03
	SP–AG	3.71 ± 0.92	3.74 ± 0.90	0.48	0.03
	FL	3.29 ± 1.06	3.34 ± 1.07	0.47	0.03

SD = standard deviation; CRF = cardiorespiratory fitness; MF = muscular fitness; SP–AG = speed and agility; FL = flexibility.

Table 4. Goodness-of-fit indices for the IFIS-C in Chinese children and adolescents.

Statistics	Boys	Girls	Children	Adolescents	Overall
χ^2	1.157	13.395	4.067	1.018	3.525
df	2	2	2	2	2
p	0.561	0.001	0.131	0.601	0.172
$\chi^2/$df	0.579	6.697	2.033	0.509	1.763
SRMR (95% CI)	0.011	0.043	0.027	0.010	0.015
RMSEA (95% CI)	0.000	0.114	0.052	0.000	0.029
GFI	0.999	0.985	0.995	0.999	0.998
CFI	1.000	0.971	0.994	1.000	0.999
NFI	0.998	0.966	0.988	0.998	0.997
TLI	1.004	0.912	0.982	1.004	0.996
IFI	1.001	0.971	0.994	1.001	0.999

Abbreviation: df = degree of freedom; SRMR = standardized root means square residual; RMSEA = root mean square error of approximation; CI = confidence interval; GFI = goodness-of-fit index; CFI = comparative fit index; NFI = normed fit index; TLI = Tucker–Lewis index; IFI = incremental fit index.

Table 5. Factor loadings, CR and AVE of the IFIS-C in Chinese children and adolescents.

Statistics	Boys	Girls	Children	Adolescents	Overall
CRF	0.78	0.69	0.76	0.73	0.74
MF	0.80	0.68	0.68	0.80	0.75
SP–AG	0.77	0.74	0.74	0.78	0.77
FL	0.47	0.47	0.34	0.49	0.43
CR	0.80	0.74	0.73	0.80	0.77
AVE	0.52	0.43	0.43	0.51	0.47

Abbreviation: CRF = cardiorespiratory fitness; MF = muscle fitness; SP–AG = speed and agility; FL = flexibility; CR = composite reliability (>0.7); AVE = average variance extracted (>0.5) [37].

4. Discussion

To the authors' knowledge, this study was the first to evaluate the reliability of the IFIS in China. Overall, the results indicate the moderate reliability of the IFIS-C in Chinese children and adolescents with weighted kappa values for different sub-measures of the IFIS-C ranging from 0.51 to 0.60. In terms of subgroups, the weighted kappa values were slightly higher in girls and children than in boys and adolescents, respectively. In addition, there was a lower reliability for overall fitness in comparison to other components of the IFIS-C, and MF showed a greater reliability in Chinese children and adolescents.

Several previous studies suggested that the IFIS has moderate reliability in children and adolescents [4,19–22]. The overall weighted kappa coefficients in this study are comparable to previously reported weighted kappa coefficients between 0.54 and 0.65 in European children and adolescents [19]. Furthermore, the weighted kappa coefficient for MF was similar to that found in Francisco's study [19]. Another study, however, reported a range of 0.52–0.67 in adolescents, which was higher than in this study (the weighted kappa of this study ranged from 0.45 to 0.56) [22]. Higher weighted kappa coefficients in children and adolescents were also reported in two other studies that used the Spanish version of the IFIS, which were 0.775 to 0.847 [20] and 0.64 to 0.80 [21], respectively. These differences may be attributed to the variability in physical activity and fitness level across study populations, as previous studies showed that well-designed physical activity can improve the perceived PF in adolescents [40]. The results of the Confirmatory Factor Analysis (CFA) indicated that the model fit was not acceptable in girls, which revealed that the IFIS-C had poor construct validity in Chinese girls. However, the reason for this might be the small sample size of girls in this study. The lower level of physical fitness in Chinese school-aged girls in comparison with that in boys may also have contributed to these results [27]. Therefore, future studies with larger sample sizes are needed to further examine the validity of the IFIS-C.

This study also showed that few children and adolescents consider their PF as "poor" or "very poor", which is consistent with previous research [21]. Sex-specific analyses also showed higher self-estimations of PF in boys than in girls, except in terms of flexibility, which was consistent with previous studies [19,21]. Potential contributors to these observed differences may be maturity status [41], morphological characteristics (different somatotypes) [42] and physiological traits [21,43]. Differences in the types of physical activity performed between boys and girls may also contribute to differences in perceived PF [22,44], as different types of physical activities can enhance various aspects of PF. For example, boys prefer ball sports that can increase strength and SP–AG fitness, while girls are more willing to participate in dance or gymnastics that can increase flexibility [44]. Despite these differences, the IFIS-C can be considered a reliable instrument for use in determining PF by sex in Chinese children and adolescents.

With regard to age, adolescents reported lower self-estimations of fitness than children in this study, which has been shown previously [4]. A decline in PF across different grades has also been reported in a large-scale study of Chinese children and adolescents [27]. Compared with other components of the IFIS-C, Chinese children and adolescents reported higher self-estimated overall fitness, which was similar to the results of previous studies in Brazilian, Spanish and Colombian adolescents [4,20,21]. Notably, existing evidence revealed that PF was closely associated with motor competence [45]. Considering that daily physical activity generally involves components of strength, speed or flexibility [4], the perception of children's and adolescents' motor performance is closely related to all of the components of PF acquired in daily physical activities [4]. Therefore, participants reported higher self-estimated overall fitness in comparison with other components of the IFIS-C.

In general, the IFIS is a reliable instrument that can be used to evaluate the overall level of PF in population-based studies (i.e., epidemiological studies), and there is a need to test the reliability and validity of the IFIS in other age subgroups (i.e., youth, young adults and old adults) in different regions of China. Although this reliability study was conducted with Chinese children and adolescents and had a large sample size, several limitations should be taken into consideration. Firstly, due to various circumstances (i.e., COVID-19 restrictions and the fact that measurements require time, facilities and equipment), it was not possible to conduct a field-based PF evaluation. Additional research, therefore, is necessary to determine the validity of the IFIS-C in Chinese children and adolescents. Secondly, validity and reliability need to be determined in other age groups to promote a national use in China. Thirdly, the sample was taken from eastern China, and reliability and validity studies should be conducted in different regions in China due to differences in PF levels in these regions.

5. Conclusions

Overall, this study showed that the IFIS-C is a reliable instrument for the assessment of PF in Chinese children and adolescents. The lower reliability of overall fitness also emphasizes the need to separately assess various subcomponents of PF in Chinese children and adolescents to gain accurate insight into the key components of PF.

Author Contributions: R.B.: conceptualization, methodology, formal analysis, writing—original draft; S.C.: conceptualization, methodology, formal analysis, writing—review and editing; K.K.: writing—review and editing, supervision; C.D.: writing—review and editing, supervision; M.L.: writing—review and editing. J.Z.: writing—review and editing; L.W.: conceptualization, writing—review and editing, supervision. All authors have read and agreed to the published version of the manuscript.

Funding: This research was funded by The National Social Science Fund of China grant number 19CTY010.

Institutional Review Board Statement: The study was conducted in accordance with the Declaration of Helsinki, and approved by the Institutional Review Board (or Ethics Committee) of the Shanghai University of Sport (protocol code: 102772021RT071 and date of 24 May 2021).

Informed Consent Statement: Informed consent was obtained from all subjects involved in the study.

Data Availability Statement: Not applicable.

Acknowledgments: Ran Bao completed this work thanks to a China Scholarship Council (CSC)—University of Newcastle joint scholarship (No. 202108310013).

Conflicts of Interest: The authors declare no potential conflicts of interest with respect to the research, authorship and/or publication of this article.

References

1. Ortega, F.B.; Ruiz, J.R.; Castillo, M.J.; Sjöström, M. Physical fitness in childhood and adolescence: A powerful marker of health. *Int. J. Obes.* **2008**, *32*, 1–11. [CrossRef]
2. Caspersen, C.J.; Powell, K.E.; Christenson, G.M. Physical activity, exercise, and physical fitness: Definitions and distinctions for health-related research. *Public Health Rep.* **1985**, *100*, 126–131.
3. Ortega, F.B.; Artero, E.G.; Ruiz, J.R.; España-Romero, V.; Jiménez-Pavón, D.; Vicente-Rodríguez, G.; Moreno, L.A.; Manios, Y.; Beghin, L.; Ottevaere, C. Physical fitness levels among European adolescents: The HELENA study. *Br. J. Sports Med.* **2011**, *45*, 20–29. [CrossRef]
4. De Moraes, A.C.F.; Vilanova-Campelo, R.C.; Torres-Leal, F.L.; Carvalho, H.B. Is self-reported physical fitness useful for estimating fitness levels in children and adolescents? A reliability and validity study. *Medicina* **2019**, *55*, 286. [CrossRef]
5. Fogelholm, M. Physical Activity, Fitness and Fatness: Relations to Mortality, Morbidity and Disease Risk Factors. A Systematic Review. *Obes. Rev. Off. J. Int. Assoc. Study Obes.* **2009**, *11*, 202–221. [CrossRef]
6. Carnethon, M.R. Prevalence and Cardiovascular Disease Correlates of Low Cardiorespiratory Fitness in Adolescents and Adults. *JAMA* **2005**, *294*, 2981. [CrossRef]
7. Ruiz, J.R.; Castro-Piñero, J.; Artero, E.G.; Ortega, F.B.; Sjöström, M.; Suni, J.; Castillo, M.J. Predictive validity of health-related fitness in youth: A systematic review. *Br. J. Sports Med.* **2009**, *43*, 909–923. [CrossRef]
8. Donnelly, J.E.; Hillman, C.H.; Castelli, D.; Etnier, J.L.; Lee, S.; Tomporowski, P.; Lambourne, K.; Szabo-Reed, A.N. Physical Activity, Fitness, Cognitive Function, and Academic Achievement in Children. *Med. Sci. Sports Exerc.* **2016**, *48*, 1197–1222. [CrossRef]
9. LaVigne, T.; Hoza, B.; Smith, A.L.; Shoulberg, E.K.; Bukowski, W. Associations Between Physical Fitness and Children's Psychological Well-Being. *J. Clin. Sport Psychol.* **2016**, *10*, 32–47. [CrossRef]
10. Alvarez-Bueno, C.; Pesce, C.; Cavero-Redondo, I.; Sánchez-López, M.; Martínez-Hortelano, J.; Martinez Vizcaino, V. The Effect of Physical Exercise Activity Interventions on Children's Cognition and Metacognition: A Systematic Review and Meta-Analysis. *J. Am. Acad. Child Adolesc. Psychiatry* **2017**, *56*, 729–738. [CrossRef]
11. Santana, C.C.A.; Azevedo, L.B.; Cattuzzo, M.T.; Hill, J.O.; Andrade, L.P.; Prado, W.L. Physical fitness and academic performance in youth: A systematic review. *Scand. J. Med. Sci. Sports* **2017**, *27*, 579–603. [CrossRef]
12. Lavie, C.J.; Milani, R.V.; O'Keefe, J.H.; Lavie, T.J. Impact of Exercise Training on Psychological Risk Factors. *Prog. Cardiovasc. Dis.* **2011**, *53*, 464–470. [CrossRef]
13. Bianco, A.; Jemni, M.; Thomas, E.; Patti, A.; Paoli, A.; Ramos Roque, J.; Palma, A.; Mammina, C.; Tabacchi, G. A systematic review to determine reliability and usefulness of the field-based test batteries for the assessment of physical fitness in adolescents—The ASSO Project. *Int. J. Occup. Med. Environ. Health* **2015**, *28*, 445–478. [CrossRef]
14. Marques, A.; Henriques-Neto, D.; Peralta, M.; Martins, J.; Gomes, F.; Popovic, S.; Masanovic, B.; Demetriou, Y.; Schlund, A.; Ihle, A. Field-Based Health-Related Physical Fitness Tests in Children and Adolescents: A Systematic Review. *Front. Pediatrics* **2021**, *9*, 155. [CrossRef]

15. Kolimechkov, S. Physical fitness assessment in children and adolescents: A systematic review. *Eur. J. Phys. Educ. Sport Sci.* **2017**, *3*, 65–79. [CrossRef]

16. Fox, K.R.; Corbin, C.B. The physical self-perception profile: Devlopment and preliminary validation. *J. Sport Exerc. Psychol.* **1989**, *11*, 408–430. [CrossRef]

17. Keith, N.R.; Clark, D.O.; Stump, T.E.; Miller, D.K.; Callahan, C.M. Validity and Reliability of the Self-Reported Physical Fitness (SRFit) Survey. *J. Phys. Act. Health* **2014**, *11*, 853–859. [CrossRef]

18. Merellano-Navarro, E.; Collado-Mateo, D.; García-Rubio, J.; Gusi, N.; Olivares, P.R. Validity of the International Fitness Scale "IFIS" in older adults. *Exp. Gerontol.* **2017**, *95*, 77–81. [CrossRef]

19. Ortega, F.B.; Ruiz, J.R.; Espana-Romero, V.; Vicente-Rodriguez, G.; Martínez-Gómez, D.; Manios, Y.; Béghin, L.; Molnar, D.; Widhalm, K.; Moreno, L.A. The International Fitness Scale (IFIS): Usefulness of self-reported fitness in youth. *Int. J. Epidemiol.* **2011**, *40*, 701–711. [CrossRef]

20. Ramírez-Vélez, R.; Cruz-Salazar, S.M.; Martínez, M.; Cadore, E.L.; Alonso-Martinez, A.M.; Correa-Bautista, J.E.; Izquierdo, M.; Ortega, F.B.; García-Hermoso, A. Construct validity and test–retest reliability of the International Fitness Scale (IFIS) in Colombian children and adolescents aged 9–17.9 years: The FUPRECOL study. *PeerJ* **2017**, *5*, e3351. [CrossRef]

21. Sánchez-López, M.; Martínez-Vizcaíno, V.; García-Hermoso, A.; Jiménez-Pavón, D.; Ortega, F. Construct validity and test–retest reliability of the International Fitness Scale (IFIS) in spanish children aged 9–12 years. *Scand. J. Med. Sci. Sports* **2015**, *25*, 543–551. [CrossRef]

22. Sánchez-Toledo, P.R.O.; Rubio, J.G.; Merellano-Navarro, E. Propiedades psicométricas de la escala "International Fitness Scale" en adolescentes chilenos. In *Retos: Nuevas Tendencias en Educación Física, Deporte y Recreación*; Federación Española de Docentes de Educación Física (FEADEF): Valladolid, Spain, 2017; pp. 23–27.

23. Ortega, F.; Sanchez-Lopez, M.; Solera-Martinez, M.; Fernandez-Sanchez, A.; Sjöström, M.; Martinez-Vizcaino, V. Self-reported and measured cardiorespiratory fitness similarly predict cardiovascular disease risk in young adults. *Scand. J. Med. Sci. Sports* **2013**, *23*, 749–757. [CrossRef]

24. Español-Moya, M.N.; Ramirez-Velez, R. Psychometric validation of the International Fitness Scale (IFIS) in Colombian youth. *Rev. Esp. Salud Publica* **2014**, *88*, 271–278. [CrossRef]

25. Romero-Gallardo, L.; Soriano-Maldonado, A.; Ocón-Hernández, O.; Acosta-Manzano, P.; Coll-Risco, I.; Borges-Cosic, M.; Ortega, F.B.; Aparicio, V.A. International Fitness Scale—IFIS: Validity and association with health-related quality of life in pregnant women. *Scand. J. Med. Sci. Sports* **2020**, *30*, 505–514. [CrossRef]

26. Álvarez-Gallardo, I.C.; Soriano-Maldonado, A.; Segura-Jiménez, V.; Carbonell-Baeza, A.; Estévez-López, F.; McVeigh, J.G.; Delgado-Fernández, M.; Ortega, F.B. International FItness Scale (IFIS): Construct validity and reliability in women with fibromyalgia: The al-Ándalus Project. *Arch. Phys. Med. Rehabil.* **2016**, *97*, 395–404. [CrossRef]

27. Zhu, Z.; Yang, Y.; Kong, Z.; Zhang, Y.; Zhuang, J. Prevalence of physical fitness in Chinese school-aged children: Findings from the 2016 Physical Activity and Fitness in China—The Youth Study. *J. Sport Health Sci.* **2017**, *6*, 395–403. [CrossRef]

28. Chen, X.; Cui, J.; Zhang, Y.; Peng, W. The association between BMI and health-related physical fitness among Chinese college students: A cross-sectional study. *BMC Public Health* **2020**, *20*, 444. [CrossRef]

29. Wang, J. The association between physical fitness and physical activity among Chinese college students. *J. Am. Coll. Health* **2019**, *67*, 602–609. [CrossRef]

30. Dong, Y.; Lau, P.W.C.; Dong, B.; Zou, Z.; Yang, Y.; Wen, B.; Ma, Y.; Hu, P.; Song, Y.; Ma, J. Trends in physical fitness, growth, and nutritional status of Chinese children and adolescents: A retrospective analysis of 1.5 million students from six successive national surveys between 1985 and 2014. *Lancet Child Adolesc. Health* **2019**, *3*, 871–880. [CrossRef]

31. Liu, Y. Promoting physical activity among Chinese youth: No time to wait. *J. Sport Health Sci.* **2017**, *6*, 248–249. [CrossRef]

32. Lu, C.; Stolk, R.P.; Sauer, P.J.J.; Sijtsma, A.; Wiersma, R.; Huang, G.; Corpeleijn, E. Factors of physical activity among Chinese children and adolescents: A systematic review. *Int. J. Behav. Nutr. Phys. Act.* **2017**, *14*, 36. [CrossRef]

33. Beaton, D.; Bombardier, C.; Guillemin, F.; Ferraz, M.B. Recommendations for the cross-cultural adaptation of health status measures. *N. Y. Am. Acad. Orthop. Surg.* **2002**, *12*, 1–9.

34. Streiner, D.L.; Norman, G.R.; Cairney, J. *Health Measurement Scales: A Practical Guide to Their Development and Use*; Oxford University Press: Cary, NC, USA, 2015.

35. Mokkink, L.B.; Prinsen, C.; Patrick, D.L.; Alonso, J.; Bouter, L.M.; de Vet, H.; Terwee, C.B.; Mokkink, L. COSMIN methodology for systematic reviews of patient-reported outcome measures (PROMs). *User Man.* **2018**, *78*, 1.

36. Cohen, J. Weighted kappa: Nominal scale agreement provision for scaled disagreement or partial credit. *Psychol. Bull.* **1968**, *70*, 213. [CrossRef]

37. Landis, J.R.; Koch, G.G. The measurement of observer agreement for categorical data. *Biometrics* **1977**, *33*, 159–174. [CrossRef]

38. Hair, J.; Anderson, R.; Babin, B.; Black, W. *Multivariate Data Analysis: A Global Perspective*; Pearson Upper Saddle River: Hoboken, NJ, USA, 2010.

39. Hu, L.T.; Bentler, P.M. Cutoff criteria for fit indexes in covariance structure analysis: Conventional criteria versus new alternatives. *Struct. Equ. Modeling Multidiscip. J.* **1999**, *6*, 1–55. [CrossRef]

40. Babic, M.J.; Morgan, P.J.; Plotnikoff, R.C.; Lonsdale, C.; White, R.L.; Lubans, D.R. Physical Activity and Physical Self-Concept in Youth: Systematic Review and Meta-Analysis. *Sports Med.* **2014**, *44*, 1589–1601. [CrossRef]

41. Mota, J.; Guerra, S.; Leandro, C.; Teixeira-Pinto, A.; Ribeiro, J.; Duarte, J. Association of maturation, sex, and body fat in cardiorespiratory fitness. *Am. J. Hum. Biol. Off. J. Hum. Biol. Counc.* **2002**, *14*, 707–712. [CrossRef]
42. Marta, C.C.; Marinho, D.A.; Barbosa, T.M.; Izquierdo, M.; Marques, M.C. Physical fitness differences between prepubescent boys and girls. *J. Strength Cond. Res.* **2012**, *26*, 1756–1766. [CrossRef]
43. Rowland, T.W. Evolution of maximal oxygen uptake in children. *Pediatric Fit.* **2007**, *50*, 200–209.
44. Gao, Z.; Lee, A.; Solmon, M.; Zhang, T. Changes in Middle School Students' Motivation Toward Physical Education Over One School Year. *J. Teach. Phys. Educ.* **2009**, *28*, 378–399. [CrossRef]
45. Cattuzzo, M.T.; dos Santos Henrique, R.; Ré, A.H.N.; de Oliveira, I.S.; Melo, B.M.; de Sousa Moura, M.; de Araújo, R.C.; Stodden, D. Motor competence and health related physical fitness in youth: A systematic review. *J. Sci. Med. Sport* **2016**, *19*, 123–129. [CrossRef]

Article

Monitoring the Role of Physical Activity in Children with Flat Feet by Assessing Subtalar Flexibility and Plantar Arch Index

Ligia Rusu [1,*,†], Mihnea Ion Marin [2,†], Michi Mihail Geambesa [1,†] and Mihai Robert Rusu [1,†]

1   Sport Medicine and Physical Therapy Department, Faculty of Physical Education and Sport, University of Craiova, 200585 Craiova, Romania; mihail.geambesa@edu.ucv.ro (M.M.G.); mihai.rusu@edu.ucv.ro (M.R.R.)
2   Industrial Engineering Department, Faculty of Mechanics, University of Craiova, 200585 Craiova, Romania; mihnea.marin@edu.ucv.ro
*   Correspondence: ligia.rusu@edu.ucv.ro; Tel.: +40-723-867-738
†   These authors contributed equally to this work.

Abstract: Flat foot is a common pediatric foot deformity which involves subtalar flexibility; it can affect the plantar arch. This study analyzes the evolution of two parameters, i.e., plantar index arch and subtalar flexibility, before and after physiotherapy and orthoses interventions, and examines the correlation between these two parameters. Methods: The study included 30 participants (17 boys, 12 girls, average age 9.37 ± 1.42 years) with bilateral flat foot. We made two groups, each with 15 subjects. Assessments of the subtalar flexibility and plantar arch index used RSScan the platform, and were undertaken at two time points. Therapeutic interventions: Group 1—short foot exercises (SFE); Group 2—SFE and insoles. Statistical analyses included Student's *t*-test, Cohen's D coefficient, Pearson and Sperman correlation. Results: Group 1—subtalar flexibility decreased for the left and right feet by 28.6% and 15.9% respectively, indicating good evolution for the left foot. For both feet, a decrease of the plantar index arch was observed. Group 2—subtalar flexibility decreased for the right and left feet by 43.4% and 37.7% respectively, indicating a good evolution for the right foot. For both feet, a decrease of plantar index arch was observed. Between groups, subtalar flexibility evolved well for Group 2; this was attributed to mixt intervention, physical therapy and orthosis. For plantar arch index, differences were not significant between the two groups. We observed an inverse correlation between subtalar flexibility and plantar arch index. Conclusions: Improvement of plantar index arch in static and dynamic situations creates the premise of a good therapeutic intervention and increases foot balance and postural control. The parameter which showed the most beneficial influence was the evolution is subtalar flexibility.

Keywords: flat foot; subtalar flexibility; plantar arch index; orthoses

Citation: Rusu, L.; Marin, M.I.; Geambesa, M.M.; Rusu, M.R. Monitoring the Role of Physical Activity in Children with Flat Feet by Assessing Subtalar Flexibility and Plantar Arch Index. *Children* **2022**, *9*, 427. https://doi.org/10.3390/children9030427

Academic Editors: Georgian Badicu, Ana Filipa Silva and Hugo Miguel Borges Sarmento

Received: 3 March 2022
Accepted: 16 March 2022
Published: 18 March 2022

Publisher's Note: MDPI stays neutral with regard to jurisdictional claims in published maps and institutional affiliations.

Copyright: © 2022 by the authors. Licensee MDPI, Basel, Switzerland. This article is an open access article distributed under the terms and conditions of the Creative Commons Attribution (CC BY) license (https://creativecommons.org/licenses/by/4.0/).

1. Introduction

Flat foot (FF) describes flatness of the medial longitudinal arch (MLA) of the foot, which is visible in standing position but also in the gait; it may be categorized as either flexible or rigid. Hyperpronation or excessive pronation of the foot refers to ankle bones being turned inward and the rest of the foot turned outward. This happens when the arch of the foot is flattened when weight is applied to it. Both aspects open a discussion about the behavior of the subtalar angle which determines subtalar flexibility in functional activities [1].

Several discussions are also ongoing about the definition of subtalar angle and its relationship with foot pathologies.

Flat foot is a pathologies which often is associated with pain and instability. The condition requires an early diagnosis to determine its etiology and design therapeutic interventions.

The subtalar joint has two parts: anterior and posterior. Subtalar joint movements are inversion–eversion from a clinical point of view; the other components (e.g., anteroposterior and mediolateral translation) are not assessable by clinical examination [2]. The subtalar joint axis has an average inclination 42° in the sagittal plane and 23° medial deviation in relation to the long axis of the foot [3].

Discussions about this angle have stemmed from the different methods used for assessing its mobility; in this context, many authors agree that inversion is 25–30° and eversion 5–10°. In practice, the relevance of this angle seems not to be so important; as such, it is necessary to focus instead on subtalar flexibility. This flexibility is connected to the body mass index (BMI). In many cases, increased flexibility seems to be correlated with a lack of anterior facet of the joint which favorizes the development of flat foot [4].

Gait analysis is one method that could be used to design physiotherapy programs in patients with flat foot (FF) due to the need to understand the behavior of the foot during gait, which involves three actions: rollover of the foot from supination to pronation on the floor, dorsal and plantar flexibility, and rotations. Subtalar joint movement can be described as rotation, translation or a combination of both; motion of this joint can be broken into a rotation about, and a translation along, the helical axis. During weight-bearing motion, all of the bones in the foot rotate around the subtalar joint axis [5]. All of these afford the foot the capacity to adapt to different conditions whereby the foot absorbs the ground force effect.

In this context, the subtalar joint has an important role, i.e., it combines the inversion with internal rotation and the eversion with external rotation. The consequence of this is that it has a possible effect under the plantar arch and in cases of collapse of the arch and midfoot during inversion, but with the arch lifted during eversion.

Flat foot may be defined by the collapse of MLA and overpronation. The condition can cause pain, gait disorders, static and balance disorders and affect the entire lower limb [6,7].

Today, the early diagnosis of FF is a challenge because clinical practices have a lot of variations regarding evaluations and therapy of the condition. Most diagnoses are based on clinical evaluations and radiography rather than functional diagnoses, which assess the behavior of joints through static and dynamic movements. For this reason, many authors do not accept this diagnostic method [6].

Many evaluations include measuring indexes like the valgus index, arch index, Staheli index, Chippaux-Smirak index, Foot posture index and Clarke angle [8–10]. Many studies have been published on the differences in evaluation results depending of the index choice. As such, there is no consensus regarding the diagnosis of FF because the reliability of these measurements can be poor.

Some authors consider the Chippaux-Smirak to be the best solution for diagnoses of FF in children [11]. Chen et al. considers that this index is the gold standard in FF evaluations for children under 8 years of age [12].

In contrast, some authors consider that clinical examinations are the best approach [13]. Fascione et al. considered that there is a limit of concordance between the Clarke angle and Chippaux-Smirak index, with regard to the plantar arch status [14]. The same aspect is relevant for Langley, who considered that the Chippaux-Smirak index makes possible more accurate predictions and diagnoses than the Staheli index [15].

First evaluations of FF and physiotherapy approaches are often based on the plantar print foot and measurements of the indexes. Clinical examinations of the longitudinal arch give information about the probability of FF occurrence; however, medical equipment make it possible to analyze the gait using pressure and force plates, as well as also baropodography [16]. To determine the foot type, arch index values are used, based on the contact area of the middle section of the plantar footprint [16]; however, this has little relevance to clinical classifications of the foot.

A new approach to diagnosing and treating FF is the development of a cluster model which makes it possible to identify the etiology and assess the foot alignment, which can influence the foot dynamic during gait and determine how physical therapy could improve this [17,18].

The relationship between subtalar angle and MLA is the subject of debate, and some authors have written about the eversion of subtalar angle and flat foot arch and vice versa [19]. Based on this, hyperpronation is a feature of FF; this is just a hypothesis, because there are no conclusive studies to date, as numerous methods exist to measure the morphological parameters of the foot, and unanimity of which is most suitable has not been reached [20].

Menz et al. measured the correlation between the results of plantar arch measurements (using force and pressure plates) and radiographyc aspects of calcaneal angle inclination, as well as correlations between this angle and navicular bone position. They observed a strong correlation between plantar arch and calcaneal angle inclination [21].

Even if the majority of the researchers and physicians consider that approaches based on the plantar arch are the most appropriate for foot classifications, many acknowledge that there are inconsistencies regarding the indexes used for foot classifications, and that clinical assessments have shortcomings.

The literature reveals many questions regarding FF, in terms of clinical and functional diagnosis, biomechanic parameters, relationships between these parameters in normal foot and in FF, etc. Therapeutic interventions and orthesis require complex evaluations and follow up.

The hypothesis of this study is that biomechanical parameters (subtalar flexibility and index arch) are relevant for monitoring the evolution of FF and response to physiotherapy and orthoses interventions. The purpose of this study is to analyze the evolution of two parameters, i.e., plantar index arch and subtalar flexibility, before and after physiotherapy and orthoses interventions. Also, this study includes an analysis of the correlation between these two parameters.

2. Study Design

This study was conducted in the Sports Medicine Department of the University of Craiova. It conformed to the guidelines of the Declaration of Helsinki, and was approved by the Ethics Committee of University of Craiova (20–28 September 2021). Written informed consent was obtained from all participants. Clinical trial registration according to the ICMJE guidelines was registered with number 243/30.09.2021.

A total of 30 participants with flat-foot were selected: 17 boys and 13 girls, with an average age of 9.37 ± 1.42 years. Both groups had bilateral flat foot degree II.

The inclusion criteria were bilateral flat foot diagnosed by clinical examination in standing position and during gait. The exclusion criteria were: (1) prior foot or ankle surgery; (2) pain in the lower extremities; (3) overweight or obese; (4) any other foot deformities; or (5) neuromuscular and neurological disorders.

We created two groups. Group 1 (15 subjects) included participants in the physiotherapy program. Group 2 (15 subjects) included participants in the physiotherapy program who also wear foot orthoses (insoles).

2.1. Evaluation

Biomechanic evaluations consisted of assessing the subtalar flexibility and plantar arch index. For both parameters, we used a force and pressure platform RSScan (Figure 1).

Data were recorded in static and dynamic positions during a gait cycle. We made three measurements and then selected the best data, as it was necessary to obtain the entire plantar surface. We asked subjects to relax and walk with their normal gait.

Statistical analyses were performed using SPSS V.2.0 package.

The aim of the statistical analyses was to identify significant differences in subtalar flexibility and plantar index arch between the two groups of patients. We used student's *t*-test, Cohen's D coefficient, Pearson and Sperman correlation.

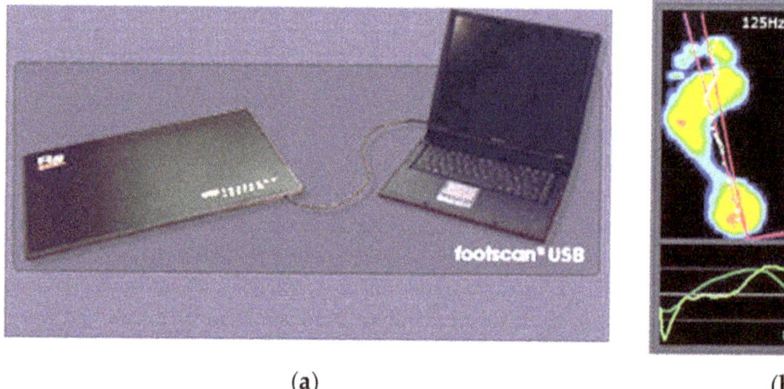

(a)   (b)

Figure 1. RSscan platform: (**a**) RSscan components; (**b**) RSscan software (Rsscan—RS software for footscan | Logemas).

2.2. Intervention

Therapeutic interventions consisted of:

For group 1—The physiotherapy program included classic exercises, e.g.,:

- strengthening of ligaments;
- strengthening of the muscles of the shins and thighs, heel bones, metatarsal fingers, ankle joint and plantar aponeurosis;
- correction of the foot arch position;

The exercises included toe bending or towel-curl exercises which mobilize the extrinsic muscles of the foot, such as the flexor digitorum longus muscle [22].

Also, we proposed short foot exercises (SFE) which are a form of sensory-motor training that activates the intrinsic muscles in the foot and actively forms the longitudinal arch and the horizontal arch [23]. A posture was maintained for 10 s followed by 5 s of relaxing; this process was repeated for a total of 30 min per session, with three sessions per week for a total of 12 weeks. The exercises [24] included:

- Single leg stance on a fixed surface (three sets, 10 repetitions followed by a 10 s break)
- Forward lean on a fixed surface (three sets, 10 repetitions followed by a 10 s break)
- Standing on one leg on an unstable surface (three sets, 10 repetitions followed by a 10 s break)
- Forward lean on an unstable surface (three sets, 10 repetitions followed by a 10 s break)
- Throwing a ball with different directions on fixed surface (three sets, 10 repetitions followed by a 10 s break)
- Throwing a ball in different directions on an unstable surface (three sets, 10 repetitions followed by a 10 s break)
- Squat on a fixed surface (three sets, 10 repetitions followed by a 10 s break)
- Jump on a fixed surface (three sets, 10 repetitions followed by a 10 s break)
- Squat on an unstable surface (three sets, 10 repetitions followed by a 10 s break)
- Jump on an unstable surface (three sets, 10 repetitions).

For group 2—We proposed the same physiotherapy program but also recommended the use insoles to improve the medial longitudinal arch (MLA). In this way, we combined foot orthotics with sensory-motor training, i.e., SFE.

Foot orthosis reduces the severity of symptoms that are secondary to increased flexibility.

We prescribed personalized, semirigid insoles, manufactured by Ortoprotetica (Bucharest, Romania) based on 3D casting.

The orthoses were manufactured based on plantar pressure evaluations using the RSScan pressure platform. They featured a semirigid thermoplastic heel cup extending to the base of the metatarsals with a full-length, perforated ethylene vinyl acetate top cover.

Each person had his/her own orthosis and was allowed to take part in normal activities. The insoles were given to the subjects, who were instructed to put them into their shoes.

The insoles were designed based upon force-plate data using the CAD-CAM (CNC) method (Figure 2a,b).

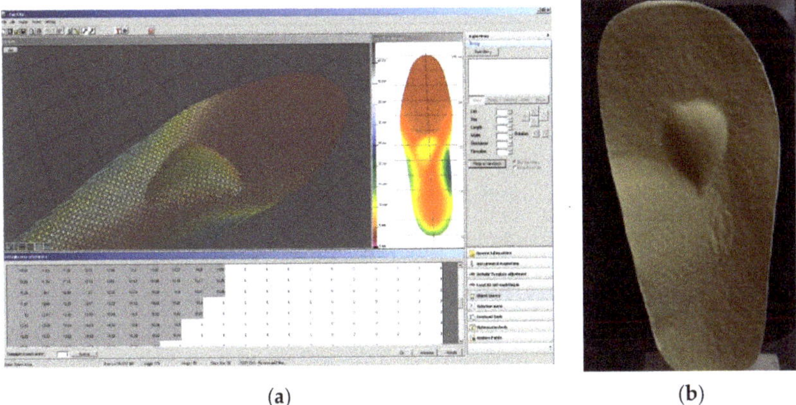

(a)   (b)

Figure 2. Insole: (**a**) CAD design of the insole; (**b**) Insole with arch support.

In this way, we ensured arch supports in the insoles.

2.3. Statistical Analysis

The aim of the statistical analysis was to identify significant differences in subtalar flexibility and plantar arch index between the two groups of patients. Statistical analysis is also used to measure differences between first and second evaluations of these parameters for each group. This was intended to assess the evolution of each group and to measure the differences between groups 1 and 2 at the second evaluation.

Data analysis based software packages were used for statistical analyses.

A database was initially created containing the experimental data from the significant aspects of this research. The recording values of the parameters were analyzed to visualize the variables, and the statistical analysis was intended to reveal significant differences between the data series for each group. Descriptive data (means SD) were reported for the entire patient cohort. Normal distribution was tested using the JB test and visual analysis of Gauss function for both parameters. We applied a *t* test for equal and unequal variances depending on the results of the Levene test for all parameters. Statistical significance was set at $p < 0.05$.

All data presented herein have been normalized.

3. Results

The average anthropometric characteristics of the 30 participants were: weight 41.8 ± 12.72 kg, height 148.7 ± 10.96 cm, and body mass index 18.84 ± 5.32 kg/m^2.

3.1. Evolution of Parameters between EV1 and EV2 for Each Group

We monitored changes in the aforementioned parameters before (EV1) and after (EV2) the physiotherapy program. The results are presented for each group, for right and left feet at EV1 and EV2.

Table 1 presents the evolution of subtalar flexibility and Figure 3 shows the average values of subtalar flexibility for group 1.

Table 1. Evolution of subtalar flexibility (Group 1, n = 15).

Subject	Subtalar Flexibility, Left Side EV1	Subtalar Flexibility, Left Side EV2	Variation of Subtalar Flexibility, Left Side %	Subtalar Flexibility, Right Side EV1	Subtalar Flexibility, Right Side EV2	Variation of Subtalar Flexibility, Right Side %
P16	7.50	5.58	25.60	12.68	9.27	26.89
P17	20.79	13.03	37.33	17.82	15.78	11.45
P18	11.45	10.77	5.94	21.42	19.86	7.28
P19	2.79	1.87	32.97	4.57	3.98	12.91
P20	24.35	23.31	4.27	11.45	10.22	10.74
P21	8.74	7.70	11.90	18.64	16.10	13.63
P22	5.86	4.72	19.45	7.28	6.75	7.28
P23	10.19	6.04	40.73	9.30	8.15	12.37
P24	9.13	2.30	74.81	10.32	9.84	4.65
P25	6.05	5.67	6.28	−17.86	−16.48	7.73
P26	4.54	3.76	17.18	3.37	2.67	19.57
P27	14.86	12.27	17.43	12.83	10.35	19.33
P28	12.66	6.15	51.42	8.41	7.34	12.72
P29	8.57	7.73	9.80	8.88	3.94	55.63
P30	9.45	2.38	74.81	5.22	4.33	17.05

P1–P30 patient number; EV 1—evaluation 1, EV2—evaluation 2.

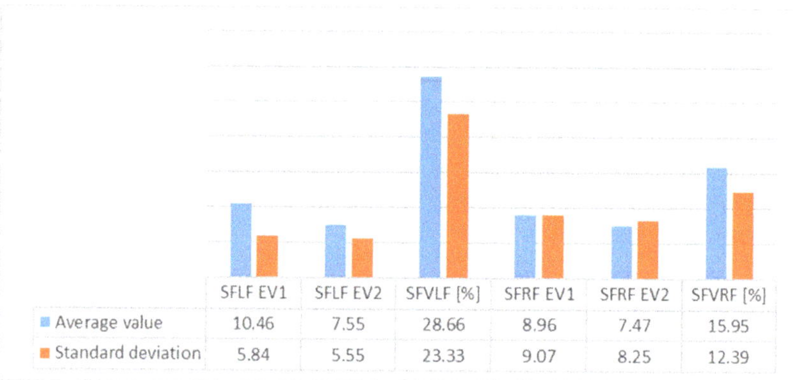

Figure 3. Variation of average values and standard deviation for subtalar flexibility for Group 1 between EV1 and EV2 (SFLF = subtalar flexibility for left foot; SFVLF = subtalar flexibility variation for left foot; SFRF = subtalar flexibility for right foot; SFVRF = subtalar flexibility variation for right foot).

Table 2 presents the evolution of plantar index arch and Figure 4 shows the average values of plantar index arch for group 1.

Table 2. Plantar index arch evolution (Group 1, n = 15).

Subject	Plantar Index Arch, Left Side EV1	Plantar Index Arch, Left Side EV2	Difference EV1–EV2	Plantar Index Arch, Right Side EV1	Plantar Index Arch, Right Side EV2	Difference EV1–EV2
P16	0.93	0.83	0.10	0.91	0.87	0.04
P17	0.51	0.41	0.10	0.57	0.45	0.12
P18	0.78	0.68	0.10	1.12	0.98	0.14
P19	0.73	0.63	0.10	0.95	0.86	0.09
P20	0.70	0.61	0.09	0.73	0.60	0.13
P21	0.79	0.7	0.09	0.73	0.62	0.11
P22	0.80	0.71	0.09	0.79	0.72	0.07

Table 2. *Cont.*

Subject	Plantar Index Arch, Left Side EV1	Plantar Index Arch, Left Side EV2	Difference EV1–EV2	Plantar Index Arch, Right Side EV1	Plantar Index Arch, Right Side EV2	Difference EV1–EV2
P23	0.39	0.30	0.09	0.32	0.29	0.03
P24	0.41	0.32	0.09	0.48	0.33	0.15
P25	0.48	0.39	0.09	0.58	0.42	0.16
P26	0.58	0.50	0.08	0.62	0.55	0.07
P27	0.76	0.68	0.08	0.66	0.56	0.10
P28	0.78	0.70	0.08	1.05	0.78	0.27
P29	0.84	0.77	0.07	0.81	0.78	0.03
P30	0.32	0.28	0.04	0.48	0.32	0.16

P1–P30 patient number; EV 1—evaluation 1, EV2—evaluation 2.

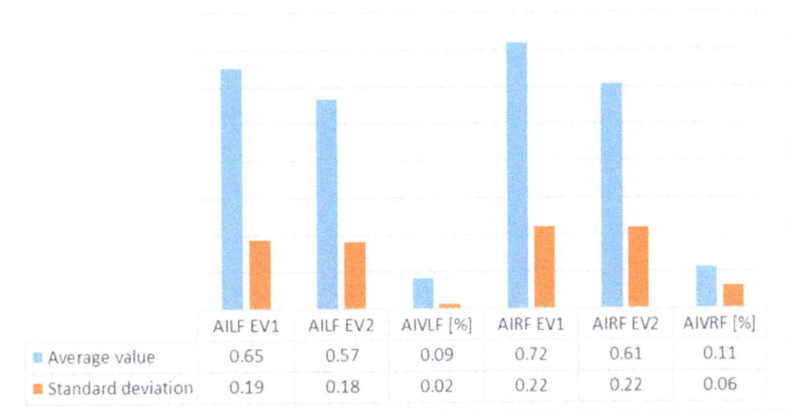

Figure 4. Variation of average values and standard deviation for plantar index arch for Group 1 between EV1 and EV2 (AILF = plantar index arch for left foot; AIVLF = plantar index arch variation for left foot; AIRF = plantar index arch for right foot; AIVRF = plantar index arch variation for right foot).

Table 3 presents the evolution of subtalar flexibility and Figure 5 shows the average values of subtalar flexibility for Group 2.

Table 3. Evolution of subtalar flexibility (Group 2, $n = 15$).

Subject	Subtalar Flexibility, Left Side EV1	Subtalar Flexibility, Left Side EV2	Variation of Subtalar Flexibility, Left Side %	Subtalar Flexibility, Right Side EV1	Subtalar Flexibility, Right Side EV2	Variation of Subtalar Flexibility, Right Side %
P1	3.78	2.23	41.01	7.87	5.72	27.32
P2	14.32	0.76	94.69	6.85	2.39	65.11
P3	14.73	12.36	16.09	3.51	1.52	56.70
P4	12.43	11.57	6.92	11.37	9.28	18.38
P5	7.64	1.95	74.48	4.88	4.67	4.30
P6	8.61	2.42	71.89	9.21	3.54	61.56
P7	15.34	13.85	9.71	8.77	7.32	16.53
P8	8.53	7.21	15.47	5.05	4.87	3.56
P9	4.70	4.64	1.28	5.74	−1.28	122.30
P10	15.29	8.65	43.43	10.33	6.83	33.88
P11	16.61	6.08	63.40	7.88	3.65	53.68
P12	7.49	6.37	14.95	12.08	0.25	97.93

Table 3. *Cont.*

Subject	Subtalar Flexibility, Left Side EV1	Subtalar Flexibility, Left Side EV2	Variation of Subtalar Flexibility, Left Side %	Subtalar Flexibility, Right Side EV1	Subtalar Flexibility, Right Side EV2	Variation of Subtalar Flexibility, Right Side %
P13	10.64	8.91	16.26	15.56	9.25	40.55
P14	1.66	0.99	40.36	8.67	5.87	32.30
P15	18.26	7.91	56.68	9.12	7.53	17.43

P1–P15 patient number; EV 1—evaluation 1, EV2—evaluation 2.

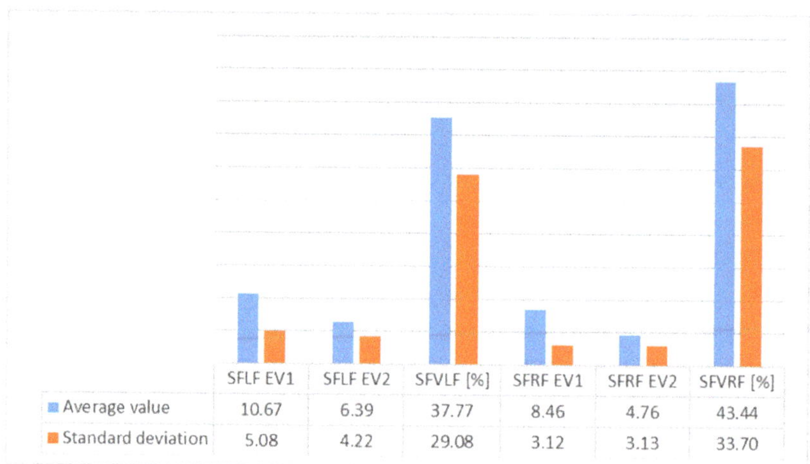

Figure 5. Variation of average values and standard deviation for subtalar flexibility for Group 2 between EV1 and EV2 (SFLF = subtalar flexibility for left foot; SFVLF = subtalar flexibility variation for left foot; SFRF = subtalar flexibility for right foot; SFVRF = subtalar flexibility variation for right foot).

Table 4 presents the evolution of plantar index arch and Figure 6 shows the average values of index arch for Group 2.

Table 4. Plantar index arch evolution (Group 2, *n* = 15).

Subject	Plantar Index Arch Left Side EV1	Plantar Index Arch Left Side EV2	Difference EV1–EV2	Plantar Index Arch Right Side EV1	Plantar Index Arch Right Side EV2	Difference EV1–EV2
P1	0.79	0.54	0.25	0.68	0.55	0.13
P2	0.87	0.64	0.23	0.81	0.71	0.10
P3	0.94	0.74	0.20	0.66	0.58	0.08
P4	0.57	0.38	0.19	0.51	0.39	0.12
P5	0.96	0.78	0.18	0.43	0.32	0.11
P6	0.79	0.62	0.17	0.84	0.79	0.05
P7	1.04	0.89	0.15	0.80	0.75	0.05
P8	0.95	0.80	0.15	0.95	0.81	0.14
P9	1.00	0.87	0.13	0.85	0.74	0.11
P10	0.68	0.56	0.12	0.57	0.44	0.13
P11	0.70	0.58	0.12	0.87	0.65	0.22
P12	0.85	0.73	0.12	0.76	0.66	0.10
P13	0.65	0.53	0.12	0.61	0.49	0.12
P14	0.64	0.53	0.11	0.69	0.54	0.15
P15	0.38	0.27	0.11	0.35	0.29	0.06

P1–P15 patient number; EV 1—evaluation 1, EV2—evaluation 2.

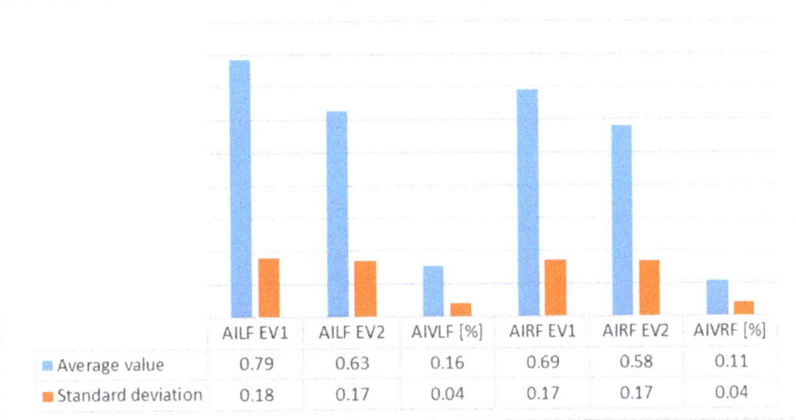

	AILF EV1	AILF EV2	AIVLF [%]	AIRF EV1	AIRF EV2	AIVRF [%]
Average value	0.79	0.63	0.16	0.69	0.58	0.11
Standard deviation	0.18	0.17	0.04	0.17	0.17	0.04

Figure 6. Variation of average values and standard deviation for plantar index arch for Group 2 between EV1 and EV2 (AILF = plantar index arch for left foot; AIVLF = plantar index arch variation for left foot; AIRF = plantar index arch for right foot; AIVRF = plantar index arch variation for right foot).

3.2. Statistical Analysis of Parameters Evolution between EV1 and EV2, for Each Group

For both groups, the student's *t*-test results and Cohen's D coefficient in Table 5.

Table 5. *p* value (student's *t*-test) * and Cohen's D coefficient ** comparing the results between EV1 and EV2.

Group	Tests	Foot Side	Subtalar Flexibility	Plantar Index Arch
Group 1	student's *t*-test	right	0.321	0.088
	student's *t*-test	left	0.086	0.110
	Cohen's D coeff.	right	0.176	0.522
	Cohen's D coeff.	left	0.528	0.472
Group 2	student's *t*-test	right	0.001	0.042
	student's *t*-test	left	0.009	0.012
	Cohen's D coeff.	right	1.225	0.677
	Cohen's D coeff.	left	0.947	0.900

significance level $p < 0.05$. $0.2 \leq$ Cohen ≤ 0.5 small size effect; $0.5 \leq$ Cohen ≤ 0.8 medium size effect; Cohen ≥ 0.8 high size effect.

3.3. Statistical Analysis of Parameters, Comparison between the Two Groups

Student's *t*-test results and Chen's D coefficient are presented in Table 6.

Table 6. *p* value (student's *t*-test) * and Cohen's D coefficient ** for parameters at EV2 comparing the results of the two groups.

Tests	Foot Side	Subtalar Flexibility	Plantar Index Arch
student's *t*-test	right	0.001	0.348
student's *t*-test	left	0.262	0.170
Cohen's D coeff.	right	0.449	2.198
Cohen's D coeff.	left	0.243	0.365

significance level $p < 0.05$. $0.2 \leq$ Cohen ≤ 0.5 small size effect; $0.5 \leq$ Cohen ≤ 0.8 medium size effect; Cohen ≥ 0.8 high size effect.

3.4. Statistical Analysis of Parameter Variations, Comparison between the Two Groups

Student's *t*-test results and Cohen's D coefficient are presented in Table 7.

Table 7. *p* value (student's *t*-test) * and Cohen's D coefficient ** for parameters variations at EV2 comparing tje two groups.

Tests	Foot Side	Subtalar Flexibility	Plantar Index Arch
student's *t*-test	right	0.004	0.495
student's *t*-test	left	0.175	0.000
Cohen's D coeff.	right	1.120	0.003
Cohen's D coeff.	left	0.357	0.148

significance level $p < 0.05$. $0.2 \leq$ Cohen ≤ 0.5 small size effect; $0.5 \leq$ Cohen ≤ 0.8 medium size effect; Cohen ≥ 0.8 high size effect.

3.5. Statistical Correlation

Correlations between the two parameters were determined using two coefficients: Pearson and Spearman. The results are presented in Table 8.

Table 8. Person and Sperman correlation coefficients.

Groups	Parameter	Pearson Coefficient	Spearman Coefficient
Group 1	EV1 subtalar flexibility/plantar index arch left foot	−0.04	0.19
Group 1	EV2 subtalar flexibility/plantar index arch left foot	0.22	0.23
Group 1	EV1 subtalar flexibility/plantar index arch right foot	0.27	0.14
Group 1	EV2 subtalar flexibility/plantar index arch right foot	0.27	0.16
Group 2	EV1 subtalar flexibility/plantar index arch left foot	−0.26	−0.16
Group 2	EV2 subtalar flexibility/plantar index arch left foot	0.03	0.017
Group 2	EV1 subtalar flexibility/plantar index arch right foot	−0.22	−0.26
Group 2	EV2 subtalar flexibility/plantar index arch right foot	−0.53	−0.54

EV 1—evaluation1; EV 2—evaluation 2.

4. Discussion

An analysis of Figures 3–8 gives information about the evolution of each parameter for both groups.

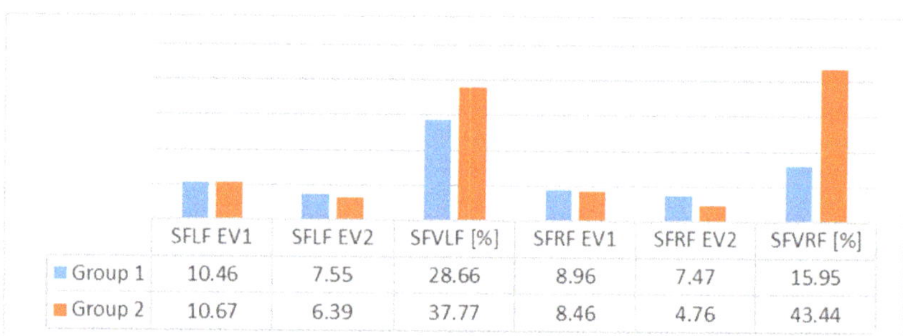

Figure 7. Average values for subtalar flexibility—comparison between the two groups (SFLF = subtalar flexibility for left foot; SFVLF = subtalar flexibility variation for left foot; SFRF = subtalar flexibility for right foot; SFVRF = subtalar flexibility variation for right foot).

For Group 1, we observed average values of subtalar flexibility (Figure 3) and noted an important decrease for the left (28.6%) and right feet (15.9%), indicating good evolution for the left foot.

The standard deviation (SD) showed homogeneity for both feet.

For Group 1, we observed in the average (Figure 4) arch index values that good evolution had occurred, as seen by decrease in the index (which, nonetheless, still remained outside out of normal values). We also observed a high decrease on the right foot, perhaps because of right dominance or because of the high degree of severity of FF in right foot among some of the subjects. The SD revealed more homogeneous values for the left foot.

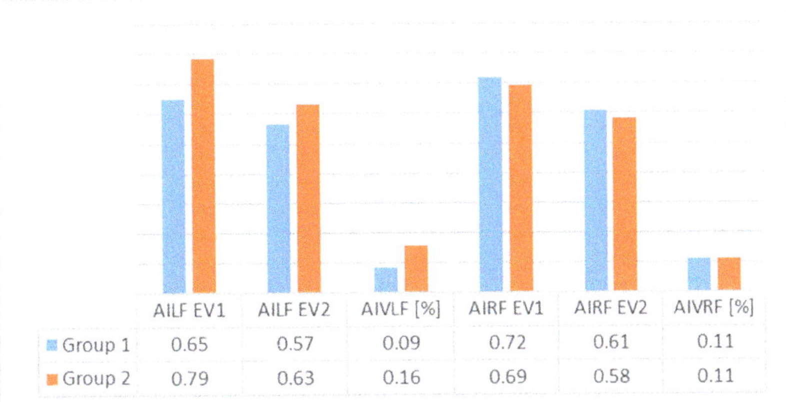

Figure 8. Average values for plantar index arch—comparison between the two groups (AILF = plantar index arch for left foot; AIVLF = plantar index arch variation for left foot; AIRF = plantar index arch for right foot; AIVRF = plantar index arch variation for right foot).

In the average subtalar flexibility values (Figure 5) for Group 2, we observed an important decrease for the right (43.4%) and left feet (37.7%), indicating good evolution for the right foot. The SD showed homogeneity for both feet.

In the average arch index values (Figure 6) for Group 2, we saw that good evolution had occurred for both feet, as evidenced by decrease in the plantar index (which, nonetheless, still remained outside out of normal values). This decrease was more pronounced for the left foot than for the right, perhaps as a result of the physical therapy and orthoses interventions. SD showed homogeneity for both feet.

In Figures 7 and 8, we can see differences in the average values of the two groups. The results indicate that the differences were greater for Group 2 for both feet.

We conclude that for Group 1, because p was more than 0.05, there was no significant difference between EV1 and EV2. For group 2, we observed a degree of high significant for subtalar flexibility and plantar index arch for both feet (Table 5).

In conclusion, the good evolution for both parameters in Group 2 was attributed to the mixed intervention, i.e., physical therapy and orthoses.

Table 6 reveals a highly significant difference regarding subtalar flexibility in the right foot ($p = 0.001$); also, for plantar index arch, the p value was significant value, indicating that at EV2, the values were not significantly different between the two groups.

In conclusion, we note that subtalar flexibility is the parameter which is most likely to influence positive therapeutic outcome.

Regarding the comparison of parameters between two groups, we observed a size effect, as assessed using the Cohen's D coefficient at EV2; the result is presented in Table 6.

A comparison regarding the variations in the parameters revealed highly significant differences between two groups regarding subtalar flexibility for the right foot ($p = 0.004$) and plantar index arch for the left foot ($p = 0.000$) (Table 7).

For other data, p values greater than 0.05 indicated the lack of significant difference between two groups (Table 7).

In conclusion, we consider that that parameters to be take into consideration when monitoring the effects of physiotherapy and orthosis interventions ought to be subtalar flexibility and plantar index arch, as the evolution of both parameters showed significant differences between the two groups. A much greater difference was observed for the left foot for plantar index arch.

We observed that was an inverse correlation for the left foot for both groups, and for the right foot with Group 2 only. This means that a decrease of subtalar flexibility could generate an increase in plantar index arch, even if the values of the two coefficients are around 0.00. We conclude therefore that there is no correlation between these two parameters (Table 8).

As many authors have underlined, classifying flat foot is a complicated process. In this context, Menz et al. discussed possible ways to place the foot in one of three situations: hyperpronation, neutral position, supination [21,25]. In our research, we tried to create an algorithm with which to assess FF in a therapeutic context. The results of our research revealed that assessments of physiotherapeutic interventions (rehabilitation) including orthoses are need to continuously monitor the program protocols. This requirement is based upon many questions searches regarding measurement methods for flat foot assessments. Analyses of the relationship between subtalar angle and MLA have been made by many authors using radiographic methods. For example, Sinclair et al. evaluate this relationship during running and observed that it was not relevant, and that the subtalar angle is not a predictor of the development of FF. Despite this, the authors observed that after 45 min of running, significant changes of subtalar angle sometimes occurred [19]. These aspects justify our research, which involved static and dynamic conditions, and are also in accordance with our results regarding subtalar flexibility and its role in monitoring the evolution of FF during physical therapy.

Bosch et al. studied the correlation between radiographic measurements of calcaneal inclination angle and plantar index arch in children. They observed that the plantar arch normalizes at around 5 years age if therapeutic intervention occurs early on, due to its effect on the calcaneal inclination angle [26].

Regarding physical therapy interventions, Wong et al. outlined the lack of a real foot classification method that could be used for clinical and therapeutic decision-making according to scientific evidence. They proposed an analysis of two aspects: clinical evaluation results and plantar index arch (calculated using AutoCAD software from footprint photographs) in 11 children that were involved in a physical therapy program. The results demonstrated that the plantar index arch values derived from plantar footprint photographs showed excellent reliability in people with varying BMI. Foot-type classification may help clinicians and researchers to subdivide sample populations to better differentiate mobility, gait, or treatment effects among foot types [27]. This study was in accordance with our results, which highlights the importance of examining the literature on the morphofunctional parameters of FF before designing rehabilitation programs.

Our approach indicated that biomechanic evaluations of FF could be used to monitor the efficacy of physical therapy interventions. This is in according with Langley et al., who observed the absence of a consensus regarding the choice of a therapeutic approach based on the morpho-functional aspects of foot [15].

Halabachi et al. also described the potential value of a physiotherapeutic strategy based on the use of appropriate footwear, foot orthoses (shoe inserts), and physical therapy including stretching and strengthening [16,28]. They concluded that are a lot of shortcomings regarding complex assessments of FF involving correlations between foot biomechanical parameters. However, they made reference to a study by Evans et al. which recommended exercise for flat feet, in the form of barefoot walking [29]. The main focus of such an exercise program is on stretching tight structures, strengthening weak components and improving proprioception and postural balance, which is in line with the approach proposed in our study.

Our intention here is to improve the literature data, assist in the design of rehabilitation programs based on evaluations, and demonstrate that SFE causes the head of the metatarsal bone to approach the heel without bending the toes in a weight-bearing state where.

These aspects are more relevant, because surgery in FF is still controversial, and little information is available regarding the efficacy of conservative management of flexible flat foot in children [30].

The results of our research underline the importance of customized management in FF, in accordance with Kachoosangy et al., who suggested that management strategies should be customized to symptomatic patients or persons with hereditary gait disorders or other comorbidities which increase the probability of growing dysfunction over time [31].

Our results suggest that orthoses interventions have to be monitored and adaptable according to the evolution of two parameters studied in this research. In this context Evans et al. concluded that until such time that the parameters of FF have become standardized, the best approach is the use of available, evidence-based management models. Those authors support ongoing studies and conclude that physical exercises for foot flat and follow-up involving the use of modified footwear are effective [29].

The results of our research demonstrate that physiotherapeutic and orthoses interventions result in improved subtalar flexibility and plantar arch index due to their ability to reduce the compensatory role of muscle groups that promote inversion and increase the role of the long perionier muscle (PL), as noted by Cho et al. [32].

At the same time, Youn et al. analyzed the subtalar joint during and after orthoses intervention using radiography, and observed that custom-made rigid foot orthoses (RFO) in children older than 6 years of age with pes planus (flat foot) can bring about significant improvements in calcaneus-related radiographic indices, and subsequently, improve talus-related radiologic indices [33].

5. Conclusions

Analyses of two parameters, i.e., subtalar flexibility and plantar arch index, allowed us to design and monitor therapeutic interventions and evaluate the difference between physiotherapy intervention alone and combined methods.

Improvements of plantar index arch in static and dynamic situations created the premise of a good therapeutic intervention, resulting in increased foot balance and postural control.

There are a lot of studies on the effects of therapeutic footwear, wedges, and insoles, but none which examines the evolution of MLA.

Sensory-motor training such as SFE is insufficient; such approaches must be extended to include analyses of biomechnic FF parameters in different conditions.

Future studies are required to investigate the long-term results of combined treatment of SFE on postural balance in subjects with a flat feet, and to evaluate the efficacy of the combination of SFE and orthoses interventions based on biomechanic analyses of FF.

We found no correlation between subtalar flexibility and plantar arch index. However, perhaps an association between these two parameters in FF could be found by examining the anatomic changes in the subtalar joint and its impact under the dynamic and passive systems of the ankle–foot complex. Such a connection could give rise to improved therapeutic methods.

Future research is need to assess and design therapeutic programs for the treatment of flat foot. This study was limited by its small group of patients and the fact that the subjects did always adhere to the physical therapy program or use the insoles.

Author Contributions: L.R.—Conceptualization, methodology, writing—original draft preparation measurements, investigation; M.I.M.—Methodology, software, visualization; M.M.G.—Data curation, investigation, software; M.R.R.—Resources, writing—original draft preparation, writing—review and editing. All authors have read and agreed to the published version of the manuscript.

Funding: This research received no external funding.

Institutional Review Board Statement: The study was conducted in accordance with the Declaration of Helsinki, and approved by Ethics Committee of University of Craiova (20–28 September 2021).

Informed Consent Statement: Informed consent was obtained from all subjects involved in the study.

Data Availability Statement: Not applicable.

Acknowledgments: Thank you to the research institute INCESA of the University of Craiova, for support us in development the research.

Conflicts of Interest: The authors declare no conflict of interest.

References

1. Nilsson, M.K.; Friis, R.; Michaelsen, M.S.; Jakobsen, P.A.; Nielsen, R.O. Classification of the height and flexibility of the medial longitudinal arch of the foot. *J. Foot Ankle Res.* **2012**, *5*, 3. [CrossRef] [PubMed]
2. Krähenbühl, N.; Horn-Lang, T.; Hintermann, B.; Knupp, M. The subtalar joint: A complex mechanism. *EFORT Open Rev.* **2017**, *2*, 309–316. [CrossRef] [PubMed]
3. Jastifer, J.R.; Gustafson, P.A. The subtalar joint: Biomechanics and functional representations in the literature. *Foot* **2014**, *24*, 203–209. [CrossRef] [PubMed]
4. Kothari, A.; Bhuva, S.; Stebbins, J.; Zavatsky, A.B.; Theologis, T. An investigation into the aetiology of flexible flat feet: The role of subtalar joint morphology. *Bone Jt. J.* **2016**, *8*, 564–568. [CrossRef]
5. Sarrafian, S.K. Biomechanics of the subtalar joint complex. *Clin. Orthop. Relat. Res.* **1993**, *290*, 17–26. [CrossRef]
6. Gonzalez-Martin, C.; Pita-Fernandez, S.; Seoane-Pillado, T.; Lopez-Calviño, B.; Pertega-Diaz, S.; GilGuillen, V. Variability between Clarke's angle and Chippaux-Smirak index for the diagnosis of flat feet Variabilidad entre el ángulo de Clarke y el índice de Chippaux- Smirak para el diagnóstico de pie plano. *Colomb. Méd.* **2017**, *48*, 25–31. [CrossRef]
7. Xiong, S.; Goonetilleke, R.S.; Witana, C.P.; Weerasinghe, T.W.; Au, E.Y. Foot arch characterization a review, a new metric, and a comparison. *J. Am. Podiatr. Med. Assoc.* **2010**, *100*, 14–24. [PubMed]
8. Staheli, L.T.; Chew, D.E.; Corbett, M. The longitudinal arch: A survey of eight hundred and eighty-two feet in normal children and adults. *J. Bone Jt. Surg. Am.* **1987**, *69*, 426.
9. Redmond, A.C.; Crosbie, J.; Ouvrier, R.A. Development and validation of a novel rating system for scoring foot posture: The Foot Posture Index. *Clin. Biomech.* **2001**, *21*, 89. [CrossRef] [PubMed]
10. Smith, L.S.; Clarke, T.E.; Hamill, C.L.; Santopietro, F. The effects of soft and semi-rigid orthoses upon rearfoot movement in running. *JAPMA* **1986**, *76*, 227. [CrossRef]
11. Chen, K.C.; Yeh, C.J.; Kuo, J.F.; Hsieh, C.L.; Yang, S.F.; Wang, C.H. Footprint analysis of flatfoot in preschool-aged children. *Eur. J. Pediatr.* **2011**, *170*, 611–617. [CrossRef] [PubMed]
12. Chen, K.C.; Tung, L.C.; Yeh, C.J.; Yang, J.F.; Kuo, J.F.; Wang, C.H. Change in flatfoot of preschool-aged children a 1-year follow-up study. *Eur. J. Pediatr.* **2013**, *172*, 255–260. [CrossRef]
13. Pita-Fernández, S.; González-Martín, C.; Seoane-Pillado, T.; LópezCalviño, B.; Pértega-Díaz, S.; Gil-Guillén, V. Validity of footprint analysis to determine flatfoot using clinical diagnosis as the gold standard in a random sample aged 40 years and older. *J. Epidemiol.* **2015**, *25*, 148–154. [CrossRef] [PubMed]
14. Fascione, J.M.; Crews, R.T.; Wrobel, J.S. Dynamic footprint measurement collection technique and intrarater reliability ink mat, paper pedography, and electronic pedography. *J. Am. Podiatr. Med. Assoc.* **2012**, *102*, 130–138.
15. Langley, B.; Cramp, M.; Morrison, S.C. Clinical measures of static foot posture do not agree. *J. Foot Ankle Res.* **2016**, *9*, 45. [CrossRef] [PubMed]
16. Rome, K.; Ashford, R.L.; Evans, A. Non-surgical interventions for paediatric pes planus. *Cochrane Database Syst. Rev.* **2010**, *7*, CD006311. [CrossRef] [PubMed]
17. Cobb, S.C.; Bazett-Jones, D.M.; Joshi, M.N.; Earl-Boehm, J.E.; James, C.R. The relationship among foot posture, core and lower extremity muscle function, and postural stability. *J Athl Train.* **2014**, *49*, 173–180. [CrossRef]
18. Tong, J.W.; Kong, P.W. Association between foot type and lower extremity injuries: Systematic literature review with meta-analysis. *J. Orthop. Sports Phys. Ther.* **2013**, *43*, 700–714. [CrossRef]
19. Sinclair, C.; Svantesson, U.; Sjöström, R.; Alricsson, M. Differences in Pes Planus and Pes Cavus subtalar eversion/inversion before and after prolonged running, using a two-dimensional digital analysis. *J. Exerc. Rehabil.* **2017**, *13*, 232–239. [CrossRef] [PubMed]
20. Williams, D.S.; McClay, I.S. Measurements used to characterize the foot and the medial longitudinal arch: Reliability and validity. *Phys. Ther.* **2000**, *80*, 864–871. [CrossRef] [PubMed]
21. Menz, H.B.; Munteanu, S.E. Validity of 3 clinical techniques for the measurement of static foot posture in older people. *J. Orthop. Sports Phys. Ther.* **2005**, *35*, 479–486. [CrossRef] [PubMed]
22. Saeki, J.; Tojima, M.; Torii, S. Clarification of functional differences between the hallux and lesser toes during the single leg stance: Immediate effects of conditioning contraction of the toe plantar flexion muscles. *J. Phys. Ther. Sci.* **2015**, *27*, 2701–2704. [CrossRef] [PubMed]
23. Jung, D.Y.; Kim, M.H.; Koh, E.K.; Kwon, O.Y.; Cynn, H.S.; Lee, W.H. A comparison in the muscle activity of the abductor hallucis and the medial longitudinal arch angle during toe curl and short foot exercises. *Phys. Ther. Sport* **2011**, *12*, 30–35. [CrossRef] [PubMed]
24. Moon, D.; Juhyeon, J. Effect of Incorporating Short-Foot Exercises in the Balance Rehabilitation of Flat Foot: A Randomized Controlled Trial. *Healthcare* **2021**, *9*, 1358. [CrossRef] [PubMed]
25. Billis, E.; Katsakiori, E.; Kapodistrias, C.; Kapreli, E. Assessment of foot posture: Correlation between different clinical techniques. *Foot* **2007**, *17*, 65–72. [CrossRef]
26. Bosch, K.; Gerß, J.; Rosenbaum, D. Development of healthy children's feet–nine-year results of a longitudinal investigation of plantar loading patterns. *Gait Posture* **2010**, *32*, 564–571. [CrossRef] [PubMed]
27. Wong, C.K.; Weil, R.; Emily de Boer, E. Standardizing Foot-Type Classification Using Arch Index Values. *Physiother. Can.* **2012**, *64*, 280–283. [CrossRef] [PubMed]

28. Halabchi, F.; Mazaheri, R.; Mirshahi, M.; Abbasian, L. Pediatric flexible flatfoot; clinical aspects and algorithmic approach. *Iran. J. Pediatr.* **2013**, *23*, 247–260. [PubMed]

29. Evans, A.M.; Rome, K.A. Cochrane review of the evidence for non-surgical interventions for flexible pediatric flat feet. *Eur. J. Phys. Rehabil. Med.* **2011**, *47*, 69–89.

30. Homayouni, K.; Karimian, H.; Naseri, M.; Mohase, N. Prevalence of Flexible Flatfoot Among School-Age Girls by using navicular drop (ND). *Shiraz E-Med. J.* **2015**, *16*, e18005. [CrossRef]

31. Kachoosangy, R.A.; Aliabadi, F. A cross-sectional study upon the Prevalence of Flat Foot: Comparison between Male and Female Primary School Students. *Iran. Rehabil. J.* **2013**, *18*, 22–24.

32. Cho, D.J.; Ahn, S.Y.; Bok, S.K. Effect of Foot Orthoses in Children With Symptomatic Flexible Flatfoot Based on Ultrasonography of the Ankle Invertor and Evertor Muscles. *Ann. Rehabil. Med.* **2021**, *45*, 459–470. [CrossRef] [PubMed]

33. Youn, K.J.; Ahn, S.Y.; Kim, B.O.; Park, I.S.; Bok, S.K. Long-Term Effect of Rigid Foot Orthosis in Children Older Than Six Years With Flexible Flat Foot. *Ann. Rehabil. Med.* **2019**, *43*, 224–229. [CrossRef] [PubMed]

MDPI

St. Alban-Anlage 66

4052 Basel

Switzerland

Tel. +41 61 683 77 34

Fax +41 61 302 89 18

www.mdpi.com

Children Editorial Office

E-mail: children@mdpi.com

www.mdpi.com/journal/children

www.ingramcontent.com/pod-product-compliance
Lightning Source LLC
LaVergne TN
LVHW071533170726
843515LV00008B/2128